Environmental Division 2013

AA003111

Core Programming Area at the 2013 AIChE Annual Meeting: Global Challenges for Engineering a Sustainable Future

San Francisco, California, USA
3 – 8 November 2013

ISBN: 978-1-63439-040-8

Printed from e-media with permission by:

Curran Associates, Inc.
57 Morehouse Lane
Red Hook, NY 12571

Some format issues inherent in the e-media version may also appear in this print version.

Printed by Curran Associates, Inc. (2014)

For permission requests, please contact AIChE
at the address below.

AIChE
3 Park Avenue
New York, NY 10016-5991

Phone: (203) 702-7660
Fax: (203) 775-5177

www.aiche.org

Additional copies of this publication are available from:

Curran Associates, Inc.
57 Morehouse Lane
Red Hook, NY 12571 USA
Phone: 845-758-0400
Fax: 845-758-2634
Email: curran@proceedings.com
Web: www.proceedings.com

TABLE OF CONTENTS

(11a) On-Site Destruction of Radioactive Oily Wastes Using Combination of Adsorption With Electrochemical Regeneration 1
Edward P. L. Roberts, Nigel W. Brown, Andrew K. Campen, David A. Wickenden

(11b) Process Chemistry Improvements in Saltcake Dissolution Modeling– A Retrospective 2
Rebecca K. Toghiani, Jeffrey S. Lindner, Laura T. Smith, Valerie Phillips

(11c) Modeling of Cold-Cap Reactions for Vitrification of Nuclear Waste Glass Based On Simultaneous Differential Scanning Calorimetry–Thermogravimetry (DSC-TGA) 3
Jaehun Chun, David Pierce, Richard Pokorny, Pavel Hrma

(11d) Technology Development for Radioactive Sludge Treatment At the Hanford Nuclear Cleanup Site 4
Thomas W. Fogwell

(11e) Noble Metal Fission Product Recovery By Electrodeposition in Ionic Liquids 5
Eriugen Gjoka, Katia Ascue, Elizabeth J. Biddinger

(29a) Economic Evaluation of Freshwater Conservation Strategies for Thermoelectric Power Plants 6
Michael E. Walker, Ranjani B. Theregowda, Iman Safari, Javad Abbasian, Hamid Arastoopour, David A. Dzombak, Ming-Kai Hsieh, David C. Miller

(29b) Modeling Of Pollutant Transport in the Nile Delta, Egypt 7
Mohamed Mostafa, Robert W. Peters

(29c) Evaluation of RO Concentrate Treatment for High Recovery Desalting of High Salinity Inland Brackish Water 8
Anditya Rahardianto, Brian C. McCool, Jose Faria, Yoram Cohen

(29d) Simulation and Sustainability Analysis for Produced Water Thermal Desalination Technologies 9
Liwen Chen, Helen H. Lou

(29e) Dispatching Electricity Generation to Increase Water Availability in Times of Drought: A Feasibility Study in Texas 10
Adam Pacsi

(29f) A Novel Method to Manage the Concentrate Disposal of Desalination Units 12
Saeid Aghahossein Shirazi, Alireza Saraeian, Jalal Rastegary, Abbas Ghassemi

(32a) Overview Of Materials In The Built Environment 13
Ron Judkoff

(32b) Energyplus Model-Based Predictive Control (EPMPC) Using Matlab/Simulink and Mle+ 14
Jie Zhao, Khee Poh Lam, B. Erik Ydstie

(32c) The Sol-Char Toilet: Concentrated Solar Power Transported By Fiber Optics for Pyrolysis of Human Waste to Biochar 15
Richard P. Fisher, Allan Lewandowski, Tesfayohanes W. Yacob, Ryan B. Mahoney, Dragan Mejic, Cori Oversby, Sara Beck, Barbara J. Ward, Josh Kearns, R. Scott Summers, Karl G. Linden, Alan W. Weimer

(32d) Building-Integrated Active Modular Phytoremediation System 16
Anna Dyson, Ahu Aydogan Akseli, Jason Vollen, Marianne Nyman, Jeffrey Bird

(32e) Integrating Technological and Ecological Design for Sustainable Buildings On a University Campus 17
Erin L. Gibbemeyer, Bhavik R. Bakshi

(32f) Energy Conservation Approach to Cooling Process By Employing Preheat Tank and Chilled Water Tanks 18
Dr. Robert W. Peters, Zhuo Li

(32g) Operation of Residential Hybrid Renewable Energy Systems: Integrating Forecasting, Optimization and Demand Response 19
Xiaonan Wang, Nael H. El-Farra, Ahmet Palazoglu

(36a) Green Chemistry, Green Engineering, Green Economy At US EPA 's Sustainable Technologies Division 21
John Leazer

(36b) Sustainability Practice in Industry 22
Joseph B. Powell

(36c) Green Engineering and Sustainability in the Chemical Engineering Curriculum 23
David R. Shonnard, David T. Allen, Nhan Nguyen, Sharon Austin

(36d) Driving Private Sector Adoption of Greener Processes and Technologies 24
Abby Lindsay

(36e) Minimization of the GHG Emissions At the Macroeconomic Level Via a Multi-Objective Input-Output Approach: A Case Study of the EU-25 Economy 25
Daniel Cortés, Antonio Ruiz-Hernández, Gonzalo Guillén-Gosálbez, Maria Llop, Roger Guimerá, Marta Sales-Pardo

(36f) Economic Analysis and Optimization of Carbon Capture Storage Pipeline Transport Network System 27
Seungkyu Dan, Byung Joon Kang, Seonghyun Cho, En Sup Yoon

(36g) Economic Development Through Entrepreneurship and Technology Innovation: Bringing Together Industry, Academia, and Government 28
Nada Marie Anid

(37a) The Carbon Capture Simulation Initiative (CCSI): Computational Tools To Accelerate Commercial Development 29
David C. Miller

(37b) The Four Dimensional Product: Integration Over Time Is the Only Way to Understand Sustainability 30
William Banholzer, Mark E. Jones
(37c) Maximization of Energy Efficiency of a Combined Heat and Power Plant 31
Thomas F. Edgar, Kody Powell, Jong Kim, Kriti Kapoor
(37d) Pan American Issues on Biofuel Sustainability 32
David R. Shonnard
(65a) Templated Ordered Mesoporou Materials for Supercapacitors and Lithium Ion Batteries 33
Jinwoo Lee
(65b) Reduced Graphene Oxide Wrapped FeS Nanocomposite for Lithium-Ion Battery Anode With Improved Performance 34
Ling Fei, Hongmei Luo
(65c) Alkanethiol-Passivated Germanium Nanowires As High-Performance Anode Materials for Lithium-Ion Batteries 35
Hsing-Yu Tuan, Fang-Wei Yuan
(65d) A Facile Microwave-Assisted Route to Co(OH)2 and Co3O4 Nanosheet for Li-Ion Battery 36
Gen Chen, Hongmei Luo
(65e) Li Adsorption and Intercalation in Graphene, Graphite and Carbon Onions By Reaxff Reactive Force Field 37
Muralikrishna Raju, Panchapakesan Ganesh, Paul R. C. Kent, Adri C. T. Van Duin
(65f) Silicon-Rich Carbon Nanofibers From Water-Based Spinning to Control Combination of Silicon and Carbon for Li-Ion Battery 38
Yong L. Joo, Yong Seok Kim, Kyung Woo Kim, Daehwan Cho, Nathaniel S. Hansen
(65g) Electrochemical Properties of Carbon-Coated SnO2-Graphene Nanocomposites for Rechargeable Lithium-Ion Batteries 39
Mahbuba Ara, Dr. Steven Salley, Simon K. Y. Ng
(65h) Multilayer Hybrid Electrodes of TiO2/Mwnt for Electrochemical Applications 40
Md. Nasim Hyder, Reza Kavian, Yang Shao-Horn, Paula T. Hammond
(65i) Facile Preparation of Mesoporous α-Fe2O3 With Unique Cocoon-Like Hollow Nanostructure for Reversible Lithium Ion Storage 41
Jian Zhu, K. Y. Simon Ng, Da Deng
(86a) Biofuel Production Wastewater Minimization and Reclaim 42
Marc Privitera
(86b) Macroscopic Water Integration Direct Recycle Applications Within Industrial City Plots 43
Sabla Alnouri, Patrick Linke, Mahmoud El-Halwagi
(86c) Predictive Modelling of a Conventional Clean Water Treatment Work 44
Eva Sorensen, Nina F. Thornhill, Folashade Akinmolayan
(86d) Bringing (Mass) Balance to California Water: A Role for Engineers in Shaping Policy 45
Maurice Hall
(86e) Regulatory and Cost Drivers Affecting Industrial Water Management 46
Brian Petty
(86f) Feasibility of Using Reclaimed Water for Thermoelectric Power Plant Cooling 47
Ashlynn S. Stillwell, Michael E. Webber
(86g) Evaluation of Effectiveness of Condensate Recovery and Its Reuse As a Make-Up Water Requirement in Chilled Water Plants At the University of Alabama At Birmingham 48
Atul Kajale, Dr. Robert W. Peters, Matthew Winslett
(101a) Advanced Treatment of Shale Gas Frac Water to Produce NPDES Quality Water 49
Robert S. Dahlin, Bill Chatterton, Behrang Pakzadeh, Austin Vaillancourt
(101b) Recovering/Reuse of Sodium Chloride from Zero Discharge Desalination Concentrate Streams 50
Alemayehu Yetayew, Lucy Mar Camacho, Malynda A. Cappelle, Tom Davis
(101c) Advances in Electrosorption By Mesoporous Carbon Materials for Water Desalination 51
Ketki Sharma, Richard T. Mayes, Jim Kiggans, Sotira Yiacoumi, David W. Depaoli, Sheng Dai, Costas Tsouris
(101d) Evaluation of Ceramic Membrane Filtration for the Removal of Dbp Precursors From Surface Water Supplies 52
Alexander P. Mathews, Steven Fann, Leila Maurmann
(101e) Efficacy of Membrane Bioreactor Process Combined With Advanced Oxidation Processes for Organic Removal in Water Reclamation and Reuse 53
Woonhoe Kim, Varadarajan Ravindran, Massoud Pirbazari
(101g) Sanitation Shallow Water and Groundwater Contaminated With Petroleum Hydrocarbons 55
Yair Cruz, Jose J. Castro, Enrique Rico
(134a) Identifying Defect Sites in Lithium-Ion Battery Materials: Local Disorder in LiVPO4F and Its Influence On Bulk Properties 56
Robert J. Messinger, Michel Ménétrier, Dany Carlier, Jean-Marcel Ateba Mba, Laurence Croguennec, Christian Masquelier, Dominique Massiot, Michaël Deschamps
(134b) Electrochemical Properties and Characterization of Li1.2Mn0.54Co0.13Ni0.13O2 Composite Cathode Powders Prepared by Ultrasonic Spray Pyrolysis 57
Alex Langrock, Sheryl H. Ehrman, Chunsheng Wang
(134c) Atomic Layer Deposition (ALD) On the Nanostructured Li-Mn-Rich Composite Li1.2Ni0.13Mn0.54Co0.13O2 Cathode Powder 58
Xiaofeng Zhang, Ilias Belharouak
(134d) Metal Fluoride/Graphene Composites for High-Performance Li-Ion Cathodes 61
Cary M. Hayner, Xin Zhao, Yue Yang Yu, Mayfair C. Kung, Harold H. Kung

(134e) Nanostructured Li2S-Graphene Composites As Cathode for High Energy Density Li-S Batteries ... 62
Kai Han, Jingmei Shen, Cary M. Hayner, Yue Yang Yu, Mayfair C. Kung, Harold H. Kung

(134f) Carbon/Sulfur Microspheres With Multi-Modal Pore Structures for Lithium-Sulfur Battery Cathodes ... 63
Cunyu Zhao, Lianjun Liu, Huilei Zhao, Ying Li

(134g) Nanofiber-Based Cathode Electrocatalysts Supported On Carbon Substrates for Lithium-Air Rechargeable Batteries ... 64
Yong Lak Joo, Jun Yin, Jangwoo Kim

(134h) Low-Temperature Dehydrogenation From Aminoborane Complexes Under Carbon Dioxide Atmospheres ... 65
Jae W. Lee, Ran Xiong, Junshe Zhang

(134i) Intermittent Electrical Energy Storage By Liquid Organic Hydrogen Carriers ... 66
Wolfgang Arlt, Karsten Müller, Daniel Teichmann

(134j) Synthesis of Nanoporous Palladium Powder With Controlled Pore and Particle Size for Hydrogen Storage Applications ... 67
Patrick Cappillino, Christopher Jones, Khalid Hattar, Blythe Clark, Michelle Hekmaty, Benjamin Jacobs, David B. Robinson

(146a) Development of Composite Hollow Fiber Membranes for Aqueous and Non-Aqueous Nanofiltration ... 68
Shi-Peng Sun, Sui Yung Chan, Tai Shung Chung

(146b) Self-Heat Recuperative Seawater Desalination Process Using Fluidized Bed ... 69
Hiroyuki Mizuno, Yasuki Kansha, Masanori Ishizuka, Atsushi Tsutsumi

(146c) Characteristics of Water and Salt Transport, Free Volume and Their Relationship With the Functional Groups of Novel Cellulose Esters ... 70
Rui Chin Ong, Bradley J. Helmer, Jos S. De Wit, Tai Shung Chung

(146d) Bioremediation of Cu(II) Contaminated Aqueous Solution Using Acinetobactor Sp ... 71
Subhajit Majumder, Gayathri Gangadhar, Smita Raghuvanshi, Suresh Gupta

(146e) Calcium Ions in Gypsum (CaSO4 2H2O) Scaling in Forward Osmosis Processes and a Novel Design Strategy to Fabricate Polybenzimidazole Anti-Scaling Hollow Fiber Membranes ... 72
Si Cong Chen, Jincai Su, Feng-Jiang Fu, Baoxia Mi, Tai-Shung Chung

(146f) New Insights Into Hydrophobic/Hydrophilic Composite Hollow Fibers for Direct Contact Membrane Distillation ... 74
Jia Ming Zhu, Lan Ying Jiang

(156a) A Study On the Survival of Oral and Faecal Bacteria in Different Beverages ... 75
Chiedozie Mbachu

(156b) Potable Water Production Via Microchannel Capacitive Desalination Technology (CDT) System Architecture ... 76
Christopher Loeb, Goran N. Jovanovic

(156c) Disinfection of Water By Adsorption Combined With Electrochemical Treatment ... 77
Syed N. Hussain, Nuria De Las Heras, Hafiz M. A. Asghar, Nigel W. Brown, Edward P. L. Roberts

(156d) Municipal Wastewater Treatment By An Osmotic Membrane Bioreactor: Use of Sodium and Magnesium Chloride As Draw Solution ... 78
Guanglei Qiu, Yen-Peng Ting

(156e) Scalability Analysis of An Open Channel UV Wastewater Disinfection Reactor Using CFD Simulation ... 79
Madhumita B. Ray, Rajib K Saha, Chao Zhang

(156f) Performance Evaluation of a Low Cost and High Rate Vertical Flow Wetland for Residential Graywater Treatment and Reuse ... 80
Zita L. T. Yu, Anditya Rahardianto, Michael K. Stenstrom, Yoram Cohen

(156g) Coagulation and Ultrafiltration of High-Alkalinity Greywater ... 81
Ted J. Amundsen, Dustin J. Zastrow, Andrew L. Wagner

(160a) Heterogeneous Chlorine Chemistry in the Atmosphere: Insights From Modeling and Experiments ... 82
Cameron Faxon, David T. Allen

(160b) Chemical and Physical Investigation of Fractal-Like Iodine Oxide Particle Formation ... 85
Shunsuke Nakao, Gookyoung Heo, Bethany Warren, Li Qi, Hiroyuki Hagino, William Carter, David Cocker

(160c) Changes in Water-Soluble Particulate Composition From Vehicles ... 86
Daniel Short, Diep Vu, Georgios Karavalakis, Thomas Durbin, Akua Asa-Awuku

(160d) Condensed-Phase Photolysis Processes in Secondary Organic Aerosol ... 87
Scott A. Epstein, Mallory L. Hinks, Sergey A. Nizkorodov

(160e) Stratocumulus Cloud Water Properties Off the California Coast During the 2011 E-PEACE Field Campaign ... 88
Armin Sorooshian, Zhen Wang, Matthew Coggon, Haflidi Jonsson

(160f) Aqueous Aerosol Chemistry: SOA Formation and Impacts On Aerosol Physical Properties ... 89
V. Faye McNeill

(160g) Effect of Humidity On Secondary Organic Aerosol (SOA) Formation From Biogenic Hydrocarbons and Nitrate Radicals ... 90
Ng Lee Ng, Christopher M. Boyd, Lu Xu

(160h) The Effect of Urban Biomass Combustion for Space Heating On PM Exposure ... 91
Dimosthenis Sarigiannis, Spyros Karakitsios, Marianthi Kermenidou, Spyridoula Nikolaki

(160i) Carbon Dioxide Control Strategies for Power Generation Using Analytic Hierarchy Process ... 94
Mohammed S. Ba-Shammakh

(192a) Fabrication of Porous Carbon Nanofibers With Adjustable Pore Sizes As Electrodes for Supercapacitors ... 95
Chau Tran, Vibha Kalra

(192b) Enhanced Energy Storage By Tunable Electrolyte Confinement in Structure-Directed CNT Arrays ... 96
Katherine T. Nicol, Dustin Zastrow, Justin J. Hill

(192c) Multifunctional Nitrogen-Rich "Brick-and-Mortar" Carbon As High Performance Supercapacitor Electrodes and Oxygen Reduction Electrocatalysts 97
Dingshan Yu, Li Wei, Yuan Chen

(192d) Multiwalled Carbon Nanotubes With Tuned Surface Functionalities for Electrochemical Energy Storage 98
Huige Wei, Hongbo Gu, Jiang Guo, Suying Wei, Zhanhu Guo

(192e) Three-Dimensional Core@Shell Nanostructured Array for Microscale Electrochemical Energy Storage 99
Yuanbing Mao, Xing Sun

(192f) Microfabricated Nickel Oxide Supercapacitors Based On High Aspect Ratio Concentric Cylindrical Electrodes 100
Andac Armutlulu, Sue Ann Bidstrup Allen, Mark G. Allen

(192g) Pretreatment Effects On Charge Storage of Early Transition-Metal Carbides and Nitrides 101
Abdoulaye Djire, Priyanka Pande, Alice E. S. Sleightholme, Aniruddha Deb, Paul G Rasmussen, James Penner-Hahn, Levi T. Thompson

(192h) Doped Transition Metal Oxide Composite Electrodes for Supercapacitor Applications 104
Prashanth Jampani Hanumantha, Karan Kadakia, Dae Ho Hong, James Poston, Manivannan Ayyakkannu, Prashant N. Kumta

(192i) Novel 3-D MnO2/Holey Graphene Nanostructure for Supercapacitor Applications With Enhanced Electrochemical Performances 105
Lixin Wang, Tiejun Meng, Mahbuba Ara, Da Deng, Simon Ng

(192j) Electrochemical Energy Storage of Magnetic Carbon Nanocomposites: Role of Magnetocapacitance and Magnetohydrodynamics 106
Jiahua Zhu, Suying Wei, Zhanhu Guo

(192k) Synthesis and Characterization of Orange Peel As Electrodes in Li-Ion Capacitors 107
Arenst Andreas Arie

(198a) Biomass Resources and Current Status Of Biofuels In China 108
Zhenhong Yuan

(198c) Air Quality Impacts of Natural Gas Production and Use 110
David T. Allen

(198d) Disasters Risks Reduction and Sustainable Development 111
Kenza Robinson

(215a) Supercritical Water Oxidation: the Next Generation Process to Incineration 112
Bushra Duri

(215b) Different Behavior of H2O2 in Aldicarb and Hexazinone Oxidized By O3/H2O2 113
Li Shaofeng

(215c) Enhancing Selectivity of Salicylic Acid Degradation By Using Molecular Imprinted Doped TiO2 Under Solar Light 114
Ziqiang Wang, Wenqiang Li, Xiang Liu

(215x) Optimal Process Conditions for Preparation of Porous Carbons and Methylene Blue Adsorption 115
C Srinivasakannan

(215ad) Integrated Bioenergy Farming System Using Communal Waste WATER – A NOVEL Approach To Carbon Mitigation 116
Ramanujam Praveen Kumar, Balasubramanian Bharathiraja, Sivasankaran Chozhavendhan, Iyyasamy Regupathi, Prukinthraj Moscow

(215d) Fabrication of Reusable Ag/BiVO4 for Water Disinfection 117
Amin Yoosefi Booshehri, Rong Xu

(215e) Visualization of Coupled Mass Transfer and Reaction Between Plasma and Liquid Phase in a Hybrid Gas-Liquid Dielectric Barrier Discharge Reactor for Water Purification 118
Xuelan Feng, Ting Shao, Binhang Yan, Wei Lu, Wentan Wang, Yi Cheng

(215ac) Removal of Phosphate and Arsenic From Aqueous Solution By Metal(hydr-)Oxide Coated Sand 119
Yuxiong Huang, Jae-Kyu Yang, Arturo A. Keller

(215m) Experimental Study On Iodine Adsorption By Silve-Exchanged Mordenite 120
Ronghong Lin, Lawrence L. Tavlarides

(215o) Study On the Wastewater Reuse By Combination of Thermophilic Anaerobic Bioreactor With Forward Osmosis (FO)/Membrane Distillation (MD) Integrated Module 121
Jin Song Zhang, Anthony G. Fane, Joseph Zheng Xin Kwan, Shuwen Goh

(215p) Removal of Oil From Oily Waters By Chitosan Microspheres: Batch Studies 123
Izabel C. Grem, Yure G. Queirós, Bruno S. Garcia, Claudia E. Mansur

(215g) Study On Supercritical Water Oxidation of Quinoline Wastewater 124
Yonghui Li, Bangxing Zheng, Jing Ma, Guiming Li, Xiuqin Dong

(215h) Photoelectrocatalytic Performance of TiO2/SnO2 Composite Thin Film Electrode 125
Yun Wei, Xingfu Zhou

(215y) Hydrothermal Pre-Treatment for Mesoporous Carbon Synthesis: Enhancement of Chemical Activation 126
Akshay Jain, M. P. Srinivasan, Rajasekhar Balasubramanian, Zhengang Liu

(215f) An Integrated Solar Photocatalysis-Organic Membrane Separation and SBR Process for the Treatment of Pharmaceutical Wastewater 127
Chunmiao Shen, Liping Xie, Yuzhong Zhang, Xiaoyao Tan, Chenxi Zou, Jun Wang

(215q) Investigating the Performance of a Continuous Electrocoagulation Unit for the Combined Removal of Nitrate and Phosphate Ions From Wastewater 131
Ahmad El-Shazly, Muhammad A. Daous

(215i) Efficient Treatment of Pharmaceutical Wastewater By Heterogeneous Fenton-Like Process With Three-Dimensional and Nano-Scaled Perovskite Based Catalyst 133
Hongying Zhao, Jinlei Cao, Yanbin Wang, Guohua Zhao
(215r) An ATR-FTIR Spectroscopic Approach for Measuring Active Functional Groups in Arsenic and Selenium Stabilization in Sulfur Environment 134
Bahngmi Jung, Bill Batchelor, Ahmed Abdel-Wahab
(215s) Degradation of Artificial Humidity Condensate in the Confined Space Using Large-Area Boron Doped Diamond Electrode 135
Hao Li, Lecheng Lei, Bin Yang
(215t) Effluent Treatment Using Hydrodynamic Cavitation: Vortex Diode As a Cavitating Device 140
R. S. Hiremath, Vinay M. Bhandari, Vivek V. Ranade
(215ae) Uniform Single Crystalline Sulphide-Based Quaternary Alloy Hexagonal Nanorings for Highly Efficient Photocatalytic Hydrogen Evolution 143
Timothy T. Y. Tan
(215j) Graphene Membranes for Water Nanofiltration and Desalination 145
Miao Yu
(215z) Ionic Liquid Mediated Electrochemical Reduction of CO2 in a Microbial Electrolysis Cell 146
Zhongjian Li, Qianli Wang, Lecheng Lei
(215aa) New Adsorbents and Material Modifications- An Approach for Ultraclean Fuels 149
Laxmi Gayatri Sorokhaibam, Vinay M. Bhandari, Vivek V. Ranade
(215v) Dynamic Modelling and Simulation of Anaerobic Digester For High Organic Strength Waste 151
Pooja Sharma, U. K. Ghosh, Amiya Kumar Ray
(215ab) Development of a Low Cost Carbon Materials and Its Functionalisation: for Waste Water Treatment 154
Chen J. Paul, Narahari Mahanta
(215u) Removal of Disinfection By-Products (DPBs) By Electro-Coagulation Process in a Continuous Flow Reactor As a Localized Water Treatment Technology 155
Cheng-Chun He, S. L. Lo
(215w) Biofiltration of Cu (II) Using Acclimated Mixed Culture 156
Suresh Gupta, Smita Raghuvanshi, Gayathri Gangadhar, Subhajit Majumder
(215k) Removing Microorganisms From Wastewater in a Painting Industry By Advanced Oxidation Processes 158
Wounjim Chang, Lourdes F. Vega, Sonia Guri
(215n) Life Cycle Assessment of the Gasplasma Process: An Innovative Waste Management Option 159
Carla Tagliaferri, Sara Evangelisti, Paola Lettieri, Chris Chapman
(215l) Kinetics Analysis for Development of a Rate Constant Estimation Model of Ultrasonic Degradation Reaction 161
Daisuke Kobayashi, Chiemi Honma, Hideyuki Matsumoto, Tomoki Takahashi, Yuichiro Shimada, Chiaki Kuroda, Katsuto Otake, Atsushi Shono
(226a) New Developments on Membranes for CO2 Separation and Capture 162
Yuanxin Chen, Kartik Ramasubramanian, Michael A. Severance, Zi Tong, Varun Vakharia, Lin Zhao, Yanan Zhao, Prabir K. Dutta, W. S. Winston Ho
(226b) Organic Solvent Nanofiltration Molecular Separations In Organic Liquids 163
Andrew G. Livingston
(226c) Sustainability Issues in Membrane Systems for Bioprocessing 164
Andrew L. Zydney
(226d) Membrane Separations As "Lever" Points in Biomass-Based Processes 165
John Pellegrino
(155e) Progresses in Membrane Integrated Systems for Water, Energy and Raw Materials Production 166
Enrico Drioli, Tai-Shung Chung, Francesca Macedonio
(225a) Fundamental Developments and Economic Feasibility of AOPs Involving Ultrasound for Environmental Remediation 167
Yusuf G (Debo) Adewuyi, Dr. Robert W. Peters
(225b) Advanced Oxidation of Estrone in Water and Wastewater 168
Shubhajit Sarkar, George Nakhla, Madhumita B. Ray
(225c) Analysis of Ozone Absorption and Reaction Kinetics in the Oxidation of Organics in Effluents in a Stirred Tank Reactor 169
Alexander P. Mathews, Kishora K. Panda
(225d) Electrochemical Wastewater Treatment: Applications and Scale-Up 170
Dieter Woisetschläger, Matthäus Siebenhofer
(225e) Case Study - In-Situ Oxidation of Ground Water Contaminated By Petroleum Hydrocarbons, Acetone and Chlorinated Hydrocarbons At a Former Industrial Facility in Northern California 171
Riffat Maqbool Qadir
(225f) Engineering Challenges of Implementing the Peroxone Process for in-Situ Oxidation of Ground Water Contaminated By Acetone and Petroleum Hydrocarbons At a Former Industrial Facility 172
Riffat Maqbool Qadir
(225g) Heterogeneous Fenton Oxidation of Synthetic Dye in Water With Surface-Modified Iron-Amended Activated Carbon 173
Jihyun R. Kim, Branden Santiano, Hyosang Kim, Eunsung Kan
(225h) Reaction Kinetics and Mechanistic Studies of Nitric Oxide Removal By Combined Aqueous Persulfate and Ferrous-EDTA Systems 174
Yusuf G (Debo) Adewuyi, Md Arif Khan

(225i) Sonication Treatment of Sump Water Containing Diesel Fuel 176
Dr. Robert W. Peters, Roxanne Bessette
(225j) UV-Based Treatment of Fluoroquinolone Antibiotics in Wastewater 177
Sebastian Snowberger, Ke He, Lee Blaney
(242a) Simultaneous Adsorption and Heterogeneous Oxidation of Endocrine Disrupting Compounds in Wastewater Using Nano Metal Catalyst-Deposited Carbon Nanotubes 178
Vincent Cleveland, Jon-Paul Bingham, Eunsung Kan
(242b) Study of Carbon-Hydrogen Bond On the Microstructure of Hydrogenated Graphite 179
Y. Zhang
(242c) Efficient Removal of Both Arsenic and Fluoride By Lanthanum Modified Carbon Nanoparticles 180
Guoting Li, Yang Yu, J. Paul Chen, Yue Ma
(242d) Efficient Removal of Fluoride By Lanthanum Modified Seaweeds Sargassum S.P 181
Guoting Li, Yang Yu, J. Paul Chen
(242e) Size-Dependent Removal of Arsenite From Aqueous Solution By Akaganeite (β-FeOOH) 182
Song-Hai Wu, Yong-Li Sun, Shao-Yi Jia, Ting-Ting Zan, Jun Wang
(242f) Removal of Ammonia From Water Using Nanoscale Zero Valent Irons Modified Fly Ash-Derived Zeolite Adsorbents 183
Hou Li'An, Liu Minmin
(242g) Drinking Water Disinfection Using Silver Nanoparticle Impregnated Activated Carbon Hybrid 184
Pritam Biswas, Rajdip Bandyopadhyaya
(245a) Implementation of Tower Gardens On An Urban University Campus 186
Dr. Robert W. Peters, Natalia Barbour, Sharyn Gaston, Mark Burns
(245b) New Developments in Implementing Green Engineering in the Chemical Engineering Curriculum 187
David R. Shonnard, David T. Allen, Nhan Nguyen, Sharon Austin
(245c) Incorporating Process Integration Methodologies Into Senior Capstone Design Courses 188
Russell F. Dunn, Kenneth A. Debelak
(245d) Assembling Engineering for Sustainable Development Programs 189
Cory D. Jensen
(337a) Ambient-Air Plasmas for Frugal and Sustainable Water Disinfection in the Developing World 190
Matthew J. Pavlovich, Connor Galleher, Toshisato Ono, Zdenko Machala, Douglas S. Clark, David B. Graves
(337c) Chemical Degradation of Xanthate By Pulse Corona Discharges 191
Katherine V. Cerda, Pedro E. Arce
(337d) Pursuing a Viable Plasma-Based Water Treatment Process: Identifying Transport Limitations and Investigating the Effect of Reactor Design On the Degradation of Bisphenol A 192
Selma Mededovic, Christopher Bellona, Fei Dai, Joshua Franclemont, Thomas Holsen, Gunnar Stratton, Chao Zhang
(337e) Enhanced Photocatalytic Degradation of Malachite Green By Visible Light Sensitive ZnO/ZrO2 Doped TiO2 Nanoparticles 193
Anuradha Shende, Rajesh V. Shende
(337f) Endogenous Hormones Degradation Using Nanotubular Oxide Layer Grown On Tiw Alloy By Electrochemical Anodization 194
Christiane A. Rodrigues, Marizilda E. Oliveira, Bruno L. Barroso, Mariany L. M. Gomes, Rodnei Bertazzoli, Maria De Lourdes L. Moraes
(337g) Flame Aerosol Synthesized Cr Incorporated TiO2 for Visible Light Photodegradation of Acetonitrile 196
Siva Nagi Reddy Inturi, Thirupathi Boningari, Makram Suidan, Panagiotis Smirniotis
(359a) Nanoparticle Emission Characterization and Management in Nanocoating Material Spray Operation 197
Rohan Uttarwar, Yinlun Huang
(359b) Innovative Room-Temperature Preparation of One-Dimensional NiO Nanowires and Their Energy and Environmental Applications 198
Xinghua Meng, K. Y. Simon Ng, Da Deng
(359c) The Direct Activation of Targeted Partial Oxidation Reactions in Liquid Phase Via Enhanced Corona Discharge Techniques From Nanofibrous Emitter Electrodes in Microreactor Channels 199
Alexandre F. T. Yokochi, Justin Pommerenck, Jordan Pommerenck, Peter Kreider, Fan Xiangru, Yousef Alanazi
(359d) Magnetic Nanocomposite Microparticles for On/Off Binding of Persistent Organic Pollutants 200
Angela Maria Gutierrez, Brad J. Newsome, Thomas D. Dziubla, J. Zach Hilt
(359e) Carbon Nanocomposite Fabrics Toward Cr(VI) Removal From Polluted Water 201
Jiahua Zhu, Suying Wei, Zhanhu Guo
(359f) Pilot Scale Evaluation of Nano-Size Adsorbent for Arsenic Contaminated Water Treatment 202
Yang Yu, J. Paul Chen, Guoting Li, Narahari Mahanta
(359g) Composition and Shape Controlled Syntheses of ZnxCd1-Xs Photocatalysts for Reducing CO2 to Methyl Formate in Methanol 203
Wanlin Jiang, Xiaohong Yin, Feng Xin, Dandan Sui
(361a) First Principles Investigation of SOx (x = 0 - 4) Chemistry On Pt (111) and Pd (111) 204
Hom Sharma, Vinit Sharma, Ashish B. Mhadeshwar, Ramamurthy Ramprasad
(361b) In-Situ Drifts Studies of Selective Catalytic Reduction of NO With C3H6 On Cu-Chabazite Catalyst 206
Richa Raj, Michael P. Harold, Vemuri Balakotaiah
(361d) Development of Low-Temperature SCR Catalysts for the Cold-Start Applications of NOx Reduction 208
Thirupathi Boningari, Padmanabha Reddy Ettireddy, Adam Kotrba, Thomas Spinks, Panagiotis Smirniotis
(361e) Effect of Mg/Al Ratios in HT Supports On NOx Storage-Reduction Processes Over Pt-BaO/Hydrotalcite(HT)-Derived Support 209
Soyeon Jeong, Seunghee Youn, Do Heui Kim

(361f) NH3-SCR Over Cu/SAPO-34 - Zeolite Acidity and Active Site Changes As a Function of Cu Loading 210
Di Wang, Li Zhang, Krishna Kamasamudram, William Epling
(361g) In-Situ Generated Hydrogen Driven Reductive Photocatalysis 211
Sravani Gullapalli, Juan C. Velazquez, Lori A. Pretzer, Michael S. Wong
(364a) Energy-Optimal Control of An Integrated UF-RO Seawater Desalination Plant 212
Larry Gao, Anditya Rahardianto, Han Gu, Panagiotis D. Christofides, Yoram Cohen
(364b) Iron Electro-Coagulation for Simultaneous Arsenic Removal and Microbial Attenuation: Mechanisms of E.Coli Removal and Inactivation 213
Caroline Delaire, Susan Amrose, Case Van Genuchten, Ashok Gadgil
(364d) Comparison of Efficiency for Impellers Commonly Used in Anoxic Denitrification 214
Richard K. Grenville, Jason J. Giacomelli, Robert W. Higbee, Todd M. Hutchinson, Christopher J Knecht, Janusz Roszczenko, Wojciech Wyczalkowski, Jack Barnes
(364c) Design of Anaerobic Fluidized Bed Reactor Technology for Microbial Sulfate Reduction in Surface and Ground Water Sources 215
Masoud Samee, Varadarajan Ravindran, Massoud Pirbazari
(374a) Simulation Tools for Nanoparticle-Based Composite Processing and Property Prediction 217
Dan S. Bolintineanu, Jeremy B. Lechman, P. Randall Schunk
(374b) Tin Oxide Nanowires and Their Hybrid Architectures for Kinetically Fast Redox Couples in Dye-Sensitized Solar Cells 218
Venkat Kalyan Vendra, Tu Nguyen, Thad Druffel, Mahendra Sunkara, Delaina A. Amos
(374c) Multiscale Sustainability Assessment in Nanocoating Material Design and Manufacturing 219
Hao Song, Rohan Uttarwar, Yinlun Huang
(374d) Continuous Nanoparticle Sizing and Characterization Via Microfluidic Interfacial Fluorescent Complexation 220
Fanxu Meng, Victor M. Ugaz
(374e) Safer Formulation Concept for Flame-Generated Engineered Nanomaterials 221
Georgios A. Sotiriou, Samuel Gass, Joel Cohen, Georgios Pyrgiotakis, Sotiris E. Pratsinis, Philip Demokritou
(374f) Green Pathways for Development of Nanostructured Aerogel Photocatalysts Effective At Both UV and Visible Range 222
Haitao Li, Sermin G. Sunol, Aydin K Sunol
(387a) Findings On Life Cycle Aquatic Ecotoxicity Impacts of Ionic Liquids 223
Amirhossein Mehrkesh, Arunprakash T. Karunanithi
(387b) Comparative Life Cycle Assessment of Vaccine Cold Chain Infrastructure in Developing Countries 224
Bahador Mousavi, Arunprakash T. Karunanithi
(387c) Hybrid Life Cycle Assessment Model of Silicon Photovoltaics 225
Yuan Yao, Yuan Chang, Eric Masanet
(387d) Root-Cause Analysis in the Design of Chemical Processes 227
Preeti Gangadharan, Liwen Chen, Helen H. Lou
(387e) Development of Product-Based Energy Intensity Metrics for the California Food Industry 228
Michael E. Walker, Margot Hutchins, Dave Lettieri, Vince Romanin, Paul Blinde, Eric Masanet
(387f) Estimations of Direct and Indirect Emissions From Mobile Air Conditioner Units in Automobile Transportation: The Cool Mile Functional Unit 229
Paul Blowers, Ibrahim Alayedi, Austin Dougless, Brian Gerwe, Adam Mesta, Caleb Rugel, Yi Yang Wang
(387g) Regional Air Quality Improvement By Scheduling of Multiple Chemical Plant Start-Ups 230
Tianxing Cai, Qiang Xu
(405a) Lean NOx Reduction By H2/CO/C3H6 Using NSR and SCR Catalysts 231
Arun S. Kota, Dan Luss, Vemuri Balakotaiah
(405b) Fast Lean-Rich Cycling for Enhanced NOx Conversion 232
Charles Perng, Michael P. Harold
(405c) Experimental and Kinetic Study of the SO2 Oxidation Reaction Over a Pt/Al2O3 Catalyst 234
Tayebeh Hamzehlouyan, Chaitanya Sampara, William Epling, Junhui Li, Ashok Kumar
(405d) Ozone Removal Using New Structure of Matter for High Performance Heterogeneous Catalysis At Short Contact Time 236
Qiang Gu, Bruce Tatarchuk
(405e) Thermal Analysis of Active Catalytic Diesel Particulate Filter Regeneration 237
Changsheng Su, Ying Yuan, James Zokoe, Matthew Henrichsen, Paul McGinn
(405f) Kinetics of Passive Soot Oxidation On a Model Un-Catalyzed and Commercial Catalyzed Diesel Particulate Filter 238
Ashok Kumar, Randy Stafford, Krishna Kamasamudram, Neal Currier, Aleksey Yezerets
(405g) Visible Light Induced Degradation and Disinfection Using Multifunctional Ag/AgCl – Activated Carbon Composite Photocatalysts 239
Joanne Gamage McEvoy, Wenquan Cui, Zisheng Zhang
(421a) Multicomponent Ppb Competitive Adsorption in the Presence of Varying Relative Humidity 240
Peng Cheng, Bruce Tatarchuk
(421b) Removal of Low Vapor Pressure Perfluoroalkanes By Activated Carbon 241
Bryan J. Schindler, James H. Buchanan, John J. Mahle, Gregory W. Peterson, T. Grant Glover
(421c) The Design and Analysis of Activated Biochar Material for the Control of Elemental Mercury Emission From Industrial Facilities Into the Global Atmospheric Pool 242
Stanley R. Shewchuk, Ajay K. Dalai, Dave Smith, Mahuya De, Ramin Azargohar

(421d) Mercury Interaction With the Fine Fraction of Coal Fly Ash in a Simulated Coal Power Plant Flue Gas Stream 243
Adam D. Jew, Erik C. Rupp, Ji-Eun Jung, Gabriela Farfan, Gordon E. Brown, Jennifer Wilcox
(421e) Mercury Sorption Technology in Coal-Fired Power Plants 245
Noah D. Meeks
(421f) Gas-Phase Mercury Removal Using Sulfur Doped Porous Carbon 246
K Suresh Kumar Reddy, Ahmed Al Shoaibi, C Srinivasakannan
(421g) Study of H2S Removal At High Temperature With 20%Fe2O3/Al2O3 Sorbent Applying Factorial Experimental Design Method 247
Ching Ying Huang, Yau Pin Chyou
(422a) Engineering Protein Delivery Systems to Protect Pyrethroids (λ-Cyhalothrin) Against Photodegradation 248
Hyoungill Lee, Pranav Thirumalai, Hao Feng
(422b) Toxicity Effects Associated With Exposure of Lung Epithelial Cells to Polymer Nanocomposites and Nano-Platelets of Silicates 249
Sushant Agarwal, Reem Eldawud, Cerasela Zoica Dinu, Rakesh K. Gupta
(422c) The Age of Carbon Nanotubes and Biomolecular Convergence for Cancer Therapeutics 250
Cerasela Zoica Dinu
(422d) Monitoring Ligand Exchange On the Surface of Gold Nanoparticles Using Isothermal Titration Calorimetry 251
Christopher L. Kitchens, Ashley E. Hart, O. Thompson Mefford, Brian A. Powell, Dan D'Unger
(422e) Regulatory Association of Cell Responses Induced By Metal and Metal Oxide Nanoparticles 252
Rong Liu, Bryan France, Saji George, Haiyuan Zhang, Tian Xia, Andre E. Nel, Kenneth Bradley, Robert Rallo, Yoram Cohen
(422f) Investigations of the Oral Uptake of Titanium Dioxide Nanoparticles Via the Buccal Mucosa 253
Birgit Teubl, Gerd Leitinger, Eleonore Fröhlich, Marc Schneider, Claus-Michael Lehr, Andreas Zimmer, Eva Roblegg
(423a) Incorporating Inherently Safer Designs to Improve Sustainability of Industrial Facilities 255
Morgan Reed, Eric Peterson
(423b) Designing Greener Plasticizers: Influence of Geometry of Central Group and Side Chains 256
Hanno C. Erythropel, Sarah Shipley, Tobin Brown, Milan Maric, Richard L. Leask, David G. Cooper
(423c) A Program for Predicting Air Pollution in Chemical, Biological, Radiological, Nuclear and Explosives (CBRNE) Disasters 258
Kwanghee Lee, Jaedeuk Park, Yi Yoon, Sungyun Her, Il Moon, Sunghyun Woo, Hyungjoon Yoon
(423d) Experimental Study and Theoretical Analysis On Decomposition Mechanism of Benzoyl Peroxide 260
Jiayu Lv, Wanghua Chen, Liping Chen
(423e) Determination of Ignition Probabilities for LNG Spills 261
Nirupama Gopalaswami
(423f) Zero Valent Silver-Based Electrode for 2, 4-Dinitrotoluene Detection 262
Jonathan C. Mbah, Kiara Moorer, Nader Vahdat, Samuel Hernandez-Rivera
(423g) Environmental Implications of Phenol Gasification in Supercritical Water 263
Chad M. Huelsman, Phillip E. Savage
(423h) Modeling the Transport of Mercury in Indoor Spaces Following a Release of Mercury Droplets 264
Anthony Gaglione, Ariel Hou, Kurt Rindfusz
(423i) The Influence of Solid Inertants On the Explosion Behavior of Dusts From Mechanical Waste Treatment Facilities 266
Kerstin Held, Harald Raupenstrauch
(427a) Metal Recovery From Spent Catalyst By Means of Biotechnological Strategies to Scale-Up Studies 271
Varsha Bisht, Bina Singh, Priyangshu M Sarma, Ajoy K. Mandal, Banwari Lal
(427b) Synthesis of P25 Doped CoFe2O4/Graphene Magnetic Catalyst and Its High-Performance Photocatalytic Activities Under Visible Light Irradiation 272
Jingwen Sun
(427c) Removal of Cr(VI) From Aqueous Solutions By Self-Flocculating Yeast 273
Lijie Chen, Fengwu Bai
(427d) Speciation of Chromium Adsorbed On Activated Carbon and Stabilization of Lime Precipitated Chromium Residue With Activated Carbon 274
Lei Mao, Shitang Tong, Mingxia Fan, Charles Q. Jia
(427e) Blue Preserve – Recovery Of Oil Spilled In Sea A Review 276
Utkarsh Maheshwari
(427f) An Eco-Friendly Technology for Metal Dissolution Using Biologically Produced Acid From Chelated Exhaust Catalyst Using Acidithiobacillus Thiooxidans and Its Kinetic Aspect 277
Bina Singh, Priyangshu M Sarma, Ajoy K. Mandal, Banwari Lal
(427h) Detoxification of Halogenated Aromatic Compounds By Anaerobic Cultures 278
Chang Ding, Shanquan Wang, Jianzhong He
(427i) Screening for Efficient Hydrocarbon Degradation By Bacterial Isolates From Crude Oil Contaminated Sites of Assam, India 279
Aloke Kumar Ghoshal
(430a) Impact of Acids On the Structure of Linde Type A Zeolites for Use in Reverse Osmosis Membranes for Wastewater Recovery 281
Heather Jamieson, Aaron Waller, Huidan Yin, Mary Laura Lind
(430b) Arsenic Rejection By Nanofiltration Membranes: Experimental Results and Model Prediction 283
Baolin Deng, Jun Fang

(430c) Fouling-Resistant Membranes for the Treatment of Flowback Water From Hydraulic Shale Fracturing: A Pilot Study 284
Daniel J. Miller, Xiaofei Huang, Hua Li, Sirirat Kasemset, Albert Lee, Dileep Agnihotri, Thomas Hayes, Donald R. Paul, Benny D. Freeman

(430d) Microbial Adhesion to Advanced Water Treatment Membrane Materials: Implications for Biofouling 285
Teresa L. Kirschling, Lauren F. Greenlee

(430e) Real-Time Monitoring Approach and System for Detection and Characterization of RO Membrane Fouling 286
John F. Thompson, Anditya Rahardianto, Yoram Cohen

(430f) A Molecular Marker Approach for Characterization of RO Membranes and Solute Transport in Spiral-Wound RO Membrane Systems 287
Sirikarn Surawanvijit, John Thompson, Andi Rahardianto, Yoram Cohen, Val Frenkel

(436a) Hydrogen Production From Water 288
Martin D. McDaniel, John G. Ekerdt

(436b) Core-Shell Photoelectrochemical Electrodes for Water Splitting 289
Qing Peng, Berc Kalanyan, Paul Hoertz, Andrew Miller, Do Han Kim, Kenneth Hanson, Leila Alibabaei, Jie Liu, Thomas Meyer, Gregory N. Parsons, Jeffrey T. Glass

(436c) Design of Photocatalysts Active in Infrared and Visible Range 290
Doh C. Lee

(436d) Efficient Photoelectrochemical Water Splitting With Si-Based Metal-Insulator-Semiconductor Photoelectrodes 291
Daniel V. Esposito, Youngmin Lee, A. Alec Talin, Thomas P. Moffat

(436e) Graphene Oxide Photocatalysts for Water Splitting and Its Upconverted Photoluminescence 292
Te-Fu Yeh, Hsisheng Teng

(436f) The Study of Cu2ZnSnS4 Nanocrystal/TiO2 Nanorod Arrays Heterojuction Photoelectrochemical Cell for Hydrogen Generation 296
Tsung-Yeh Ho, Liang-Yih Chen

(436g) Development of New Generation of Sub-Nanometer Catalysts for Sustainable Energy Applications 297
Qiyuan Wu, Peichuan Shen, Shen Zhao, Girish Ramakrishnan, Dong Su, Yan Li, Alexander Orlov

(436h) ZrO2 Incorporated ZnO/TiO2 Mesoporous Photocatalyst for H2 Generation 298
Abdulmenan Hussein, Rajesh Shende

(444a) Novel Catalysts and Reactor Concepts for Sustainable Processes: the Use of LCA Data in Process Design 299
Polina Yaseneva, Alexei Lapkin

(444b) Techno-Economic and Environmental Assessment of Ionic Liquid-Based Biorefineries 301
Sara Giarola, Nilay Shah

(444c) Optimal Simultaneous Production of Bio-Ibutene and Bioethanol From Switchgrass 303
Mariano Martin, Ignacio E. Grossmann

(444d) Synthesis and Design of Biorefinery Processing Networks With Uncertainty and Sustainability Analysis 305
Peam Cheali, Krist V. Gernaey, Gürkan Sin

(444e) Ecosystems As Unit Operations for Innovative and Sustainable Process Design 307
Bhavik R. Bakshi, Robert A. Urban, Guy Ziv

(444f) Process Analysis of Continuous Catalytic Gasification As a Waste to Energy Alternative 308
Charles F. Tillie, Melissa J. Riedthaler, Shreya Adhikari, Stephen A. Reeves, Jorge E. Gatica

(477a) Remediation of Diesel-Contaminated Soil With a Biostimulation and Bioaugmentation Joint Process 309
Xiaosen Li, Diannan Lu, Liu Zheng

(477b) Electrically Enhanced Contaminant Degradation in Sediment Caps 310
Ruiling Zhang

(477c) pH-Dependent Transport of Energy By-Products in the Subsurface 311
Valentina Prigiobbe, Marc Hesse, Steven Bryant

(477d) Experimental Investigation Using a Contaminant-Monitoring Sytem in Storm Drains: A Dichotomous Preliminary Risk Assessment 313
Dana Lackey, Robert W. Peters

(477e) Innovative Polymer Quantification and Management At a Large Scale Remediation Project 317
Scott Grieco, Adam Redding, Nonnie Lim

(477f) Agile Multiple Water Network Designs for Soluble Contaminant Transport Identification and Site Remediation 318
Tianxing Cai, Qiang Xu

(477g) Modeling Electrokinetic-Soil Cleaning By Spatial Averaging 319
Rocio Tijaro-Rojas, Pedro E. Arce

(477h) Dynamics, Source Diagnosis, and Risk Assessment of Persistent Organic Pollutants in Water and Sediments of the Urban Aquatic Bodies 320
Rahul Kumar

(477i) Mass Transfer and Reaction Kinetics Limitations in Hazardous Waste (Organic and Metal) Treatment 323
Ramesh Chawla, Jude Ighere, Jacob Mlusu, Anthony Farrell

(477j) A Simulation Of Weathering Of Colombian Crude Oils In The Colombian Caribbean Sea 324
Juan Ramirez, Aura Merlano, Juan Guerrero-Gallego, Daniel Peláez, Juan Hernandez, Andres Osorio, Alejandro Molina

(477k) Increasing Strength of Thickened Oil Sand Tailings with Geopolymers 327
Yuming Xu, Shahrad Karimi

(484a) Selective Adsorption of CO2 On Flexible Ni/Dbm-Bpy MOF 328
Christopher Matranga, Jeffrey Culp, Fan Shi

(484b) Amine Impregnated N-Doped Mesoporous Carbon for Selective CO2 Capture 330
John To, Jiajun He, Brannon Gary, Daniel Stack, Jennifer Wilcox, Zhenan Bao
(484c) Effect of Alumina Doping On Structure and Stability of CaO Based Sorbents for High Temperature CO2 Capture 331
Krishna Reddy Gunugunuri, Steven Qullin, Panagiotis Smirniotis
(484d) Enhanced Performance of Polymeric Amines Supported On Heteroatom-Incorporated Silica in CO2 Adsorption/Desorption 332
Yasutaka Kuwahara, John R. Copeland, Carsten Sievers, Takashi Kamegawa, Hiromi Yamashita, Christopher W. Jones
(484e) Hydrophobic Polymeric Solvents for the Selective Absorption of CO2 From Warm Gas Streams That Also Contain H2 and H2O 333
Robert M. Enick, David Luebke, Hunaid Nulwala, Samantha Warman, Craig Stevenson, Peter Koronaios
(484f) Synthesis and Characteristics of Amine Functionalized Biopolymers for CO2 Capture 334
Junpei Fujiki, Katsunori Yogo
(484g) Removal of Sulfur Compounds in Methane Or Nitrogen By Nano-Sized Magnesium Oxide and Its Composite 335
Chang Ha Lee, Yo-Han Kim, Jae Jeong Kim, Vu Anh Tuan
(485a) Multimedia Environmental Distribution of Nanomaterials 336
H. Haven Liu, Yoram Cohen
(485b) Origin, Cure and Control of Nanosilver Toxicity 337
Georgios A. Sotiriou, Kakeru Fujiwara, Sotiris E. Pratsinis
(485c) Bacterial Colonization of Surfaces Displaying Adhered Silver Nanoparticles 338
Stacy M. Wirth, Gregory V. Lowry, Robert D. Tilton
(485d) Carbon Nanotubes Induced Cellular Biomeahanic Changes Is Depended On Treatment Time 339
Chenbo Dong, Reem Eidawud, Michael L. Kashon, David Lowry, Linda M. Sargent, Cerasela Zoica Dinu
(485e) A Simple Kinetic Model Describing Nanoparticle Interactions With a Tethered Lipid Bilayer 340
Ying Liu, Robert M. Worden
(485f) Effect of Protein Corona On Nanoparticle Cellular Uptake 341
Shikha Nangia, Wenjuan Jiang, Haarika Kamani
(485g) Lessons From Nature: Analogies Between Protein and Nanoparticle Interactions With Lipid Membranes 342
Geoffrey Bothun, Christopher Bobba, Aihong Xi, Christopher L. Kitchens
(485h) Reaction-Diffusion Model Describing Antioxidant Depletion in Polyethylene-Clay Nanocomposites 343
Iftekhar Ahmad, Grace Hsuan, Christopher Li, Richard Cairncross
(499a) SiC Photoanodes for Solar Water Splitting 344
Christopher Bohn, F Sharifi, M. G. Kang, Veronika Szalai
(499b) Advanced Electrocatalysts From Organic Solution Synthesis 345
Chao Wang
(499c) Nanostructured Ti –Fe2O3/Cu2O Bilayered Thin Films for Photoelectrochemical Water Splitting 346
Dipika Sharma, Sumant Upadhyay, Surbhi Choudhary, Vibha Rani Satsangi, Rohit Shrivastav, Sahab Dass
(499d) Synthesis and Characterization of Mn2O3-ZnO Nanocomposites for Hydrogen Generation Via Photoelectrochemical and Photocatalytic Splitting of Water 348
Nirupama Singh, Surbhi Choudhary, Vibha Rani Satsangi, Sahab Dass, Rohit Shrivastav
(499e) High Performance Fluorine Doped Oxygen Evolution Reaction (OER) Electro-Catalysts for PEM Based Water Electrolysis 349
Karan Kadakia, Moni Kanchan Datta, Oleg Velikokhatnyi, Prashant N. Kumta
(499f) Multiwalled Carbon Nanotubes Drive the Activity of Metal@Oxide Core-Shell Catalysts in Modular Nanocomposites 350
Matteo Cargnello
(499g) Separation of Sulphur Dioxide and Oxygen in Thermochemical Hydrogen Production 351
Rachael H. Elder, Guanghu He, Denis J. Cumming, Ray W. K. Allen
(499h) Chemisorption, Physisorption, and Hysteresis of Hydrogen On Carbon Nanotubes 352
Seyedhamed Barghi, Theodore T. Tsotsis, Muhammad Sahimi
(529a) Lignocellulosic Biomass Pyrolysis Product Comparison From a Micropyrolyzer, Batch and Continuous Reactor Systems 353
Rene Garrido, Nicole L. Hammer, Alexander Zmiewski, Justinus A. Satrio
(529b) Spatio-Temporal Carbohydrate Characterization of Wood Particles During Fast Pyrolysis 355
Alex D. Paulsen, Blake Hough, Daniel Schartz, Jim Pfaendtner, Paul J. Dauenhauer
(529c) Numerical Investigation of Multiphase Dynamic Effects in Catalytic Upgrading of Biomass Pyrolysis Vapor 356
Jesse S. Capecelatro, Olivier Desjardins, Perrine Pepiot, Mark Jarvis, Thomas D. Foust
(529d) Catalytic Effects of MgSO4 and ZnCl2 On Fast Pyrolysis of Pine for the Production of Bio-Oils 357
Liang Yuan, Yun-Quan Liu, Dong-Can Lv, Duo Wang
(529e) Pyrolysis of Alkali Pre-Treated Biomass and Biomass Constituents 358
Paige Case, Elizabeth Stemmler, William J. Desisto, M. Clayton Wheeler
(529f) Study On the Aging Mechanism of Pyrolysis Bio-Oil 359
Jiajia Meng, David Tilotta, Sushil Adhikari, Steve Kelley, Sunkyu Park
(529g) Effects of Post-Condensation Centrifugal Filtration On Pyrolysis Oil 360
Anandavalli Varadarajan, Keisha B. Walters
(529h) Transportable Fast-Pyrolysis Process for Distributed Conversion of Waste Biomass to Renewable Liquid Fuels 361
Paul E. Yelvington, Nicholas Schwartz, Dustin Zastrow, Ted Amundsen

(529i) Biomasspyrolysisrefinery: Product Stability of Intermediates and Products 362
Nikolaus Schwaiger, Roland Feiner, Hannes Pucher, Lisa Steiner, Lisa Ellmaier, Peter Pucher, Matthäus Siebenhofer
(529j) Biomass Gasification: Effect of Sulfur Compounds On Catalytic Tar Removal Activity Using Nickel-Clay Catalysts 364
Joesph Vutukuri, Prashanth Buchireddy, R. Mark Bricka, John Guillory, Rakesh Bajpai, Mark E. Zappi
(539a) Study of the Tcep-Zeolite System: Kinetics and Surface Chemistry Analysis Investigations 365
Scott Grieco, Bandaru V. Ramarao
(539b) Adsorption and Degradation of Perfluorooctanoic Acid 366
Catherine B. Almquist, Jason A. Berberich
(539c) Removal of Arsenic in Water With An Adsorbent Derived From Waste Concrete Sludge 367
Takeshi Sasaki, Masayuki Watanabe, Teruhisa Hongo, Atsushi Iizuka, Akihiro Yamasaki
(539d) Effect of Organic Anion On Selective Adsorption 368
Len Foong Koong, Koon Fung Lam, John Barford, Gordon McKay
(539e) Optimization of Commercial Para-Xylene (PX) Simulated-Moving-Bed (SMB) Simulation Using Extended Langmuir Isotherm 369
Thu Thi Dang, Young-Il Lim
(539f) Synthesis of Mesoporous Carbons: Oxidative Modification and Hydrothermal Carbonization for Efficient Chemical Activation 370
Akshay Jain, M. P. Srinivasan, Rajasekhar Balasubramanian
(539g) Biosorption of Heavy Metals Using An Acid Activated Walnut Shell 371
Guillermo Andrade-Espinosa, Maria Selene Berber-Mendoza, Luis Armando Bernal-Jácome, Nahúm Andrés Medellín-Castillo
(540a) Biomimetic Properties of Carbon Nanotubes In Vivo 372
Juan C. Villegas, Rafael Valiente, Lidia Rodriguez-Fernández, Jesus González, Mónica L. Fanarraga
(540b) Differential Analysis Of Single Wall Carbon Nanotubes Cellular Uptake Mechanism 373
Reem Eldawud, Chenbo Dong, Linda M. Sargent, Yon Rojanasakul, Cerasela Zoica Dinu
(540c) Development of in-Vivo Toxicity Screening Benchmarks for Complex Engineered Nanomaterials 374
Sharlee Mahoney, Michelle Najera, Qing Bai, Edward Burton, Götz Veser
(540d) NLRP3 Inflammasome Activation Induced By Long Aspect Ratio Engineered Nanomaterials: Role of Oxidative Stress 375
Bingbing Sun, Tian Xia
(540e) Effects of Aerosolized Sub-Micron Particles On The Interfacial Properties of Lung Surfactant Models 376
Amir M. Farnoud, Jennifer Fiegel
(540f) Carbon Nanotubes Induce Invasion of Human Mesothelial Cells Through Matrix Metalloproteinase-2 377
Warangkana Lohcharoenkal, Cerasela Zoica Dinu, Todd Stueckle, Liying Wang, Yon Rojanasakul
(540g) Neoplastic-Like Transformation Ability of Carbon Nanotubes On Small Airway Epithelial Cells: Linking Toxicogenomic Signatures and Lung Cancer Hallmarks 378
Todd Stueckle, Anurag Mishra, Raymond Derk, Terence Meighan, Vince Castranova, Liying Wang, Yon Rojanasakul
(546a) Controllability of Complex Networks for Sustainable System Dynamics 379
Pahola T. Benavides, Urmila D. Diwekar, Heriberto Cabezas
(546b) Sustainable Process Decision Making With Greenscope 380
Raymond L. Smith, Gerardo J. Ruiz-Mercado, Michael A. Gonzalez
(546c) Innovative Chemical Engineering Applications to Control Air Emissions From Oil Refineries 381
Dr. Chandra Nair
(546d) Direct and Indirect Emissions From Mobil Air Conditioner Units in Transportation By Truck and Rail 382
Paul Blowers, Austin Dougless, Brian Gerwe, Adam Mesta, Caleb Rugel
(546e) Multi-Objective Optimization of International Economies Via Multi-Regional Input-Output Analysis: Application to the US Economy 383
Janire Pascual, Gonzalo Guillén, Laureano Jimenez, Daniel Cortès
(546f) Urban Systems and Sustainability: Comparative Analysis of Multiple Metropolitan Statistical Areas in Ohio, USA 384
Alejandra Gonzalez-Mejia, Tarsha Eason, Heriberto Cabezas, Makram Suidan
(546g) Environmental Impact Minimization of Vaccine Supply Chain for Developing World 385
Bahador Mousavi, Arunprakash T. Karunanithi
(564a) Strategies in the Sustainable Design of Active Pharmaceutical Ingredients for Solvent Recovery Systems 386
Mariano J. Savelski, C. Stewart Slater, Eduardo J. Cavanagh
(564b) Towards a Complete Solution of Allocation in Life Cycle Inventories 387
Rebecca Hanes, Nathan Cruze, Prem K. Goel, Bhavik R. Bakshi
(564c) Hybrid Techno-Economic Modeling Tool for Greener Chemicals Supply Chains 388
Yuan Yao, Yuan Chang, Fred Thwaites, Eric Masanet
(564d) Design Decision-Making Under Uncertainty to Identify Sustainable Integrated Forest Biorefinery Strategies 389
Shabnam Sanaei, Paul R. Stuart
(564e) MILP-Based Clustering Method for Multi-Objective Optimization: Application to Environmental Problems 390
Diego Gabriel Oliva, Gonzalo Guillén-Gosálbez, Laureano Jiménez-Esteller
(564f) Emergy Method Analysis On Sludge Residues For Use From Waste-WATER Treatment Plant 391
Natalia Andrea Cano, Dario Gallego, Hector Ivan Velásquez
(592a) Catalytic Fast Pyrolysis of Forest Thinnings in a Bubbling Fluidized Bed Reactor With ZSM-5 Catalysts 394
Ville Paasikallio, Foster Agblevor

(592b) Effects of Reactor Configurations and Catalyst Properties for Vapor Upgrading in HZSM5 Catalyzed Pyrolysis Vapor Upgrading 395
Shaolong Wan, Christopher Waters, Abhishek Gumidyala, Rolf Jentoft, Lance Lobban, Steven Crossley, Daniel E. Resasco, R. G. Mallinson
(592c) CFD Modeling of the Catalytic Pyrolysis of Biomass With Zeolite in Fluidized Bed 397
Dan Sun, Konstantinos Papadikis, Sai Gu
(592d) Deoxygenation of Bio-Oil Intermediates: Examination of Methane As a Direct Reducing Agent 398
Aaron A. Oberg, Dennis J. Miller
(592e) Fast-Hydropyrolysis and Catalytic Hydrodeoxygenation for Conversion of Biomass to Liquid Fuel 399
Vinod Kumar Venkatakrishnan, John Degenstein, W. Nicholas Delgass, Rakesh Agrawal, Fabio H. Ribeiro
(592f) Low-Oxygen Bio-Crude From Hydropyrolysis in a Pressurized Fluid-Bed Reactor 400
John R. Carpenter, Josh Hlebak, Justin Farmer, David C. Dayton
(592g) Catalytic Fast Pyrolysis of Isotopically Labeled Biomass and Plastic Blends 401
Christina Dorado, Charles Mullen, Akwasi A. Boateng
(592h) Mild Pyrolysis of P3HB/Switchgrass Blends for the Production of Bio-Oil Enriched With Crotonic Acid 402
Charles A. Mullen, Akwasi A. Boateng, Christina Dorado, Dirk Schweitzer, Kevin Sparks, Kristi Snell
(600a) Mixed-Conductor Supported Ferrites for Chemical Looping Conversion of Methane 403
Yanguang Chen, Fanxing Li
(600b) A Novel Mixed Metallic Ni Based Oxygen Carrier for Chemical Looping Combustion 404
Hugo De Lasa, Mohammad R. Quddus, Mohammad Mozahar Hossain
(600c) Low-Cost Oxygen Carrier Materials for Chemical-Looping Combustion of Biomass 406
Georg Schwebel, Gonzalo Vilela Perez, Pavleta Knutsson, Henrik Leion
(600d) Fe and Mn-Based Oxygen Carriers for Chemical Looping Combustion 407
Saurabh Bhavsar, Amey More, Götz Veser
(600e) Experimental and Model Comparison of Mixed Metal Ferrites for H2 Production Using Chemical Looping 408
Victoria J. Aston, Brian W. Evanko, Christopher L. Muhich, Jonathan Scheffe, Alan W. Weimer
(600f) Sol-Gel Derived Fe2O3-Based, Al2O3 Supported Oxygen Carriers for Hydrogen Production Via a Chemical Looping Cycle 409
Nur Sena Yüzbasi, Qasim Imtiaz, Christoph R. Müller
(600g) NiO/Ce-γAl2O3 Oxygen Carrier for Chemical-Looping Combustion: Reactivity and Stability 411
Shamseldin Mohamed, Shaikh Razzak, Mohammad R. Quddus, Hugo I. De Lasa, Mohammad M. Hossain
(619a) Nanoscale Interfacial Design Concept Towards High Flux CO2 Separation Maintaining High Selectivity and Longevity 412
Lakshmi Suhasini Kocherlakota, Tiep Hoang Pham, René M. Overney
(619b) Investigation of Water Quantification in Ionic Liquids Utilizing Near Infrared Absorption of Water 413
Daniel Fagnant, George S. Goff, Wolfgang Runde, Joan F. Brennecke
(619c) Mixed Gas Sorption of CO2/CH4 Mixtures in PIM-1 and PTMSP Membranes: Experiments and Modeling 414
Ondrej Vopicka, Maria Grazia De Angelis, Giulio C. Sarti, Michael D. Guiver
(619d) Diffusive Separation of Propylene/Propane With ZIF-8 Membranes 416
Nobuo Hara, Miki Yoshimune, Hideyuki Negishi, Kenji Haraya, Shigeki Hara, Takeo Yamaguchi
(619e) Simulation of Propylene/ Propane Plasticization in 6FDA-Based Polyimide Membranes 417
Sadiye Halitoglu-Velioglu, M. Goktug Ahunbay, S. Birgül Tantekin-Ersolmaz
(641a) Algal Biorefinery: Production of Biocrude Oil & Byproducts 419
Harvind Kumar Reddy, Thinesh Selvaratnam, Tapaswy Muppaneni, Nagamany Nirmalakhandan, Tanner Schaub, Barry Dungen, Nilusha Sudasinghe Appuhamilage, Peter Lammers, Wayne Voorhies, Shuguang Deng
(641b) Biodiesel Production Through Microwave Assisted Transesterification of Microbial Cells 420
Yi Cui
(641c) Design Optimization By Response Surface Methodology for Continuous Fermentative Production of 1,3 Propanediol From Waste Glycerol By Product of Biodiesel Processes 421
Baishali Kanjilal, Iman Noshadi, Nicholas Intoci, Matthew Dowding, William Hale, Brittany Brendel, Ranjan Srivastava, Richard Parnas
(641d) Review of Bio-Jet Fuel Conversion Pathways 422
Wei-Cheng Wang, Ling Tao
(641e) Pellets From Lignocellulosic Materials: Briquetting Process Evaluation 423
Yuri Gonzalez, Sahra Rincon, Gabriel Camargo, Fabio Sierra, Carlos A Forero N
(641f) Pyrolytic Liquid Yields From Bamboo Species (Bambusa Oldhamii and Bamboo Angustifolia) 424
Y. S. Pliego-Bravo, Lidia Cervantes- Pesado, Eusebio Bolaños-Reynoso, Pedro Gonzalez-Garcia
(641g) The Impact of Ethanol and Iso-Butanol Blends on Regulated Emissions, Air Toxics, and Particle Emissions from SIDI Vehicles 427
Georgios Karavalakis, Daniel Short, Diep Vu, Thomas Durbin, Akua Asa-Awuku
(642a) Recovery of Phosphorus From Sewage Sludge and Sewage Sludge Ashes With the Thermo-Reductive Recophos Process 429
Andreas Schönberg, Kasra Samiei, Hannes Kern, Harald Raupenstrauch
(642b) Data Availability and Management for Process Sustainability Evaluation: Greenscope 432
Rajib Mukherjee, Gerardo J. Ruiz-Mercado, Raymond L. Smith, Michael A. Gonzalez
(642c) Development of a Sustainable Phase Transfer Catalysed Continuous Process for a Pharmaceutical Intermediate 433
Soo Khean Teoh, Kanicha Sa-Ei, Mohammad Salih Noorulameen, Yi Ling Ng, Paul N. Sharratt, Katherine Wheelhouse, Andrew Rutter

(642d) Innovative Sustainable Materials & Processes 434
Tom Enright, Adela Goredema
(642e) Dry Granulation of Blast Furnace Slag for Heat Recovery 435
Markus Kofler, Klaus Doschek, Harald Raupenstrauch
(642f) Advances in Technology for Producing Alumina From Complex Resources in China 439
Yifei Zhang
(664a) Evaluation of Spray Dried Oxygen Carriers Based On Manganese Ore and Ca(OH)2 for Chemical-Looping With Oxygen Uncoupling 440
Sebastian Sundqvist, Tobias Mattisson, Henrik Leion, Anders Lyngfelt
(664b) Solid Fuel Conversion Via Chemical-Looping With Oxygen Uncoupling (CLOU) Using Co-Precipitation Derived Cu-Mn Oxygen Carriers 442
Davood Hosseini, Qasim Imtiaz, Agnieszka Kierzkowska, Christoph R. Müller
(664c) Using (MnxFe1-x)2SiO5 As Oxygen Carriers for Chemical-Looping With Oxygen Uncoupling (CLOU) 443
Tobias Mattisson, Dazheng Jing, Jasper Van Noyen, Magnus Ryden, Anders Lyngfelt
(664d) CO2 Utilization Via Chemical Looping Process 445
Vladimir V. Galvita, Hilde Poelman, Guy B. Marin
(664e) The Synthesis of Highly Efficient Ca-Based CO2 Sorbents Containing a Hierarchical Pore Structure 448
Marcin Broda, Christoph R. Müller
(664f) Li4SiO4-Based Novel Solid Fluidizable Sorbent for CO2 Capture: 450
Muhammad B. I. Chowdhury, Hugo De Lasa, Mohammad R. Quddus
(677a) Study Of Hydroden Production By Photofermentation From Different Organic ACID As Substrate 451
Thamayne V. Oliveira, Felipe Segismundo Oliveira, Thiago M. Sousa, Juliana S. Ferreira, Fabiana R. X. Batista, Vicelma L Cardoso
(677b) Influence Of pH, Ammonium and Iron On The Hydrogen Production By Microbial Consortium 452
Betania B. Romão, Julia G. Santos, Felipe Segismundo Oliveira, Juliana S. Ferreira, Fabiana R. X. Batista, Vicelma L Cardoso
(677c) Monitoring of Gaseous Sulfur Species in a Modified Sulfur-Iodine Thermochemical Water Splitting Cycle in a Hydrophobic Ionic Liquid 453
Kevin Caple
(677d) The Water Splitting Kinetics of Two-Step Solar Thermochemical Process With CeO2 454
Darwin Arifin, Anthony H. McDaniel, Alan W. Weimer
(677e) Oxidation/Reduction Kinetics of Fe Particles in a ZiO2/ SiO2 Fluidized Bed 455
Benjamin Greek, Nicholas Auyeung, Luca Imponenti, James Klausner
(677f) Hydrogen Fueling Stations: Optimization of Fill-Up Process 456
Fernando Olmos, Vasilios Manousiouthakis
(704a) Sustainable Pathways for Recovering Materials and Energy From Waste 457
Debalina Sengupta, Raymond L. Smith, Chun C Lee, Sudhakar Takkellapati
(704b) Simultaneous Reaction and Purification Process for Continuous Production of High Quality Biodiesel From Acid Oils 458
Naomi Shibasaki-Kitakawa, Toru Ihara, Toshikuni Yonemoto
(704c) Algal Biomass and Lipid Production From Wastewater Using a Phagotrophic Microalga 461
Cong Li
(704d) Novel Biomass-Supported Sorbents for Coal Combustion Desulfurization 462
Sheng Chu, Geoffrey Tompsett, Jin Yang, Cheng Zhu, T. J. Mountziaris, Paul Dauenhauer
(704e) Fuel Quality Roadmap for the Kingdom of Saudi Arabia Domestic Transportation Fuel 463
Saeed S. Alalloush, Nasiru M. Tukur
(704f) Torrefied Biomass: Superior Fuel for Co-Firing in Pulverized Coal Fired Power Generation Facilities 464
Prashanth Buchireddy, John Guillory, Mark Zappi
(704g) The Osmotic Heat Engine: The Promise of High Power Density Closed Loop Pressure Retarded Osmosis 465
Daniel Anastasio, Jason T. Arena, Jeffrey R. McCutcheon
(704h) A Participatory Pilot Study On Awareness of Biomass Pyrolysis: Feedback From Two Upper-Cumberland Counties 466
Jessica D. Murillo, Joseph J. Biernacki, Lachelle Norris
(727a) Effect of Pore and Particle Size of Oxygen Carriers On Chemical-Looping Combustion and Reforming 467
Lu Han, Zhiquan Zhou, George M. Bollas
(727b) Thermal Evaluation of Chemical Looping Process for Hydrogen Production Using Bivalent Metal Added Iron Oxide As An Oxygen Carrier 469
Wonchul Cho, Kyoungsoo Kang, Chusik Park, Seong Uk Jeong, Doyeon Lee, Myung Won Seo, Chang Hee Kim, Ki-Kwang Bae, Sang Done Kim
(727c) Modeling of Cu Oxidation in Air Reactors of Chemical Looping Combustion Systems 470
Eli A. Goldstein, Reginald E. Mitchell, Turgut M. Gür
(727d) An Overview of Chemical-Looping Reduction in Fixed-Bed and Fluidized Bed Reactors Focused On Oxygen Carrier Utilization and Reactor Efficiency 471
Zhiquan Zhou, Lu Han, Shoucheng Du, Ari Fischer, George M. Bollas
(727e) Modeling A Scaled 300 kWth Circulating Fluidized Bed Reactor With Barracuda 473
Matthew A. Hamilton, Joann S. Lighty
(727f) The Ohio State University Chemical Looping Processes: 25 Kwth Sub-Pilot Performance Studies for Natural Gas Conversion and 250 Kwth Pilot Unit Developments 475
Mandar Kathe, Andrew Tong, Liang Zeng, Dawei Wang, Samuel Bayham, Omar McGiveron, L. S. Fan

(727g) Combustion Of Polyurethane By Chemical Looping Process In A Moving Bed Reactor Using Iron-Based Oxygen Carrier 476
Young Ku, Ping-Chin Chiu, Hsuan-Chih Wu, Yao-Hsuan Tseng, Yu-Lin Kuo

(735a) Detailed Kinetic Modeling of Selective Catalytic Reduction of NO With C3H6 On Cu-Chabazite Monolithic Catalyst 477
Richa Raj, Michael P. Harold, Vemuri Balakotaiah

(735b) A Detailed Microkinetic Model for SO2 Oxidation From Diesel Engine Exhaust 479
Hom Sharma, Ashish B. Mhadeshwar, Ramamurthy Ramprasad

(735c) Reaction Characteristics of Sncr De-NOx in Moderate Temperature Range Based Urea Spiked Hydrazine Reductant 481
Hui Chen, Liu Hong, De-Zhen Chen

(735d) Kinetic Model Study in a Photo-CREC Reactor for Photocatalytic Hydrogen Production 482
Hugo De Lasa, Salvador Escobedo, Benito Serrano

(735e) Abatement of Diesel Soot Under Various Catalyst/Soot Contact Conditions 485
Changsheng Su

(735f) Gklj;lklk 486
X. Chanda Nagarajan

(735g) System-Wide Modeling of HCCI Engine With After-Treatment 488
Pratheeba Chanda Nagarajan, Preeti Aghalayam

(735h) A New Approach for NOx Control Based On the Chelate Absorption-Biofilm Electrode Integrated System 492
Yinfeng Xia, Bailong Xu, Ya Zhou, Wei Li

(735i) Preparation and Catalytic Performance of Au/3DOM Co3O4 for the Oxidative Removal of CO and BTX 493
Yuxi Liu, Jiguang Deng, Hongxing Dai

(735j) Kinetic Mechanism Investigation On NOx Removal With Hydrazine Hydrate At Moderate to High Temperatures 495
Liu Hong, De-Zhen Chen, Du Wang

(764a) Sulfolenes As Sustainable Substitutes for Dimethyl Sulfoxide 496
Esteban E. Ureña-Benavides, Afrah J. Boigny, Zachary J. Randall, Pamela Pollet, Charles L. Liotta, Charles A. Eckert, Jason S. Fisk, Bruce S. Holden

(764b) Production of Ethylene and Methanol From Dry Natural Gas and Crude Oil: A Deep Dive Sustainability Assessment 497
Venkata K. K. Upadhyayula, Eric Masanet, Joshua Warren, Das Sujit, Diane J. Graziano, Mathew Riddle, Guenter Conzelmann

(764c) Process Ecoefficiency of Methanol and Dmc Production From CO2: A Comparative Analysis 498
Rita M. B. Alves, Camila Ribeiro, José L. Medeiros, Paulo L. A. Coutinho, Ofélia Araújo

(764d) An Innovative Process for the Production of Polymer Grade Propylene From Propane 499
Vivek Julka, Sagar B. Gadewar

(764e) Gas to Olefins: Sustainable Development of Shale Gas? 500
Heejae Huh, Benjamin John Davis

(764f) Polyol-Ester Synthesis Via Transesterification of Jatropha-Based Methyl Ester With Trimethylolpropane 501
Suzana Yusup, Ruzaimah Nik Mohamad Kamil, Mustakimah Mohamed

(764g) Studies On the Stability Behavior of a Depolymerization Reaction Based On Modified Denbigh Scheme for the Reclamation of Waste Tires 503
Abhishek Dutta, Denis Constales, Guido Beny, Kal Renganathan Sharma

(774a) A Novel Fluidized Bed Hydrator for Steam Reactivation of Calcium-Based CO2 Sorbent 504
Alan Wang, Dawei Wang, Niranjani Deshpande, Nihar Phalak, William Wang, L. S. Fan

(774b) Dynamic Modeling of a Solid-Sorbent CO2 Capture System 505
Srinivasarao Modekurti, Debangsu Bhattacharyya, Stephen E. Zitney, David C. Miller

(774c) Comparative Environmental Life Cycle Assessment of CO2 Capture Systems Based On Ionic Liquid and Monoethanolamine Solvents 506
Reza Farahipour, Arunprakash T. Karunanithi

(774d) Whole-Chain CCS System Modelling: Enabling Technology to Help Accelerate Commercialisation and Manage Technology Risk 507
Mark Matzopoulos, Alfredo Ramos

(774e) Optimization of CO2 Capture, Utilization and Sequestration (CCUS) Supply Chain Networks 508
M. M. Faruque Hasan, Fani Boukouvala, Christodoulos A. Floudas

(808a) CO2 Capture By Gas Hydrate Formation: Combined Effect of Thermodynamic and Kinetic Promoters 510
Nagu Daraboina, Nicolas Von Solms

(808b) CO2/Brine/Rock Interactions Under CO2 Sequestration Conditions 511
Yee Soong, Bret H. Howard, S. W. Hedges, Dustin McIntyre, Robert P. Warzinski, Gino Irdi, Igor Haljasmaa, Dustin Crandall

(808c) Novel CO2 Capture and Carbonation Process With Waste Concrete 514
Daiki Shuto, Akihiro Yamasaki, Teruhisa Hongo, Atsushi Iizuka

(808d) Direct Capture and Conversion of CO2 From Flue Gas Via Ex-Situ Mineral Carbonation Using Chemical and Biological Catalysts 515
Edward J. Swanson, Patrick V. Brady, Ah-Hyung Alissa Park

(808e) A Green Approach to Mitigate CO2 Using Mixed Microbial Culture 516
Smita Raghuvanshi, Suresh Gupta, Somesh Mishra

Author Index

(11a) On-Site Destruction of Radioactive Oily Wastes Using Combination of Adsorption With Electrochemical Regeneration

Roberts, E. P. L., University of Manchester
Brown, N. W., Arvia Technology Ltd
Campen, A. K., Arvia Technology Ltd
Wickenden, D. A.,
Originally presented on: 11/3/2013 15:30:00 - 15:55:00

Intermediate level waste (ILW) organic liquids and high alpha oils are difficult to dispose of via existing routes and are typically stored on site, classified as orphan wastes. Taking these sites to care and maintenance requires these wastes to be retrieved and treated to assure either passive storage as an immobilised waste or destruction of the waste. In this study adsorption coupled with electrochemical regeneration has been investigated as a method of treatment of contaminated oils. The process results in the complete destruction of the organic phase where the radioactivity is transferred to liquid and solid secondary wastes that can then be processed using existing authorised on-site waste- treatment facilities. Following successful laboratory and pilot scale trials, a demonstrator unit was commissioned at the Magnox Trawsfynydd decommissioning site in the UK to destroy 10 L of radioactive oils, comprising high alpha/ILW oils. Over 99% of the emulsified oil was removed and destroyed with the majority of activity (80–90%) being transferred to the aqueous phase. Secondary wastes were disposed of via existing routes with the majority being disposed of via the sites active effluent treatment plant. The regeneration energy required to destroy a litre of oil was 42.5 kWh. This on-site treatment approach eliminates the risks and cost associated with transporting the active waste oils off site for incineration or other treatment.

(11b) Process Chemistry Improvements in Saltcake Dissolution Modeling- A Retrospective

Toghiani, R. K., Mississippi State University
Lindner, J. S., Mississippi State University
Smith, L. T., Mississippi State University
Phillips, V.,
Originally presented on: 11/3/2013 15:55:00 - 16:20:00

In 1997, the Process Chemistry for Waste Processing group was formed at the Diagnostic Instrumentation and Analysis Laboratory (DIAL; now the Institute for Clean Energy Technology) at Mississippi State University. Initial efforts of the group focused on working with Hanford site personnel to identify process chemistry needs to improve the reliability and accuracy of predictions of saltcake retrieval strategies using the Environmental Simulation Program (ESP, OLI Systems, Inc.). At the 222-S Laboratory at Hanford, Herting[1]judiciously selected saltcake core samples from tanks with differing waste inventory and performed 'saltcake dissolution' testing over the next few years. Based on the results of these dissolution experiments, simulation of the experiments using the ESP software allowed chemical systems to be identified where the predictive behavior and experimental measurements differed significantly. These select chemical systems were then examined in the ICET laboratory, with solubility experiments conducted under the high ionic strength conditions present in the Hanford waste tanks. The gathered data for all chemical systems examined in the ICET laboratory have now been incorporated into the Mixed Solvent Electrolyte (MSE) database within ESP. A comparison of the early predictions with the predictions now possible within the MSE framework of ESP for these saltcakes of differing composition will be presented.

---

[1] Herting, D.L., Edmonson, D.W., "Saltcake Dissolution FY 1998 Status Report", HNF-3437, Rev. O, 1998.

(11c) Modeling of Cold-Cap Reactions for Vitrification of Nuclear Waste Glass Based On Simultaneous Differential Scanning Calorimetry-Thermogravimetry (DSC-TGA)

Chun, J., Pacific Northwest National Laboratory
Pierce, D.,
Pokorny, R., Institute of Chemical technology in Prague
Hrma, P.,
Originally presented on: 11/3/2013 16:20:00 - 16:45:00

During the waste-glass vitrification, multiple and overlapping gas-evolving reactions called cold-cap reactions take place, leading to the feed-to-glass conversion. The cold-cap reactions involve mass losses and foaming which significantly influence the mass and heat transfers, and eventually affect the melting rate in the vitrification. We studied the cold-cap reactions of a representative waste glass feed using DSC-TGA and developed a kinetic model. The model is based on the assumption that the reactions are nearly independent and can be approximated by an nth order kinetic equation. To separate the reaction heat from both the heat associated with the heat capacity of the feed and experimental artifacts, we employed the run/rerun method, which enabled us to define the degree of conversion based on the reaction heat and to estimate the heat capacity of the reacting feed. We obtained the kinetic parameters using the Kissinger method combined with least squares analysis. The resulting mathematical simulation of the cold-cap reactions provides a key element for the development of an advanced cold-cap model.

(11d) Technology Development for Radioactive Sludge Treatment At the Hanford Nuclear Cleanup Site

Fogwell, Ph.D., P.E., T. W., Fogwell Consulting
Originally presented on: 11/3/2013 16:45:00 - 17:10:00

Title: Technology Development for Radioactive Sludge Treatment at the Hanford Nuclear Cleanup Site

Highly radioactive sludge resulting from the storage of degraded spent nuclear fuel has been consolidated in Engineered Containers (ECs) in the 105-K West Storage Basin located on the Hanford site near the Columbia River in Washington State. CH2M Hill Plateau Remediation Company (CHPRC) is proceeding with a project to retrieve the sludge, place it in Sludge Transport and Storage Containers (STSCs) and store those filled containers within the T Plant Canyon facility on the Hanford Site Central Plateau (Phase 1). Retrieval and transfer of the sludge material will enable removal of the 105-K West Basin and allow remediation of the subsurface contamination plumes under the basin. The U.S. Department of Energy (DOE) is developing technology to treat and dispose of this K Basins sludge (Phase 2) as Remote Handled Transuranic Waste (RH TRU) at the Waste Isolation Pilot Plant (WIPP) located in New Mexico. A formal approach is being taken for the development of the technology, which entails development in specific stages according to US Department of Energy guidelines for such projects. This process is presented in detail.

The K Basin sludge currently contains uranium metal which reacts with water present in the stored slurry, generating hydrogen and other byproducts. The established transportation and disposal requirements require the transformation of the K Basins sludge to a chemically stable, liquid-free, packaged waste form. The Treatment and Packaging Project includes removal of the containerized sludge from T Plant, the treatment of the sludge as required, and packaging of all the sludge into a form that is certifiable for transportation to and disposal at WIPP. Completion of this scope will require construction and operation of a Sludge Treatment and Packaging Facility (STPF), which could be either a completely new facility or a modification of an existing Hanford Site facility. The U.S. Department of Energy's Richland Field Office (DOE-RL) has approved a recommended technical approach. The baseline process is broken down into the following main process steps: (1) STSC transfer from T Plant to the Sludge Treatment and Packaging Facility (STPF). (2) Retrieval of sludge from the STSCs and transfer to the Receipt and Reaction Tank (RRT). (3) Preparation for immobilization by oxidation for those batches that require it and concentration by evaporating water at about atmospheric pressure in the RRT. (4) Immobilization by using additives to eliminate free liquids and packaging of the treated sludge into drums. (5) Inspection and handling of the filled drums prior to transfer to a separate storage and shipping facility. (6) Handling of vapor, condensate, and other waste streams generated by the process. Each of these steps is discussed in the paper, together with the current state of progress in developing the technology and requirements for continued development.

(11e) Noble Metal Fission Product Recovery By Electrodeposition in Ionic Liquids

Gjoka, E.,
Ascue, K., KTH-Royal Institute of Technology
Biddinger, E. J., City College of New York
Originally presented on: 11/3/2013 17:10:00 - 17:35:00

U.S. nuclear power plants "temporarily" house more than 57,000 tons of spent fuel. This spent fuel is composed of unreacted fuel and, radioactive and non-radioactive fission products. Nuclear energy processes can become more sustainable by recycling the non-reacted fuel and separating the fission products by half-life classification. This also significantly reduces the volume of waste that requires long-term sequestration.

Ionic liquids have been proposed as non-flammable, minimally volatile, selective extractant/extractant solvents for the separation of fission products from the aqueous solution resulting from the acid digestion of spent fuel rods. The extracted fission products must be further separated from the ionic liquid. This can be done by electrodeposition from the ionic liquid. Ionic liquids can be used as both the solvent and electrolyte in electrochemical systems. Many ionic liquids have wide electrochemical potential windows, making electrodeposition from most metals, including many fission products, possible.

In the work presented here, the electrodeposition of noble metals in ionic liquids will be reported. Noble metals are a small, but valuable component of the fission product waste stream and can be used as model systems to other radioactive fission product systems. The electrodeposition of noble metals including Pd, Ru and Rh will be reported on. The impact of the ionic liquid selection and the electrochemical conditions on the electrodeposition efficiency, deposit composition and morphology has been investigated.

(29a) Economic Evaluation of Freshwater Conservation Strategies for Thermoelectric Power Plants

Walker, M. E., Illinois Institute of Technology
Theregowda, R. B., Carnegie Mellon University
Safari, I., Illinois Institute of Technology
Abbasian, J., Illinois Institute of Technology
Arastoopour, H., Illinois Institute of Technology
Dzombak, D. A., Carnegie Mellon University
Hsieh, M. K., Carnegie Mellon University
Miller, D. C., U.S. Department Of Energy, National Energy Technology Laboratory
Originally presented on: 11/3/2013 15:30:00 - 15:50:00

Thermoelectric power plants withdraw and consume large volumes of freshwater. In parts of the country where freshwater resources are scarce, this demand can be problematic. One option for conserving freshwater is to utilize degraded water sources as a replacement cooling medium for thermoelectric power plants. However, utilizing degraded water in cooling systems can be problematic; increased concentrations of particulates and salt forming species can lead to increased fouling on heat exchanger surfaces. A degradation of performance in the heat exchanger will translate to a loss of thermal efficiency and may require the plant to perform more frequent maintenance routines; it is therefore necessary to treat the degraded water source before introducing it to the system.

This presentation discusses a comprehensive methodology for the determination of the total cost of degraded water use in thermoelectric plants, including the cost of treatment, additional fuel use, additional downtime and lost power production that results. This methodology is then applied via a novel, cross-disciplinary case study to (1) estimate the freshwater conservation cost associated with secondary municipal wastewater use in the recirculating cooling loop of a thermoelectric power plant; (2) evaluate the freshwater conservation cost associated with secondary municipal wastewater against that of dry-air cooling technology and the U.S. Department of Energy's 2020 target; and (3) to determine the breakeven differential purchase price of freshwater to secondary municipal wastewater, beyond which the use of secondary municipal wastewater is economically beneficial.

(29b) Modeling Of Pollutant Transport in the Nile Delta, Egypt

Mostafa, M., University of Alabama at Birmingham
Peters, R. W., University of Alabama at Birmingham
Originally presented on: 11/3/2013 15:50:00 - 16:10:00

MODELING OF POLLUTANT TRANSPORT IN THE NILE DELTA, EGYPT

Mohamed Mostafa, Department of Civil, Construction and Environmental Engineering, University of Alabama at Birmingham, Birmingham, AL and Robert W. Peters, Department of Civil, Construction and Environmental Engineering, University of Alabama at Birmingham, Birmingham, AL

Egypt is an arid country with high temperatures and low rainfall, and depends mainly on the Nile River as its source of fresh water, where it supplies Egypt with more than 95% of its diverse water needs. The Nile River divides into two branches (Damietta and Rosetta) forming the Nile delta. Both branches flow into the Mediterranean Sea. The Nile delta extends nearly 200 km (124.27 mi) from north to south and over 300 km (186.4mi) from west to east. This research is being performed on the Nile delta. The primary objective of this paper is to develop a dynamic model of Nile delta pollution by using MATLAB software. The model is calibrated based on field and historical data. Problem concerning the water quality on the Nile delta involve toxic dumping, and water quality degradation. The lengths of Damietta and Rosetta branches are approximately 245 km (152.2 miles), and 239 km (148.5 miles), respectively.

The Nile delta receives pollution loadings from industrial, agricultural, and domestic activities which are located along its path. This model can be useful in assessing the downstream distance affected by pollutant releases and will aid as a decision tool on restrictions to be imposed on industrial, agricultural, and domestic activities. The analytical model is also a reliable tool for water quality assessment and management in affected areas, and can be used to predict water quality and will aid in the estimation of potential environmental damage. The investigation started with sample collection on December 2012 and *will last until August* 2013. Water samples are collected every 15 days *along the Nile Delta*, and from drains discharging into the Nile delta. Water samples are analyzed for different parameters during the winter, spring, and *summer* seasons. The environmental water quality parameters include pH, chemical oxygen demand (COD), biochemical oxygen demand (BOD), *total organic carbon* (TOC), total suspended solids (TSS), total dissolved solids (TDS), dissolved oxygen (D.O), and temperature. The main sources of pollution are studied to identify the most appropriate ways to reduce pollution and improve water quality in the Nile delta.

(29c) Evaluation of RO Concentrate Treatment for High Recovery Desalting of High Salinity Inland Brackish Water

Rahardianto, A., University of California, Los Angeles
McCool, B. C., University of California, Los Angeles
Faria, J.,
Cohen, Y., University of California, Los Angeles
Originally presented on: 11/3/2013 16:10:00 - 16:30:00

Development of new water sources is crucial in order to address the crisis of dwindling water supplies around the globe. Reverse osmosis (RO) desalination is a technology that is now in common practice for developing new water sources via desalting of seawater, brackish surface water and groundwater, as well as municipal wastewater. While RO is considered a mature technology, its application for desalination of inland brackish water remains challenging due to the high costs associated with residual RO concentrate disposal and membrane mineral scaling. The above challenge is particularly daunting in California's San Joaquin Valley with respect to the critical need for high-salinity agricultural drainage (AD) water management. Accordingly, in the present study, the technical feasibility of primary RO (PRO) desalting and its integration with PRO concentrate treatment was evaluated, via both experimental field tests and process analysis, to assess its potential use in enabling recovery enhancement via secondary RO (SRO) desalting. Two concentrate treatment options were considered: precipitation softening (PS) and a new chemical enhanced seeded precipitation (CESP) process. In the CESP process, initial partial lime treatment provides adsorptive removal of residual antiscalant from the PRO concentrate. This enables subsequent concentrate desupersaturation (with respect to the major scaling salt calcium sulfate) to occur via seeded gypsum precipitation, unimpeded by residual antiscalant. The desupersaturated PRO concentrate (~ 34% reduction in SIg from ~ 1.7 to ~ 1.1) can then be further desalted in an SRO step up to the limit that is feasible with antiscalant dosing for scale control. In order to assess the cost of various AD water desalination management options, on the basis of treated feed volume, a comparison was undertaken of the total cost (including brine disposal for all options) for: (a) PRO-CESP-SRO and PRO-PS-SRO at the maximum recovery limit of 93%, and (b) RO desalting up to the maximum attainable recovery of 62% without concentrate demineralization. Process and economic analysis of water salinity management options for the AD water source considered suggested that, when considering the value of the desalted product water and accounting for the disposal cost of residual brine, PRO-CESP-SRO desalting is the most economically attractive option. The break-even AD water desalination management cost (i.e., zero net cost) for PRO-CESP-SRO, for the present cost estimates, was at a product water value of 0.67–0.74 \$/m3 which is lower by 22–29% compared to that of the PRO-PS-SRO process (0.95 \$/m3). Admittedly, RO desalting (along with disposal of residual brine) can be more economical when brine disposal cost is below about ~ 0.44 \$/m3 brine, but this is unlikely to be a feasible or acceptable water management option given the environmental challenge of disposal and loss of valuable water resource.

(29d) Simulation and Sustainability Analysis for Produced Water Thermal Desalination Technologies

Chen, L., Lamar University
Lou, H. H., Lamar University
Originally presented on: 11/3/2013 16:30:00 - 16:50:00

Simulation and Sustainability Analysis for Produced Water Thermal Desalination Technologies

Abstract

Produced water is the largest waste stream generated in oil and gas industries. Characterized by high salinity, environmental concerns and the prospect of beneficial uses have driven intensive research into produced water management. Dominant thermal desalination technologies include Multi-stage Flash (MSF) and Multi-effect Distillation (MED). In this work, several multi-stage MSF and MED processes are simulated using Aspen Plus based on the plant data, and comparisons are made on multiple dimensions. In technological level, water recovery and gain output ratio (GOR) are the major indicators. For economic comparison, capital and operational costs (CAPEX & OPEX) are collected. The Waste Reduction Algorithm (WAR) is utilized for the evaluation of their environmental impacts.

(29e) Dispatching Electricity Generation to Increase Water Availability in Times of Drought: A Feasibility Study in Texas

Pacsi, A.,

Originally presented on: 11/3/2013 16:50:00 - 17:10:00

Over the past several decades, the frequency and intensity of droughts has increased throughout the United States, and this trend is predicted to continue. The use of water as a coolant fluid in thermoelectric power generation accounts for 40% of freshwater withdrawals and 3% of consumption in the United States, which leaves these facilities vulnerable during times of decreased water availability. Thus, drought can stress both the water and electricity generation resources within an affected region.

This work seeks to characterize the feasibility, cost, and air quality implications of using the grid as a virtual water pipeline to "deliver" increased water availability in regions with high drought using a case study of eastern Texas during the summer of 2006. This idea involves shifting power generation from parts of the grid which are under extreme drought conditions to areas with relatively more water availability, changing the magnitude and spatial extent of consumptions and withdrawals of freshwater resources for power plant cooling. This approach could be rapidly implemented using existing power plants and transmission lines and could respond to spatial changes in the location of drought during other periods. Changing the spatial location of power generation based on water availability will impact the cost of electricity generation by shifting generation to facilities which are lower on the dispatch order (and thus more expensive to operate) and may affect regional air quality by changing the location and magnitude of air quality pollutant precursors such as nitrogen oxides (NO_x) and sulfur dioxide (SO_2), which are related to regional ozone and fine particulate matter (PM) concentrations. Texas offers an interesting test bed for the interconnections of water and electricity resources since the state has its own self-contained electric grid (ERCOT) that has a fuel-mix that is a reasonable approximation of electricity resources within the United States as whole. In addition, Texas has 23 self-contained river basins that drain to the sea, reducing complications in modeling associated with downstream water users from other states. Also, projected population growth within the state of Texas will likely add further the strain on the electric grid and water resources in the next few decades.

This work is the first to demonstrate that ERCOT has the existing capacity to switch ~10% of its base case generation from a highly-drought stricken area to other parts of the grid. For the June 2006 episode, water-withdrawing electricity generation was re-dispatched away from areas classified as being under extreme and exceptional drought by the U.S. Drought Monitor, which were in South Texas during this period. While drought is a complex hydrological phenomenon and water planning decisions within this framework would likely be more complex than assumed in this work, the case study offers insight into the resiliency of the grid to respond to localized drought conditions. In addition, scenarios in which water consumption in severe drought regions was also limited were examined to limit the exportation of water shortages to borderline drought regions. The June 2006 episode is important for air quality planning in Texas as it was the period chosen for the State Implementation Plan (SIP) filed with the Environmental Protection Agency in response to non-attainment of the 8-hour ozone standards in the Dallas-Fort Worth area.

Coupled models of ERCOT power generation (Power World Simulator), power plant water withdrawals and consumption, and regional air quality (Comprehensive Air Quality Model with Extensions, CAMX) were utilized in this work. Factors for water consumption and use as well as air quality precursor emissions were developed for individual power plants within ERCOT on a per kWh basis so that these parameters changed with generation at the facility. A base case model that minimized the total cost of electricity generation in ERCOT without water-based constraints was utilized in order to compare the impacts of four water-consumption based scenarios. The first scenario involved turning off all water-withdrawing electricity generation in extreme and exceptional drought regions in South Texas and re-dispatching electricity generation from that region to other power plants on a least cost basis. Other scenarios in which water consumption in severe drought regions (the next most intense US Drought Monitor classification) was constrained to be equal to the base case, to be reduced by 5%, and to be reduced by 10% were also modeled in the work.

The cost of the drought-based re-dispatching scenarios was compared to the base case operation of the grid to determine the financial impacts of this strategy to increase water availability. The grid-based re-dispatching modeling indicated that water consumption for thermoelectric power generation in South Texas could be forgone during the 2006 episode, creating enough water-availability in the region for the personal use of 360,000 Texans. The cost of this strategy was compared to the installation and use of air cooling technologies at the subset of power plants in extreme and exceptional drought regions as well as to a pipeline project in Texas. The use of drought-based grid dispatching was found to be generally more expensive than either of these strategies but had the advantage of being able to respond to seasonal changes in the location of the most severe drought within ERCOT. Future work and costing strategies could potentially improve the economics of grid re-dispatching. In addition, drought-based grid re-dispatching could be used as a parallel strategy while more long-term strategies, such as pipelines or air-cooling, were implemented.

Drought-based re-dispatching of power generation also changed the spatial location of power plant emissions of ozone and PM precursors within Texas in addition to changing patterns of water use within the state. Impacts on 8-hour ozone and average PM concentration, which are the metrics related to federal air quality standards, were generally modeled to be small. The areas with the greatest net air quality benefits were generally down-wind from the South Texas drought area from which power generation was shifted. The locations with the greatest air quality dis-benefit were near coal-fired facilities with increased power generation during the episode.

(29f) A Novel Method to Manage the Concentrate Disposal of Desalination Units

Aghahossein Shirazi, S., New Mexico State University
Saraeian, A.,
Rastegary, J., New Mexico State University
Ghassemi, A., New Mexico State University, Institute for Energy and the Environment/WERC
Originally presented on: 11/3/2013 17:10:00 - 17:30:00

Scarcity of clean water is a great dilemma today and needs better management. About 1.2 billion people are living in areas where water is physically scarce and also 1.6 billion encounter economic water shortage. Population growth, climate change, and inappropriate use of available resources are the chief factors that worsen the current condition. Given that the amount of water on the earth has remained quite constant over time, it is crucial to desalinate saline and brackish waters. However, cost and concentrate stream of desalination units are two major issues that have restricted expansion of inland water desalination technologies. The salinity of concentrate stream which is sent to disposals increases with each succeeding separation step and causes environmental issues resulting in confining practical deployment of desalination units. Cultivation of microalgae in concentrate disposal is a new method which could contribute to solving these problems inasmuch as microalgae have been considered as excellent candidates being capable of removing ions and salts existing in concentrate stream. In addition, the derived dry biomass could be used as a feedstock for biofuel production. To examine this idea, a full factorial experiment with completely random design (CRD) was conducted aiming to investigate the growth of two strains of marine algae, *Dunaliella tertiolecta* and *Nannochloropsis oculata,* in three different media including concentrate, artificial seawater medium, and a 50-50 combination of these two under 16 hours of illumination in 25°C. Furthermore, Nitrate, Phosphate, Sodium, and potassium removal were measured everyday to characterize the trend of removing ions and salts from media. Results revealed that both strains of algae can grow fairly well in concentrate and have significant effect on removing salts and ions from concentrate disposal.

(32a) Overview Of Materials In The Built Environment
Judkoff, R.,
Originally presented on: 11/4/2013 8:30:00 - 8:52:00

Materials advances could help to reduce the energy and environmental impacts of buildings. Globally, buildings consume 30%-40% of primary energy and account for 25%-33% of CO_2 emissions. This talk explores some foundational concepts about how energy is used in buildings, and also highlights areas where materials advances would be most beneficial from an energy perspective.

(32b) Energyplus Model-Based Predictive Control (EPMPC) Using Matlab/Simulink and Mle+

Zhao, J., Center for Building Performance and Diagnostics
Lam, K. P., Center for Building Performance and Diagnostics
Ydstie, B. E., Carnegie Mellon University
Originally presented on: 11/4/2013 8:52:00 - 9:14:00

This paper proposes the EnergyPlus Model-based Predictive Control (EPMPC) system to control under floor air distribution (UFAD) in a multi-zone two stories open office building – The Center for Sustainable Landscapes – at Phipps Conservatory in the city of Pittsburgh. The building is equipped with a central air-handling unit with ground source heat pump as the cooling and heating source. The UFAD is used for the open office, conference rooms and other occupied spaces. Ceiling-based air distribution system is used for service spaces, such as restrooms, mechanical rooms, and storage rooms. The ASHRAE 90.1-2007 baseline and design phase whole building models are created using DesignBuilder and EnergyPlus program under the "Design-Build-Operate Energy Information Modeling (DBO-EIM)" infrastructure. The key of the DBO-EIM infrastructure is to use the detailed design phase energy model throughout the entire building life cycle by validating and adapting model assumptions during construction, commissioning and operation periods. Therefore, the model is expected to be "accurate" and easy-to-use for model-based building control systems in the daily operation to optimize energy and comfort. In this paper, the design phase model is applied in both the baseline and the EPMPC control. The EPMC controller includes both continuous and discrete variables and it is not possible to use optimization algorithms that rely on gradient search. In order to solve the optimization problem we developed an exhaustive search optimization algorithm within the Matlab/Simulink environment via maximum likelihood estimation+, co-simulation tool. In two weeks of winter and summer simulation using *typical meteorological year 3* weather file, comparing to the baseline logic control, which is also implemented using the same co-simulation method, the EPMPC can save HVAC operation energy by 18.9%. The occupant *predicted mean vote* value is as good as the baseline if not better. The benefits of using the EnergyPlus model over the simplified first principle or identified linear model include: (1) it can model complex buildings and systems; (2) it can be continuously updated with little cost and effort; (3) it requires less mathematics skills to manage in daily operation once the system has been developed.

(32c) The Sol-Char Toilet: Concentrated Solar Power Transported By Fiber Optics for Pyrolysis of Human Waste to Biochar

Fisher, R. P., University of Colorado at Boulder
Lewandowski, A., National Renewable Energy Lab
Yacob, T. W., University of Colorado at Boulder
Mahoney, R. B., University of Colorado at Boulder
Mejic, D., University of Colorado at Boulder
Oversby, C., University of Colorado at Boulder
Beck, S., University of Colorado at Boulder
Ward, B. J., University of Colorado at Boulder
Kearns, J., University of Colorado at Boulder
Summers, R. S., University of Colorado
Linden, K. G., University of Colorado at Boulder
Weimer, A. W., University of Colorado at Boulder
Originally presented on: 11/4/2013 9:14:00 - 9:36:00

Currently, 2.6 billion people on the planet do not have access to adequate sanitation and many communities in the developing world have neither the facilities nor the resources to properly treat human waste, leading to open defecation posing enormous public health risks as well as issues related to individual dignity and safety. Sanitation is not a problem unique to developing countries – modern sanitation processes are also unsustainable due to their often massive requirements for energy and water. The Sol-Char Toilet, a solar waste treatment technology developed at the University of Colorado Boulder, takes advantage of our most abundant renewable resource: the sun. In the Sol-Char Toilet, concentrated solar power (CSP) is transmitted through fiber optic cables to drive pyrolysis of human waste, converting a hazardous and unusable material to biochar. Biochar is a carbonized solid product of thermal biomass decomposition consisting largely of condensed aromatic zones that when applied as a soil amendment imparts agronomic benefits and is recalcitrant over a long timescale (10^2 – 10^3years). Biochar presents a solution relevant to the world today: reducing greenhouse gas emissions by sequestering carbon in biochar for several hundreds of years or even millennia. The over-reaching goal of the project is to develop a cost-effective, self-contained, and sustainable sanitation process by which human waste can be converted to useful products without connection to water, sewer or electricity.

Sunlight is concentrated to nearly 2000 Suns by a parabolic mirror, directed into a fiber optic bundle, and routed to a solar hood that directly irradiates a pyrolysis reactor. Two identical reactor vessels are switched between a pyrolysis location under the solar hood and a toilet location via a rotating carousel mechanism, thus achieving a semi-continuous pyrolysis process during daylight hours. The Sol-Char Toilet prototype represents a novel application of CSP to the areas of environmental, sanitation, and reaction engineering. This presentation will focus on the design and operation of the 1 kW Sol-Char Toilet prototype designed and constructed at the University of Colorado at Boulder.

(32d) Building-Integrated Active Modular Phytoremediation System

Dyson, A., Rensselaer Polytechnic Institute/Center for Architecture Science and Ecology
Aydogan Akseli, A., Rensselaer Polytechnic Institute
Vollen, J., Rensselaer Polytechnic Institute
Nyman, M., Rensselaer Polytechnic Institute
Bird, J.,
Originally presented on: 11/4/2013 9:36:00 - 9:58:00

Contemporary construction materials and mechanical building ventilation systems types are complicit in the degradation of indoor air quality (IAQ) and have become principal contributors to public health concerns in developed countries. These challenges are further exacerbated by the increasing cost of energy that is required to maintain healthy levels of IAQ in these buildings. Addressing these issues will require not only innovative technologies and approaches, but also deep collaborations across several disciplines.

This research stems from an interdisciplinary framework for the development of a building-integrated phytoremediation system using bio-mechanical techniques to 'scrub' toxins from the air and maintain desired psychometric profiles for IAQ by integrating modules containing hydroponic-supported plant materials into heating, ventilating and air conditioning (HVAC) systems. This research is multi-dimensional and interdisciplinary in scope, with the precise goal of integrating lab-based testing protocols to reiterative architectural systems design parameters.

There are three scopes that will be presented within this research: 1.Parametric framework for information exchange across disciplines; 2.Acqusition of scientific data on the sorption of VOCs (*e.g.,* formaldehyde and benzene) to growing media and granular activated carbon, removal efficiency of the selected VOCs from air utilizing the building-integrated phytoremediation system and the effect of phytoremediation as a bio-mechanical technique to remove formaldehyde and benzene from indoor air and; 3. Co-design of an integrated bio-mechanical HVAC system at building-scale. A series of experiments were undertaken to determine the performance parameters for architectural integration, including: the toxin removal capacity of various growing media and plant species; the impact of module morphology on contaminant concentration levels and air flow rates; and the rates of remediation by constituent components within plant leaves and the root rhizosphere. The conclusions from these experiments show that the building-integrated active modular phytoremediation system (AMPS) could have a significant impact in contributing to remediation strategies addressing poor IAQ in buildings and energy reduction through improved HVAC performance.

(32e) Integrating Technological and Ecological Design for Sustainable Buildings On a University Campus

Gibbemeyer, E. L., The Ohio State University
Bakshi, B. R., The Ohio State University
Originally presented on: 11/4/2013 9:58:00 - 10:20:00

Importance

The Ohio State University has pledged to become carbon neutral by 2050, an ambitious and currently out of reach goal. Technological systems often find the task of carbon neutrality to be challenging and frequently impossible. Therefore, we look to nature and ecosystems as examples of systems that have maintained sustainable levels of carbon, as well any other resource we care to examine. Thus, instead of merely optimizing a technological system to have minimum carbon emissions, we also consider the surrounding ecosystem and optimize the entire system to minimize carbon in what we call "techno-ecological networks." For example, in optimizing a building, our technological system, we can optimize it to be as close to carbon neutrality as possible; however, by simply planting additional trees outside of it, we can further reduce its carbon footprint by accounting for the carbon these trees will sequester. Including the ecosystem in our network allows us to reach a smaller carbon footprint that was previously infeasible when only looking at the technological components of our system.

Originality

While there have been many studies evaluating and optimizing the Ecological Footprints of buildings, ecosystems are usually not included in the evaluation. This means that any optimized solutions found in these studies could be further improved by also including the surrounding ecosystems. We also plan to include behavioral variables such as choosing a thermostat set point. Previously, we have incorporated these ideas into a simplified residential system; in this project, our techno-ecological network will be the real system of a segment of the OSU Columbus Campus that will include buildings and their surrounding green space. This real system will allow us to continue testing the effects of including ecological and behavioral variables into this optimization framework of minimizing carbon footprint, water footprint, and cost.

Methods Employed

We will use software developed by the DOE to model our campus buildings, and the models will be verified using data collected by the OSU Energy Services and Sustainability office. We then will begin making changes to technological variables to see what retrofitting options would lower our carbon and water footprints, as well as adding in the ecological variables through biophysical models including InVEST and CENTURY. Once we have developed all the different variables, we will use optimization software to solve for scenarios that minimize carbon, water, and cost.

Results

Because our optimization is multi-objective, we will not simply obtain a single solution; we will find a Pareto optimal set of solutions. However, we anticipate some variables to lead to "win-win" scenarios, such as installing a rainwater capture system to lower water footprint and also lower water utility costs. In finding "win-win" scenarios, we hope to find opportunities for changes that can be made and allow this study to turn into a true "living laboratory" where students can become more ecologically conscious. Nonetheless, for other variables, we will require a judgment call on the relative importance of our objectives if we want to be able to select a "best" scenario that we could implement on campus.

(32f) Energy Conservation Approach to Cooling Process By Employing Preheat Tank and Chilled Water Tanks

Peters, D. R. W., University of Alabama at Birmingham
Li, Z., University of Alabama at Birmingham
Originally presented on: 11/4/2013 10:20:00 - 10:42:00

Energy conservation has been a hot topic worldwide. With the rapid growth of population and urbanization, the energy demand is increasing annually. According to the 2005 *Annual Energy Review* compiled by the Energy Information Administration (EIA) within the U.S. Department of Energy (DOE), the energy consumption in the United States including transportation, residential, and commercial end uses, shows a trend of increasing during the period of 1970 to 2005. Over the years from 1970 to 2004, total commercial end uses have an annual growth rate of 2.6 percent, and electricity consumption accounts for 76.2% of total commercial energy consumption. The HVAC system is an important process to provide comfortable temperatures, but demanding a large amount of energy and mainly contributing to the peak electricity consumption. The chilled water tank is a technology that can dramatically reduce the peak energy consumption with installation in traditional air conditioning systems, especially for the hot and humid climates. The preheat tank is a device used for heat recovery purposes that absorbs the heat from air flows to decrease the air temperature entering the chilled water tank and enhance the performance of cooling with even less energy requirements.

Our research utilizes the preheat tank and chilled water tank in commercial and conventional air conditioning systems to achieve energy conservation opportunities in Birmingham, Alabama. The preheat tank was installed to exchange heat with hot air, and the heated water can be used for lavatory and other purposes. The chilled water tank was set up to provide cooling with less energy consumption. The Generalized Fluid System Simulation Program (GFSSP) software was utilized to analyze the air conditioning system involving the heat recovery tank and chilled water tank. In addition, experiments were conducted in a small building at the University of Alabama at Birmingham campus with an installed small preheat tank and chilled water tank, respectively, to provide data needed for software programming. The energy consumption of the modified HVAC system was monitored to check the energy conservation rate. Some factors that may affect the performance of the modified system, include ambient temperature, humidity, water temperature in the tank, air conditioning set temperature, and dew point. The variables were recorded and analyzed in regression models to determine which factor(s) are significant for energy conservation. Moreover, covariance analysis was performed to determine the relationship between these factors. The results are expected to provide energy conservation for air conditioning systems, especially during the peak hours of energy consumption in hot and humid summer climates which is the typical weather condition in Birmingham, Alabama. Finally, the payback period was estimated by comparing energy bill savings and capital investment. The research can be extended to other southern states that experience hot and humid climates, such as Texas, Arizona, Georgia, and Florida.

(32g) Operation of Residential Hybrid Renewable Energy Systems: Integrating Forecasting, Optimization and Demand Response

Wang, X.,
El-Farra, N. H., University of California, Davis
Palazoglu, A., UC Davis
Originally presented on: 11/4/2013 10:42:00 - 11:04:00

Hybrid renewable energy systems, which consist of solar, wind and other energy generation and storage units, have been widely researched in recent years owing to the reduction in traditional fossil resources and increase in the world-wide energy demand. This paper focuses on the optimal design and operation of a standalone community-scale hybrid renewable energy system comprising photovoltaic (PV) cells, wind turbines, diesel generators and storage units such as batteries. A detailed modeling of each system component is presented. The demand-side management as well as the performance of the generation system are considered to guarantee that the electricity demand of the consumer is satisfied and the overall system economic and environmental cost is minimized.

To optimally allocate electricity generation among different units, a precise prediction of the future electricity consumption and the available solar and wind resources is necessary. Both the day-ahead and real-time predictions are made to provide sufficient information for further optimization. For the day-ahead prediction, historical demand as well as solar radiation and wind speed data are used as the basis for the classification of day types, which are determined by the temperature, humidity, seasonality and other related characteristics. During the day-ahead operational management, a general electricity usage and a generation profile are developed to realize the preliminary dispatch strategy. For the real-time prediction, autoregressive and polynomial fitting methods are adopted to facilitate time-series analysis and forecasting. The online forecasting model is able to consider the updated data and modify the overall prediction and optimization.

On the demand side, a demand-responsive scheme is integrated with the operation of power generation. Among all the residential electricity usage, some are rigid in demand which cannot be changed or shifted such as lighting and cooking while others are flexible loads including washing machines, clothes dryers and vacuum cleaners, which are the key factors to be considered to manage the demand side. These loads can be arranged to reshape demand profiles to match renewable output better.

In the optimization stage, the method of model predictive control (MPC) is adopted by considering real-time optimization to minimize multiple objectives over a receding horizon, the most significant of which is to minimize the generation cost within the time window considered. The environmental influences are also accounted for by the indicators of equivalent cost. As constraints, each energy resource has its maximum generation capacity, as well as minimum output level, which decides the operation range of the generation unit. Limits of ramping rate of each resource are also posed to avoid damage caused by excessively rapid change of power output. The formulation is simplified to a convex optimization problem. At each time interval, only the first set of power references calculated by the optimization are implemented. The optimal allocation between different resources drives the generation to meet electricity requirement of the demand side, with minimum operating cost and environmental impact. Different strategies to realize the receding-horizon method are applied and the results are compared. It turns out that the length of the moving horizon has a critical impact on the optimal allocation of electricity generation among the available resources. Proper selection of the horizon allows for an efficient and optimal operation of the energy system.

The strength of the proposed strategy is that it is able to achieve global optimization over a certain time period and at the same time the users are able to obtain sufficient information of the future conditions and thus make rational decisions about electricity management. Compared with traditional operation of energy systems, the on-line process that combines real-time prediction, optimization and demand-responsive scheme will improve system efficiency and take better advantage of all available resources.

We will show several simulation and optimization case studies that demonstrate the effectiveness and applicability of the proposed receding horizon scheme. Future work to connect the current system to the electrical smart-grid to realize large-scale application will also be discussed.

(36a) Green Chemistry, Green Engineering, Green Economy At US EPA 's Sustainable Technologies Division

Leazer, J., U.S. Environmental Protection Agency
Originally presented on: 11/4/2013 8:35:00 - 8:50:00

Green Chemistry and Green Engineering Research in the sustainable Technologies Division of USEPA's National Risk Management Research Laboratory is focused on reducing the adverse human health and environmental impacts of the synthesis of economically important chemicals and chemical processes. Our research focuses on several categories including:

1) Sustainable Materials Management, which accounts for ecological scarcity and resource efficiency;

2) Green Chemistry, which focuses on the minimization/elimination of human health and environmental impacts as well as resource efficiency;

3) Green Engineering, which focuses on energy and resource conservation;

4) Sustainability Measurement, which allows us to determine whether a system is becoming more sustainable or not;

5) Life Cycle Assessment, ensuring minimal impact across the entire life cycle of a product/good/service and

6) Supply Chain Design, which looks at the optimization and increased sustainability of the entire supply chain of a product.

These research areas have the common goal of green production of economically viable chemicals and chemical processes.

(36b) Sustainability Practice in Industry

Powell, J. B., Shell Westhollow Technology Center
Originally presented on: 11/4/2013 8:50:00 - 9:05:00

This seminar will address how sustainability is implemented in industry, with real world examples from the energy and chemical processing industries. Topics will include making the case for sustainability, sustainability metrics and reporting, the importance of considering sustainability early during new process or product development, and how industry may be organized to incorporate sustainability principles into decision making. The importance of external stakeholder engagement in project planning and execution will be addressed.

In industry, sustainability is viewed as a key enabler for license to operate, and can be seen as a driver for much of the new innovation and development in our profession. Competency in life cycle inventory and overall environmental footprint and social impact assessment has become increasingly important in assessing preferred opportunities for technology development and business deployment. Demand for excellence in safe and responsible implementation and performance will continue strong for the developed world, and dissemination of best technologies and practices to developing economies will be critical for addressing global sustainability needs and environmental challenges.

(36c) Green Engineering and Sustainability in the Chemical Engineering Curriculum

Shonnard, D. R., Michigan Technological University
Allen, D. T., The University of Texas at Austin
Nguyen, N., U.S. Environmental Protection Agency
Austin, S., Chemical Engineering Branch, US EPA
Originally presented on: 11/4/2013 9:05:00 - 9:20:00

A green economy will be well-served by an educated chemical engineering workforce firmly grounded in green and sustainable engineering concepts. The goal of this presentation is to outline updated green engineering and sustainable engineering content for undergraduate and graduate chemical engineering education. The work of chemical engineers is central to the development of sustainable chemical and energy production systems and to meet global human needs. Incorporating environmental issues and sustainability into the design of engineered products, processes, and systems generally means evaluating environmental, economic, and social performance. This presentation will summarize two approaches to incorporate green and sustainable engineering into undergraduate chemical engineering education; **i.** a stand-alone elective course, and **ii.** modules for required core chemical engineering courses. A detailed overview of chapter-by-chapter content will also be presented with tie-ins to core courses.

(36d) Driving Private Sector Adoption of Greener Processes and Technologies

Lindsay, A., Author: Abby Lindsay, American University, Doctoral Candidate

Originally presented on: 11/4/2013 9:40:00 - 9:55:00

In addition to advanced environmental technologies and improved chemical processes, a green economy will require widespread adoption by the private sector. This presentation will outline ways in which greener practices are spread to a wider number of companies through focusing on experiences throughout Latin America. Most larger companies with extensive resources and access to financing have realized the environmental and cost benefits to reducing energy and water use, decreasing pollution, and increasing water and materials efficiency. Micro, small, and medium-sized enterprises (MSMEs) can benefit just as much, but often lack the awareness, financial resources, or skills to increase their efficiency. Given that the majority of companies are MSMEs, reducing the resource-use and pollution intensity of these companies is critical to a green economy. This presentation will look at factors that drive the diffusion of environmental processes and technologies, and determinants required for successful adoption, such as adequate engineering education systems, availability of financing, and policy incentives.

(36e) Minimization of the GHG Emissions At the Macroeconomic Level Via a Multi-Objective Input-Output Approach: A Case Study of the EU-25 Economy

Cortés, D., University Rovira i Virgili

Ruiz-Hernández, A., University Rovira i Virgili

Guillén-Gosálbez, G., Universitat Politécnica de Catalunya

Llop, M., Universitat Rovira i Virgili

Guimerá, R., Universitat Rovira i Virgili

Sales-Pardo, M., Universitat Rovira i Virgili

Originally presented on: 11/4/2013 9:55:00 - 10:10:00

Minimization of the GHG emissions at the macroeconomic level via a multi-objective input-output approach: A case study of the EU-25 economy

ABSTRACT

D. Cortés-Borda[a], A. Ruiz-Hernández[a], G. Guillén-Gosálbez[a], M. Llop[b], R. Guimerà, M. Sales-Pardo.

a.Departament d'Enginyeria Química, Universitat Rovira i Virgili, Av. Països Catalans, 26, Tarragona E-43007, Spain.

b.Centre de Recerca en Economia Industrial i Economia Pública (CREIP), Universitat Rovira i Virgili, Av. Universitat 1, 43204 Reus, Spain.

Global warming has grown significantly in the recent past due to the increasing atmospheric concentration of anthropogenic greenhouse gasses (GHGs). As a result, worldwide environmental policies are placing global warming mitigation as a top priority in the nations' environmental agendas. That is, most of the National governments are adopting policies that aim to minimize GHG emissions.

Unfortunately, in today's globalized markets designing effective environmental policies is challenging because international trade establishes channels through which impacts are imported and exported among nations, overshadowing the main source of pollution into an intricate trading network. Therefore, the design of these policies requires robust environmental assessment methods capable of capturing the complexity of the existing international supply chains of goods/services .

Methods based on extended environmental input-output (EEIO) models are particularly suitable to address global warming at a global scale, since they can assess macroeconomic connections between productive sectors of different nations and the associated environmental impact. This information can be used to allocate the corresponding environmental responsibility. EEIO models are flexible, transparent and accurate enough to perform life cycle assessment (LCA) studies. Hence, LCA-based EEIO models translate the economic output of a given economy into a tangible environmental impact.

By using macroeconomic EEIO models, we can identify a wide range of alternatives for environmental savings, as it is possible to reduce each sector's output (and therefore its associated GHGs emissions) to a different extent through taxation and/or economic penalties. In this context, systematic optimization tools are especially helpful to identify the most promising alternatives considering economic and environmental concerns.

We have developed in this work a systematic approach of this type for reducing the environmental impact of an economy that is based on the combined use of EEIO models and multi-objective linear programming techniques. This tool has been applied to the European Union (EU-25) economy in order to find solutions that simultaneously maximize the total output and minimize the GHGs emissions. Hence, in our approach the problem of identifying key economic sectors that should be given higher priority is posed in mathematical terms as a bi-criteria LP problem that seeks to optimize simultaneously the total European gross domestic product and the associated life cycle GHGs simultaneously. Our study makes use of the E3IOT database, which covers 487 EU economic sectors.

Numerical results show that substantial reductions in global warming potential can be achieved by imposing taxes on sectors such as energy, mineral extraction, and meat and its derived products. By applying this wide-scope assessment, we find potential alternatives for reducing GHG emissions (that are not obvious at first sight) that can assist policy-making.

(36f) Economic Analysis and Optimization of Carbon Capture Storage Pipeline Transport Network System

Dan, S., Seoul National University
Kang, B. J.,
Cho, S.,
Yoon, E. S., Seoul National University
Originally presented on: 11/4/2013 10:10:00 - 10:25:00

Carbon Capture & Sequestration (CCS) is the effective means that can be used to reduce CO2 emissions. Among the CCS network system, a pipeline transport network is the least cost element of the full CCS network system, but the most planning part during the development of the CCS network system.

The research presented here based on the physical properties of CO2 and Co2 containing mixtures calculated. The methodology of this research consists of four steps: Optimal CO2 pipeline design, CCS transport network system modeling, Cost evaluation for model, Selection of the optimal CCS transport network system.

Objective of this research is to find the optimal proposal of CCS pipeline transport network system. This research proposes the optimal engineering CO2pipeline design and CCS pipeline network design for transporting CO2 using HYSYS 7.3 simulator.

(36g) Economic Development Through Entrepreneurship and Technology Innovation: Bringing Together Industry, Academia, and Government

Anid, N. M., New York Institute of Technology

Originally presented on: 11/4/2013 10:25:00 - 10:40:00

The School of Engineering and Computing Sciences (SoECS) of New York Institute of Technology (NYIT) is launching a new Entrepreneurship and Technology Innovation Center (ETIC). The ETIC will consolidate and expand the School's ongoing industry-academic partnerships, foster innovation and promote collaborations between industry, the academic community, professional organizations, and government. The Center will support the NY region's economic development by focusing on three critical areas Health Analytics & Bioengineering, IT & Cybersecurity, and Energy and Green technologies. This paper will discuss a case study of a green economic development initiative that involves chemical engineers: the creation of a microgrid pilot installation that utilizes various storage technologies. Process economics and financing via a public private partnership and seed government funding will be presented. The goal is to infuse the engineering curriculum with real-world problems, and to show how project-based learning, teamwork, and a collaborative environment and direct contact with practitioners, instill in students the key competency and entrepreneurship skills required in today's rapidly evolving, globalized economy.

(37a) The Carbon Capture Simulation Initiative (CCSI): Computational Tools To Accelerate Commercial Development

Miller, D. C., U.S. Department Of Energy, National Energy Technology Laboratory

Originally presented on: 11/4/2013 8:30:00 - 8:55:00

The U.S. Department of Energy's Carbon Capture Simulation Initiative (CCSI) is developing, demonstrating and deploying advanced computational tools and models to accelerate the development of next generation technologies, specifically the development of cost effective carbon capture technologies. This CCSI Toolset will (1) enable promising concepts to be evaluated more quickly based on an optimized process, (2) reduce the time for design and troubleshooting by integrating process and device scale simulations to better predict performance and more effectively resolve issues, (3) quantify technical risk to enable more focused scale up activities, and finally, (4) stabilize the costs during commercial deployment by having a more comprehensive understanding of the process and underlying behavior of the system. This paper will provide an overview of the program and highlight recent accomplishments applying the tools to potential carbon capture technologies.

(37b) The Four Dimensional Product: Integration Over Time Is the Only Way to Understand Sustainability

Banholzer, W., Dow Chemical Company
Jones, M. E., The Dow Chemical Company
Originally presented on: 11/4/2013 8:55:00 - 9:20:00

Sustainability is a question of time, integrating knowledge of production, use and end-of-life. Far too often, hype has surpassed reality, boundary conditions have been neglected and actions taken that don't truly lead to a sustainable future. Singular focus on attributes like renewable content or biodegradability gives the wrong answer. Quality engineering forms the foundation for any meaningful discussion of sustainability, providing insights into energy and material flows, energy return, and thermodynamic constraints. The chemical industry is increasing resource and energy efficiency through innovation in both products and processes. Our mastery of materials science enables unique energy producing and energy saving products, products that improve sustainability by being significantly better than the next best alternatives during use.

(37c) Maximization of Energy Efficiency of a Combined Heat and Power Plant

Edgar, T. F., The University of Texas at Austin
Powell, K., The University of Texas at Austin
Kim, J., The University of Texas at Austin
Kapoor, K., The University of Texas at Austin
Originally presented on: 11/4/2013 9:20:00 - 9:45:00

Energy efficiency is receiving increased attention as a way to reduce the use of fossil fuels and the resulting production of greenhouse gases. Automation, process control, and real-time optimization are key technologies to operate plants in the most efficient way. A combined heat and power plant at the University of Texas at Austin has been studied in order to reduce its carbon footprint as well as reduce costs. The steam power plant has been modeled using steady state mass and energy balances. Chillers have also been modeled so that their efficiencies can be maximized. There is also a thermal storage tank for chilled water to be used during periods of high demand. Forecasting of loads as well as the effect of weather have been incorporated into the real-time optimization approach. The effects of time of day pricing of power and use of demand response techniques to flatten demand curves are discussed. Usage of thermal energy storage along with automatic control gives additional degrees of freedom to deal with dynamic power conditions.

(37d) Pan American Issues on Biofuel Sustainability

Shonnard, D. R., Michigan Technological University
Originally presented on: 11/4/2013 9:45:00 - 10:10:00

Biofuels are in rapid development in many Pan American countries, yet the nature and extent of impacts on social and environmental systems from large-scale biofuel development are not well understood. This talk will explore sustainability issues across a broad set of impacts and in different Pan American countries and will begin to outline important research gaps. A future research agenda to overcome these knowledge gaps would help establish a new and growing biofuels industry onto a sustainable path.

(65a) Templated Ordered Mesoporou Materials for Supercapacitors and Lithium Ion Batteries

Lee, J., Pohang University of Science and Engineering

Originally presented on: 11/4/2013 8:30:00 - 9:00:00

For high performance supercapacitor and Li-ion battery electrodes, various mesoporous metal oxide materials have been synthesized by employing lab-synthesized block copolymers as a structure-directing agent or ordered mesoporous silica materials as hard templates. An ordered mesoporous WO_{3-X} material with high electrical conductivity was employed for use as a supercapacitor electrode in 2.0 M H_2SO_4. The electrode made of mesoporous WO_{3-X} exhibited a high rate capability and an excellent capacitance (366 Fcm^{-2}, 639 Fcm^{-3}), which were attributed to the large ordered mesopores, high electrical conductivity, and high material density. Fe_3O_4 nanocrystals confined in mesocellular carbon foam (MSU-F-C) are synthesized by 'Host-Guest' approach and tested as anode for lithium-ion batteries (LIBs). 13-27 nm magnetite Fe3O4 nanocrystals are formed inside ~30 nm pores of mesocellular carbon matrix. Fe_3O_4/MSU-F-C (45 wt% Fe_3O_4) exhibited 1007 mA h g^{-1} of first charge (Li de-insertion) capacity at 0.1 A g^{-1} and 111% of capacity retention at 150th cycle, and retains 37% capacity at 7 A g^{-1} (~7 C rate). For the first time, we synthesized mesostructured LTO/carbon composite (Lithium titanate/carbon) with large (> 15 nm) and uniform pores for high performance LIB anodes via a block copolymer assembly. Sn-mesostructured carbon nanocomposites have been synthesized by simple self-assembly of block copolymers with Tin and carbon precursors.

(65b) Reduced Graphene Oxide Wrapped FeS Nanocomposite for Lithium-Ion Battery Anode With Improved Performance

Fei, L., New Mexico State University
Luo, H., New Mexico State University
Originally presented on: 11/4/2013 9:00:00 - 9:15:00

The fast growing business of portable electronics (e.g. computers, mobile phones, cameras) has stimulated the rapid development of lithium-ion batteries (LIBs) due to their high energy density, flexible and lightweight design, and longer lifespan than conventional battery technologies. Graphite is the most popular commercial anode. However, its low theoretical capacity (372 mAh/g) renders it hard to satisfy the increasing market demands. Recently, metal sulfides with high theoretical capacity are widely studied as another promising class of anode materials with excellent electrochemical performance.

Here, we report FeS based nanocomposite as anode for lithium-ion batteries. FeS nanoparticles wrapped in reduced graphene oxide (RGO) are produced via a facile direct-precipitation approach. The resulting nanocomposite FeS@RGO structure has better lithium ion storage properties exceeding those of FeS prepared without RGO sheets. The nanocomposite delivered an initial discharge capacity of 1357 mAh/g and a charge capacity of 1116 mAh/g, giving the first cycle coulombic efficiency of 82 %. The composite shows a very stable cyclic performance with its capacity remaining as high as 978 mAh/g even after 40 cycles at a current density of 100 mA/g. This value is almost three times the capacity of commercial graphite anode, and is comparable to those reported in literature. At current density of 300 mA/g, the composite delivered discharge capacity of 927, 690, and 618 mAh/g in the first, 10th, and 30th discharge cycles, respectively; while FeS suffered rapid capacity decline, only delivering 885, 518, and 382 mAh/g in the first, 10th, and 30th discharge cycles. Electrochemical impedance demonstrate that FeS@RGO has a lower electrochemical resistance than that of FeS. The enhanced electrochemical performance is attributed to the robust sheet-wrapped structure and the synergetic effects between FeS and RGO sheets, such as increased conductivity, shortened lithium ion diffusion path, and the effective prevention of polysulfide dissolution.

(65c) Alkanethiol-Passivated Germanium Nanowires As High-Performance Anode Materials for Lithium-Ion Batteries

Tuan, H. Y., National Tsing Hua University
Yuan, F. W., National Tsing Hua University
Originally presented on: 11/4/2013 9:15:00 - 9:30:00

Rechargeable lithium-ion batteries have been widely used in portable electronic devices such as cell phones, laptop computers, and digital cameras, and have been considered as an electrochemical energy storage system for powering hybrid electric vehicles (HEV) and zero-emission vehicles (ZEV). It is necessary to develop lithium-ion batteries having a higher energy density and a higher power density to meet the increasing energy demand. Germanium (Ge) has emerged as a promising anode materials due to its high theoretical capacity of 1384 mAh/g - around four times greater than commercial graphite anode (372 mAh/g). However, the practical usage of Ge as an anode material is hindered by dramatic volume changes (~300%) during the insertion/extraction process of lithium ions, causing crack and pulverization, and loss of electrode contact.

Ge nanowires can effectively accommodate the volume changes, tolerate relaxed mechanical strain, provide channels for efficient electron transport, and alleviate pulverization. Although direct incorporation of Ge nanowires into slurry coating process shows improved electrochemical performance relative to microparticles, the anode performance decayed rapidly within a few tens of cycles. In this talk, we demonstrate that dodecanethiol monolayer passivation on Ge nanowire surface can significantly enhance their anode performance in lithium-ion batteries. The dodecanethiol-passivated Ge nanowires exhibit an excellent electrochemical performance with a reversible specific capacity of 1130 mAh/g at 0.1 C rate over 100 cycles, which is above 80% of the Ge theoretical capacity. The functionalized Ge nanowires show high-rate capability having charge and discharge capacities of ~555 mAh/g at rates as high as 11 C. An aluminum pouch type lithium cell using a $LiFePO_4$ cathode was also assembled to provide larger current (~30 mA) for uses on light-emitting-diodes (LEDs) and audio devices. Organic monolayer-coated Ge nanowires represent promising Ge-C anodes with very low carbon content (~2-3 wt %) for high capacity, high-rate lithium-ion batteries and are readily compatible with commercial slurry-coating process for cell fabrication.

(65d) A Facile Microwave-Assisted Route to Co(OH)2 and Co3O4 Nanosheet for Li-Ion Battery

Chen, G., New Mexico State University
Luo, H., New Mexico State University
Originally presented on: 11/4/2013 9:30:00 - 9:45:00

Cobalt hydroxide ($Co(OH)_2$), as one of the most important functional inorganic materials, can be used in the field of catalysis and electrochemical electrodes. The previous report revealed that $Co(OH)_2$ is a promising alternative conversion-based anode materials for lithium-ion batteries (LIBs) due to its high capacity. Unfortunately, like other conversion-based anode materials, the implementation in LIBs is greatly hampered by its poor cycling performance. One of the most promising strategies to tackle this obstacle is to make $Co(OH)_2$ into sheet-like structure to improve the structural stability in the electrochemical reactions. Compared with the bulk $Co(OH)_2$, $Co(OH)_2$nanosheet displays obvious advantages due to large surface area, highly oriented growth direction, tunable interlayer space for the de/insertion of lithium ions.

In this work, a facile microwave-assisted synthetic route has been successfully developed for preparing hexagonal $Co(OH)_2$ nanosheets with average width of 2 μm and thickness of 100 nm. $Co(OH)_2$ can further convert to porous Co_3O_4 nanosheets *via* thermal decomposition. Their phases, structures and morphologies were investigated by X-ray diffraction (XRD), atomic force microscopy (AFM), and transmission electron microscopy (TEM). Both $Co(OH)_2$ and Co_3O_4 nanosheets can serve as potential candidates for anodes of Li-ion battery. The electrochemical study revealed that $Co(OH)_2$ and Co_3O_4 nanosheets delivered a reversible capacity of 600 and 700 mAh g^{-1} after 40 cycles, respectively. Cyclic voltammetry (CV) curves also confirmed the relative stability of the as-synthesized electrode materials. This effective microwave-assisted route may be a promising approach for preparing other transition metal hydroxide/oxides for energy applications.

(65e) Li Adsorption and Intercalation in Graphene, Graphite and Carbon Onions By Reaxff Reactive Force Field

Raju, M., The Pennsylvania State University
Ganesh, P., Oak Ridge National Laboratory
Kent, P. R. C., Oak Ridge National Laboratory
van Duin, A. C. T., Pennsylvania State University
Originally presented on: 11/4/2013 9:45:00 - 10:00:00

A ReaxFF reactive force field has been developed to describe Li/C systems. The ReaxFF force-field parameters for Li-C systems were optimized against a density-functional theory (DFT) (with van der Waals corrections) based training set, containing a collection of data (energies, geometries, charges etc.) related to adsorption and diffusion of Li on graphene and in graphite. ReaxFF reproduces accurately the adsorption energies for Li on hollow, bridge and top sites on pristine graphene, formation energy of stage I compounds in graphite and intralayer Li migration barriers in graphite at separations of 3.35 Å and 3.7 Å. To validate this force field we have compared it against DFT calculations for Li diffusion in the direction perpendicular to the graphene sheet for graphene with single and double-vacancy. We performed grand canonical Monte Carlo simulations to obtain ground states for Li intercalation in graphite and find that ReaxFF describes the in-plane Li ordering and interlayer separations for stage I and II compounds in good agreement with previous DFT studies and experiment. The voltage profile for the Li-graphite system as a function of Li concentration obtained by ReaxFF matches experiment and DFT data. We study Li intercalation in presence of vacancy defects in graphite and find that the presence of vacancy defects increase the ratio of Li/C and shifts the voltage profile downwards, both in proportion to the number of vacancy defects. Li intercalation and diffusion kinetics in Carbon Onions was investigated for potential applications in batteries and electrochemical capacitors and we report a different loading behavior/voltage profile for these Carbon Onions as compared to graphitic systems. The Voltage vs Li concentration for Carbon Onions has a 'saw tooth' shape whereas the voltage profile for graphitic systems consists of plateaus.

(65f) Silicon-Rich Carbon Nanofibers From Water-Based Spinning to Control Combination of Silicon and Carbon for Li-Ion Battery

Joo, Y. L., Cornell University
Kim, Y. S., Cornell University
Kim, K. W., Cornell University
Cho, D.,
Hansen, N. S., Cornell University
Originally presented on: 11/4/2013 10:00:00 - 10:15:00

Silicon-rich carbon nanofibers from water-based spinning to control combination of silicon and carbon for Li-ion battery

Yong Seok Kim, Kyoung Woo Kim, Daehwan Cho, Nathaniel S. Hansen, and Yong Lak Joo*

School of Chemical and Biomolecular Engineering, Cornell University, Ithaca 14853, USA

ABSTRACT: Carbon (C) nanofibers with high silicon (Si) loading (> 65 wt%) were fabricated from water–based electrospinning of aqueous polyvinyl alcohol/Si nanoparticle solution for Li–ion battery anode applications. Our Si-rich C (SRC) nanofibers showed much facilitated charge transport features and increased activities because of continuous one–dimensional (1–D) carbon backbone structure, leading to superior battery performance to that of bare silicon nanoparticles. The presence of carbon as 1–D nanofibers could not only mitigate volume expansion of silicon but also avoid formation unstable solid electrolyte interface onto the surface of silicon. Our study on the optimum combination of carbon and silicon in the nanofibers for their electrochemical properties and battery performance revealed that the SRC nanofibers with 72.8 wt% Si content (27.2 wt% C) exhibits an adequate balance between high energy capacity of Si nanoparticles and dimensional stability with effective charge transport by carbon nanofibers, leading to an outstanding cycle life to maintain 1076 mAh/g normalized by total electrode mass and high Coulombic efficiency of about 99% for fifty cycles. Such scalable Si-rich C nanofibers made from water-based electrospinning using polyvinyl alcohol should be greatly contributed to a cost–effective development of high performance negative battery electrodes.

(65g) Electrochemical Properties of Carbon-Coated SnO2-Graphene Nanocomposites for Rechargeable Lithium-Ion Batteries

Ara, M., Wayne State University
Salley, D. S., Wayne State University
Ng, S. K. Y., Wayne State University
Originally presented on: 11/4/2013 10:15:00 - 10:30:00

SnO_2 nanoparticles were carbon-coated using two different methods and then dispersed on the reduced graphene oxide nanosheets prepared from modified Hummer's method. Carbon-coating was performed by a solvothermal method from glucose precursor material or by carbonization in Polyacrylonitrile (PAN) based carbon fibers.The later was prepared by dispersing SnO_2 nanoparticles in N-Methyl-2-pyrrolidone and then the dried powder was stabilized in air at 270°C for 30 minutes followed by the heat treatment at 900°C for 2 hours. The coated nanocomposites were characterized by X-ray diffraction, scanning electron microscopy (SEM), transmission electron microscopy (TEM), and Raman spectroscopy. From TEM observations, it was found that the SnO_2 nanoparticles were 2-6 nm in diameter and were uniformly distributed on the reduced graphene oxide nanosheets. Electrochemical properties were evaluated by galvanostatic charge-discharge cycling. Carbon-coated SnO_2-reduced graphene oxide nanocomposites showed higher capacity and more stable electrochemical performances than non-coated SnO_2-reduced graphene oxide nanocomposites. Carbon coating enhances the conductivity of the electrode and it reduces the stress developed by the volume changes of the SnO_2 nanoparticles during charge/discharge cycles. The synergistic effect of the amorphous carbon, SnO_2 nanoparticles and reduced graphene oxide nanosheets is also responsible for the superior specific capacity of carbon-coated anode material. Excellent rate capability and cyclic performances were also demonstrated with higher current densities of 500 and 1000 mA/g.

(65h) Multilayer Hybrid Electrodes of TiO2/Mwnt for Electrochemical Applications

Hyder, M. N., Massachusetts Institute of Technology
Kavian, R.,
Shao-Horn, Y., Massachusetts Institute of Technology
Hammond, P. T., Massachusetts Institute of Technology
Originally presented on: 11/4/2013 10:30:00 - 10:45:00

Rational design of next-generation electrochemical devices requires novel electrode materials and architectures to achieve high-energy, high-power density to bridge the gap between battery and capacitors. Conventional electrode fabrication processes such as doctor blading allows limited control over the electrode thickness and structure at the nanoscale, in addition to the incorporation of insulating binder materials that limits device performance. To address these challenges, we have engineered a route for the synthesis of highly stable, sub-8nm TiO_2 nanoparticles that were incorporated with acid-functionalized multi-walled carbon nanotubes (MWNT). Using electrostatic layer-by-layer nanofabrication, we assembled binder-free hybrid electrode of TiO_2/MWNT to obtain a synergistic effect from high electronic conductivity and excellent charge storage capacity. These thin film electrodes with highly controllable thicknesses shows well developed mesopores with finely dispersed, non-agglomerated TiO2 nanoparticles on MWNTs. Electrochemical measurements show high charge storage capacity (>150 mAh/$g_{electrode}$ at 0.1 A/$g_{electrode}$) with excellent cycling stability during charging and discharging up to 200 cycles indicating their promise as thin-film negative electrodes for future Li storage applications.https://aiche.confex.com/aiche/2013/t5/papers/index.cgi?username=341047&password=343273

(65i) Facile Preparation of Mesoporous α -Fe2O3 With Unique Cocoon-Like Hollow Nanostructure for Reversible Lithium Ion Storage

Zhu, J., Wayne State University
Ng, K. Y. S., Wayne State University
Deng, D., Wayne State University
Originally presented on: 11/4/2013 10:45:00 - 11:00:00

After 20 years of development since the first commercialization of rechargeable lithium ion batteries by Sony in 1991, carbon is still exclusively used as anode materials for lithium ion batteries. However, the theoretical capacity (372mAh g^{-1}) of graphite is almost achieved now and it becomes a bottleneck to further increase lithium-ion storage capacity using carbon. Carbon-alternative materials with higher capacity are currently being extensively explored to meet the increasing demand for energy in mobile electronic devices, electric vehicles and energy storage in grid. Transition metal oxides receive increasing attention as carbon-alternatives because of their much higher theoretical capacity. For example, α-Fe_2O_3 with a theoretical capacity of 1007 mAh g^{-1} is a promising candidate. Compared to other transition metal oxides, α-Fe_2O_3 is outstanding in terms of low cost, abundance, ease of fabrication, and environmental benignity. However, the poor cyclability caused by volume change during the insertion /extraction of Li ions and poor conductivity is the main challenge. Our strategy is to design and tailor α-Fe_2O_3 at nanoscale with structure control to improve their electrochemical performance. In this presentation, the strategy to facilely prepare mesoporous α-Fe2O3 with unique cocoon-like hollow nanostructure as anode materials will be reported. The formation mechanism is revealed by extensive characterization by FESEM, TEM and XRD for samples prepared under different controlled conditions. The improved electrochemical performance in reversible lithium ion storage will be discussed.

(86a) Biofuel Production Wastewater Minimization and Reclaim
Privitera, M., PreProcess, Inc
Originally presented on: 11/4/2013 8:30:00 - 8:52:00

Biofuel feedstock is wet. The fuel specifications understandably have very low moisture limits. The mass balance must be complete and the disposition of the water in the feedstock must be resolved. Regulatory requirements make for very challenging low level constituent control in order to release separated biofuel water streams.

As reaction technologies develop to allow for the conversion of more and more waste feedstocks, the control and disposition of each constituent that is separated from the product becomes more and more of a challenge.

Some waste water constituents are easily controlled and agglomerated into a reasonable production waste flow. Some constituents require unique separation systems to insure the management of levels and the concentrations.

Biofuel production also presents the challenge of multi phase fluid characteristics. The process systems employed also present the opportunity to reclaim and re-use the separated streams as the mass balance for the system is developed and put into practice.

(86b) Macroscopic Water Integration Direct Recycle Applications Within Industrial City Plots

Alnouri, S.,
Linke, P., Texas A&M University at Qatar
El-Halwagi, M., Texas A&M University
Originally presented on: 11/4/2013 8:52:00 - 9:14:00

The development of efficient strategies for managing water usage and handling wastewater discharge in industrial zones have become imperative, as a result of the intensified environmental regulations, and the scarcity of naturally-occurring freshwater reserves. In order to assist in the design process of water systems in complex industrial regions, a macroscopic optimization-based framework for water networks within an industrial city is employed. The methodology considers direct recycling strategies for wastewater re-use amongst multiple processing facilities. The primary goal is to achieve attractive water integration schemes amongst all plants operating within a given industrial city arrangement, compared to the situation of having separate water processing systems associated with individual plants.

Targeting methods that are commonly utilized for determining minimum freshwater usage and wastewater discharge were applied to an industrial area in which several water consuming and producing processes all run in parallel. Moreover, industrial city layouts have been captured in a sense that would allow for water source-sink mapping, whilst taking into consideration service corridors, and any barriers or infeasible routing scenarios. The shortest paths that would allow effective source-sink matching amongst available water streams were all extracted using Dijkstra's Algorithm, which in turn solves a single source-to-destination shortest path problem in weighted graphs. This greatly facilitates finding optimum distances that can be used for water piping layouts and construction. In doing so, water networks that aim to achieve economical piping systems at low cost within any given industrial city arrangement can be developed, whilst managing to reduce deviations from minimum fresh and waste targets.

A case study has been carried out as a demonstration to the proposed work. A total of 6 plants were assumed to be operating within the industrial city plot, with an overall number of 6 water sources and 6 water sinks distributed amongst the individual plants. Shared freshwater and wastewater mains were assumed to and from all plants. Several cases have been investigated for single and multiple contaminant scenarios that utilize different locations for fresh and waste mains. Efficient water allocation strategies that employ shortest source-to-sink routes were obtained.

(86c) Predictive Modelling of a Conventional Clean Water Treatment Work

Sorensen, E., University College London
Thornhill, N. F., University College London
Akinmolayan, F.,
Originally presented on: 11/4/2013 9:14:00 - 9:36:00

Adequate and safe water is important for human health and well-being, economic production, and sustainable development, and failure to ensure the safety of drinking water may expose the community to the risk of waterborne and infectious diseases. The water industry in the United Kingdom is regarded as one of the most heavily regulated industries, and water utilities are faced with increasingly tighter targets for the quality of the water received at customers' taps. While the principles of drinking water treatment have been established for more than a century, water treatment plants have become increasingly complex due to the introduction of the more stringent regulations. These developments have further increased the need for optimum functionality of the water treatment plants to remain competitive by minimising risks.

Mathematical models are essential to describe, predict and control the complicated interactions between different parts of a clean water treatment process, and thereby facilitate a more accurate prediction of a plant's operational status and help to identify capacities for improvements and risk minimisation, a concept which is well understood within the process industry but not yet established within the water treatment industry. However, most modelling work within water operations has so far been based on empirical observations or hydraulic modelling, rather than mathematically describable relationships of the process. By accurately modelling the controlled flows of water from source of supply through treatment units such as flocculators, filters, chlorinators, intermediate storage vessels and pumps through to treated-water delivery, simulations can predict the behaviour of continuous and sequence process controls and their effects on water flows, pressures and hold-ups in vessels. This will in turn lead to better water quality, cost reduction and a more sustainable performance of the water treatment plant.

The objective of this work is the development of a novel dynamic mathematical model representing a complete conventional water treatment plant. Established data from the literature for single units have been used to verify the performance of each unit individually, and are then combined to enable the simulation of a complete clean water treatment work. Based on real industrial data from one of the United Kingdom's largest water treatment companies, we will demonstrate how this model allows the characterisation and quantification of changes in process parameters in one operational unit, and their effects on the overall treatment process, thus allowing plant wide operational optimisation for risk mitigation.

(86d) Bringing (Mass) Balance to California Water: A Role for Engineers in Shaping Policy

Hall, M., The Nature Conservancy

Originally presented on: 11/4/2013 9:36:00 - 9:58:00

The process of making public policy, from the recognition of the need, to the final negotiation in the wee hours of the morning before a vote, is a messy and inexact process – and often an uncomfortable domain for scientists and engineers. However, when it comes to resource management, technology issues, and the infrastructure that supports our society, there is an urgent need for the organized vision, subject matter expertise, and constant reality check that engineers (and like-minded technologists and scientists) bring to the table.

The Nature Conservancy's California Water Program has added significant engineering and physical science expertise to their team to improve water management and policy in California. In addition to the obvious subject matter expertise this adds, we are learning that these skill sets also bring increased capacity to educate stakeholders and decision-makers and the ability to shape influential demonstration projects, which are very powerful tools in finding common ground, building trust, and dispelling myths that obstruct progress toward better water management.

We will describe some of the huge disconnects in current water policy that demand our urgent attention and some of our experiences and approaches that illustrate why engineering and science engagement is so critical to improving our policies.

When it comes to sustainable water management, you just can't get there without a mass balance. Who can tell that story more credibly than an engineer?

(86e) Regulatory and Cost Drivers Affecting Industrial Water Management

Petty, B., Geosyntec Consultants, Inc.

Originally presented on: 11/4/2013 9:58:00 - 10:20:00

Based solely on the costs to purchase or develop water supplies, water is a relatively inexpensive resource. The costs increase considerably when accounting for the tangible costs related to environmental compliance, treatment, and discharge. More stringent discharge limits are being imposed for numerous chemicals such as 1,4-dioxane, chromium(VI), perchlorate, and many others. Stringent discharge limits are also being imposed for naturally-occurring substances such as selenium, arsenic, and nitrate. The discharge limits can be set to below receiving water background levels, and treatment can be impracticable or costly. This presentation will provide recent discharge limit developments and associated costs.

Water costs increase even more when accounting for the intangible costs related to regulatory scrutiny, negative public perception, and increased risk of lawsuits. Examples of ways these can impact businesses include regulatory fines, risk of shutdowns, loss in market share, and civil liability. This presentation will illustrate how a "total quality management" framework can be used to assess these costs.

Examples of ways to mitigate the additional costs and regulatory issues include water footprinting, pollution prevention techniques, beneficial reuse, and closed-loop systems. These types of water management techniques are much more appealing if the tangible and intangible costs of water are considered, which will be illustrated by relevant case studies.

(86f) Feasibility of Using Reclaimed Water for Thermoelectric Power Plant Cooling

Stillwell, A. S., The University of Texas at Austin
Webber, M. E., The University of Texas at Austin
Originally presented on: 11/4/2013 10:20:00 - 10:42:00

Reclaimed water – treated effluent from municipal wastewater treatment plants – is a source of non-potable water with the opportunity to offset freshwater use for cooling thermoelectric power plants. We performed an analysis of the geographic, technologic, and economic feasibility of using reclaimed water for power plant cooling in both retrofit and forward-looking applications, applying our methodology to the state of Texas. This methodology employs spatial analysis, optimization, multiple linear regression, and geographic calculations to evaluate areas of suitability. In retrofit applications, results show that use of reclaimed water for cooling was best suited for small- to medium-sized power plants near urban areas operating closed-loop cooling systems. Estimated annual cost of retrofit was dominated by the purchase of reclaimed water in our model; however, pipeline construction cost was determined to be a significant variable in estimating power generation. In forward-looking applications, pipeline construction cost, water scarcity, and reclaimed water flow rate were found to contribute to future suitability of reclaimed water use at power plants. Our analysis revealed that reclaimed water can be a geographically, technologically, and economically feasible resource to cool thermoelectric power plants.

(86g) Evaluation of Effectiveness of Condensate Recovery and Its Reuse As a Make-Up Water Requirement in Chilled Water Plants At the University of Alabama At Birmingham

Kajale, A., University of Alabama at Birmingham
Peters, D. R. W., University of Alabama at Birmingham
Winslett, M., University of Alabama at Birmingham
Originally presented on: 11/4/2013 10:42:00 - 11:04:00

Collection of air handling unit (AHU) condensate has become an essential part of the new building design codes; many of the new buildings being constructed throughout the world are incorporating this technique in their buildings. However, installation of condensate recovery systems in the existing building is still not very popular amongst building/facility operators. The majority of the federal buildings, universities, and other institutional buildings have already installed CRSs in their buildings. Although, the weather conditions in Alabama are very favorable for a successful condensate recovery system, implementation of these projects is surprisingly unpopular. This article discusses a case study of Condensate Recovery Systems installed on the various campus buildings of the University of Alabama at Birmingham (UAB). This article highlights the need and importance of CRSs in commercial, industrial, and Institutional buildings throughout the state of Alabama, given that majority of the southeastern counties in the state are more vulnerable to drought conditions. It is therefore necessary, to increase awareness of water conservation amongst building/facility operators in Alabama. This article also discusses how capturing and recycling condensate water helps in achieving water and energy conservation goals. Condensate Recovery Systems have been installed in eight campus buildings at the University of Alabama at Birmingham (UAB) and the collected condensate water is supplied to the central chilled water plant to meet the make-up water requirements. The quality and temperature of the condensate water has an advantage of reuse in chilled water plants. Total water consumption of the chilled water plant in year 2012 was 310,199 CCF which is reduced by 0.23% compared with that of year 2011. The UAB Facilities Management Department seeks to reduce the make-up water requirement by recycling condensate water in to the chilled water plants. In this study, the condensate water collected in 3 different buildings has been continuously monitored using water-level data-loggers at an interval of one minute frequency to estimate annual recovery of condensate water from each air handling unit. More than 2 billion gallons of condensate is collected and recycled from June 2012 through April 2013 out of which 1.67 billion gallons of condensate water was collected and recycled during cooling season of 2012. The primary goal of this research is to evaluate the feasibility of condensate collection systems at the Birmingham, Alabama location by analyzing effects of indoor and outdoor operating conditions on production of condensate which includes weather parameters, air supply rate, the areal size of the building, and tonnage of the system. A regression model has been also developed to predict the amount of condensate water from air handling units which can be used to determine the feasibility of future condensate collection system around the UAB campus.

(101a) Advanced Treatment of Shale Gas Frac Water to Produce NPDES Quality Water

Dahlin, R. S., Southern Research Institute
Chatterton, B.,
Pakzadeh, B.,
Vaillancourt, A., Southern Research Institute
Originally presented on: 11/4/2013 12:30:00 - 12:52:00

Advanced Treatment of Shale Gas Frac Water to Produce NPDES Quality Water

Bob Dahlin, Bill Chatterton, Behrang Pakzadeh, Austin Vaillancourt

Southern Research Institute, 5201 International Drive, Durham, NC, 27712

Natural gas extraction by fracking produces two types of wastewater that must be managed:Â 1) Flowback water is produced when the fracking is complete and pressure is released.Â Approximately 10% to 30% of the water used for fracking flows out of the well over a period of about ten days, and 2) Produced water, a mixture of water that was trapped underground with the gas deposit and frac water, flows out of the well with the natural gas over the several years life of the well at a rate of 100 to 1,300 gallons per day.

All wastewaters produced by drilling and fracking are regulated by the Clean Water Act (CWA) and are subject to the technology-based regulations (40 CFR Part 435, Subpart C) which apply to onshore facilities ? engaged in the production, field exploration, drilling, well completion and well treatment in the oil and gas extraction industry.?Â The effluent guidelines at 40 CFR 435, Subpart C establish best practicable control technology currently available (BPT) requirements for onshore facilities and prohibits the discharge of waste water pollutants into navigable waters from any source associated with production, field exploration, drilling, well completion or well treatment.Â

Southern Research Institute, with project partners M2 Water Treatment, Inc. and BKT United, has been awarded a contract by RPSEA (Research Partnership to Secure Energy for America) to develop an innovative, cost-effective, and robust approach for treatment of shale gas frac water to produce National Pollution Discharge Elimination System (NPDES) quality water for discharge and/or reuse. The approach comprises optimizing the combination of four technologies, two for frac water treatment and two for treatment and disposal of residues (high solid slurry and membrane concentrate) from the frac water treatment.Â The frac water treatment technologies are Â magnetic ballast clarification (MBC) for removal of total suspended solids (TSS), metals, andÂ naturally occurring radioactive materials (NORMs) , and vortex-generating and nano filtration membranes, for removal of TSS and/or total dissolved solids (TDS).Â The residue treatment/disposal technologies are hydrogel adsorbent for metals, NORMs, and trace element removal, and precipitation, solidification and stabilization.Â

The project began in May of 2013 and is anticipated to continue for a period of 24 months.Â The project tasks include bench-scale simulation, onsite field research/testing, and a feasibility analysis.Â The objective of the bench-scale simulations is to develop a method to allow for a quick evaluation as to whether waters from a specific well are amenable to treatment with MBC, membranes, and hydrogel media or precipitation/solidification/stabilization.Â The objective of the onsite field research/testing is to provide a linkage from the bench-scale simulations to what can be expected in the field.Â The objective of the feasibility analysis is to bring together the technical, economic and environmental data to assess the advantages and disadvantages of the proposed integrations compared to conventional technologies.Â If selected, Southern Research plans to present the detailed analysis and results from the bench-scale testing and any available preliminary data from the onsite field testing.

(101b) Recovering/Reuse of Sodium Chloride from Zero Discharge Desalination Concentrate Streams

Yetayew, A.,
Camacho, L. M., University of Texas at El Paso
Cappelle, M. A., University of Texas at El Paso
Davis, T., University of Texas at El Paso
Originally presented on: 11/4/2013 12:52:00 - 13:14:00

Zero Discharge Desalination (ZDD) technology is able to achieve up to 98% recovery by desalinating brackish groundwater. ZDD is typically comprised of a reverse osmosis (RO) or nanofiltration (NF) system and a unique variation of electrodialysis called electrodialysis metathesis (EDM). The ZDD technology increases the yield of fresh water from water supplies that contain enough $CaSO_4$ to cause precipitation problems when the water is treated with RO or NF alone. EDM can remove $CaSO_4$ and other troublesome salts, using NaCl as the exchangeable ions. NaCl recovery and return to the EDM system is important to advance the development of the EDM technology. Electrodialysis with monovalent selective anion and cation exchange membranes was explored to selectively remove sodium chloride from a feed stream saturated with calcium sulfate. Recovered sodium chloride was reuse in the EDM process. Experiments with simulated solutions and extended tests with the actual EDM concentrate streams were performed. Very good selectivity was observed for chloride over sulfate. The performance of the process with actual EDM concentrate streams shows great promise for rapid transfer to the pilot scale.

(101c) Advances in Electrosorption By Mesoporous Carbon Materials for Water Desalination

Sharma, K., Georgia Institute of Technology
Mayes, R. T., Oak Ridge National Laboratory
Kiggans, J., Oak Ridge National Laboratory
Yiacoumi, S., Georgia Institute of Technology
DePaoli, D. W., Oak Ridge National Laboratory
Dai, S., Oak Ridge National Laboratory
Tsouris, C., Oak Ridge National Laboratory
Originally presented on: 11/4/2013 13:14:00 - 13:36:00

Electrosorption of ions from aqueous solutions by porous carbon electrodes has been investigated for the desalination of saline water through the capacitive deionization (CDI) process. Mesoporous carbon materials of 10-nm average pore size, synthesized at Oak Ridge National Laboratory (ORNL), have been proven suitable for CDI because of their high surface area associated with an optimum pore size for ion transport. To study ion transport in mesoporous carbon electrodes, we employed the neutron imaging technique, which is a non-destructive in-situ method for providing information on the local concentration of ions in the electrodes as a function of time. Neutron imaging of ions in mesoporous carbon was conducted to determine the effective diffusivity of ions under different conditions of applied potential. It was also found that an oscillating voltage component superimposed on the applied direct-current (dc) voltage that is usually applied in capacitive deionization enhances the rates of adsorption and desorption of ions. Using neutron imaging, it was determined that the effective diffusion coefficient increased by 35% when the oscillating potential was superimposed onto the dc potential of 1.2 V. Thus, modulation of the electrical signal applied in CDI processes can be used to enhance the kinetics of CDI and increase the competitiveness of this process in real water-treatment and reuse applications.

(101d) Evaluation of Ceramic Membrane Filtration for the Removal of Dbp Precursors From Surface Water Supplies

Mathews, A. P., Kansas State University
Fann, S.,
Maurmann, L.,
Originally presented on: 11/4/2013 13:36:00 - 13:58:00

The presence of high concentrations of natural organic matter (NOM) in surface water supplies makes many such resources unusable as drinking water resources due to the formation of disinfection byproducts (DBPs) during disinfection of the water for pathogen control. Chlorine, which is an inexpensive and effective oxidant, has been traditionally used for disinfection in a majority of water treatment facilities. The presence of numerous chlorinated compounds that have adverse human health impacts from chlorination, have resulted in new regulations that set maximum limits on total trihalomethane (TTHM) and haloacetic acids (HAA5) in potable waters. Alternative disinfectants such as chloramine, ozone, etc., are being used at some facilities. But, they may have other problems such as growth of nitrifying bacteria in the distribution system, and the lack of disinfectant residual in the distribution system and the potential for recontamination. As such, the removal of DBP precursors from water supplies using membranes has received considerable attention in the last decade. Polymeric membrane filters using microfiltration (MF), ultrafiltration (UF), and nanofiltration (NF) have been studied for the removal of DBP precursors from water suppliers. Nanofiltration, although effective, is expensive to operate. Fouling and flux decline are the major problems with NF membranes. Considerable advances have been made in the manufacture of ceramic membranes (CMFs) in recent years. CMFs are resistant to chemicals, pH variations, can withstand high transmembrane pressures, and provide a robust platform for the removal of contaminants from water supplies and wastewaters. Bench-scale and pilot studies were conducted using 15 kDa to 3 kDa ceramic membrane filters for the removal of DBP precursors from a surface water supply at a Navy installation water treatment facility. The NOM content was analyzed for using measures such as TOC, and trihalomethane potential (THMFP). Size exclusion chromatography was used to characterize the molecular size distribution of the organic matter in the water. The membrane flux, TMP values, and the data for NOM from bench-scale pilot CMF studies will be presented.

(101e) Efficacy of Membrane Bioreactor Process Combined With Advanced Oxidation Processes for Organic Removal in Water Reclamation and Reuse

Kim, W., University of Southern California
Ravindran, V., University of Southern California
Pirbazari, M., University of Southern California
Originally presented on: 11/4/2013 13:58:00 - 14:20:00

An increasing number of public owned wastewater treatment and water reclamation utilities and several municipalities have recognized the importance of recycled water as a new water source, particularly in arid areas of the United States. These utilities have developed a wide variety of options for water reuse, and have considered water recycling water for indirect potable reuse, or discharge into surface waters or infiltration into subsurface environments for augmenting potential water supply sources. They have considered membrane technologies such as microfiltration (MF) followed by nanofiltration (NF) or reverse osmosis (RO) for this purpose. In the present context, our study compares the performance of membrane bioreactor (MBR) process using ultrafiltration (UF) membranes integrated with the application of powder activated carbon (PAC) as adsorbent.

The major concerns of the U.S. Environmental Protection Agency and local regulatory agencies are health problems arising from the introduction of pathogens (protozoa, bacteria, viruses and other parasites), anthropogenic organic or inorganic chemicals into groundwater or surface water sources. Additionally, organic matter or natural organic matter (NOM) already present in drinking water or those formed during wastewater treatment due to decomposition of organic matter could significantly affect water quality in terms of dissolved organic carbon (DOC) concentrations, or can act as precursors to the formation of disinfection byproducts (DBPs) such as trihalomethanes, haloacetic acids, aldehydes, ketones and others, many of which are well known carcinogens. Therefore, the proposed environmental regulations in the State of California (more stringent than those of the US EPA) include a goal of achieving TOC levels not exceeding 1 mg/L originating from reclaimed waters.

The purpose of this study was to evaluate the performance of the MBR process regarding the fate of organics in the treated effluent, when reclaimed water is used as the feed. The intention was to compare the performance of the MBR system (using the UF membrane) with those of other advanced membrane treatment systems such as NF or RO processes, that have been reported in several earlier studies. The present work considered the application of advanced oxidation processes (AOPs) such as ozonation, ozone-hydrogen peroxide oxidation and UV-hydrogen peroxide oxidation to render the organic matter more easily amenable to biodegradation in the MBR system. These evaluations also included the evaluation of permeate fluxes and effluent quality pertaining to the MBR system.

Extensive characterization of organics (or DOC) in the reclaimed water and the treated effluent

(MBR) effluent was necessary for the following reasons. Quality issues for residual compounds arise from their resistance to sorption and biodegradation. Some of these organic constituents exhibit potentials for forming DBPs during disinfection. Hydrophilic constituents are poorly sorbed on adsorbents and cannot be easily removed. Large molecular weight compounds are resistant to degradation, while low molecular weight compounds and polar compounds are sometimes not well rejected during membrane filtration (NF or RO processes), and some have significant health concerns. Hence RO filtration and size fractionation using resins is important for organics classification.

The organics characterization strategies require state-of-the-art analytical techniques including chromatographic and spectroscopic methods. These techniques focused on several aspects including the bulk organic carbon characteristics including hydrophobicity/hydrophilicity, molecular weight or size distributions, elemental analysis, organic carbon functional groups, potential formation of trace organics and DBPs, and identification of fluoropores. The gas chromatography/mass spectrometry (GC/MS) analysis provided the qualitative and quantitative assessments of the trace organics and DBPs present. The RO filtration and resin adsorption methods determine the hydrophobicity and polarity of the constituents. The size exclusion chromatography (SEC) methods provided details on molecular weight and size distributions of the product, while the carbon-13 nuclear magnetic resonance spectroscopy (^{13}C-NMR) after size separation techniques showed the functional group characteristics of the DOC as well as their polar and non-polar fractions. The UV absorbance (UVA) measurements provide information on the natural of NOM present and the types of products formed after the application of AOPs. These advanced techniques for organics characterization provided an in-depth evaluation of the MBR system with reference to the effluent characteristics and removals of various organic constituents.

(101g) Sanitation Shallow Water and Groundwater Contaminated With Petroleum Hydrocarbons

Cruz, Y.,
Castro, J. J.,
Rico, E.,
Originally presented on: 11/4/2013 14:20:00 - 14:42:00

In this work a study was made of the biodegradation of water contaminated synthetically with diesel and oil , and a water sample taken from a actual-estate of the Valley of Mexico, having this problem; using a microbial consortium containing microorganisms of the genus Pseudomonas. The involvement of native microorganisms in the biodegradation, when implementing a bioremediation process must be carefully evaluated since their presence also affects the efficiency of biodegradation, reaching strengthen it in the best case. It was carried out to characterize the actual water and preparing synthetic solutions with salts essential for bacterial growth. Adapting the bacterial consortium was performed by supplying the necessary essential salts and adjusting the optimum pH and temperature of development. The biodegradation of hydrocarbon substrates was followed via COD, respirometry, CO_2 absorption, UV-vis spectroscopy,IR/ATR infrared and gas chromatography-FID. Biodegradation profiles and bacterial growth curves are obtained. Degradation efficiencies were obtained 90 to 98 percent, varying the nature of the hydrocarbon substrate used in a treatment period of 5 days.

(134a) Identifying Defect Sites in Lithium-Ion Battery Materials: Local Disorder in LiVPO4F and Its Influence On Bulk Properties

Messinger, R. J., CNRS
Menetrier, M., CNRS
Carlier, D.,
Ateba Mba, J. M., CNRS
Croguennec, L.,
Masquelier, C.,
Massiot, D.,
Deschamps, M.,
Originally presented on: 11/4/2013 12:30:00 - 12:45:00

The transition metal fluorophosphate $LiVPO_4F$ is a promising electrode material for lithium-ion batteries that is capable of operating both as a high-voltage positive electrode (4.25 V, $V^{III/IV}$) and a negative electrode (1.8 V, $V^{II/III}$).[1,2] While its Tavorite-like structure has been solved by a combination of X-ray and neutron diffraction, solid-state nuclear magnetic resonance (NMR) measurements reveal a significant extent of local disorder that has not been observed at longer length scales with scattering techniques. In particular, solid-state single-pulse ^7Li NMR spectra acquired under conditions of ultra-fast (64 kHz) magic-angle-spinning (MAS) reveal that $LiVPO_4F$ exhibits several additional ^7Li NMR signals of unknown origin that can account for up to 20% of the total lithium content. Electrochemical measurements in Swagelok-type cells with lithium metal as the negative electrode establish that materials with different populations of unknown lithium sites exhibit different cycling behavior, including differing extents of irreversible capacity and polarization. 2D $^7Li\{^7Li\}$ dipolar correlation NMR measurements establish the sub-nanometer proximities of the unknown lithium sites with respect to the $LiVPO_4F$ site, proving unambiguously that the unknown sites are local defects within the $LiVPO_4F$ structure and are not associated with separate impurity phases. In combination with magnetic measurements, variable-temperature NMR measurements reveal that the paramagnetic ^7Li NMR shifts of the different lithium sites vary linearly with the bulk magnetic susceptibility, further confirming that the sites are defects within the $LiVPO_4F$ structure. The results provide a general strategy for identifying lithium defect sites in lithium-ion battery materials and highlight the importance of understanding local disorder and its relationship to macroscopic material and device properties.

[1]J.-M. Ateba Mba, C. Masquelier, E. Suard, L. Croguennec, *Chem. Mater.* **2012**, *24*, 1223 – 1234.

[2]J.-M. Ateba Mba, L. Croguennec, N.I. Basir, J. Barker, C. Masquelier, *J. Electrochem. Soc.* **2012**, *159*, A1171-A1175.

(134b) Electrochemical Properties and Characterization of Li1.2Mn0.54Co0.13Ni0.13O2 Composite Cathode Powders Prepared by Ultrasonic Spray Pyrolysis

Langrock, A., University of Maryland
Ehrman, S. H., University of Maryland
Wang, C., University of Maryland
Originally presented on: 11/4/2013 12:45:00 - 13:00:00

Lithium-rich layered oxide $Li_{1.2}Mn_{0.54}Co_{0.13}Ni_{0.13}O_2$ (LMO) composite cathode powders are prepared by ultrasonic spray pyrolysis. The crystal structure and morphology of the prepared samples are characterized with X-ray diffraction and scanning electron microscope. The spherical shape and porous nanostructure of the as prepared powders are maintained after post-treatment at temperatures between 600 and 800°C. Grain growth in the powders increases with increasing post-treatment temperatures. The BET surface area of the composite cathode powders post-treated at 800°C is 12 m^2 g^{-1}. Discharge capacity of the cathode powders post-treated at 800°C decreases from 278 mAh g^{-1} to 208 mAh g^{-1} by the 50th cycle, the capacity retention being 75%. The post-treatment temperature strongly affected the electrochemical properties of the composite cathode powders. Composite cathode powders post-treated at high temperatures with high amounts of the Li_2MnO_3 phase had good cycle properties as well as large initial charge/discharge capacities.

(134c) Atomic Layer Deposition (ALD) On the Nanostructured Li-Mn-Rich Composite Li1.2Ni0.13Mn0.54Co0.13O2 Cathode Powder

Zhang, X., Institute of Nuclear and New Energy Technology, Tsinghua University
Belharouak, I.,
Originally presented on: 11/4/2013 13:00:00 - 13:15:00

Atomic Layer Deposition (ALD) on the Nanostructured Li-Mn-rich Composite $Li_{1.2}Ni_{0.13}Mn_{0.54}Co_{0.13}O_2$Cathode Powder

Xiaofeng Zhang,[a] Ilias Belharouak,[a,*]

[a] Chemical Sciences and Engineering Division, Argonne National Laboratory, 9700 S. Cass Ave, Argonne, IL 60439, USA, Email: belharouak@anl.gov(Corresponding author)

Li-Mn-rich (LMR) composite cathode materials $Li_{1+x}M_{1-x}O_2$ (M = Ni, Mn and Co) are of particular interest for plug-in hybrid electric vehicles (PHEVs), in which a high-energy battery is always desired for a higher mileage per charge.[1, 2] In the composite structure, the layered, the spinel and rocksalt structure are compatible at nano-scales, that many composites with different electrochemical properties can be synthesized by simply tuning the Li content in the materials. However, these materials would suffer from voltage decay and capacity fading with cycles, which significantly deteriorate their energy densities and life time for practical applications. In this type of materials, Li content is essential to the arrangement of the transition metal atoms in the composites and also to the structural stability against undesired reactions in a real electrochemical cell. It has been demonstrated that the occurrence of layered-spinel transformation induces the voltage decay. It is because the Li intercalation/deintercalation reactions may preferentially occurs at the tetrahedral site other than octahedral site, which facilitates the spinel formation.[1, 2] Further study by HR-TEM indicated that the composite particle surface could lose the layered feature after a few cycles, and the first-principle calculation suggested a defect-spinel structure forming on the surface due to the migration of transition metal ions into the Li layers.[3] Therefore, the surface chemistry and structure stability appear to be essential to limit the undesired reactions and phase transformation in the composite materials. Some achievements have been made *via* developing new synthesis processes, new chemistry or surface modifications (e.g., coating), which possibly altered the short-range and long-range ordering of transition metal atoms (Ni and Mn) near the surface, which lead to an improvement in the battery performance.[4-8]

Atomic layer deposition (ALD) is a well-established method to grow ultrathin films on flat surfaces and powders in atomic thickness using sequential, self-limiting surface reactions.[9-14] It has been demonstrated that the surface film can react as a surface barrier between the cathode active material and the electrolyte, which would reduce the loss of active materials in the electrode. [9-14] Consequently, less capacity loss is achieved for the coated cells as compared with the pristine materials. The well-established ALD process normally is using trimethylaluminium (TMA) and H_2O as precursors forming an Al_2O_3 surface film. Experimental results showed that the coated $LiCoO_2$ powders exhibited a capacity retention of 89% after 120 charge–discharge cycles in the 3.3–4.5 V vs. Li/Li+ range. [12, 15-17] The ALD surface coating can also be applied to battery electrodes at a low temperature <150 °C, which would not degrade the electrodes (e.g., melting of polymer binders) in this temperature range. It is reported that Al_2O_3-coated $LiCoO_2$ electrodes display superior electrochemical performance in comparison to the uncoated $LiCoO_2$ electrodes. However, Liu et al observed that the ALD surface coated Al_2O_3 could be lithiated forming passivation $LiAlO_2$ glass on the materials surface using in-situ TEM [18], which provided new insight into the electrochemical characteristic of the Al_2O_3surface film against Li.

In this paper, we are aiming for studying the importance of a conformal surface protection film using atomic layer deposition (ALD) technique on the high-capacity $Li_{1.2}Ni_{0.13}Mn_{0.54}Co_{0.13}O_2$ cathode materials in Li-ion batteries, which can be potentially significantly improved in their electrochemical performance. More precisely, we evaluated the effect of coating species, coating thickness and morphology on the rate performance, cycling life and high temperature performance of the high-capacity $Li_{1.2}Ni_{0.13}Mn_{0.54}Co_{0.13}O_2$ cathode materials in half-cell and full-cell configurations.[19] We found that an conformal Al_2O_3 thin film could be deposited on the powders using 10 ALD cycles of TMA and H_2O at 150 °C. TiO_2 coating was prepared using 20 ALD cycles of titanium isopropoxide (TTIP) and H_2O at 150 °C, which displayed a particulate morphology. Nevertheless, the surface coated TiO_2 film was electrochemically active, which could be lithiated forming Li_xTiO_2 as detected by X-ray photoelectron spectroscopy (XPS). The electrochemical reactivity of TiO_2 will cause degradation of the surface film, leading to capacity fading of the cell. Nevertheless, it was also found that the Al_2O_3 surface coating was not electrochemically active after repetitive cycles, which was contrary to the early report by Lui et al that Al_2O_3 can be lithiated into $LiAlO_2$. [18] To further explorate the phenomena, a systematic study on the ALD coating of Al_2O_3 and $LiAlO_2$was performed.

Refereances:

[1] C.S. Johnson, N. Li, J.T. Vaughey, S.A. Hackney, M.M. Thackeray, Electrochem. Commun., 7 (2005) 528-536.

[2] C.S. Johnson, N.C. Li, C. Lefief, J.T. Vaughey, M.M. Thackeray, Chem. Mater., 20 (2008) 6095-6106.

[3] B. Xu, C.R. Fell, M.F. Chi, Y.S. Meng, Energy & Environmental Science, 4 (2011) 2223-2233.

[4] J. Liu, B. Reeja-Jayan, A. Manthiram, Journal of Physical Chemistry C, 114 (2010) 9528-9533.

[5] S.H. Kang, M.M. Thackeray, J. Electrochem. Soc., 155 (2008) A269-A275.

[6] J.R. Croy, M. Balasubramaniam, D. Kim, S.-H. Kang, M.M. Thackeray, Chem. Mater., 23 (2012) 5415-5424.

[7] J.R. Croy, S.H. Kang, M. Balasubramanian, M.M. Thackeray, Electrochem. Commun., 13 (2011) 1063-1066.

[8] D. Kim, S.H. Kang, M. Balasubramanian, C.S. Johnson, Electrochem. Commun., 12 (2010) 1618-1621.

[9] M.Q. Snyder, S. Trebukhova, B. Ravdel, M.C. Wheeler, J. DiCarlo, C.P. Tripp, W.J. DeSisto, Mater Res Soc Symp P, 972 (2006) 331-338.

[10] M.Q. Snyder, S.A. Trebukhova, B. Ravdel, M.C. Wheeler, J. DiCarlo, C.P. Tripp, W.J. DeSisto, J. Power Sources, 165 (2007) 379-385.

[11] S.M. George, Chem. Rev., 110 (2010) 111-131.

[12] Y.S. Jung, A.S. Cavanagh, A.C. Dillon, M.D. Groner, S.M. George, S.H. Lee, J. Electrochem. Soc., 157 (2010) A75-A81.

[13] L.A. Riley, A.S. Cavanagh, S.M. George, S.H. Lee, A.C. Dillon, Electrochem. Solid-State Lett., 14 (2011) A29-A31.

[14] L.A. Riley, S. Van Ana, A.S. Cavanagh, Y.F. Yan, S.M. George, P. Liu, A.C. Dillon, S.H. Lee, J. Power Sources, 196 (2011) 3317-3324.

[15] Y.S. Jung, A.S. Cavanagh, L.A. Riley, S.H. Kang, A.C. Dillon, M.D. Groner, S.M. George, S.H. Lee, Advanced Materials, 22 (2010) 2172-2176.

[16] I.D. Scott, Y.S. Jung, A.S. Cavanagh, Y.F. An, A.C. Dillon, S.M. George, S.H. Lee, Nano Lett., 11 (2011) 414-418.

[17] Y.S. Jung, P. Lu, A.S. Cavanagh, C. Ban, G.H. Kim, S.H. Lee, S.M. George, S.J. Harris, A.C. Dillon, Advanced Energy Materials, 3 (2012) 213-219.

[18] Y. Liu, N.S. Hudak, D.L. Huber, S.J. Limmer, J.P. Sullivan, J.Y. Huang, Nano Lett., 11 (2011) 4188–4194.

[19] X. Zhang, I. Belharouak, L. Li, Y. Lei, J.W. Elam, A. Nie, X. Chen, R. Shahbazian-Yassar, R.L. Axelbaum, Advanced Energy Materials, (2013). 10.1002/aenm.201300269

(134d) Metal Fluoride/Graphene Composites for High-Performance Li-Ion Cathodes

Hayner, C. M., Northwestern University
Zhao, X., Northwestern University
Yu, Y. Y., Northwestern University
Kung, M. C., Northwestern University
Kung, H. H., Northwestern University
Originally presented on: 11/4/2013 13:15:00 - 13:30:00

Rechargeable batteries with improved energy and power density have many potential, high impact applications including advanced portable electronics and electric vehicles. However, most 'next-generation' materials capable of providing improved performance suffer from rapid capacity degradation or severe loss of capacity when rapidly discharged. Graphene, a unique two-dimensional material composed of atomically thin carbon sheets, has shown potential to address unsatisfactory rate capability, limited cycling performance and abrupt failure of these high capacity material. Therefore, we have been exploring graphene-based architectures, allowing us to exploit graphene's unique combination of high surface area, high in-plane electrical conductivity, excellent tensile modulus and mechanical durability.

Metal fluorides, including iron trifluoride (FeF_3), have the potential of both high energy delivery and high cyclability. We have recently developed a stable nanostructured iron fluoride (FeF_3)/graphene composite cathode. A facile route that combines co-assembly and photothermal reduction was developed to synthesize free-standing, flexible FeF_3/graphene papers. The papers contained a uniform dispersion of FeF_3 nanoparticles (< 40 nm) and open ion diffusion channels in the porous, conducting network of graphene sheets that resulted in a flexible paper cathode with high charge storage capacity, rate, and cycling performance, without the need for other carbon additives or binder. The composites demonstrate promising applications as cathodes in high-energy density Li-ion batteries. In a free-standing form, they showed a high storage capacity of >400 mAh/g and improved cycling performance compared to bare FeF_3 particles. In addition, we have investigated a novel mixed metal fluoride cathode system, containing tunable manganese and iron fluoride, for which characterization and electrochemical performance will be presented.

(134e) Nanostructured Li2S-Graphene Composites As Cathode for High Energy Density Li-S Batteries

Han, K., Northwestern University
Shen, J., Northwestern University
Hayner, C. M., Northwestern University
Yu, Y. Y., Northwestern University
Kung, M. C., Northwestern University
Kung, H. H., Northwestern University
Originally presented on: 11/4/2013 13:30:00 - 13:45:00

Lithium-ion batteries, although a promising technology for energy storage, is reaching their energy density limitation, making it difficult to fulfill the requirements of emerging applications such as electric vehicles and smart grid. Lithium-sulfur batteries are one of the most attractive systems for next generation high energy storage beyond Li-ion batteries. Conventional Li-S batteries mainly use elemental sulfur as the starting cathode material due to its high theoretical capacity of ~1675 mAh/g and high energy density of ~2500 Wh/kg based on the redox reaction $S_8 + 16Li \rightleftharpoons 8Li_2S$. Sulfur is also low cost, abundant, and of low toxicity. However, significant hurdles such as rapid capacity fade and poor rate capability have to be overcome before practical application can be realized. Two important underlying reasons for the poor performance of this battery are the insulating nature of sulfur and the dissolution of the intermediate product polysulfide into the electrolyte resulting in redox shuttling. Although extensive research efforts have been expanded to solve these issues, most studies are limited to the simple cell configuration of sulfur as cathode and metallic lithium as anode. For safety and ease of operation, introducing Li into the cell in the form of Li_2S is more desirable than Li metal. Moreover, Li_2S, with a theoretical capacity of ~1166 mAh/g, could be paired with a lithium-metal-free anode such as Si or Sn.

Here, we report the use of a nanostructured Li_2S-graphene composite as cathode material. Li_2S nanoparticles (20~50nm) are imbedded in pocket-like structures in a three-dimensional graphene/reduced graphene oxide via a facile, wet chemical method followed by thermal treatment. Structural characterization by XRD and SEM provide clear evidence for the unique structure. The constructed 3-D graphene network could enable efficient use of Li_2S active material by providing excellent electrical contacts, offer long-term stability against loss of polysulfide and accommodate volume changes during charge and discharge processes. The as-prepared Li_2S-graphene nanocomposites exhibit high initial reversible capacity and good cycling performance. The effect of lithium polysulfide as additive of the electrolyte is also investigated, especially with respect to its contribution to the overall charge storage capacity. The structural and electrochemical characterization results of the nanocomposites will be presented.

(134f) Carbon/Sulfur Microspheres With Multi-Modal Pore Structures for Lithium-Sulfur Battery Cathodes

Zhao, C., University of Wisconsin-Milwaukee
Liu, L., University of Wisconsin-Milwaukee
Zhao, H., University of Wisconsin-Milwaukee
Li, Y., University of Wisconsin-Milwaukee
Originally presented on: 11/4/2013 13:45:00 - 14:00:00

Sulfur has received increasing attention as a cathode material for lithium-sulfur batteries due to its high theoretical specific capacity of 1675 mAh g-1, which is five times higher than that of conventional lithium-ion batteries. However, poor electronic conductivity of sulfur, dissolution of the lithium polysulfide intermediates and large volume expansion of sulfur during cycling limit the commercialization of Li-S batteries. In this work, we have demonstrated a cathode design with a novel structure made of multi-modal porous carbon microspheres. The tunable porous structure of sulfur-infiltrated carbon nanocomposites allows easy access to Li ion transport, inhibits loss of active sulfur material, and provides void space for sulfur expansion. An initial capacity of 1289 mAh g-1 at 1C and a Coulombic efficiency of 90% over 200 cycles are achieved.

(134g) Nanofiber-Based Cathode Electrocatalysts Supported On Carbon Substrates for Lithium-Air Rechargeable Batteries

Joo, Y. L., Cornell University
Yin, J., Cornell University
Kim, J.,
Originally presented on: 11/4/2013 14:00:00 - 14:15:00

Nanofiber-Based Cathode Electrocatalysts Supported on Carbon Substrates for Lithium-Air Rechargeable Batteries

*Jun Yin[a], Jangwoo Kim[a], and Yong Lak Joo[a]**

[a] School of Chemical and Biomolecular Engineering, Cornell University, Ithaca, New York 14853

In terms of usable energy density rivaling that of gasoline, rechargeable lithium-air batteries have attracted increasing interests presently. However, the practical capacity was highly affected by the effective catalyst materials in the air cathode in Li-air batteries. The demand for effective catalyst materials in the air cathode requires the design and synthesis multifunction catalysts for power density improvement. In this presentation, several types of metal oxide nanofibers (e.g., ZnO, Co_3O_4 etc) were synthesized by water-based electrospinning of highly loaded precursor solutions, followed by thermal treatment and their use as the cathode catalysts in single-cell measurements was described. In the meantime, the correlation between the various carbon-based substrates and the metal oxide fiber catalysts and how they synergistically influence the electrocatalytic activity will also be discussed. The metal oxide nanofibers are characterized using an array of techniques such as SEM, XRD, XPS, BET and Raman to determine their morphology and structural properties. Charging – discharging evaluation, cyclic voltammetric measurements in organic electrolytes were carried out to assess the mechanistic details as well.

(134h) Low-Temperature Dehydrogenation From Aminoborane Complexes Under Carbon Dioxide Atmospheres

Lee, J. W., The City College of New York
Xiong, R., CCNY
Zhang, J., The City College of New York
Originally presented on: 11/4/2013 14:15:00 - 14:30:00

Ammonia borane (AB) is a chemical hydride and one of attractive solid-state media for on-board hydrogen storage due to high hydrogen contents (19.6 wt. %) and low-temperature dehydrogenation compared to other metal hydrides. The hydrogen release from AB can be accelerated at the working temperature (~ 85 °C) of polymer electrolyte membrane fuel cell. However, this talk shows that CO_2-treatment can accelerate hydrogen release of AB polymer complex, polyaminoborane (PAB). Polyaminoborane (PAB) is the spent fuel of ammonia borane (AB, NH_3BH_3) thermolysis with around 1 mole hydrogen release per mol AB. Dehydrogenation of PAB by thermolysis generally commences at temperatures above 120 °C. However, we find gaseous CO_2 environments dramatically enhance the thermolytic dehydrogenation of polyaminoborane (PAB) at 85 - 90 °C. By exposing PAB to 2.07 MPa CO_2, 5.48 wt % hydrogen based on the mass of ammonia borane (AB) desorbs in less than 1 h, while 4 h is required to generate 5.04 wt % hydrogen by maintaining PAB at 120 °C without using CO_2. The total hydrogen desorbed from the initial AB is 11.38 wt % under the CO_2 atmosphere at 85 °C, including the hydrogen release during the transition from AB to PAB. The main contributing factors to the facile hydrogen release are the promoting effect of newly formed formic acid and the exothermic effect of the reaction between PAB and CO_2. We will explain the possible promoting mechanism by various spectroscopic and elemental analyses.

(134i) Intermittent Electrical Energy Storage By Liquid Organic Hydrogen Carriers

Arlt, W., University of Erlangen-Nuremberg
Muller, K., University of Erlangen-Nuremberg
Teichmann, D., University of Erlangen-Nuremberg
Originally presented on: 11/4/2013 14:30:00 - 14:45:00

The intermittent delivery of energy from sources like photovoltaics or wind demands a large and long-term storage of energy. Liquid Organic Hydrogen Carriers like "Carbazol" address this topic using the infrastructure of liquids like domestic fuel oil. In Germany, the situation is far more complicated, because the country decided to phase out nuclear energy in the year 2022. The lecture highlights the situation in the German state of Bavaria and the planning.

The lecture reports the progress of Liquid Organic Hydrogen Carriers (LOHC). A demo-project is under construction at a condo in the city of Erlangen/Bavaria. Important data are safety, efficiency of electrical energy storage and robust construction. The purpose of this demo is to investigate

1. The ratio between storage and heat demand of the building
2. The number of storage and delivery actions
3. The robustness of the process in a public area

The time frame is 2 years from beginning of 2014 on.

(134j) Synthesis of Nanoporous Palladium Powder With Controlled Pore and Particle Size for Hydrogen Storage Applications

Cappillino, P., Sandia National Laboratories
Jones, C.,
Hattar, K., Sandia National Laboratories
Clark, B.,
Hekmaty, M.,
Jacobs, B.,
Robinson, D. B., Sandia National Laboratories
Originally presented on: 11/4/2013 14:45:00 - 15:00:00

Synthesis of Nanoporous Palladium Powder with Controlled Pore and Particle Size for Hydrogen Storage Applications

by Patrick J. Cappillino, Christopher G. Jones, Khalid M. Hattar, Blythe G. Clark, Michelle A. Hekmaty, Benjamin W. Jacobs, and David B. Robinson

Materials capable of rapidly storing and delivering hydrogen are currently in demand for numerous energy applications.Â The high surface area exhibited by nanoporous palladium has the potential to greatly improve its kinetics of hydrogen charge and discharge *versus* nonporous metal while retaining favorable operating temperatures, pressures, and volumetric capacity.Â In addition, the void space imparted by porosity should accommodate the volumetric expansion of palladium upon hydriding, mitigating plastic deformation and improving cycle life.

By chemically reducing palladium salts in the presence of various surfactants, we are able to synthesize palladium having size-tunable mesopores from a few to tens of nm, in a scalable fashion.Â Our results suggest that Pd nanoparticles form and sinter around micelles present in the aqueous media.Â Changing the chemical composition of the surfactant affects the size of these micelles and ultimately determines the pore dimensions.Â We demonstrate that pore size affects not only surface area, but also pore thermal stability under vacuum, and in the presence of hydrogen gas.

In addition to pore geometry, particle size and shape are important factors in determining gas flow characteristics. Using nonporous copper particles as sacrificial reductants for palladium salts, in the presence of surfactants, we are able to synthesize micron-sized particles of uniform size and shape.Â This is in contrast to continuous aggregates and films that are obtained using soluble chemical reductants or planar electrodes.

We will present details of the synthesis of nanoporous palladium, demonstrating control of both pore and particle size.Â Bulk measurements of hydrogen storage properties and pore characteristics, as well as microscopic measurements, including *in situ* heated-stage TEM under vacuum and in the presence of hydrogen, will be included, as well as details on the kinetics of hydriding/dehydriding.

Sandia National Laboratories is a multi-program laboratory managed and operated by Sandia Corporation, a wholly owned subsidiary of Lockheed Martin Corporation, for the U.S. Department of Energy's National Nuclear Security Administration under contract DE-AC04-94AL85000.

(146a) Development of Composite Hollow Fiber Membranes for Aqueous and Non-Aqueous Nanofiltration

Sun, S. P., National University of Singapore
Chan, S. Y., National University of Singapore
Chung, T. S., National University of Singapore
Originally presented on: 11/4/2013 12:30:00 - 12:55:00

Nanofiltration (NF) has been growing rapidly in pharmaceutical separation, wastewater treatment, seawater desalination and many other applications. This presentation focuses on the new developments of composite NF hollow fiber membranes from both co-extrusion and interfacial polymerization (IP). Firstly, a highly water permeable dual-layer NF hollow fiber membrane fabricated by the simultaneous co-extrusion of Torlon® polyamide-imide (PAI) and cellulose acetate (CA) polymer solutions through a triple-orifice spinneret in a dry-jet wet phase inversion process. As compared to conventional membranes, the dual-layer hollow fiber membranes developed here have a unique structure which consists of (1) controllable thin selective outer layer; (2) innovatively designed interface, which is finely connected between outer and inner layers, does not contribute significant transport resistance while still maintains good mechanical strength; (3) fully porous substructure with reduced transport resistance. The cross-section struction and the nanopores of dual-layer hollow fiber membranes were molecularly designed by controlling the phase inversion process with the aid of various non-solvent additives into the polymer solutions. The newly developed NF dual-layer hollow fiber membrane with methanol as additive has a relatively high pure water permeability of 11.93 $l\ m^{-2}bar^{-1}h^{-1}$ with a mean effective pore radius of 0.63 nm and a more than 96.5% rejection of Na_2SO_4.

Secondly, a novel thin-film composite NF membrane was fabricated by IP of hyperbranched polyethyleneimine (HPEI) and isophthaloyl chloride (IPC) on a dual-layer hollow fiber membrane. The substrate possesses a unique cross-section structure comprising a layer full of macrovoids located in the middle and sandwiched by two thin spongy-like layers. Such a sandwich-like sublayer structure provides minimal transport resistance and sufficient mechanical strength for water permeation under high pressures. After IP, the NF membrane possesses a negatively charged substrate and a positively charged selective layer with a mean effective pore radius of 0.36 nm, molecular weight cut off of 500 Da, and pure water permeability of 4.9 $lm^{-2}bar^{-1}h^{-1}$. Due to this double-repulsion effect, together with the steric-hindrance and the solute electro-neutrality effects, by adjusting the pH of cephalexin aqueous solution to modify the ionization states of this zwitterionic molecule, the NF membrane shows high rejections over a wide pH range. Furthermore, the latest preliminary data shows the membrane is very stable in harsh organic solvents such as dimethyl sulfoxide. Therefore, the NF membrane may potentially be useful to reduce waste, recycle valuable pharmaceuticals and reuse solvents for the pharmaceutical industry.

(146b) Self-Heat Recuperative Seawater Desalination Process Using Fluidized Bed

Mizuno, H., The University of Tokyo
Kansha, Y., The University of Tokyo
Ishizuka, M., The University of Tokyo
Tsutsumi, A., The University of Tokyo
Originally presented on: 11/4/2013 12:55:00 - 13:20:00

Recently, it is presumed that there will be a serious shortage of water in several years later due to the high growth rate of population and industrial development. To alleviate the water shortage, more attention has been paid to desalination technology. One of the typical desalination processes is a multi-stage flash desalination. This process is categorized as thermal desalination process because it produces pure water by evaporation. The thermal process has the following two major advantages; seawater quality does not seriously affect the operation condition and product water salinity is low enough for industrial usage. However, this process consumes a lot of energy, thus the remarkable energy saving technology of thermal process is in great demand. So far, to reduce the energy consumption, we proposed a novel thermal desalination based on self-heat recuperation, which is a method for energy saving. By a process simulator, it was revealed that the proposed desalination process energy consumption is reduced to 1/4 of conventional thermal desalination process. Furthermore, it was elucidated that the specific energy consumption decreased as the recovery ratio (= the product water flow rate / the feed water flow rate) increased. However, in the thermal desalination process with high recovery ratio, a problem of scale deposition, which leads to a reduction of heat transfer efficiency in heat exchanger occurs.

In this research we introduce self-heat recuperative seawater desalination process with fluidized bed for solving the problem. In the desalination process, the fluidized bed is employed as an evaporator. It might be seen that this process has the following advantages; good heat transfer properties and less scale deposition because of water evaporation taking place on heated particles in fluidized bed evaporator.

In addition, we investigated the influence on the specific energy consumption and feasibility to the society of the proposed desalination process.

(146c) Characteristics of Water and Salt Transport, Free Volume and Their Relationship With the Functional Groups of Novel Cellulose Esters

Ong, R. C., National University of Singapore
Helmer, B. J., Eastman Chemical Company
de Wit, J. S., Eastman Chemical Company
Chung, T. S., National University of Singapore
Originally presented on: 11/4/2013 13:20:00 - 13:45:00

Fundamental transport properties including salt and water partition coefficients, permeability and diffusivity of various newly synthesized cellulose esters were evaluated in this study in order to investigate the relationship among them as a function of molecular structure. Dense flat cellulose ester films were prepared and studied so that the influence of membrane fabrication, morphology and processing history on transport properties can be minimized. Annihilation lifetime spectroscopy (PALS) was employed to characterize the free volume of these cellulose esters. It was found that the transport and free volume properties can be correlated with the functional groups and their content in cellulose esters. A high hydroxyl content leads to a high hydrophilicity but results in a high water and salt partition coefficient. Hydrophilic cellulose esters suffer from a low water/salt selectivity. On the other hand, an increase in bulky and hydrophobic propionyl or butyryl groups generally enhances cellulose esters' selectivity but brings down the permeability. However, a very high content of bulky functional group causes a drop in selectivity due to poor chain packing and the enlarged free volume as confirmed by PALS. Experimental results suggest that cellulose esters with moderate content of both hydroxyl and bulky side groups have the best diffusivity selectivity and may be suitable as membrane materials for salty water separation. An upper bound relationship between solubility selectivity and water partition coefficient was also observed in these cellulose esters as suggested in previous literatures for other polymers.

(146d) Bioremediation of Cu(II) Contaminated Aqueous Solution Using Acinetobacter Sp

Majumder, S., Birla Institute of Technology & Science, Pilani
Gangadhar, G.,
Raghuvanshi, S., Birla Institute of Technology & Science, Pilani
Gupta, S., Birla Institute of Technology & Science, Pilani
Originally presented on: 11/4/2013 13:45:00 - 14:10:00

Bioremediation of metal wastes using enriched mixed consortium of microbial species is a very promising and viable alternative in environmental biotechnology. Present study deals with the removal of divalent copper [Cu(II)] compound from aqueous solution using *Acinetobacter sp.* (GenBank Accession Number: HQ538658.1), an indigenous bacterial strain which is isolated from enriched mixed culture. Enrichment of mixed culture is carried for a period of 6 days. At the end of the enrichment period, the biomass concentration obtained is 0.0187 g/L of solution. The effect of different operating conditions such as contact time, initial pH and amount of nutrient dosage on bioremediation of Cu(II) are studied in a batch mode. The effect of contact time is studied for an initial Cu(II) concentration ranging from 60-150 mg/L. Increase in initial Cu(II) concentration from 60 mg/L to 150 mg/L shows better removal of Cu(II). The effect of initial pH is also studied by varying the solution pH in the range of 2 - 12. The maximum biomass growth obtained is 0.112 g/L at pH = 7.5 for initial Cu(II) concentration of 150 mg/L. The maximum percentage removal of Cu(II) is obtained for the initial Cu(II) concentration of 60 mg/L at the solution pH value of 7.5. Percentage removal of Cu(II) value is also compared with reported value of percentage removal of Cu(II). This study also shows that the microbial growth for Cu(II) removal is significantly affected by nutrient dosage. The optimum biomass growth obtained is 0.0248 g/L at double of the standard composition of MSM.

Keywords: *Acinetobacter sp., Divalent copper; Bioremediation; Acclimated mixed culture, MSM.*

(146e) Calcium Ions in Gypsum ($CaSO4·2H2O$) Scaling in Forward Osmosis Processes and a Novel Design Strategy to Fabricate Polybenzimidazole Anti-Scaling Hollow Fiber Membranes

Chen, S. C.,

Su, J., National University of Singapore

Fu, F. J.,

Mi, B., The George Washington University

Chung, T. S., National University of Singapore

Originally presented on: 11/4/2013 14:10:00 - 14:35:00

Calcium ions in gypsum ($CaSO_4$°¤$2H_2O$) scaling in forward osmosis processes and a novel design strategy to fabricate polybenzimidazole anti-scaling hollow fiber membranes

Si Cong Chen[a,b], Jincai Su[b], Feng-Jiang Fu[b] , Baoxia Mi[c], Tai-Shung Chung[a,b,]*

[a] NUS Graduate School for Integrative Sciences & Engineering (NGS), National University of Singapore, 28 Medical Drive, Singapore 117456, Singapore

[b] Department of Chemical and Biomolecular Engineering, National University of Singapore, 4 Engineering Drive 4, Singapore 117576, Singapore

[c] Department of Civil and Environmental Engineering, University of Maryland, College Park, MD 20742, USA

* Corresponding author: chencts@nus.edu.sg

Abstract

Gypsum ($CaSO_4$°¤$2H_2O$) scaling phenomena have been studied on hollow fiber membranes with different physicochemical properties in forward osmosis processes. Three types of hollow fiber membranes were demonstrated i.e., (1) cellulose acetate (CA), (2) polybenzimidazole (PBI)-polyethersulfone (PES) and (3) PBI-polyacrylonitrile (PAN) - polyhedral oligomeric silsesquioxane (POSS). For the first time, we have found that calcium ions not only dominate but also take a leading role on gypsum ($CaSO_4$°¤$2H_2O$) scaling through ionic interactions with the membrane surface. A high level of calcium ions in scaling solutions can initiate gypsum crystallization more effectively than a high level of sulfate ions. Based on this finding, for the first time, a novel strategy to molecularly engineer PBI hollow fiber membranes with superior anti-scaling properties has been proposed and demonstrated. As shown in Figure 1c, the newly designed PBI-PAN (POSS) membranes have a positively charged and smooth membrane surface at the nanoscale level but possess a ridge and valley structure at the micrometer level. It shows slightly positive charge at a low pH and exhibits minimal scaling of only 5% flux reduction in contrast to CA membranes with 63% flux reduction. The adhesion force measured by atomic force microscopy (AFM) confirms a strong repulsive force between gypsum particles and PBI (Figure 1b) at pH values lower than the isoelectric point of 5.7. The repulsion force leads to the anti-fouling properties, while the novel ridge and valley morphology in the micrometer scale facilitates surface turbulence flow and effectively remove foulants.

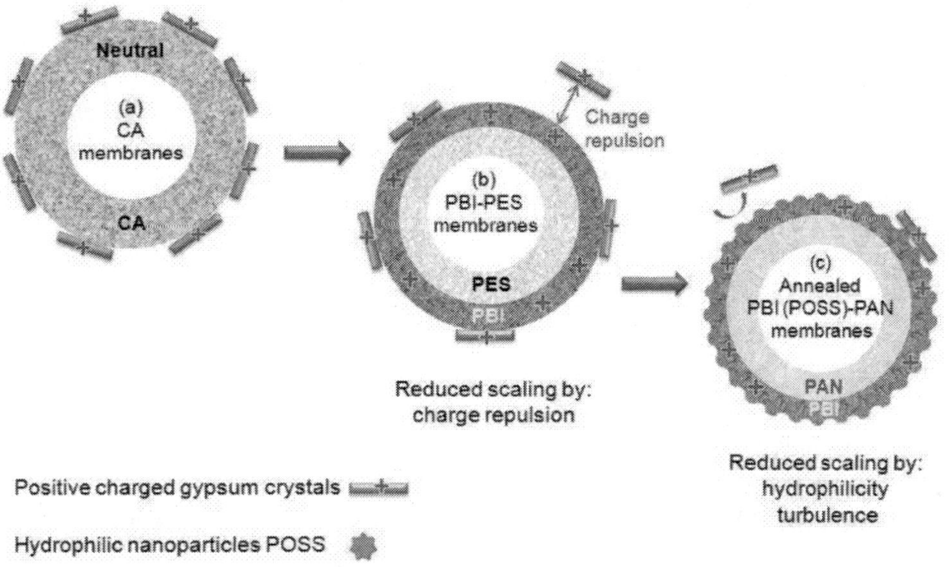

Figure 1: Fouling mechanisms of gypsum scaling on membranes

(146f) New Insights Into Hydrophobic/Hydrophilic Composite Hollow Fibers for Direct Contact Membrane Distillation

Zhu, J. M., Central South University
Jiang, L. Y., Central South University
Originally presented on: 11/4/2013 14:35:00 - 15:00:00

Hydrophobic/hydrophilic composite membrane is a new design of membrane in terms of structures and materials for DCMD. With dry-wet phase inversion process, a novel type of hydrophobic/hydrophilic hollow fiber membranes was fabricated in this research. The outer layer was made from polyvinylidene fluoride (PVDF), while the inner layer PVDF and polyvinyl alcohol (PVA) blend. Effects of dope composition, PVA/PVDF blending ratio and coagulants on the morphological, physiochemical and mechanical characteristics of the membranes thus generated were investigated. The pore size and porosity in respective layer of the dual-layer membranes were determined mainly by data obtained from SEM and corresponding software processing. The DCMD performance observed in experiments was compared with that obtained by mass and heat transport modeling of both single-layer and dual-layer DCMD membrane. Analysis relating the membrane structural features and the above comparison revealed that the inner layer characteristics in terms of both material hydrophilicity and pore size should be optimized to give a better membrane performance. The maximum flux in DCMD obtained was around 24.4kgm^{-2}h^{-1} at 65℃.

(156a) A Study On the Survival of Oral and Faecal Bacteria in Different Beverages

Mbachu, C., Middlesex University London

Originally presented on: 11/4/2013 15:15:00 - 15:37:00

Bacterial contamination of beverages has been a major concern and can cause many health issues. *Escherichia coli, Enterococci*, Coli forms and *Pseudomonas aeruginosa* are common known bacteria that can be found in beverages such as fruit juices and bottled water and these organisms can thrive well depending on the conditions (e.g. temperature and pH). These beverages sold at all public places and roadside shops and they can become a health concern depending on handling and storage or if the manufacturing process is unsafe. In this research, five different beverages (spring water, fountain water, fruit juice, lucozade and coke) were inoculated with *Pseudomonas aeruginosa* and *Escherichia coli* and survival rate monitored over time under two different temperature (4oC and 22oC) conditions. In this, the effect of storage time and temperature of different beverages on the survival growth of bacteria were studied to assess for the potential risk of food borne illnesses on different beverages. Analysis of Variance (ANOVA) of the data was conducted to determine and assess the differences in growth rate and for comparison purposes and to know which condition was significant. It was found that temperature was significant in the growth rate of *Escherichia coli* and both temperature and type of beverage used for the study was significant in the growth rate and survival of *Pseudomonas aeruginosa*.

(156b) Potable Water Production Via Microchannel Capacitive Desalination Technology (CDT) System Architecture

Loeb, C., Oregon State University
Jovanovic, G. N., Oregon State University
Originally presented on: 11/4/2013 15:37:00 - 15:59:00

As clean water availability diminishes throughout the world, an inexpensive, reliable, and scalable system is required to produce potable water from the world's vast supply of seawater. Taking advantage of the short diffusion lengths inherent in microchannel architectures, the research team at Oregon State University and MTEK Energy Solution Inc. is developing a parallel flow-through system capable of removing and segregating ionic species from concentrated salt water.

The application of a voltage potential to high surface area electrodes quickly removes ions from the salt-water feed to the electrode surface, allowing clean water to exit the CDT system while ions remain immobile. Segregated side channels quickly and efficiently elute concentrated brine solution during periods of unapplied electrical potential. Clean water production will be achieved through a series arrangement of CDT cells.

Validated computer simulations have provided the necessary insight to dramatically improve system design, operating conditions, and fluid flow relative to our demonstration prototype. The microchannel CDT system utilizes common materials of construction and fabrication processes available at the Microproducts Breakthrough Institute (MBI) in Corvallis, Oregon to quickly modify for enhanced performance. Increased throughput of seawater is readily achieved by 'numbering-up' parallel microchannel CDT architectures to match water production requirements. The CDT system allows for tremendous end-user customization that makes the CDT system the optimal alternative for any required process metrics, space limitations, and location.

(156c) Disinfection of Water By Adsorption Combined With Electrochemical Treatment

Hussain, S. N., University of Manchester
de Las Heras, N.,
Asghar, H. M. A., University of the Punjab
Brown, N. W., Arvia Technology Ltd
Roberts, E. P. L., University of Manchester
Originally presented on: 11/4/2013 15:59:00 - 16:21:00

The process of adsorption using graphite based adsorbents with electrochemical regeneration was developed at the University of Manchester for the removal of low concentrations of organic impurities from water. The technology employs a graphite intercalation compound (GIC) adsorbent that can be electrochemically regenerated. This process has significant potential application for water recycling since it does not involve the addition of chemicals and no solid or liquid wastes are generated. Safe storage of water for recycling often requires that the water is disinfected. To date there have been no studies of the effect of the adsorption / electrochemical regeneration process on any microorganisms present in the water. The aim of the present work was to investigate the removal and disinfection of microorganisms by adsorption on a GIC adsorbent followed by electrochemical treatment. The adsorption of an enteropathogenic bacterium, *Escherichia coli*, was performed using GIC particles followed by electrochemical treatment under a range of experimental conditions in an air agitated sequential batch reactor. The adsorption of *E. coli* cells on the GIC adsorbent was found to be a fast process and was capable of removing >99% *E. coli* cells from solution after 10 min with a ca. $6.5\log_{10}$ reduction in the CFU concentration. The electrochemical treatment gave 100% regeneration efficiency showing that the capacity of the GIC adsorbent did not diminish over five adsorption and regeneration cycles . An ~8-$\log_{10}$ reduction of *E. coli* was achieved on each cycle. During electrochemical regeneration significant indirect electrochemical disinfection of *E. coli* present in the supernatant solution was observed. The process of adsorption using GIC adsorbents with electrochemical regeneration is capable of disinfecting a high concentration of E. *coli* at low current density and low energy consumption. The most likely mechanisms for electrochemical disinfection include electrochlorination, direct electrochemical attack and pH effects. In the absence of chloride, both the adsorption of *E. coli* on the GIC and the electrochemical regeneration were found to be less effective for water disinfection. In spite of this, effective disinfection was achieved in the absence of chloride by the appropriate selection of operating conditions. This observation indiates that the process of disinfection by adsorption with electrochemical regeneration could be employed for disinfection without the formation of chlorinated disinfection by–products.

(156d) Municipal Wastewater Treatment By An Osmotic Membrane Bioreactor: Use of Sodium and Magnesium Chloride As Draw Solution

Qiu, G.,
Ting, Y. P., National University of Singapore
Originally presented on: 11/4/2013 16:21:00 - 16:43:00

Osmotic membrane bioreactor (OMBR) technology, which integrates forward osmosis (FO) and biological treatment process, has received increasing attention. In this study, an OMBR was developed for municipal wastewater treatment, using two draw solutes (DS), i.e. $MgCl_2$ and NaCl. The potential and limitations of this innovative treatment process was investigated with respect to pollutant removal efficiencies, flux stabilities and membrane fouling. During a 100-day operation, high removal of organic matter (98%), nitrogen (98%) and PO_4^{3+}-P (99%) were achieved with both DS, with the effluent TOC, NH_4^+-N, total nitrogen (TN) and PO_4^{3-}-P being lower than 3.0 mg/L, 0.5 mg/L, 1.5 mg/L and 0.1 mg/L, respectively. Significant aerobic denitrification (up to 98%) was observed in the OMBR, owing to the high rejection of the FO membrane and the prolonged retention of NO_2^--N and NO_3^--N within the bioreactor which facilitated their aerobic removal. However, salt leakage from the DS and the retention of electrolytes from the feed wastewater resulted in significant salt accumulation in the bioreactor, which not only reduced the water flux considerably but also adversely affected pollutant removal efficiencies. Nitrification was affected acutely with both the DS. Marginal inhibition of denitrification was also observed with NaCl. However, for both DS, TOC removal was not evidently affected. Compared to $MgCl_2$, NaCl gave 20% higher initial water flux but resulted in more significant water flux decline (about twice), due to a more severe reverse transport of NaCl as the DS (about 6.5 times). Membrane fouling was controlled by periodically soaking and washing with tap water, and resulted in only 5% reduction in water flux decline over 60 days. Different rejection and transportation behaviors of NH_4^+-N, NO_2^--N and NO_3^--N were also observed with different DS, which was found to be governed by electrostatic interactions and the Gibbs-Donnan effect.

(156e) Scalability Analysis of An Open Channel UV Wastewater Disinfection Reactor Using CFD Simulation

Ray, M. B., The University of Western Ontario
Saha, R. K., University of Western Ontario
Zhang, C., University of Western Ontario
Originally presented on: 11/4/2013 16:43:00 - 17:05:00

The scalability of an open channel ultra-violet (UV) wastewater disinfection reactor is numerically investigated using a computational fluid dynamics (CFD) model. The CFD model consists of volume of fluid (VOF) method for capturing water-air interface, Lagrangian particle tracking for calculating microbial particle trajectory and the discrete ordinate (DO) model for resolving the UV intensity field in the reactor. The realizable k-ϵ turbulence model was adopted to simulate the turbulent flow, and the enhanced wall treatment was used to resolve the turbulence near the walls. Different scaling down procedures were tested to establish similarity of the hydrodynamics and the radiation field between the scale model and the full size reactor. To ensure similarity of the hydrodynamics, Froude number and the Reynolds number were taken into consideration. Scaling methodology for the UV radiation field was formulated to ensure the same reduction equivalent dose (RED) values can be achieved from the full size and scaled down reactors. A parametric study was performed in terms of the flow rate in the reactor and water UV transmittance to characterize the range of scalability. Initially, the model was validated using experimental data on free-surface profile. The CFD model predicts that the open channel reactor is scalable, thus allowing one to predict the large scale behavior from the small scale laboratory test.

(156f) Performance Evaluation of a Low Cost and High Rate Vertical Flow Wetland for Residential Graywater Treatment and Reuse

Yu, Z. L. T.,
Rahardianto, A., University of California, Los Angeles
Stenstrom, M. K., University of California, Los Angeles
Cohen, Y., University of California, Los Angeles
Originally presented on: 11/4/2013 17:05:00 - 17:27:00

Water shortages in urban environments across the US and around the world have sparked increased interests in the development of sustainable water practices and technologies. Despite increased sustainability efforts, water reuse in the United States accounts for only ~1% of the nation's total water supplies. Water reclamation and reuse through centralized suppliers (i.e. municipalities or water agencies) represent a major challenge given the lack of infrastructure and operational costs for treatment and distribution system for reclaimed wastewater and its reuse. One possible approach that addresses, in part, the above challenges is to utilize point-of-use (i.e. onsite) treatment to generate non-potable water source for onsite reuse. Such onsite reuse activity will expand the nation's water recycling efforts driven by end-users or small communities (e.g. multifamily homes) and augment the existing water resources portfolio. Accordingly, it is argued that on-site treatment and reuse of graywater is a viable option in residential and non-residential premises. The key challenge of encouraging such a practice is the lack of affordable and simple treatment technology suitable for residential users. In order to meet the above challenge, the present study presents a low cost, high rate, compact vertical flow wetland (CVFW) system for graywater treatment. The developed small footprint graywater treatment (~1m^2) was field tested in a single family home in Los Angeles with the treated water being used for irrigation. The kinetics of the graywater treatment system was characterized using a composite model for the vertical wetland configuration and system operation was accordingly optimized. Results from the field study demonstrated that the CFVW system is capable of removing up to 90% of total organic carbon and up to 99% of turbidity. The present approach has a potential for deployment in multifamily homes for onsite treatment and reuse of graywater with an estimated return on investment within 2 years.

(156g) Coagulation and Ultrafiltration of High-Alkalinity Greywater

Amundsen, T. J., Mainstream Engineering Corporation
Zastrow, D. J., Mainstream Engineering Corporation
Wagner, A. L., Mainstream Engineering Corporation
Originally presented on: 11/4/2013 17:27:00 - 17:49:00

Water is perhaps the most essential natural resource. Unfortunately, fresh water is becoming increasingly scarce, and with the world population continuing to grow, the need for improved methods of water utilization, treatment and recycling is becoming ever more pressing. Greywater is produced by a number of household activities including dishwashing, showering, bathing and laundry. Water from these sources is characterized by the presence of a variety of organic and inorganic impurities such as soap, oil, grease and dissolved solids. Greywater produced during dishwashing poses a particularly difficult problem for treatment processes due to a relatively large concentration of pollutants, particularly soap. The soap solubilizes nonpolar oils and greases in the water while simultaneously increasing the alkalinity of the water. Many treatment processes are very sensitive to pH and require either neutral or acidic pH to operate effectively. Traditional coagulants based on iron or aluminum do not perform well in an alkaline environment and can require permeate post-processing to remove dissolved metals. We investigated the use of novel electrode materials in an electrocoagulation process to treat dishwashing water without any pH adjustment. The coagulated water was then further purified by ultrafiltration. We report the impact of the pre-coagulation on membrane performance and longevity.

(160a) Heterogeneous Chlorine Chemistry in the Atmosphere: Insights From Modeling and Experiments

Faxon, C., University of Texas at Austin

Allen, D. T., The University of Texas at Austin

Originally presented on: 11/4/2013 15:15:00 - 15:30:00

Heterogeneous Chlorine Chemistry in the Atmosphere: Insights from Modeling and Experiments.

Background

To date, numerous studies using both computational and experimental methods have shown that the presence of gas phase atomic chlorine radicals ($Cl^?$) can influence tropospheric photochemistry. Specifically, $Cl^?$ participates in the ozone production cycle in a manner analogous to the hydroxyl radical ($OH^?$), and elevated concentrations can lead to increases in ozone production rates. $Cl^?$ radicals are produced from the photolysis of a gas phase precursor such as Cl_2 or HOCl. Impacts on ozone production resulting from anthropogenic point source emissions of these gas phase $Cl^?$ precursors tend to be highly localized due to rapid rates of photolysis during the day.

Aside from direct emission from anthropogenic sources, gas phase chlorine species can be produced through heterogeneous mechanisms involving chloride aerosols. One such species is nitryl chloride ($ClNO_2$), which was initially observed in ambient air in 2006 during a study along the Gulf Coast of the U.S. Additional observations have also been made in Boulder, CO, Alberta, Canada and off the coast of California, near the Los Angeles Basin. Multiple studies investigating $ClNO_2$ production have determined that the responsible mechanism is the reactive uptake of N_2O_5 onto aqueous chloride aerosols. N_2O_5, which is formed from the gas phase reaction between NO_2 and NO_3, photolyzes rapidly and therefore only builds up to significant concentrations at night. The nocturnal conversion of N_2O_5 into $ClNO_2$ results in the rapid production of $Cl^?$ when $ClNO_2$ photolyzes at sunrise. Through gas phase reactions initiated by $Cl^?$, this can increase the rate of oxidation for volatile organic compounds (VOCs) and thus increase the rate of ozone production. The heterogeneous uptake and dissociation of N_2O_5 has been determined to be the rate limiting step, and the full mechanism is shown below.

$$NO_{2,g} + NO_{3,g} \leftrightarrow N_2O_{5,g} \quad (1)$$

$$N_2O_{5,g} \leftrightarrow N_2O_{5,aq} \quad (2)$$

$$N_2O_{5,aq} \leftrightarrow NO_2{}^+{}_{aq} + NO_3{}^-{}_{aq} \quad (3)$$

$$NO_2{}^+{}_{aq} + H_2O_{aq} \rightarrow H_3O^+{}_{aq} + HNO_{3aq} \quad (4a)$$

$$NO_2{}^+{}_{aq} + Cl^- \rightarrow ClNO_2 + H_2O_{aq} \quad (4b)$$

$$NO_3{}^-{}_{aq} + H^+ \leftrightarrow HNO_3{}^+{}_{aq} \quad (5)$$

Another heterogeneous mechanism of interest involves the production of Cl_2 from chloride aerosol particles. Previous studies have suggested a direct reaction between O_3 and particulate chloride. Other studies have suggested a surface reaction between particulate chloride and $OH^?$. These mechanisms were formulated in response to experimental observations of gas phase Cl_2 when aqueous NaCl and O_3 were present together.

The reaction has been reported to occur both in the presence of light as well as in the dark. The mechanism for the reaction between $OH^{?}$ and particulate chloride can be represented as follows, where 2OH???Cl^{-} represents an intermediate species that forms at the particle surface.

$$OH_g + Cl^{?-}_{aq,surface} \rightarrow OH???Cl^{?-}_{aq,surface} \quad (6)$$

$$2OH???Cl^{-}_{aq,surface} \rightarrow Cl_{2,g} + 2OH^{-}_{aq} \quad (7)$$

Cl_2 concentrations up to 150 ppt$_v$ have been detected in marine air at night, suggesting a nocturnal 330 ppt$_v$ day^{-1} Cl_2 source. Additionally, detection of Cl_2 in both coastal marine and coastal urban air has been reported. Therefore, the determination of an exact mechanism and quantification of the rate would be useful in predicting the impact Cl_2 production from chloride aerosols on tropospheric chemistry. This is particularly relevant in coastal regions where chloride particles are typically present in large quantities. The mechanism involved in reactions 6 ? 7 is the focus of the experimental results presented here.

Methods and Discussion

This work presents both computational and experimental investigation of the heterogeneous processes described above. An overview of a sensitivity analysis of the physical parameters involved in the mechanism described by reactions 1 ? 5 will be discussed. The parameters investigated in the analysis include the reactive uptake coefficient, $ClNO_2$ yield, particle surface area, and gas phase concentrations of VOCs and NO_x. The sensitivity analysis was carried out through photochemical box modeling using the SAPRC software and the Carbon Bond 5 mechanism. The focus of this sensitivity analysis was on the effects of variations in parameter values on the production of $ClNO_2$ and subsequent impacts on ozone production. The results obtained from simulations using various parameter value combinations were compared to a base case modeling scenario in which all heterogeneous reactions were absent. In order to quantify differences in O_3 production, the peak O_3 concentration from each simulation was used as a point of comparison. An analysis of parameter values reaching the upper limits reported by the literature was undertaken.

Results from the sensitivity analysis indicate that $ClNO_2$ chemistry can potentially change peak O_3 concentrations by -10.5% ? 27%. The availability of NO_x was found to play an important role. The presence of additional NO_x during nocturnal hours resulted in greater production of N_2O_5 and thus higher concentrations of $ClNO_2$. However, NO_x emissions after sunrise reduced the impact of N_2O_5-$ClNO_2$ chemistry on O_3 production. Additionally, the presence of large concentrations of VOCs that are highly reactive with the NO_3 radical significantly reduced the impact of N_2O_5-$ClNO_2$ chemistry on ozone production. Figure 1 shows the combined influence of variations in the reactive uptake (γ) and $ClNO_2$ yield (Y_{ClNO2}) on peak ozone concentrations as compared to the base case scenario. Further discussion of details on the impacts of $ClNO_2$ chemistry with respect to O_3 production and NO_x cycling within the atmosphere will be included in the presentation.

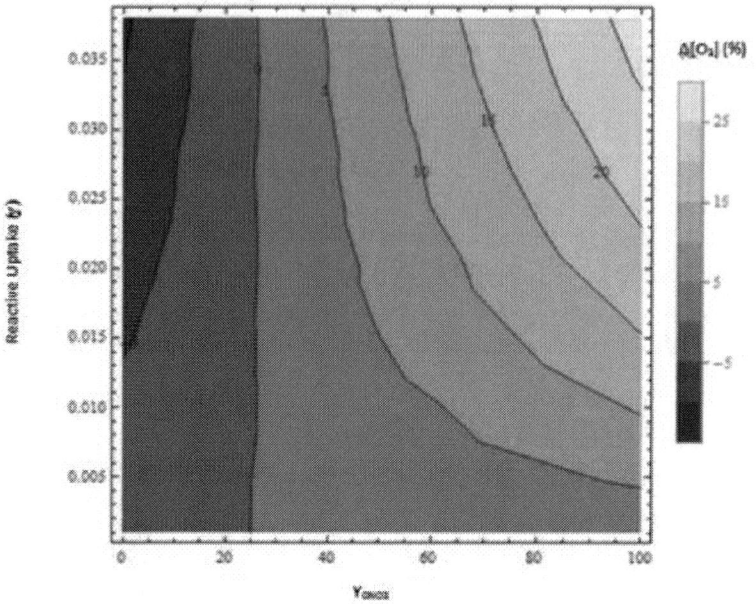

Figure 1: Percentage peak O_3 increases in over the base case scenario (no heterogeneous reactions) for various combinations of the $ClNO_2$ yield and reactive uptake parameter values.

Experimental work investigating the heterogeneous mechanism described in reactions 6 ? 7 will also be covered. Specifically, results from environmental chamber experiments involving the exposure of NaCl aerosol particles to various gas mixtures representative of conditions present in the atmosphere, including HO_X, NO_X and O_3 will be presented. Experiments were carried out in a 10 m^3 reaction chamber equipped with UV lights which induce photochemistry. Chemical Ionization Mass Spectrometry (CI-MS) utilizing iodide as the reagent ion was used to quantify and identify gas phase species produced by the mechanism. Aerosol population size distributions were monitored using a Scanning Electrical Mobility System (SEMS).

Experiments were carried out by generating and injecting NaCl aerosol particles into the chamber along with photolytic precursors of the hydroxyl radical. The mixture was then exposed to UV radiation in order to initiate photochemistry and hydroxyl radical production. Resulting gas phase Cl_2 concentrations were monitored using iodide CI-MS. Various initial conditions were used to test the impact of several factors on the overall efficiency of the mechanism in producing gas phase chlorine. One factor is the impact of variations in relative humidity levels on the efficacy of the mechanism. Another factor is the variation of the composition of the bulk aerosol. The impact of the presence or absence of elevated O_3 concentrations is also examined. Resulting variations in overall Cl_2 production due to these factors will be compared and discussed.

(160b) Chemical and Physical Investigation of Fractal-Like Iodine Oxide Particle Formation

Nakao, S., Colorado State University
Heo, G.,
Warren, B.,
Qi, L., National Research Center for Environmental Analysis and Measurement
Hagino, H.,
Carter, W.,
Cocker, D. III,
Originally presented on: 11/4/2013 15:30:00 - 15:45:00

Particulate matters in the atmosphere as well as in engineering processes are often non-spherical. Non-sphericity poses technical challenges especially in the system where chemical reactions continuously occur and thus particles' chemical composition, size, number, morphology, and mass change accordingly.

Atmospheric iodine oxide particle (IOP) formation is an example of systems with evolving fractal-like particles. Atmospheric chemistry of iodine influences the oxidizing capacity of the atmosphere (e.g., catalytic loss of tropospheric ozone). Furthermore, atmospheric iodine could be a significant precursor of newly formed aerosol in coastal areas. Previous experimental studies investigated IOP formation typically without NOx; however, recent studies point out IOP formation occurs in polluted environment with elevated NOx concentrations as well.

This study advanced the understanding of the chemical and physical processes leading to atmospheric IOP formation through the application of on-line particle effective density analysis (with approximately one-minute time resolution) to take into account the evolution of particle shape. IOP was produced via CH3I photolysis under varied O3 and NOx concentrations. An on-line chemical analysis showed minor contributions from nitrogen containing species to IOP, suggesting that nitrogen containing iodine species act as gas-phase reservoir of iodine. Real-time particle density measurement allowed determination of mass-based yield of IOP. Yield was nearly one (meaning the mass of IOP was equal to the mass of reacted CH3I) when there is sufficiently high ozone concentration for the formation of I2Ox species regardless of NOx concentration; on the other hand, particle yield approached zero when ozone formation was suppressed by the presence of NO.

Finally, a predictive model for IOP formation is developed by extending a gas-phase chemical reaction mechanism, SAPRC-07, to include iodine chemistry. IOP formation and O3 decay were reasonably predicted in the absence of NOx. However, when NOx was present, IOP was underpredicted due to build-up of reservoir species (e.g., I2 and IONO2) within the model, which suggests that further improvements are needed in describing reactions involving NOx in the iodine mechanism.

(160c) Changes in Water-Soluble Particulate Composition From Vehicles

Short, D., UC Riverside
Vu, D., UC Riverside
Karavalakis, G., UC Riverside
Durbin, T., UC Riverside
Asa-Awuku, A., UC Riverside
Originally presented on: 11/4/2013 15:45:00 - 16:00:00

Particle number, size, and composition information is important for constraining aerosol effects on air quality, climate, and health. The composition of particles, especially from vehicular sources, may contain insoluble material that may modify particle nucleating properties. Changes in fuel properties are known to modify criteria pollutants and particulate matter mass, size, and number. In this work we summarize findings as they pertain to the water-soluble and insoluble composition of particles. In field measuremnts and controlled laboratory studies, a water-based condensation particle counter (CPC) and a butanol-based CPC measure particle number concentration. Both instruments were coupled with a Scanning Mobility Particle Sizer (SMPS) and particle number and size data were recorded. Real time particle insoluble mass fractions are estimated with the SMPS data sets; theoretical soluble fractions are calculated from ideal hygroscopicty single parameter values. This is the first time that this experimental method has been employed and used to infer online insoluble fractions. The results show that near-roadway emissions contain water insoluble and black carbon components. The emissions of different ethanol and butanol gasoline blends are also explored for light-duty vehicles on a light-duty dynaometer chassis. Laboratory results indicate that soluble vehicular components are strongly correlated with vehicle driving conditions.

(160d) Condensed-Phase Photolysis Processes in Secondary Organic Aerosol

Epstein, S. A., Carnegie Mellon University
Hinks, M. L., University of California Irvine
Nizkorodov, S. A., University of California, Irvine
Originally presented on: 11/4/2013 16:00:00 - 16:15:00

Exposing secondary organic aerosol (SOA) to ultra-violet (UV) radiation at atmospherically relevant wavelengths leads to changes in aerosol composition and volatility. The carbonyls, peroxides, nitrates, and other multifunctional compounds typically present in SOA undergo direct photolysis when exposed to UV radiation. These direct photolysis rates can be competitive with atmospheric oxidation by OH and other gas-phase oxidants. Much of the previous research aimed at studying the photolysis of SOA involved exposing both particles and their surrounding vapors to UV light. These photolabile vapors may produce semi-volatile products with vapor pressures that induce condensation onto adjacent particles. These bulk SOA photolysis studies, typically performed in reaction chambers, do not permit for the decoupling of gas-phase and condensed-phase photochemistry. We have designed and built a system where SOA mixtures are stripped of their vapors after passing through three, one-meter denuders and then photolyzed in a UV-transparent flow cell. The denuder system leads to an order of magnitude reduction in vapor concentration. We employ an Aerodyne HR-ToF-AMS to monitor changes in particle composition during photolysis. Experiments designed to age SOA formed from biogenic precursors will reveal how aerosol oxidation state influences the rate of particle photolysis. Additional experiments with SOA seeds of varying compositions will illustrate the dependence of photolysis rate on seed composition.

(160e) Stratocumulus Cloud Water Properties Off the California Coast During the 2011 E-PEACE Field Campaign

Sorooshian, A., University of Arizona
Wang, Z., University of Arizona
Coggon, M.,
Jonsson, H., Center for Interdisciplinary Remotely Piloted Aircraft Studies
Originally presented on: 11/4/2013 16:15:00 - 16:30:00

This study reports on cloud water chemical and pH measurements off the California coast during the July-August 2011 Eastern Pacific Emitted Aerosol Cloud Experiment (E-PEACE). Eighty two cloud water samples were collected by the Center for Interdisciplinary Remotely-Piloted Aircraft Studies (CIRPAS) Twin Otter in boundary layer stratocumulus clouds impacted to varying degrees by marine-derived emissions, ship exhaust, and continental emissions. Cloud water pH ranged between 2.92 and 7.58, with an average of 4.46. Peak pH values were observed north of San Francisco, simultaneous with the highest concentrations of Si, B, and Cs. One case flight shows how air mass trajectories transported over land covered with vegetation and soil resulted in the highest cloud water pH (~ 7.5) and Si concentration (air equivalent of 4.39 μg m^{-3}) during E-PEACE. These results point to an important role for basic particle types (e.g. soil dust and fly ash) originating from continental sources in impacting stratocumulus clouds in the study region. The lowest pH values were observed south of San Francisco in an area with extensive ship traffic. These samples exhibited the highest concentrations of sulfate, nitrate, V, Fe, Al, P, Cd, Ti, Sb, and Mn, which are linked largely to cargo and tanker ships. Marine biogenic emissions were responsible for the close association between a number of species including chloride, methanesulfonate, Na, Se, K, Rb, Sr, and I. The highest levels of Fe, Mn, Ti, and P were observed in the area of most extensive ship traffic rather than near the northern part of the region impacted by continental sources, suggestive of the role of ship emissions in providing a source for these ocean micronutrients and aqueous-phase reaction catalysts to stratocumulus cloud drops.

(160f) Aqueous Aerosol Chemistry: SOA Formation and Impacts On Aerosol Physical Properties

McNeill, V. F., Columbia University

Originally presented on: 11/4/2013 16:30:00 - 16:45:00

Organic chemistry in aerosol and cloud water has been recognized as a source of secondary organic aerosol material (SOA) from volatile, water-soluble precursors. I will give an overview of the state of the science re: known mechanisms for aqueous SOA formation from a laboratory perspective. I will discuss the application of a numerical model of coupled gas and aqueous aerosol chemistry, GAMMA (McNeill et al. Environ. Sci. Technol. 2012) that was developed my group, to study the formation of secondary organic aerosol material in aerosol water and the associated changes in aerosol physical properties.

(160g) Effect of Humidity On Secondary Organic Aerosol (SOA) Formation From Biogenic Hydrocarbons and Nitrate Radicals

Ng, N. L., Georgia Institute of Technology
Boyd, C. M., University of South Carolina
Xu, L., Georgia Institute of Technology
Originally presented on: 11/4/2013 16:45:00 - 17:00:00

The reactions of biogenic volatile organic compounds (BVOCs) with nitrate radicals (NO_3) represent a direct link of anthropogenic and biogenic emissions. Half of the aerosol lifetime is in the dark where NO_3 radicals and ozone can be the prevalent oxidants. Compared to SOA formation from ozonolysis and photooxidation chemistry, SOA formation and evolution involving NO_3 radicals have received far less study. Water is ubiquitous in the atmosphere and water vapor can affect the mechanisms, chemical composition, and physical and chemical properties of aerosols formed. This can be especially important for nighttime chemistry (where NO_3 radicals can dominate) where ambient RH is high. Here we report a series of experiments to study SOA formation from the reaction of different biogenic hydrocarbons (isoprene, a-pinene, and b-caryophyllene) with NO_3 radicals. Experiments are performed in the newly constructed Georgia Tech indoor chamber facility, which consists of two 10 m^3 Teflon chambers suspended in a temperature controlled enclosure (4-40 ^oC). The experiments are performed in the dark using thermal decomposition of N_2O_5 as a source of NO3 radicals. Experiments are conducted under both dry and humid conditions: dry (RH< 5%) and 80% RH. Aerosol growth and composition are monitored by Scanning Mobility Particle Sizer (SMPS) and High Resolution Time-of-Flight Aerosol Mass Spectrometer (HR-ToF-AMS). The SOA yields (and parameterizations) for these experiments will be presented and the effect of water vapor on SOA formation mechanism and composition will be discussed.

(160h) The Effect of Urban Biomass Combustion for Space Heating On PM Exposure

Sarigiannis, D., Aristotle University of Thessaloniki
Karakitsios, S., Aristotle University of Thessaloniki
Kermenidou, M.,
Nikolaki, S., Aristotle University of Thessaloniki
Originally presented on: 11/4/2013 17:00:00 - 17:15:00

Particulate matter (PM) is nowadays one of the most serious urban environmental pressures to public health. PM can be directly emitted from a variety of sources, both natural and anthropogenic, or generated by atmospheric reactions, leading to a complex mixture of solid particles and liquid droplets with different size and chemical composition, met in both outdoor and indoor environments. Macro and micro environmental concentrations of PM are affected by seasonality, reflecting both changes in the prevalent meteorological conditions as well as PM emission patterns, traffic emissions and domestic heating being dominant among them. The concentrations of PM10 at urban sites increase during the winter due to the combination of strictly anthropogenic urban emissions such as traffic and domestic heating, and meteorological conditions such as the prevalence of anticyclonic scenarios which limit air mass circulation from the more polluted urban sites towards the regional environment. Over the last couple of years, the use of biomass as heating source was allowed in Greece as a CO_2-neutral means of space heating in the large metropolitan areas of Athens and Thessaloniki affecting more than half of the country's population. At the same time the use of light heating diesel was heavily taxed. In the same period Greece faces a financial crisis with significant repercussions on the average household income. This combination resulted in reduced traffic loads but excessive biomass use for domestic heating during wintertime. In this context, the current study deals with the assessment of the seasonal variability of PM exposure in the city of Thessaloniki.

A combination of measured and modeled data of outdoor and indoor PM10 and PM2.5 concentrations was fed into an integrative exposure assessment framework. Indoor concentrations were estimated with the computational tool INTERA. The latter is based on a multi-zonal mass-balance model, that takes into account the major processes governing particles concentration i.e. emissions, deposition, indoor/outdoor exchange rate, and outdoor infiltration. Exposure to PM was estimated based on the indoor/outdoor concentration data for two clusters of exposed individuals, based on whether they live in a house using a fireplace or not. Different microenvironments were treated in different way in terms of time spent within them, and time-weighted factors were used, based on the time-activity data. By using detailed activity patterns and linking them to specific microenvironments, our methodology takes stock of an additional factor influencing the exposure-effect continuum, the inhalation rate. Different types of activities demand different levels of effort which correspond to different inhalation rates.

The measurements campaign included the assessment of outdoor and indoor air quality and the evaluation of biomass use (in open fireplaces) for domestic heating. With regard to the outdoor measurements, PM10 and

PM2.5 samplers were installed at 3.5 m above ground, located at one traffic and one background site. The traffic site is surrounded by several streets with typical city heavy-traffic levels and the background site is located in an urban environment without direct traffic emissions in accordance to the EU Air Quality Directive spectifications. The measurements lasted for six months, covering the transition from the warm to the cold period and back (average daily temperatures 23 and 4 ^{0}C respectively).

Parallel measurements of indoor air quality (for PM10 and PM2.5) were carried out in 30 houses allocated close to the traffic and the urban background station respectively, so as to validate INTERA model estimates. PM2.5 and PM10 samples were collected using six low-flow air samplers (ENCO PM, TCR TECORA, Italy). The samplers used sampling heads meeting the EN 12341 (PM10) and EN 14907 (PM2.5) standards, and operated at a flow-rate of 38 L/min, with a collection time of 24 h per sample. Samples were collected on PTFE membranes filters with PMP supporting ring (PALL Life Sciences, Ø 47mm, pore size 2 μm, USA). Teflon filters were analyzed gravimetrically for particle mass concentrations using an electronic microbalance with a sensitivity of ± 1 μg after 24-h equilibration at a temperature between 20 °C and 23 °C and a relative humidity (RH) between 30 and 40%. Each filter was weighed at least three times before and after sampling, and the net mass was obtained by subtracting the average of the pre-sampling weights from the average of the post-sampling weights. Differences among replicate weightings were <5 μg for the blanks and the samples. Prior to the start of the sampling campaign, the flow rate of the PM2.5 and PM10 samplers was calibrated. Field blank filters were also collected and used to correct for background concentrations or influences from handling and transport.

The same PM samplers were used for 24 h sampling of PM10 and PM2.5 in 30 indoor locations, with and without the operation of fireplace.

Ambient air data showed a significant increase of PM10 (from 30.1 to 73.1 μg/m^3) and PM2.5 (from 19.4 to 62.7 μg/m^3) concentrations during the transition from the warm to the cold period of the year 2012, in contrast to the previous year (2011) when this transition was accompanied by a smaller increase by 12 μg/m^3 for both PM10 and PM2.5. Between the two years, significant differences exist in PM emission patterns; in 2012 the contribution of traffic is reduced, while during the colder period the contribution of biomass heating increases up to 34%. The results of the indoor measurements revealed that PM concentrations are close to the ones of the outdoor air if no strong emission source (e.g smoking indoors or combustion) is present. In any other case (i.e use of fireplace), the concentrations tend to rise significantly. Similar results were obtained also for the the days of the cold period, when the fireplaces were burning for 2-5 hours daily. In this case, average daily concentrations tended to be similar or higher than the ones outside, with an average daily increase of about 10 μg/m^3 and 14 μg/m^3 for PM2.5 and PM10 respectively compared to the houses where the fireplace was not in use. However, this variability is subjected to many parameters, such as the daily duration of fireplace usage, the type of wood used and the burning conditions. Very close agreement was noticed between the measured and the modeled concentrations.

Personal exposure variability follows a pattern that is strongly influenced by indoor air concentrations; this was expected considering that almost 85% of exposed individuals activities are performed indoors and 55% of time is spent in the home environment (where an additional contribution to the overall personal exposure came from the use of open fireplaces or wood stoves). A very interesting result is that for 2012, exposure to PM (mean of 47 and 42.7 μg/m^3 for PM10 and PM2.5 respectively) for population living in houses not equipped with fireplaces, is higher than the one for people living in a house equipped with fireplace for year 2011 (mean of 39.6 and 34.6 μg/m^3 for PM10 and PM2.5 respectively) during the cold season. For individuals exposed to indoor biomass burning in 2012, average daily exposure is up to 50.1 and 46.8 μg/m^3 for PM10 and PM2.5 respectively. During the warm period, exposure is clearly higher in 2011, due to the higher ambient air concentrations, the difference being prominent for PM 2.5 (32.1 μg/m^3 for 2011 instead of 20.9 μg/m^3 for 2012). Beyond average exposure, intra-day variability of exposure was also investigated. For a typical intra-day winter time profile of an exposed individual living in a house with fireplace, during most of the day, exposure follows the indoors concentration profile; the latter is highly dependent on ambient air concentrations, which in turn is affected by the intensity of outdoor emission sources, mainly traffic during the day time and domestic heating during the evening and night hours. However, Figure 5 shows that actual PM intake (as expressed by the grey line that represents the personal exposure concentration modified by the activity relevant inhalation coefficient), occurs mostly during the daytime; during sleep inhalation rate is lower, thus, despite the elevated outdoor concentrations, actual intake is lower than the one estimated by the monitoring ambient and indoor air concentrations. Tracking intra-day variability, shows that peaks of ambient PM concentrations do not necessarily reflect peaks of exposure, since the timing and the location where people are performing their activities is also important. In this way late night peaks of ambient PM concentrations (related to domestic heating emissions) do not correspond to peaks of actual intake (especially for people not operating a fireplace), since people stay indoors, most of the time sleeping (thus inhalation rate is low) in an environment with minimized indoor/outdoor interaction. On the contrary, peaks of exposure (and potential intake) are related to specific activities performed outdoor (e.g. commuting by any transportation means) or indoors (use of open fireplace), verifying the need for targeted policy interventions.

(160i) Carbon Dioxide Control Strategies for Power Generation Using Analytic Hierarchy Process

Ba-Shammakh, M. S., King Fahd University of Petroleum & Minerals
Originally presented on: 11/4/2013 17:15:00 - 17:30:00

Climate change is one of the greatest and probably most challenging environmental, social and economical threats facing the world this century. Human activities have altered the chemical composition of the atmosphere through the buildup of significant quantities of greenhouse gases (GHGs), which remain in the atmosphere for long periods of time and intensify the natural greenhouse effect. Increasing concentrations of greenhouse gases, mainly CO_2, are likely to accelerate the rate of climate change.

Power generation is a major contributor to the climate change issue. The purpose of this study is to evaluate several available technologies to control CO2 emissions in power generation. Analytic hierarchy process (AHP) technique is used as a decision support tool to find the best technology under multi-criteria. These criteria are: cost, efficiency and lifetime or duration. The technique is illustrated in a case study. The results show that fuel balancing will be recommended for CO_2 reduction if the cost is the criteria of selection. Chemical looping Combustion technology is recommended if the efficiency is the criteria of selection. The current paper covers the set of criteria weights considered typical for power plants. The results presented here are illustrative and user defined weighting is required to make this study valuable for a specific group of users.

(192a) Fabrication of Porous Carbon Nanofibers With Adjustable Pore Sizes As Electrodes for Supercapacitors

Tran, C., Drexel University
Kalra, V., Drexel University
Originally presented on: 11/4/2013 15:15:00 - 15:45:00

We report a facile method for obtaining extremely high surface area and uniformly porous carbon nanofibers for supercapacitors. Blends of polyacrylonitrile and a sacrificial polymer at different compositions have been electrospun into non-woven nanofiber mats with diameters in the range of 200–400 nm. Electrospun nanofiber mats are then subjected to carbonization to obtain porous carbon nanofibers (CNFs) as polyacrylonitrile converts to carbon and the sacrificial polymer decomposes out creating intra-fiber pores. Resultant porous CNFs exhibit specific surface area of up to 1600 m^2 g^{-1} with a large fraction of mesopores (2–4 nm). No additional chemical or physical activation process was used. We demonstrate the tunability of the pore sizes within CNFs by varying the composition. The non-woven fiber mats of porous CNFs are studied as free-standing electrode materials for supercapacitors eliminating the need for polymeric binding agents. Electrochemical measurements showed large specific gravimetric and volumetric capacitances of up to 210 F g^{-1} and 60 F cm^{-3} in 1 M H_2SO_4 at a high cyclic voltammetry scan rate of 100 mV s^{-1} due to the large fraction of mesopores. These materials retain 75% performance at a large current density of 20 A g^{-1} indicating excellent power handling capability.

(192b) Enhanced Energy Storage By Tunable Electrolyte Confinement in Structure-Directed CNT Arrays

Nicol, K. T., Mainstream Engineering Corporation
Zastrow, D., Mainstream Engineering Corporation
Hill, J. J., Mainstream Engineering Corporation
Originally presented on: 11/4/2013 15:45:00 - 16:00:00

The demand for power and energy dense storage devices is growing rapidly. While electrochemical double layer capacitors (EDLCs) offer high power density, many believe that high energy density is also obtainable through precise tuning of the electrolyte, electrode and structure of the overall cell. Multiwalled carbon nanotubes (MWNTs) are a promising electrode material for EDLCs because of their high electrical conductivity, chemical stability and exceptionally high surface area. Until single walled carbon nanotubes can be produced as solely conductive structures, dense and low-diameter MWNTs offer the best properties with regard to carbon-based EDLC electrode materials. When fabricated in anodic aluminum oxide templates, a MWNT is grown within each nanopore, resulting in an ultra-high density aligned array. This method produces a carbon electrode material with a significantly increased active surface area over commercially available materials. The energy density and capacitance of these materials can be further improved by tuning the size of the electrolyte domain around each MWNT. This is accomplished through partial and selective removal of the template, thereby confining the electrolyte domain. In typical EDLCs, the diffuse part of the double layer does not contribute to charge storage. However, we have observed an apparent charge storage enhancement in domain-confined MWNT electrodes. We will also discuss the tunable approach with regard to optimizing energy storage independent of the choice of REDOX couple or solvent.

(192c) Multifunctional Nitrogen-Rich "Brick-and-Mortar" Carbon As High Performance Supercapacitor Electrodes and Oxygen Reduction Electrocatalysts

Yu, D.,
Wei, L., Nanyang Technological University
Chen, Y., Nanyang Technological University
Originally presented on: 11/4/2013 16:00:00 - 16:15:00

Novel porous carbon materials with excellent electronic, chemical and structural properties and high nitrogen content are desirable for many applications. Here, we show the design and synthesis of a new multifunctional porous carbon material with unique “point to plane” architecture through a simple but effective activation-free procedure. Porous carbon derived from a new type of copolymer poly (vinylidene chloride-co-acrylonitrile) serves as the nitrogen-rich “mortar”. Reduced graphene oxide layers work as “bricks” with an aim to provide an open nanoscale scaffold and connect porous carbon, as well as modulate the ratio between mesopores and micropores. This new material has large surface area (957 m^2/g), high nitrogen content (6.6 at %), excellent conductivity (up to 5.1 S/cm), and favorable hierarchical meso- and microporosity. Benefiting from these intriguing features, this material shows ultrahigh specific capacitance at 361 F/g in aqueous electrolyte, far superior to that of many carbonaceous materials reported so far. The as-assembled asymmetric supercapacitors with the designed carbon material as negative electrodes and porous cobalt oxide nanorods as positive electrodes deliver a high-energy of 50.1 Wh/kg with a cell voltage of up to 1.6 V. Further, this material shows a superb electrocatalytic activity for oxygen reduction reaction in alkaline medium comparable to that of 20 wt% platinum-carbon electrodes, with much better methanol tolerant and long-term durability. We expect that this uniquely designed nitrogen-rich porous carbon material with its simple and scalable synthesis method will have great potentials for various applications in energy storage, energy conversion and catalysis.

(192d) Multiwalled Carbon Nanotubes With Tuned Surface Functionalities for Electrochemical Energy Storage

Wei, H., Lamar University
Gu, H., Lamar University
Guo, J., Lamar University
Wei, S., Lamar University
Guo, Z., Lamar University
Originally presented on: 11/4/2013 16:15:00 - 16:30:00

In the paper, surface functionalities effect on the electrochemical properties and energy storage properties of multiwalled nanotubes (MWNTs) were investigated via two different functionalized MWNTs, one modified with ether and carboxylic acid functional groups (denoted as MWNTs-EC), and the other with ester functional groups (denoted as MWNTs-E) as the supercapacitor materials. The functionalities onto the nanotubes were confirmed by Raman spectra and caused a minor enhancement in the band gap obtained from the Uv-vis spectra. Cyclic voltammetry results reveal that the capacitance of the MWNTs-EC, and MWNTs-E, was significantly increased by 41.5% and 123.2%, respectively, when compared to the as received MWNTs at lower scan rates. The energy density is also found to increase by 36.1 and 130.0% for the former and the latter, respectively, than that of the as received MWNTs-As in the galvanostatic charge/discharge measurement. The increases in both the capacitance and the energy density might arise from the enhanced surface wettability and larger pseudocapacitance contributions as confirmed by the electrochemical impedance spectroscopy. An equivalent circuit was proposed to disclose the nature of the capacitive behavior of the electrodes. The cyclic stability studies show an excellent long term cycling stability of the MWNTs, which retain 100.0% of the charge storage or discharge capacity even after 1000 charge-discharge cycles.

(192e) Three-Dimensional Core@Shell Nanostructured Array for Microscale Electrochemical Energy Storage

Mao, Y., Washington State University
Sun, X., Kansas State University
Originally presented on: 11/4/2013 16:30:00 - 16:45:00

Microelectromechanical systems (MEMS) based autonomous and biomedical devices play an ever increasing role in our society by monitoring the structural health of major civil works, the environmental presence of toxic or explosive compounds, and the physical health of human patients. Integration of robust and high power energy storage devices within these systems is important for reliable MEMS with necessary burst energy for data relay and dense components by reducing bulky external power sources. To overcome the shortcomings of conventional two-dimensional battery and supercapacitor configurations while developing more powerful energy storage devices, we focus on the development of three-dimensional architectures to take advantage of the vertical dimension-height. In this study, we designed a novel architecture composed of branched nanowire arrays, i.e. forest of nanotrees, to further boost electrochemical performance over one-dimensional nanowire arrays, inspired by branched trees with larger surface area to capture more sunlight. The prepared three-dimensional ZnO@MnO_2 core@shell branched nanowire arrays exhibit five times higher areal capacitance, better rate performance and smaller inner resistance than their nanowire array counterparts. These novel three-dimensional architectures offer promising designs for powering microelectronics and other autonomous devices on exceptionally small geometric scales.

(192f) Microfabricated Nickel Oxide Supercapacitors Based On High Aspect Ratio Concentric Cylindrical Electrodes

Armutlulu, A., Georgia Institute of Technology
Bidstrup Allen, S. A., Georgia Institute of Technology
Allen, M. G., Georgia Institute of Technology
Originally presented on: 11/4/2013 16:45:00 - 17:00:00

Supercapacitors are known as high-power energy storage devices, which narrow the application gap between the batteries and traditional electrostatic capacitors. As in the case of their electrostatic counterparts, they possess rapid charge and discharge capabilities with high degree of reversibility. Yet they can also store more energy per unit mass and/or volume which makes them superior to traditional capacitors and thus, yields a more attractive candidate for energy storage systems to be utilized in portable electronic devices. Supercapacitors can store energy electrostatically via separated charges in a Helmholtz double-layer that forms at the electrode/electrolyte interface, as well as electrochemically based on a phenomenon called pseudocapacitance that relies on reversible Faradaic charge transfer between the electrolyte and the surface of the electrode. In both energy storage mechanisms, electrode surface area plays a crucial role. In this study, our focus is on MEMS-enabled nickel oxide (NiO) based supercapacitors comprising high surface area 3D microelectrodes. The fabrication process consists of four major steps: formation of the array of high aspect ratio SU-8 pillars via backside exposure, deposition of alternating Cu and Ni layers on the SU-8 pillars by a sequential robotic electroplating process, mechanical and chemical etching of sacrificial Cu layers to obtain high surface area Ni structures in the form of concentric cylinders, and lastly conformal electrodeposition of the active material, i.e. nickel hydroxide ($Ni(OH)_2$), on the high surface area Ni backbone. Following the electrodeposition process of the $Ni(OH)_2$ layer, the structure is heated up to 300 °C in order to convert $Ni(OH)_2$ film to NiO, and the fabrication process is finalized. Performance tests for the resulting structures are carried out via cyclic voltammetry in 1 M KOH solution. Devices are cycled in excess of 500 times, and no significant capacitance degradation is observed. The areal and specific capacitance values of the supercapacitors are estimated to be 270 mF/cm^2 and 56 F/g, respectively.

(192g) Pretreatment Effects On Charge Storage of Early Transition-Metal Carbides and Nitrides

Djire, A., University of Michigan

Pande, P., University of Michigan

Sleightholme, A. E. S., University of Michigan

Deb, A., University of Michigan

Rasmussen, P. G., University of Michigan

Penner-Hahn, J., University of Michigan

Thompson, L. T., University of Michigan

Originally presented on: 11/4/2013 17:00:00 - 17:15:00

Pretreatment Effects on Charge Storage of Early Transition-Metal Carbides and Nitrides

Abdoulaye Djire[a], Priyanka Pande[a], Aniruddha Deb[b], Alice E. S. Sleightholme[a], Paul Rasmussen[ac], James Penner-Hahn[b], Levi T Thompson[ac]*

[a] Department of Chemical Engineering

[b] Department of Chemistry

[c] Hydrogen Energy Technology Laboratory

University of Michigan, Ann Arbor, MI_- 48109-2130

E-mail: ltt@umich.edu

Early transition-metal carbides and nitrides are promising candidates for use in supercapacitor electrodes due to their high electronic conductivities, high surface areas (can exceed 200 m^2/g), good electrochemical stabilities and high capacitance [1,2]. For example, the capacitance for VN has been reported to be as high as 1340 Fg^{-1} in aqueous KOH [3]. This high capacitance has been attributed to pseudocapacitive charge storage involving fast near-surface redox reactions. We observed that pretreatment of the surface to remove the passivation layer significantly improved the capacitances and electrochemical stabilities of early transition-metal carbides and nitrides in aqueous electrolytes. In this paper we present results from electrochemical characterization including cyclic voltammetry (CV), chronopotentiometry, and electrochemical impedance spectroscopy (EIS), and surface characterization using x-ray photoelectron spectroscopy (XPS) to explain changes in surface chemistry and electrochemical properties caused by pretreatment.

The early transition-metal carbides and nitrides of V, W and Mo were synthesized via temperature-programmed-reaction of their oxide precursors with 15% CH_4/H_2 or anhydrous NH_3 followed by passivation in 1% O_2/He at room temperature to form a oxygen-rich passivation layer preventing bulk oxidation on exposure to air [1]. Characterization of the structural properties was performed using nitrogen physisorption (BET surface area) and X-ray diffraction. Prior to the electrochemical characterization, the electrodes were pretreated in 0.3M NaOH aqueous solution for 1 min then rinsed with acetone, ethanol, and ultrapure water [4]. The CV was used to establish the stability windows and capacitances for these materials. The capacitance was deconvoluted into double-layer capacitance and pseudocapacitance by means of CV and EIS. Chronopotentiometry was used for further stability measurements and to determine changes in surface activities caused by pretreatment.

As shown in Figure 1, the specific capacitance for VN increased by 57% after pretreatment. This enhancement in performance suggests significant improvement in material electrochemical activities. We believe that pretreatment altered the charge storage for the carbides and nitrides. We also observed good electrochemical stability after successive cycling of the material for several thousand cycles. In general, the treated materials were more stable than the passivated materials.

Figure 2 shows the open circuit potential (OCP) or rest potentials for the passivated and treated VN in acidic medium. The OCP for the treated material is lower than that for the passivated material. We attributed this to changes in surface activities due to pretreatment. Using the Nernst equation, we found that the treated VN had a higher pseudocapacitance than the passivated VN. This pseudocapacitive charge-storage mechanism is being investigated using XPS and *in-situ* x-ray absorption spectroscopy.

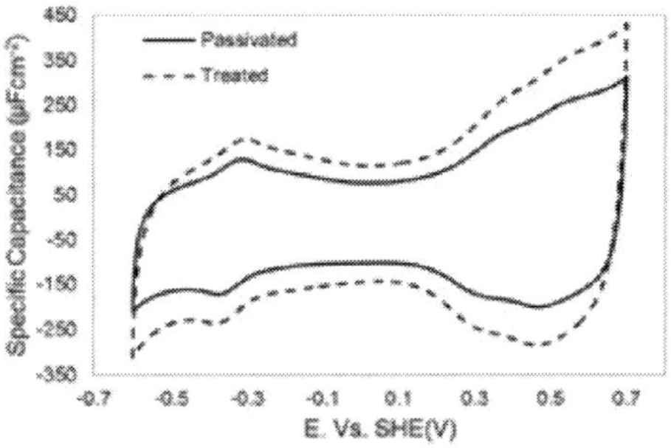

Figure 1: CV of passivated and treated VN in 0.1M H_2SO_4 at 50 mVs^{-1}.

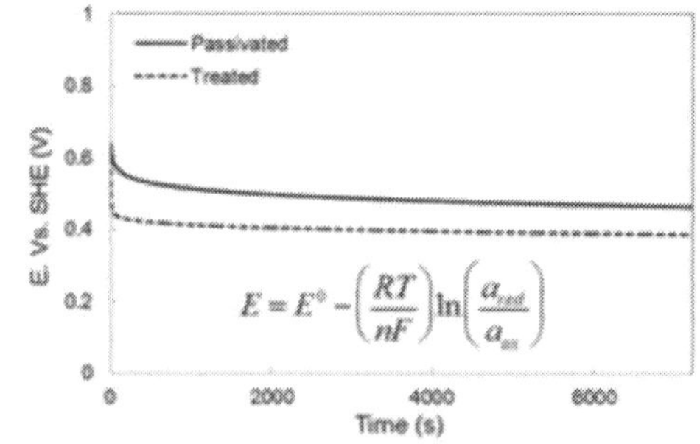

Figure 2: Rest potential of passivated and treated VN in 0.1M H_2SO_4

The physical and electrochemical properties will be described in terms of a proposed mechanism for psuedocapacitive charge storage.

References

(1) Cladridge, J. B.; York, A. P. E.; Brungs, A. J.; Green Malcolm L. H.; *Chem. Mater.* **2000**, 12, 132.

(2) Wixom, M.R.; Tarnowski, D. J.; Parker, J. M.; Lee , J.Q.; Chen, P. ?L.; Song, I.; Thompson, L. T.; *Mat. Res. Soc. Symp. Proc.* **1998,** 496, 643.

(3) Choi, D.; Kumta, P. N.; *Electrochem. Solid-State Lett.* **2005,** 8, 8, A418.

(4) Weidman, M.C.; Elposito, D.V.; Hsu, I. J.; Chen J. G.; J. Electrochem Soc.**2010**, 157, 12, F179

(192h) Doped Transition Metal Oxide Composite Electrodes for Supercapacitor Applications

Jampani Hanumantha, P., University of Pittsburgh
Kadakia, K., University of Pittsburgh
Hong, D. H., University of Pittsburgh
Poston, J., National Energy Technology Laboratory, U.S. Department of Energy
Ayyakkannu, M., National Energy Technology Laboratory
Kumta, P. N., University of Pittsburgh
Originally presented on: 11/4/2013 17:15:00 - 17:30:00

There is a dire need to provide a sustainable solution to the growing global demand for energy. It is this incessant demand that is the driving force fueling various current and emerging technologies including lithium ion batteries, supercapacitors and fuel cells. Supercapacitors occupy a unique place among these energy storage devices on account of their high power density and long cyclability. Supercapacitors are electrochemical charge storage devices capable of delivering large bursts of charge over a very short period of time. It is due to this unique ability that supercapacitors assume a very important role in the growing area of alternative energy storage systems. High surface area carbons, hydrated ruthenium oxide and MnO_2 are among the primary materials being considered for supercapacitor applications. Bulk oxide materials are traditionally hamstrung on account of their limited electrical conductivity for various electronic and electrochemical applications. In order to gain a fundamental understanding of the charge storage mechanisms in oxide materials, we deposited films of undoped transition metal oxides and doped transition metal oxides on current collectors and used various materials and electrochemical characterization techniques including X-ray photo-electron spectroscopy (XPS), cyclic voltammetry and electrochemical impedance spectroscopy to obtain a first-hand understanding of the underlying mechanisms. Effect of coating the various oxides on vertically aligned carbon nanotubes was also examined. Results of these studies will be presented and discussed.

(192i) Novel 3-D MnO2/Holey Graphene Nanostructure for Supercapacitor Applications With Enhanced Electrochemical Performances

Wang, L., Wayne State University
Meng, T., Wayne State University
Ara, M., Wayne State University
Deng, D., Wayne State University
Ng, S., Wayne State University
Originally presented on: 11/4/2013 17:30:00 - 17:45:00

As an excellent energy storage device with high-power density and long life cycle, supercapacitor has been intensely investigated in many aspects, such as electrode materials, electrode structure engineering, electrolyte compositions, separator improvements, as well as device design. Among them, one of the most urgent shortcuts is the around 10 times lower energy density compared to the traditional Li-ion batteries. This drawback has dramatically confined the applications of supercapacitors in electrical vehicles, intermittent energy storages and other occasions where both the power density and energy density are required. However, many researchers have been working on improving the energy density without significantly scarifying power density. These researches include increasing the voltage windows by employing organic electrolyte with larger voltage openings, applications of thinner and yet still mechanically strong separators, integrate materials with higher intrinsic capacitance, and so on. Despite all the effort on supercapacitor research, the current state-of-art on energy density still needs to be improved to meet the fast-growing application demanding in the emerging and potential applications. Therefore, we propose a mild method of preparing a novel 3-D MnO_2/holey graphene composite for supercapacitor electrodes. The nano MnO_2 nanorods in the composites serves as a capacitance booster due to its high intrinct capacitance and spacer to prevent 2-D graphene sheets from stacking back together. At the same time, graphene also serves both as active material to provide large surface area for charge storage and conductive agent to collect electrons generated by the surface redox reactions of semiconductive MnO_2. Holes on graphene sheets enhance the efficient accessibility of electrolyte to the surface of active materials. With this novel 3-D composite, an increasement of around 30% to 50% in capacitance is expected.

(192j) Electrochemical Energy Storage of Magnetic Carbon Nanocomposites: Role of Magnetocapacitance and Magnetohydrodynamics

Zhu, J., Lamar University
Wei, S., Lamar University
Guo, Z., Lamar University
Originally presented on: 11/4/2013 17:45:00 - 18:00:00

Magnetic field has been used in different chemical processes to improve the quality of electrodeposited films, yield of photoredox products, and oxidative reactivity of catalysts and *etc*. In this work, a small external magnetic field will be imposed on a supercapacitor unit trying to improve its electrochemical energy density. Magnetic microtubular carbon nanocomposite fabrics decorated with uniformly distributed metal oxides (iron, cobalt and nickel oxides) nanoparticles will be fabricated via both microwave-assisted and conventional annealing processes. These fabrics could serve as flexible electrodes for supercapacitors to deliver high energy density and exhibit excellent cycling stability. The effect of metal oxide species, annealing method and applied magnetic field on the capacitance of each nanocomposite material will be tested using cyclic voltammetry and galvanostatic charge/discharge measurements. The corresponding mechanism will be investigated using electrochemical impedance spectroscopy. Cycling stability of each composite with and without magnetic field will be also comparatively studied.

(192k) Synthesis and Characterization of Orange Peel As Electrodes in Li-Ion Capacitors

Arie, A. A., Parahyangan University

Originally presented on: 11/4/2013 18:00:00 - 18:15:00

Orange peel was utilized as the raw material for activated carbon-based electrodes in Li-ion capacitors. The carbon samples weres synthesized
using a chemical activation process followed by a surface modification method. The surface modification method showed no significant effect on the specific surface area, however it changed the surface property of the carbon samples. Experimental results demonstrated that the electrochemical characteristics of un-treated activated carbons had a better specific capacitance compared with those obtained by surface modification method. Based on our results, orange peel waste can potentially be used as a cheap raw material for the mass production of activated carbons as electrodes in Li-ion capacitors.

(198a) Biomass Resources and Current Status Of Biofuels In China

Yuan, Z., Guangzhou Institute of Energy Conversion, Chinese Academy of Sciences
Originally presented on: 11/4/2013 15:15:00 - 16:15:00

With the increase in population and development of industry, energy crisis and environment pollution caused by the fossil energy system have turned out to be two serious problems in the world. In China, the situations appear to be more severe according to the sharp rise in the importation of petroleum in the recent ten years(Fig.1). Biomass is a promising energy source which will be one of the imperative components contributing to the future energy system due to its renewability, cleanness and CO_2 neutral. Laws concerning renewable energy have been issued and some projects were funded in order to promote bioenergy development in China. In this paper, potential biomass resources in China are analyzed(Table 1), current status of biofuels including biodiesel, cellulosic ethanol, liquid alkane and butanol are summarized. Relevant laws and rules are also introduced. Furthermore, development tendencies of biomass energies are predicted.

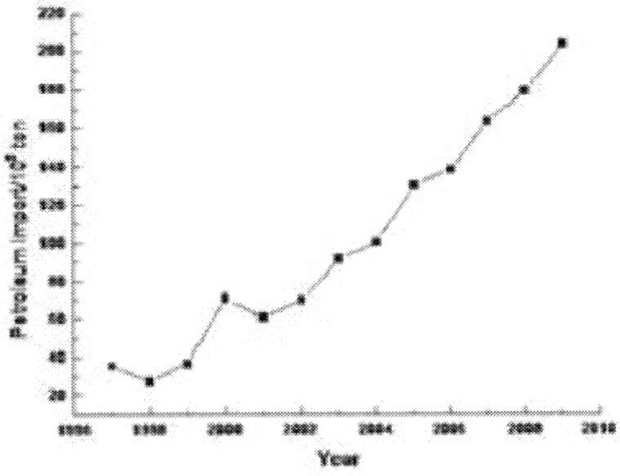

Fig.1 Petroleum importation in China from 1997 to 2009

Table 1. Potential analysis of biomass resources in China (unit: 10^8 tons of standard coal)

		2006	2010	2020	2030	2050
Biomass resources in existance		5.0	5.0	5.0	5.0	5.0
	In application	2.2	2.0	1.8	1.5	1.0
	Available	2.8	3.0	3.2	3.5	4.0
Growth in biomass resources		0.0	0.3	2.3	4.7	5.8
	Agricultural and forestry wastes		0.2	1.4	3.0	4.0

	Planting of energy crops		0.1	0.4	0.7	0.8
	Planting on marginal soils		0.0	0.5	1.0	1.0
Total		**5.0**	**5.3**	**7.3**	**9.7**	**10.8**

(198c) Air Quality Impacts of Natural Gas Production and Use

Allen, D. T., The University of Texas at Austin

Originally presented on: 11/4/2013 17:15:00 - 18:15:00

Natural gas production is increasing dramatically, enabled by technologies such as horizontal drilling and hydraulic fracturing of shale formations. The implications of this transformation in the production of natural gas and associated hydrocarbon liquids will clearly be profound in the energy sector. Much less appreciated, however, is the way in which the widespread availability of natural gas and natural gas liquids will likely transform electricity generation, chemical manufacturing, and agriculture. These changes will have complex environmental footprints. As a case study of these footprints, this presentation will examine air quality impacts associated with natural gas production and use. Data and modeling on emissions and impacts of photochemically active air pollutants, toxic air pollutants and greenhouse gas emissions will be described. In addition to addressing the direct atmospheric impacts of expanded natural gas production, indirect effects will also be described, using a case study of coupled natural gas production and electricity generation. Widespread availability of shale gas, and limited capacities for transporting shale gas to global markets, can drive down natural gas prices, which in turn, can impact the use of natural gas in electricity generation. Natural gas production in Texas and the Texas grid will be used to examine these indirect consequences of expanded natural gas availability.

(198d) Disasters Risks Reduction and Sustainable Development

Robinson, K.,

Originally presented on: 11/4/2013 18:15:00 - 19:15:00

Although the issue of disasters reduction was not included in the Millennium Development Goals (MDGs) framework, it is now evident and widely recognized that disasters, in particular water disasters have made the MDGs attainment more difficult and costly and hampered poverty reduction. They have destroyed human, social and physical capital and derailed social and economic development with long lasting impacts on economic growth. At the same time, the drivers of disasters have not been addressed properly so far. Extreme weather events will increased during the next 15 years and beyond. Within this context, Post-2015 Development Agenda will have to consider the dimension of disasters risks reduction as across-cutting issue that also requires a long-term planning perspective and integration across sectors.

Urban dwellers will increase in numbers and mega-cities are expected to emerge. Particularly vulnerable, however, are the poorest dwellers especially women.

The presentation will address these issues and introduce the United Nations Plan of Action on Disaster Risk Reduction for Resilience including a gender perspective.

(215a) Supercritical Water Oxidation: the Next Generation Process to Incineration

Duri, B.,

SUPERCRITICAL WATER OXIDATION AS THE NEXT GENERATION IN HAZARDOUS WASTE TREATMENT

Bushra Al-Duri[1][*], Faihan Alsoqyani[1] and Iain Kings[1]

[1] School of Chemical Engineering, University of Birmingham, Edgbaston – Birmingham B15 2TT, UNITED KINGDOM. Email: B.Al-Duri@bham.ac.uk

Supercritical water oxidation (SCWO) is an advanced technology, which represents a suitable alternative to incineration for the destruction of chemically stable wastes. SCWO is an intense, efficient and highly rapid process that converts hazardous hydrocarbons to benign gases and liquid water, with no emissions (dioxins, NO_X and SO_X), and no ashes; eliminating the problem of landfill. Compared to wet air oxidation it is multifold more rapid and more efficient as it takes place in a one-phase environment, based on the special properties of supercritical water [1].

Nitrogen (N) containing hydrocarbons are frequently encountered in wastes. Their abatement is specifically challenging due to formation of the recalcitrant ammonia the reason why, this work investigates the SCWO of N, N dimethyl formamide (DMF); and aliphatic hydrocarbon found in chemical industries. This work focuses on the finding that under the same operating conditions alcohol addition improves the conversion of organics and nitrogen towards benign gases[2]. SCWO of DMF in the presence of isopropyl alcohol (IPA) is investigated under a range of temperatures (673 – 800*K*), oxidant ratios (0.8 – 3*SR*), initial DMF concentrations (5 – 20*mM*) and different IPA/DMF ratios. Results are presented in the form percentage organics removal, and ammonium yield and total liquid nitrogen yield. It shows the positive effect of alcohol addition on the SCWO process.

The reaction kinetics is also investigated, showing that the activation energy of 65 kJ/mol is significantly lower in the presence of IPA than without (155 kJ/mol). The reaction rate constants and global rate expression, showing the reaction order is also given by:

-rA=3.15x107 exp64,9958.314TTOCO20.297 $-r_A = 3.15x10^7 \exp\left(\frac{64,995}{8.3147}\right)[TOC][O_2]^{0.297}$

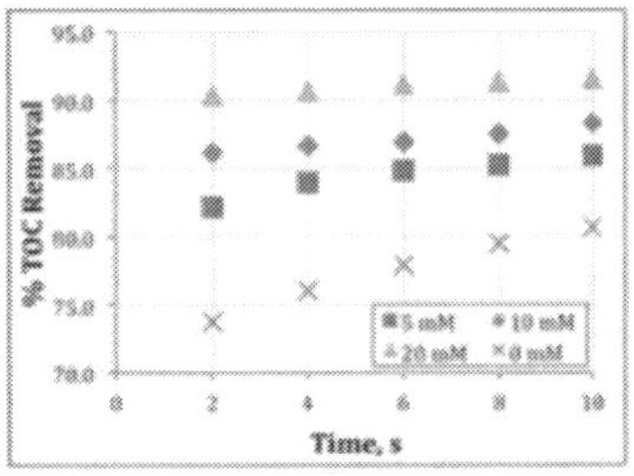

Figure 1: SCWO of 10mM DMF at 673K, 1SR adding 5, 10 and 20mM IPA.

[1] Verianshyah B. & Kim J. D.; *Journal of Environmental Sciences*, **19** (2007) 513–522.

[2] Oe T; Suzugaki H; Naruse, I; Quitain A. T; Daimon H & Fujie K; *Ind. Eng. Chem. Res.*, **46 No. 11** (2007) 3566–3573.

---

[*] Corresponding author.

(215b) Different Behavior of H2O2 in Aldicarb and Hexazinone Oxidized By O3/H2O2

Shaofeng, L.,
Shaofeng, L.,
Shaofeng, L.,

Different Behavior of H_2O_2 in Aldicarb and Hexazinone Oxidized by O_3/H_2O_2

Abstract: The degradation of hexazinone and aldicarb by direct ozonation and an advanced oxidation process (AOP) of O_3/H_2O_2 were investigated in this study with the focus on the oxidation mechanism by identifying the hydrogen peroxide consumption during the oxidation process of the two chemicals. The results show H_2O_2 could enhance the removal rate of the triazine herbicide hexazinone, and it consumes along with the variation of removal rate in the light of different pH levels. The addition of H_2O_2 contributes little to the removal of the thiocarbamate herbicide aldicarb and the H_2O_2 content kept constant throughout the degradation process. Tert-butyl alcohol (TBA) effectively scavenges the OH radical effectively for hexazinone, but had no effect on the removal rate of aldicarb. Aldicarb removal mainly attribute to direct oxidation of the ozone molecule in both O_3 (97%) and O_3/H_2O_2 (96.76%) systems .Moreover, sole O_3 could hardly oxidize hexazinone whereas OH radicals contribute respectively 74.7% and 97.5% of removal in O_3 system and O_3/H_2O_2 AOP. All of these findings which are contrast to the conclusions of formers suggest that the mechanism of OH radical generation and the chain reaction in O_3/H_2O_2 AOP should be further discussed.

Key words: hexazinone; aldicarb; O_3/H_2O_2; direct and indirect oxidation; H_2O_2 consumption

(215c) Enhancing Selectivity of Salicylic Acid Degradation By Using Molecular Imprinted Doped TiO2 Under Solar Light

Wang, Z., Jiangnan University
Li, W., Jiangnan University
Liu, X., Jiangnan University

The doped TiO_2 photocatalysts have been widely applied for the wastewater treatment because of high photocatalytic activity and chemical stability under solar light. However, poor selectivity of doped TiO_2 photocatalysis is unfavorable to photocatalytic removal of highly toxic low-level organic pollutants in polluted waters in the presence of other less toxic high-level pollutants. A new strategy of increasing this selectivity is the surface modification of doped TiO_2 via coating a thin layer of molecular imprinted polymer (MIP), which provides molecular recognition ability toward the target molecules. Firstly, we prepared a doped TiO_2 sample which has high photocatalytic activity under simulated solar light in this paper. Then, o-phenylenediamine was taken as the functional monomer and salicylic acid was taken as the target molecular. The polymerization was initiated photocatalytically by UV light illumination on doped TiO_2 without using any other chemical initiator or cross-linker, causing in situ coating of the imprinted layer onto the surface of doped TiO_2. MIP-doped-TiO_2 photocatalyst samples, which had the special interaction between the target salicylic acid and the footprints polymer via the functional groups (-NH_2, -OH and –COOH), were prepared. Lastly, an aqueous solution of salicylic acid (20mg/L, target pollutant) and phenol (100mg/L, nontarget pollutant) was chosen as the simulated wastewater system, and MIP-doped-TiO_2 was adopted as the photocatalyst of selective degradation of salicylic acid. The results indicated that the degradation rate of salicylic acid could be increased 70% by using MIP-doped-TiO_2, comparing with neat doped TiO_2, under simulated solar light. The degradation rate of phenol was decreased 90% under the same conditions. The selectivity of salicylic acid degradation by using MIP-Doped-TiO_2 under simulated solar light was remarkably increased.

(215x) Optimal Process Conditions for Preparation of Porous Carbons and Methylene Blue Adsorption

Srinivasakannan, C., The Petroleum Institute

Porous carbon is the most popular adsorbent that is being widely used in variety of industries for different applications. The present work focuses on the process optimization of porous carbon preparation from date seed pits. The Response Surface Methodology (RSM) technique with Box-Behnken Method (BBM) was utilized with the response variables being the yield and BET surface area. The influencing parameters selected were activation temperature, Impregnation ratio (IR), and activation time. The optimum conditions were identified to be an activation temperature of 500 ºC, IR of 2 and activation time of 75 min with the resulting yield and BET surface area being 46 % and 838 m^2/g, with corresponding average pore volume of 1.07 cc/g and average pore diameter of 1.69 nm. It also confirms that the RSM approach is most suitable for optimization of process conditions for preparation of porous carbon.

In order to assess the suitability of the carbon for adsorption of macromolecules, methylene blue (MB) adsorption studies were assessed by varying the initial concentration and the adsorption temperature. The monolayer adsorption of MB using the Langmuir adsorption isotherm was estimated to be 345 mg/g for optimized process condition, while the maximum MB adsorption capacity of 455 mg/g, corresponds to the porous carbon with highest of BET surface area. The net positive charge on surface of porous carbon decreased with increase in pH, resulting in an increase in MB uptake, due to the reduction in repulsion between the porous carbon and MB.

(215ad) Integrated Bioenergy Farming System Using Communal Waste WATER - A NOVEL Approach To Carbon Mitigation

Praveen Kumar, R., National Institute of Technology, Surathkal
Bharathiraja, B.,
Chozhavendhan, S.,
Regupathi, I.,
Moscow, P.,

Increasing demand and depleting resources leads to the search of new and clean energy source, on the other hand treatment and reuse of waste water generated becomes a big challenge. Researchers around the world are working to sort out this problem and to find suitable solution. This paper discusses about the possibilities in integrated energy farming for small communities using communal waste water as a source to produce alternate bioenergy. Communal waste water can be categorized into back and grey water. Black water is the waste water generated from toilets, it contains very high amount of organic load with high BOD and COD, as a result it can be used to produce Biogas and thereby electricity using biogas. Grey water contains low amount of BOD, generally inorganic load is higher in grey water. This waste water was secondarily treated (primary settling and secondary biological processing via trickling filter). This secondarily treated grey water is rich in nitrogen, phosphorus and other nutrients; hence it needs a tertiary treatment to eliminate these compounds. In this study tertiary treatment was carried out using the microalgae *Botryococcus braunii* grown in laboratory, the microalga *Botryococcus braunii* contains high amount of oil which can be used for production of biodiesel. Experiments were carried out in laboratory scale Erlenmeyer flasks which was filled with secondarily treated grey water inoculated with microalgae *Botryococcus braunii*. The pH of the grey water was maintained under 8.0 in room temperature, 12 h light/12 h darkness cycles were maintained, light intensity has been maintained higher than 1.2 ± 0.2 klux. The compounds initially present in treated grey water like nitrogen and phosphorus is reduced considerably and it is used for landscaping purpose. The results show that the oil productivity of algae using communal waste water was higher than that of fresh water. This study reveals that an integrated energy farming system can be a potential source for energy in small communities.

(215d) Fabrication of Reusable Ag/BiVO4 for Water Disinfection

Yoosefi Booshehri, A., NTU
Xu, R., NTU

In this study photocatalytic disinfection of *Escherichia coli* in water under visible light (λ>400) was studied. Bismuth Vanadate with deposited silver particles (Ag/BiVO4) was fabricated and used as a photocatalyst. Silver nano particles were deposited on the surface of BiVO4 by simple photo-deposition technique. Ag/BiVO4 catalysts were characterized by X-ray diffraction (XRD), scanning electronic microscopy (SEM), energy dispersive spectrometer (EDS). Silver NPs on the surface of BiVO4 enhance its activity significantly. It was found that a concentration of 1 mg/mL of 8% Ag/BiVO4 disinfect water completely after 3 hours. Silver nano particles on the BiVO4 were found out to be very stable and the catalyst sustained its activity in frequent usage. Due to very good precipitation performance of Ag/BiVO4, it is a very good candidate for usage water purification systems. Tests were done both in dark and under light to ensure the antibacterial activity is due to photocatalytic activity and not because of silver.

(215e) Visualization of Coupled Mass Transfer and Reaction Between Plasma and Liquid Phase in a Hybrid Gas-Liquid Dielectric Barrier Discharge Reactor for Water Purification

Feng, X.,
Shao, T., Tsinghua University
Yan, B., Tsinghua University
Lu, W., Tsinghua University
Wang, W., Tsinghua University
Cheng, Y., Tsinghua University

The interaction of cold plasma with liquid water has been widely applied as an integrated approach of advanced oxidation processes (AOPs) for water purification. A novel approach, i.e., the reactive planar laser induced fluorescence (reactive-PLIF) technique, was proposed to visualize the coupled process of mass transfer of chemically active species (O_3, H_2O_2, •OH, •H, •O, etc.) and instantaneous oxidation reaction between plasma and liquid phase in a hybrid gas-liquid dielectric barrier discharge (DBD) reactor, by quantitatively recording the dynamic change of concentration field of fluorescence dye (e.g., Rhodamine B) in the liquid layer during its decoloration process. Firstly, the decoloration dynamics were revealed under different discharge voltage in oxygen atmosphere. The increase of the discharge voltage can give rise to the increase of the density of DBD plasma filaments, which impacted on the surface of the liquid layer. As a result, more reactive species (O_3, H_2O_2, •OH and other radicals detected by optical emission spectroscopy) were generated, together with the intensification of the convective transport in the liquid layer. The induced flow patterns were confirmed by particle image velocimetry (PIV) measurements, indicating that the convective transport dominated the distribution of the reactive species. The above results demonstrated that the significant enhancement of the Rhodamine B decoloration efficiency with the increase of the discharge voltage can be attributed to the simultaneous intensification of the mass transfer and reaction. Secondly, the decoloration dynamics in the discharge atmosphere of inert gas (e.g., argon) were revealed in comparison with the oxygen discharge at the same discharge voltage. The argon discharge showed denser DBD filaments so as to yield higher intensity of radicals (•OH, •H, •O, etc.) and H_2O_2 concentration. However, the Rhodamine B decoloration efficiency in the oxygen discharge was remarkably higher than the argon discharge because of O_3 formation, which indicated that O_3 plays more important role in Rhodamine B degradation due to the reactive species. This study provided straightforward analysis on the mechanism of the interaction between discharge plasma and polluted water with focus on the fundamental plasma physics and plasma chemistry, which will be beneficial to develop highly-efficient cold plasma technique for AOPs.

(215ac) Removal of Phosphate and Arsenic From Aqueous Solution By Metal(hydr-)Oxide Coated Sand

Huang, Y., University of California, Santa Barbara
Yang, J. K.,
Keller, A. A., University of California, Santa Barbara

Phosphate and arsenic are generally found as contaminants in water bodies due to various anthropogenic sources, such as mining activity, discharges of industrial and domestic wastewater as well as agricultural drainage. Both of them can cause wide-spread environmental problems: phosphate can contribute to eutrophication in aqueous systems as well as posting a great threat to aquatic environments, and arsenic has adverse effect on the entire biota, which is known as a primary concern to humans exposed to such systems. Since phosphate and arsenic have same chemical structure as well as have similar chemical behavior, we developed cheap and easily available adsorbents such as iron and manganese oxide-coated sands to treat them efficiently. In this study, the effects of the solution pH and sand sizes on the coating efficiency and mineral type of both iron and manganese during the preparation of manganese-coated sand (MCS), iron-coated sand (ICS) and iron and manganese-coated sand (IMCS) as well as on the removal of phosphate and arsenic were evaluated. Iron and manganese-coated sand can achieve 100% removal of phosphate and 90% removal rate on arsenic in certain environmental conditions. Removal of phosphate and arsenic under different conditions were also evaluated to provide the optimized solution.

(215m) Experimental Study On Iodine Adsorption By Silve-Exchanged Mordenite

Lin, R., Syracuse University
Tavlarides, L. L., Syracuse University

Iodine-129 is one of several radioactive iodine isotopes formed in fission reactions. Due to its extremely long half-life of about 16×10^6 years, effective removal and immobilization of iodine-129 species from spent nuclear fuels represents one of the critical barriers to nuclear waste management. The state-of-the-art of iodine capture from spent fuel reprocessing off-gases is adsorption by silver-exchanged mordenite. Development of sorbents and sorption processes for off-gas treatment requires advanced predictive modeling tools for simulating dynamic adsorption processes. Development of modeling tools further requires fundamental experimental adsorption equilibrium and kinetic data. In this work, a continuous flow adsorption system was developed for single-pellet gas adsorption studies. Experiments are in progress to gather equilibrium and kinetics data for iodine adsorption on silver-exchanged mordenite. Experiments are being conducted at 150-200 ^{o}C and at varying iodine concentration below 100ppm. In this presentation, design of the experiment and preliminary adsorption equilibrium and kinetics results will be discussed.

(215o) Study On the Wastewater Reuse By Combination of Thermophilic Anaerobic Bioreactor With Forward Osmosis (FO)/Membrane Distillation (MD) Integrated Module

Zhang, J. S., Singapore Membrane Technology Centre(SMTC)

Fane, A. G., Nanyang Technological University

Kwan, J. Z. X.,

Goh, S.,

Study on the wastewater reuse by combination of thermophilic anaerobic bioreactor with Forward Osmosis (FO)/Membrane Distillation (MD) integrated module

Zhang Jinsong,* Kwan Zheng Xin Joseph, Goh Shuwen, Anthony Gordon Fane*

Singapore Membrane Technology Centre(SMTC),Nanyang Environment and Water Research Institute(NEWRI),Nanyang Technological University,1 Cleantech Loop, CleanTech One, #06-08, Singapore 637141

*Corresponding author: *jszhang@ntu.edu.sg, Tel: 65-65921835, Fax: 65- 67910756*

Abstract

The organic loading and temperature of wastewater from industries such as chemical, petroleum and pharmaceutical are typically high. These characteristics make them potential candidates for anaerobic bioprocess, which is able to process high strength wastewater and generate energy in the form of methane. Due to the high temperature of wastewater, applying thermophilic anaerobic digestion has additional benefits over mesophilic digestion, including a higher COD removal efficiency, faster degradation rate, and lower methane solubility in wastewater. However, due to the incomplete degradation of organics, the effluent from the anaerobic bioprocess usually contains a high concentration of organics, which makes it difficult to reuse or dispose according to the discharge standard.

In this study, a thermophilic upflow anaerobic sludge blanket (UASB) combined with the forward osmosis (FO)/membrane distillation (MD) integrated module has been demonstrated for the first time. The schematic diagram of anaerobic bioprocess with FO/MD module is shown in Figure 1. The MD process uses microporous hydrophobic membranes that allow water vapour to pass through the dry pores, while retaining the non-volatile organics and inorganics. If the MD membrane was directly used to treat the UASB effluent, serious biofouling, scaling and wetting will occur. The FO process is a membrane separation process which uses semi-permeable membranes that allows water molecules to pass through while retaining bigger size solutes. The driving force for water transport is an osmotic pressure gradient achieved by using an osmotic ?draw solution' (DS). The FO/MD integrated module captures the complementary advantages provided by FO and MD in a single step. By applying the FO membrane process at the anaerobic bioprocess effluent side, nearly all the organics and inorganics can be rejected at a low fouling rate. This makes the FO membrane a perfect barrier to prevent MD fouling and wetting. High temperature feed water will also mitigate the concentration polarization effects usually observed in FO processes, resulting in enhanced FO membrane flux. The draw solution is simultaneously regenerated in-situ via the MD process, which completely retains the solute (such as NaCl) in DS while allowing water to diffuse across the membrane into the permeate reservoir. The FO/MD module allows us to achieve a product that is of distilled water quality and ready for reuse in a single step.

Our experiments have shown that the anaerobic bioreactor with FO/MD module had a high rejection for

TOC, with a calculated average of 99.55 ± 0.24 %. TOC levels in the permeate remained fairly constant below 0.5 mg/L. Permeate conductivity was stable at 5 μS cm^{-1} , which is comparable to the RO permeate of wastewater reclamation. The average composition of the biogas generated from the UASB process was found to be 67.4 % methane, with the highest percentage being 83.6 %. Optimally, the bioprocess is able to produce a maximum of 0.35 L methane for every gram of chemical oxygen demand (COD) that is degraded.

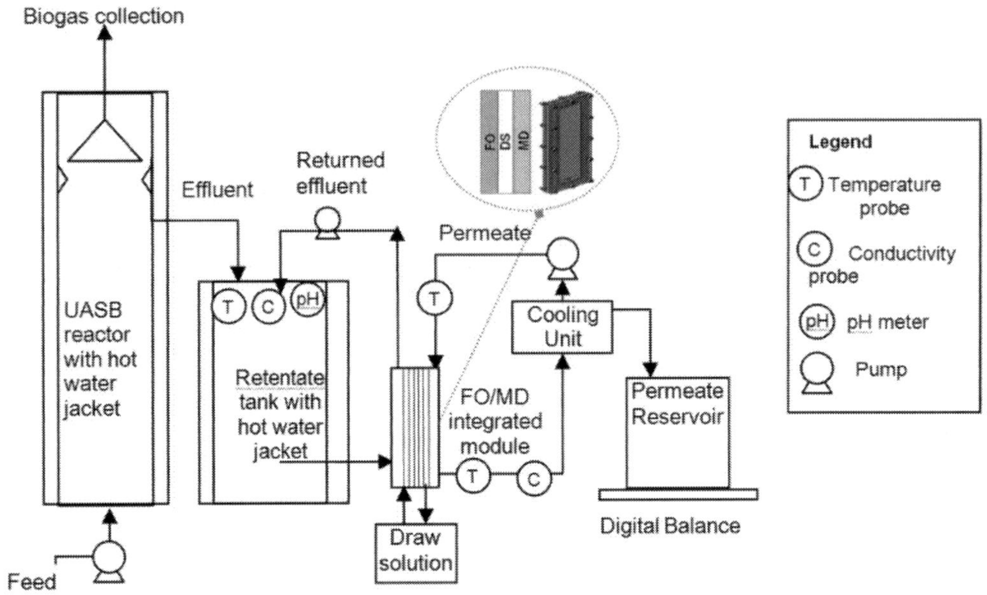

Figure 1 Schematic diagram of side-stream UASB-FO/MD system

In conclusion, the integrated process of combining the anaerobic bioprocess with the FO/MD module is a state-of-art technology and a viable alternative to wastewater reclamation. This technology appears to be particularly well suited for high-strength industrial waste streams reclamation due to the potential to achieve high quality water product with lower energy consumption and higher methane production.

Keywords: Thermophilic, Upflow Anaerobic Sludge Blanket (UASB), Forward Osmosis (FO), Membrane Distillation (MD), FO/MD integrated module

(215p) Removal of Oil From Oily Waters By Chitosan Microspheres: Batch Studies

Grem, I. C., Federal University of Rio de Janeiro
Queiros, Y. G.,
Garcia, B. S.,
Mansur, C. E., Federal University of Rio de Janeiro

One of the mayor problems of the petrochemical industry is the great amount of wastewater produced. The purification of the wastewater is an important global problem, mainly due to the restricted quantities of water that can be safely used directly and the high cost of purification installations. In recent years suitable technologies for treatment of wastewaters and other effluents has attracted great interest, due to more rigorous laws and regulations. Some of the common pollutants are phenols, dyes, detergents, insecticides and heavy metals. The nature of pollutants depends on the source of generation and varies from place to place. These pollutants are often toxic and cause adverse affects on human and animal life if presente above certain concentration levels. Unlike free or floating oil spilled in the sea, lakes or rivers, most of industrial wastewaters contain oil-in-water emulsions among their basic contaminants. Among several chemical and physical methods, adsorption process is one of the effective methods widely used in wastewater systems. Chitosan is a natural polymer that has been used in a variety of practical fields including wastewater management. Chitosan among its various properties, presents adsorptive capacity. Crosslinked chitosans have been used to adsorb mainly reactive dyes, some cations and anions, and other contaminants, but apparently no mayor studies have been done to adsorb oil from wastewater using chitosan.

In this presentation we will discuss the efficience of chitosan microspheres as adsorbents in the removal of crude oil from synthetic oily waters. For this, bacth studies were done.

First of all, we synthesized the chitin from residues of shrimp in order to obtain chitosan. Chitosan is produced by the deacetilation of chitin. The molar mass and acetylation degree of the chitosan obtained were determined by rheology and NMR H^1, respectively. We prepared chitosan microspheres crosslinked with sodium tripolyphasphate. For this, chitosan was dissolved in a acid acetic solution and the solution was slowly dropped into a tripolyphosphate solution The microspheres were analised by optical microscopy, scanning electron microscopy and ASAP. We also prepared the synthetic oily water from the dispersion of oil in water under high shear rates. After that, we made some batch adsorption studies. We studied the effect of adsorbent mass, pH, oil concentration, temperature and time. The amount of oil in these experiments was measurement by a fluorimeter, with n-hexane being used as the oil-extraction solvent. These studies were realized in a shaker with a constant stirring of 100 cicles/minutes.

With the results obtained we could observe the capacity and efficiency of the material, as well as the best conditions in which it should be used.

(215g) Study On Supercritical Water Oxidation of Quinoline Wastewater

Li, Y., Tianjin University
Zheng, B.,
Ma, J., Tianjin University
Li, G., Tianjin University
Dong, X., Tianjin University

Complex N-containing organic wastewater, which is hazard and will cause serious environment problems, is increasingly generated from industry. Quinoline is a typical N-containing compound, which is highly refractory. Supercritical water oxidation (SCWO) is a forthcoming technology for the treatment of wastewater containing non-biodegradable compounds. In this paper, the supercritical water oxidation of quinoline was investigated with hydrogen peroxide as oxidant.

The operating conditions of SCWO process were studied by the response surface methodology (RSM) and central composite design (CCD). The effects of temperature, pressure, residence time and excess oxidation ratio on the organics removal were investigated. The optimum condition for the wastewater treatment was obtained. The optimum temperature, pressure, residence time and excess oxidation ratio were found to be 441℃, 25MPa, 60s and 2.48 respectively.

The liquid and gas products were analyzed after experiment. The carbon of quinoline was mainly oxidized into carbon dioxide and small amount of carbon monoxide. Nitrogen atom of quimoline was mainly transformed into ammonia, nitrogen and little amount of N-containing compounds. It was found that increasing the temperature was benefit for the generation of gas and ammonia. Pressure had little effect on the content of products. The increase of the residence time and the excess oxidation could benefit the conversion of carbon monoxide into carbon dioxide. Quinoline oxidation in supercritical water was investigated over the temperature range of 440~460℃and the pressure range of 25~27MPa, with the residence time of 60~80s and excess oxidation ratio of 2~3. At these conditions, 70~80% of carbon were oxidized into carbon dioxide, while 10~15% of carbon was oxidized into carbon monoxide. 25~30% of nitrogen was converted into ammonia and 55~60% of nitrogen was changed into nitrogen. The TOC removal was between 94% and 98%.

(215h) Photoelectrocatalytic Performance of TiO2/SnO2 Composite Thin Film Electrode

Wei, Y.,
Zhou, X., Nanjing University of Technology

TiO_2/SnO_2 composite thin film was coated on one side of Ti/SnO_2- Sb-CeO_2 electrode by a simple sol-gel process route. The immobilization of catalyst as a film on a suitable substrate is the reasonable way to minimize post-treatment and increase the possibilities of real application.The TiO_2/SnO_2 composite thin film electrode have been employed for electrochemically assisted photocatalytic degradation of methyl blue. The photoelectric decomposition of methyl blue (MB) was used as a model system to investigate the effection of photocatalysis(PC), electrocatalysis(EC)and photoelectrocatalysis (PEC). The photocatalytic activity can be further enhanced by electric-field-assisted technology. The results reveal that the synergetic decomposition efficiency of photoelectrocatalysis is greater than that of the combine of photocatalysis and electrocatalysis. Moreover, the key factors affecting the photoelectrocatalytic efficiency were discussed, including the applied bias potential, light intensity and initial concentration of MB. Additionally, the assumptive mechanism for the enhanced photocatalytic activity of TiO_2 with SnO_2-doped were proposed.

(215y) Hydrothermal Pre-Treatment for Mesoporous Carbon Synthesis: Enhancement of Chemical Activation

Jain, A., National University of Singapore
Srinivasan, M. P., National University of Singapore
Balasubramanian, R., National University of Singapore
Liu, Z.,

A hydrothermal pre-treatment process is used to enhance activation of biomass during the production of mesoporous activated carbon. $ZnCl_2$ as the activating agent was incorporated into the biomass (coconut shells) during a hydrothermal pre-treatment step. The pre-treatment was followed by pyrolysis accompanied by physical activation. The resultant mesoporous activated carbons possessed higher total surface area and a greater degree of mesoporosity compared to the biomass that was pyrolysed without hydrothermal treatment. The cause of higher mesoporosity is inferred to be the more conducive environment for accessibility of $ZnCl_2$ to the biomass provided by the hydrothermal treatment due to reduced diffusion resistance. In addition, the hydrothermal environment facilitated generation of oxygen-containing functional groups that contributed to enhanced activity of $ZnCl_2$. Up to 67% increase in the mesoporous surface area has been achieved with the inclusion of the pre-treatment step. Analogously, the amount of $ZnCl_2$ required to deliver the same performance in the absence of the pretreatment step is an additional 50%. The effect of hydrothermal treatment with and without $ZnCl_2$ along with sequence of $ZnCl_2$ addition has been investigated in order to differentiate between the hydrothermal effect and incorporation of $ZnCl_2$ during hydrothermal treatment. The mesoporous activated carbons were tested for adsorption of textile dyes and possessed high adsorption capacities of up to 526 mg/g and 630 mg/g for methylene blue and erythrosine red, respectively. The incorporation of the hydrothermal pre-treatment is an important step in developing processes that make efficient and effective use of activating agents.

(215f) An Integrated Solar Photocatalysis - Organic Membrane Separation and SBR Process for the Treatment of Pharmaceutical Wastewater

Shen, C., Tianjin Polytechnic University
Xie, L., Tianjin Polytechnic University
Zhang, Y., Tianjin Polytechnic University
Tan, X., Tianjin Polytechnic University
Zou, C.,
Wang, J.,

1. Introduction

Due to strong oxidation ability, no secondary pollution and mild reaction conditions [1], photocatalysis treatment of pharmaceutical wastewater with refractory organic is becoming attractive recently, and related research have been carried out widely. There is, however, little information about the solar photocatalysis degradation of pharmaceutical wastewater available, and the majority of published solar photocatalysis researches are focused on the compound parabolic collector(CPC) reactor with immobilization of TiO_2 on solid supports, which exists mass transfer limitation compared with suspension photocatalytsis, to eliminate the difficult separation and recycling of TiO_2.

Based on high separation efficiency and easy operation for organic membrane separation, the paper firstly introduced a novel type of three-phase fluidized bed equipment of solar photocatalysis-organic membrane separation (the equipment) which TiO_2 can be recycled, then the combined process of the equipment with SBR (Sequential Batch Reactor) for pharmaceutical wastewater treatment were investigated.

2. Experimental

2.1 Introduction of the equipment

The equipment consists of photocatalytic reaction zone (PRZ) and membrane separation zone (MSZ), which PRZ is comprised of bottom water-tank, casing tube reactor, CPC, upper water-tank and PRZ aeration device, there also are wastewater inlet and sludge outlet in bottom tank, UV lamp (main wavelength of 254 nm, power of 15 W) which is set in the centre of tube reactor. MSZ is comprised of membrane separator shell, submerged membrane module and MSZ aeration device. The water outlet is connected with membrane module. Membrane separator is connected with upper and bottom water-tank by connecting pipe and circulating pipe with air rising jet, respectively. CPC and tube reactor were facing the equator with the certain angle to horizontal plane, which equals local latitude.

Wastewater entering into bottom water-tank rapidly mixes with TiO_2 to form three phase fluidized bed under PRZ aeration flow. When the wastewater, containing air bubble and TiO_2, flows through shell sides of tube reactor, photocatalytic degradation reaction happens under UV irradiation. Air bubbles escape out of the wastewater when degraded wastewater flows into upper water-tank. Wastewater with only TiO_2 flow back bottom water-tank by circulating pipe again after they come into membrane separator by connecting pipe. Obviously, the wastewater in the equipment is always in circulating flow state. TiO_2 is intercepted when the wastewater in membrane separator is passing through membrane module and the intercepted TiO_2 flows back to the bottom water-tank by circulating flow. Air rinsing makes deposited TiO_2 suspend again to keep TiO_2 suspension concentration invariable [2].

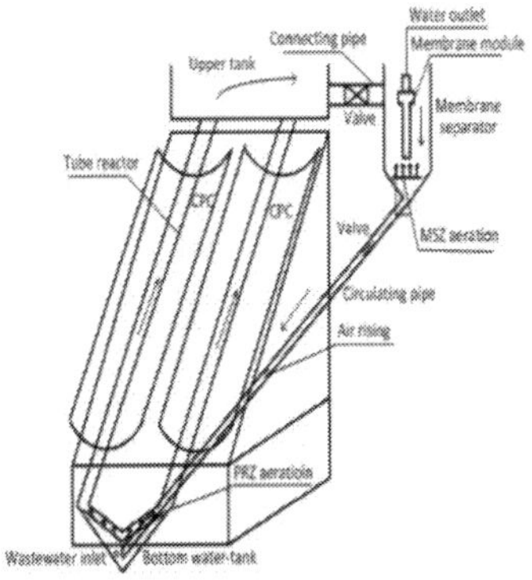

Fig. 1. Solar photocatalysis-membrane separation equipment

2.2 Experimental methods

The pharmaceutical wastewater was first pre-treated by flocculation with Poly Aluminium Chloride (PAC), and then the pre-treated wastewater obtained under the appropriate flocculation conditions was diluted 50 times and used to photocatalytic degradation experiment. The photocatalyzed wastewater was also treated by SBR under the aeration time of 8 h, anaerobic stir of 5 h and hydraulicretention time of 15 h.

3 Results and Discussion

3.1 Pretreatment of pharmaceutical wastewater

After flocculation pre-treatment with PAC, the COD_{Cr}, chromaticity and turbidity of pharmaceutical wastewater decrease from 294500 mg/L, 4100, 95NTU to 266000 mg/L, 3150, 31NTU respectively, and the appropriate PAC dosage is 400 mg/L. Although the photocatalytic degradation rate of the pre-treated wastewater always increases with its dilution ratio, the appropriate dilution ratio can be regarded as 50 in the paper.

3.2 Effect of solar UV intensity and UV light sources

The more the total UV light intensity, whether under only solar or the double light sources of solar and UV lamp, the higher the photocatalytic degradation rate of the wastewater. Besides, there is more obvious influence of UV lamp on the increase of photocatalytic degradation rate under cloudy day than clear day, which indicates that UV lamp made the equipment still keep high degradation rate in cloudy day.

3.3 Effect of the wastewater pH value

Generally, the photocatalytic degradation rate of the wastewater increase with the decrease of the wastewater pH value. Nevertheless, the rate under pH value of 3.0 is lower than that of 5.0. Therefore, the appropriate pH value of the pharmaceutical wastewater is 5.0.

3.4 Photocatalytic processes comparison of laden and commercial TiO_2

Air rinsing makes TiO_2 deposited in recycle pipe suspend again to keep the suspension concentration of laden and commercial TiO_2 0.715g/L, 0.710g/L respectively, which the concentrations of two kinds TiO_2 almost equal each other. However, the photocatalytic degradation rate of laden TiO_2 for the wastewater is obvious higher than that of commercial TiO_2.

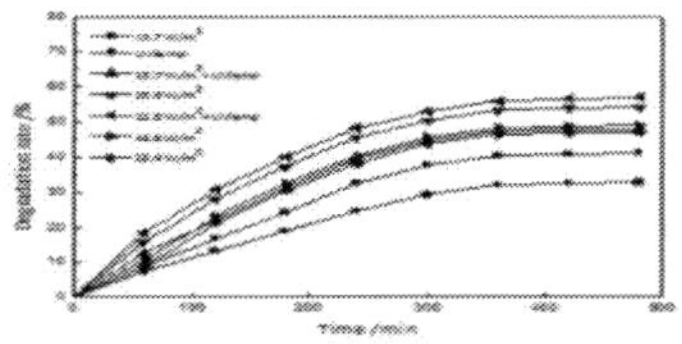

Fig.2. Effect of solar UV intensity and UV light sources

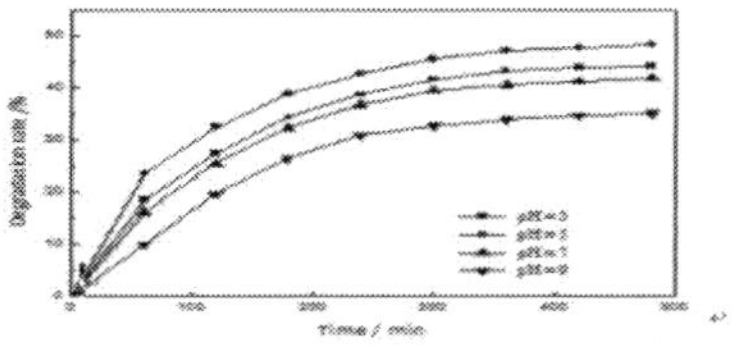

Fig.3. Effect of the wastewater pH value

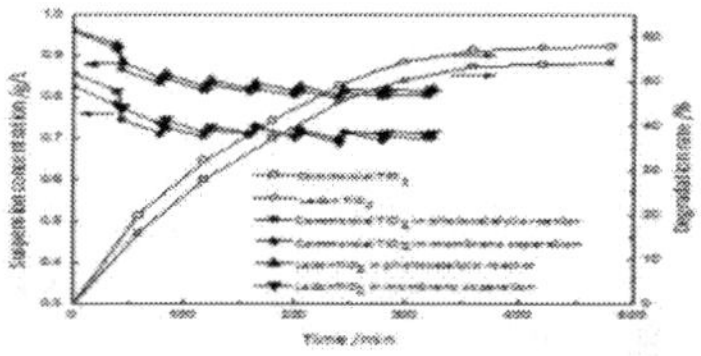

Fig.4. Photocatalysis process comparison of laden and commercial TiO_2

3.5 Treatment by combined solar photocatalysis and SBR process

After single solar photocatalytic degradation for 6 hours under cloud day, the B/C value of pharmaceutical wastewater increases from 0.29 to 0.37, which indicates that biochemical degradability of the wastewater is improved evidently, and the COD_{Cr}, TOC, BOD_5 also decline greatly. In the SBR stage, the pollutants in the wastewater are removed further, and the SBR effluent meets with the Second Chinese Effluent Standard. It is feasible to treat pharmaceutical wastewater by combined solar photocatalysis - membrane separation and SBR processes.

Table .1. Results of treatment by combined photocatalysis-membrane separation and SBR process

Index	Photocatalysis			SBR			Total Removal Rate /%
	Inflow /mg L^{-1}	Effluent /mg L^{-1}	Removal Rate /%	Inflow /mg L^{-1}	Effluent /mg L^{-1}	Removal Rate /%	
COD_{Cr}	5320	2771	47.9	2771	98	96.4	98.2
TOC	1480	750	49.3	750	57	92.4	96.1
BOD_5	1542	1025	33.5	1025	158	84.5	89.8
pH	5.0	6.5		6.5	7.5		

4 Conclusion

Combined process of solar photocatalysis-organic membrane separation and SBR can effectively treat pharmaceutical wastewater, and UV lamp in the equipment makes the combined process still effectively work even in cloudy day.

Acknowledgements

This work was sponsored by Tianjin Research Program Key Projects of Apply Base and Frontier technology (11JCZDJC24900) and Open Fund of State Key Laboratory of Hollow Fiber Membrane Materials and Processes (201022).

Literature Cited

[1] Marcela Boroski, Angela Claudia Rodrigues, Juliana Carla Garcia,Luiz Carlos Sampaio, Jorge Nozaki1, Noboru Hioka. ***J Hazardous Materials*** 162, 448 (2009)

[2] CHANG Qing, XIE Liping, FEI Xuening, DAI Xiaohong, JIANG Yuanguang. ***J Sichuan University(Engineering Science Edition)*** **44**, 249 (2012)

(215q) Investigating the Performance of a Continuous Electrocoagulation Unit for the Combined Removal of Nitrate and Phosphate Ions From Wastewater

El-Shazly, A., King Abdulaziz University
Daous, M. A., King Abdulaziz University

The presence of excess nutrient materials such as nitrates and/or phosphates in wastewater causes the well known eutrophication phenomenon, which is the depletion of oxygen in such water as a result of algae growth due to the presence of such nutrients in high concentrations. Nitrates are relatively stable and highly soluble ions with low potential for co-precipitation or adsorption. Such properties make treatment of these ions for their removal from wastewater difficult, complicated, and expensive. Existing methods of removing nitrate from wastewater include ion exchange, biological decomposition, chemical treatment, reverse osmosis, electrodialysis, and catalytic denitrification. Although Ion exchange is a very efficient process, it is however fairly high in capital and operating costs with undesirable high residual constituents such as chlorides and bicarbonates in the treated water, which must be removed prior to consumption. Biological decomposition is a stable and extremely effective process in reducing nitrate by nearly 100% without using any chemicals. Unfortunately, this process is generally time consuming, limited in the temperature range, very costly and requires extensive maintenance. Therefore, in most cases it is utilized only in treating waste water of sufficiently high original nitrate concentrations.

Current employed phosphorus removal techniques include chemical treatments such as adsorption, chemical precipitation, ion exchange and electrodialysis, hybrid systems based on fly-ash adsorption and membrane filtration, and electrocoagulation. Among these methods, adsorption, and chemical precipitation are the most widely used for phosphate removal. Phosphate removal from aqueous streams is based on the conversion of soluble phosphate to an insoluble solid phase, which can be separated from water by means of sedimentation or filtration. In wastewater applications, the most common and successful methods to precipitate phosphate involve the use of dissolved cations, such as Al^{3+}, Ca^{2+}, Fe^{3+} and to a lesser extent Fe^{2+}. It was found that when iron and aluminum are present in water, $FePO_4$ and $AlPO_4$form at a low pH range below 6.5, while at a higher pH range (above 6.5) iron and aluminum increasingly convert to oxides and hydroxides. However, precipitation of phosphate with calcium as apatites and hydroxyapatites at higher pH is more ideal for phosphate removal.

During the past two decades Wastewater treatment using electrochemical technologies has gained prominence. It has found wide applications in water treatment and metal recovery from wastewaters produced in various industries, such as tannery, electroplating, diary, textile processing, oil and oil refineries. In certain wastewater treatment applications, such as those involving refractory pollutants, electrochemical technologies are in many instances considered the best treatment choice.

Various types of reactors have found applications in electrochemical wastewater treatment processes. These include basic reactors, such as tank cells, plate and frame cells, and rotating cells, as well as, complicated three-dimensional reactor systems like fluidized bed, packed bed cell, or porous carbon packing cells. Rotating cathode cells are used to enhance mass transfer from the bulk fluid to the electrode surface and to remove the deposited metal powders from the cathode. A pump cell is another variant of a rotating cathode cell, which uses a static anode and a rotating disk cathode with a narrow spacing between the electrodes that allow for the flow of the effluent wastewater stream. Dissolved metals are electrically collected and scraped as powders. Lately, a continuous flow reactor based on a bipolar electrocoagulation-oxidation combined with electroflotation (ECEO–EF) was developed and tested for the removal of phosphate and ammonia under different operational conditions of pH, voltage, and detention time.

In this work we investigated the performance of an electrocoagulation reactor for the removal of phosphate and nitrate ions from wastewater using monopolar vertical iron electrodes. Experimental results show that phosphate and nitrate removal efficiencies are improved by increasing the solution flow rate and the current density while decreased by increasing (N/P) ions ratio and by increasing the initial concentrations of these ions. Experimental results further show that the electrocoagulation process can be described by a first order rate equation for the removal of both phosphate and nitrate ions. A correlation for the effect of solution flow rate on the rate of mass transfer for both nitrate and phosphate ions were also developed, which can be highly useful for the design of such electrocoagulation reactor for phosphate and nitrate mixed ions removal from industrial wastewater.

(215i) Efficient Treatment of Pharmaceutical Wastewater By Heterogeneous Fenton-Like Process With Three-Dimensional and Nano-Scaled Perovskite Based Catalyst

Zhao, H., Tongji University
Cao, J.,
Wang, Y., Tongji University
Zhao, G., Tongji University

Heterogeneous Fenton-like method is a developing and promising advanced oxidation technology for the treatment of non-biodegradable organic pollutants. The studies about exploring heterogeneous catalysts suitable for Fenton-like process have been widely investigated. The general requirements for heterogeneous catalysts in Fenton-like system are: high catalytic oxidation efficiency and H_2O_2 utilization, minimum metal leaching, and effectiveness at extended pH range. Aiming at these developments for heterogeneous Fenton-like system, we proposed a novel $BiFeO_3$ (BFO)-based catalyst. $BiFeO_3$ (BFO), as a typical ABO_3 compound, has stable perovskite-type structure in liquid phase and Fe element, which is a powerful candidate for catalytic activation of H_2O_2 with high stability. Considering the catalytic ability of simple BFO was not so strong especially in pure Fenton process to degrade persistent and complex pollutants, thus, in this work we developed the sol-gel method to fabricate a novel BFO/carbon aerogel (CA) as heterogeneous Fenton-like catalyst. Through BET analysis, it was found that BFO/CA nano-scaled composite presented a specific surface area of 404 $m^2\ g^{-1}$, being considerably larger than that of pure nano BFO (14 $m^2\ g^{-1}$). Apart from high surface area, nano BFO/CA catalyst possessed three-dimensional (3D) network structure, in which BFO active sites not only simply placed on CA surface, but grown along with the 3D structure of CA. The Fenton oxidation reactivity with this novel catalyst was evaluated by its reducibility through TPR measurement and the amount of generated OH radicals through HPLC analysis. Moreover, the catalytic activity was tested in degrading pharmaceutical wastewater. Here, ketoprofen (KP), recognized as a refractory and hazardous component in pharmaceutical wastewater, was selected as a target pollutant to be removed. As a result, this novel nano BFO/CA catalyst exhibited excellent Fenton-like activity towards the KP degradation, retained its efficient catalytic performance in wide pH range of 3-7, and had limited iron leaching (0.59 mg L^{-1}) even in acidic condition during whole reaction. Further investigations will be focus on improving the Fenton reaction kinetics of KP oxidation by optimizing the ratio of BFO/CA: H_2O_2: KP and operating conditions such as temperature. For respective, this promising Fenton-like composite catalyst has great potential for practical treatment of persistent and complex pollutants, since its high efficiency, stability and friendly working condition. Moreover, this developed coupled system, not only offers a new idea to improve the activity of catalysts, but also inspires us to design new advanced oxidation system for practical refractory treatment.

(215r) An ATR-FTIR Spectroscopic Approach for Measuring Active Functional Groups in Arsenic and Selenium Stabilization in Sulfur Environment

Jung, B., Texas A&M University at Qatar
Batchelor, B., Texas A&M University
Abdel-Wahab, A., Texas A&M University at Qatar

Arsenic and selenium are widely distributed in the environment and are found in wastewater discharge from oil and gas industries. These sources must be controlled to reduce adverse effects of these contaminants on human health and the environment. Arsenic and selenium are present in water in several different forms depending on the pH and oxidation potential of the water. Reductive stabilization would immobilize arsenic or selenium by reducing it to lower oxidation state and precipitate it as low-solubility solids. This study is to investigate the reduction of arsenic and selenium and subsequent precipitation of elemental arsenic and selenium in sulfidic solutions. Se(0) is relatively immobile and has low toxicity, while Se(IV) and Se(VI) are toxic. The residuals containing Se(0) will be stable and highly resistant to releasing Se when disposed. Arsenic sulfide precipitates in sulfate-reducing conditions at low pH has been reported. The overall goal of this project is to develop fundamental information on the behavior of reductive stabilization of arsenic and selenium to stable elemental (solid) forms in sulfidic environments. In this study, the rapid-scan attenuated total reflectance (ATR) Fourier transform infrared (FTIR) spectroscopy was utilized to investigate the oxidation states of arsenic and selenium in sulfur solutions. The experimental data of ATR-FTIR indicated that rapid-scan FTIR is an effective method for quantifying the change of significant functional groups regarding As and Se in liquid phase and solid phase. The effect of pH or different reducing agents on the significant functional groups were studied using ATR-FTIR with the observation of As or Se amount that involved in the solid formation and macroscopic examination (Scanning electron microscopy with energy-dispersive X-ray spectroscopy).

(215s) Degradation of Artificial Humidity Condensate in the Confined Space Using Large-Area Boron Doped Diamond Electrode

Li, H.,

Lei, L., Zhejiang University

Yang, B.,

Degradation of artificial humidity condensate in the confined space using large-area boron doped diamond electrode

Hao Li, Lecheng Lei, and Bin Yang*

(keyangb@zju.edu.cn)

Key Laboratory of Biomass Chemical Engineering of Ministry of Education, Department of Chemical and Biological Engineering, Zhejiang University, Hangzhou 310027, China

Key words: BDD electrode; Ta substrate; electro-catalytic oxidation; water reuse

The confined space is a particular location that is cut off from the atmosphere or without any gas exchange with the outside. It will be a large cost if the water is provided only by means of external supplements when human being lives there for a long time, therefore water reuse is the most realistic and economical way to produce drinking water. Humidity condensate generated by the body mainly from breathing, sweating, skin evaporation and so on is of a relatively good quality which can be used as drinking water after treatment[1]. It has been reported that humidity condensate has a major composition of small molecule alcohol, acid, and inorganic salt, and the total organic concentration are relatively low[2]. The methods of activated carbon adsorption and the medium temperature catalytic oxidation are employed to treat humidity condensate. However, a large amount of adsorbents are consumed for the former method and while the energy consumption for the later method is relatively high. Therefore, we consider of choosing the electro-catalytic oxidation
technology to remove organic pollutants in water efficiently, economically and reliably. The key to this technology is to fabricate high quality of anode. The boron doped diamond (BDD) electrode has been proved as a candidate owing to a wide potential window, low background current, high electrochemical stability and corrosion resistance[3, 4].

Due to the limitation of space in the confined space, the treatment facilities should be compact and easy to install and operate. Herein, we devoted to fabricating large-area BDD electrode. BDD films are usually deposited on the substrate of Si, Ti, Nb, Ta, W and so on. The substrate of Si is most widely used at present, however it is brittle and its conductivity is poor[5]; Ti is a much cheaper material with good electrical conductivity, but the adhesion of diamond film between Ti is poor due to the formation of TiC transition layer[6]. In particular, Ti substrate show serious deformation when fabricating large-area diamond films (see **Figure 1a** and **1b**). The substrate materials of Nb, Ta and W can be utilized, because they have good mechanical strength, high electrical conductivity and good chemical stability. In the present study, large-area BDD films (the size of tantalum plate: ¦µ100°Á1mm) are grown on Ta substrates (see **Figure 1c** and **1d**) using hot filament chemical vapor deposition. The surface of the Ta/BDD electrode is smooth and uniform distribution, and there don't have any deformation on the electrode. On the basis of some literatures[7], the main composition of artificial humidity condensate is shown in **Table 1**.

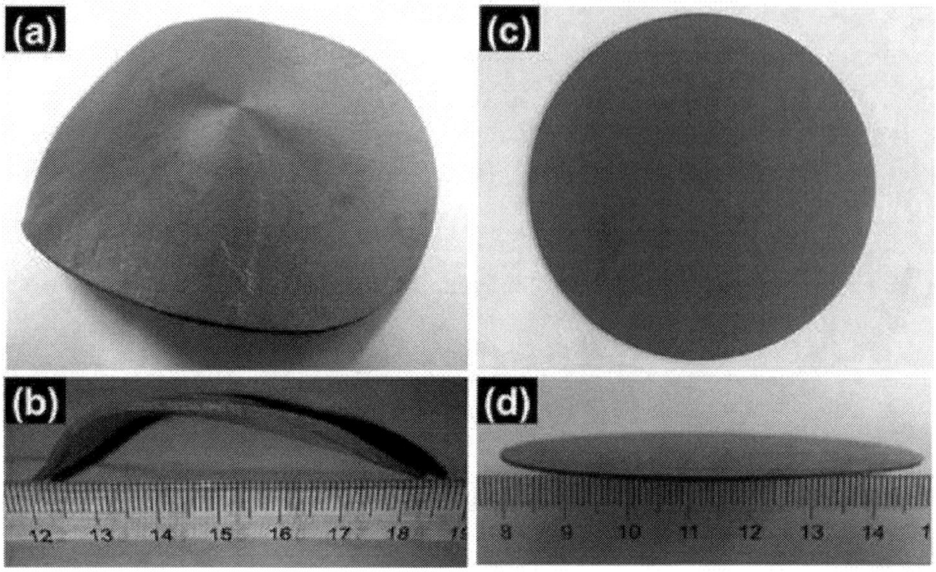

Figure 1. The BDD electrode deposited on Ti (a: top view; b: front view) and Ta (c: top view; d: front view) substrates

Table 1 The composition of artificial humidity condensate

Component	Concentration(mg/L)	Component	Concentration(mg/L)
ethanol	160	formaldehyde	10
acetic acid	60	acetone	5
ethylene glycol	30	caprolactam	5
isopropanol	30	sodium chloride	16.5
propylene glycol	25	potassium sulfate	11.2
methanol	20	sodium sulfate	26.3
formic acid	15	ammonium carbonate	112.9

Figure 2 shows the SEM picture of the surface morphology of BDD electrode. It is obvious that the substrate is totally covered by the clear-cut facet diamond crystals with hexangular, triangular and rectangular habits in the diameter range from 1 to 3μm. The XRD profile (see **Figure 3**) reveals that a diamond (111) facet preferential orientation. Moreover the diffraction peaks of TaC and Ta_2C are observed

which are interlayers formed during the deposition process. Raman spectrum and XPS are further investigated to identify the structure of BDD electrodes. It is also shown that diamonds have a compact structure, homogeneous composition and high levels of content.

Figure 2. The SEM image of the surface morphology of BDD film

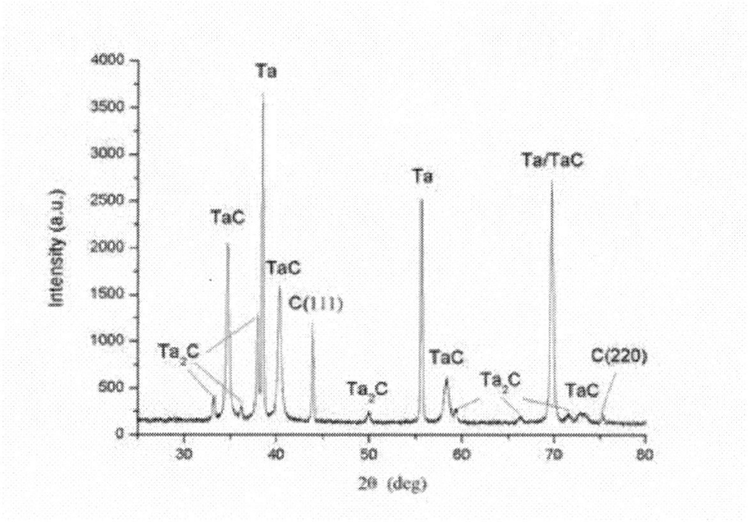

Figure 3. The XRD diagram of the surface of BDD film deposited on Ta substrate

200 mL artificial humidity condensate was treated through circular manner with the velocity of 50mL/min by using the BDD electrode as anode and the stainless steel plate as cathode. The final TOC removal efficiencies against the operating time for different current densities are shown in **Figure 4**. The removal efficiency for each current reaches up to 95%. The residual treated solution was further identified by HPLC. It reveals that the degradation products are composed of oxalic acid, formic acid and acetic acid, and among them oxalic acid is the dominant component. The operating time are approximately 120, 90, 70 min, respectively in order to obtain the TOC removal efficiency of 80% in the current density of 5, 10, 15 mA/cm^2 (the corresponding power is 3.8, 9.9, 17.8 W). Accordingly, the energy consumptions are 7.6, 14.8, 20.8 J respectively. Although the energy consumptions in these three conditions are all relatively low, the current density of 10mA/cm^2 is optimal with the consideration of treated time together.

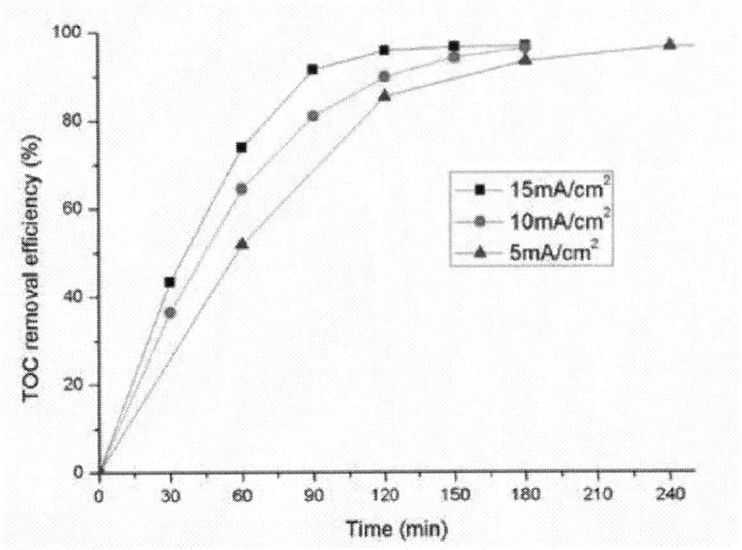

Figure 4. Influence of the current density on the TOC removal efficiency during degradation of 200mL artificial humidity condensate

One of the biggest problems of using electro-catalytic oxidation technology is the generation of large amount of hydrogen gas. When the space is filled with hydrogen, it will induce combustion, explosion or other safety hidden troubles, so hydrogen must be cleaned clearly in time. The generation of hydrogen under operating condition is listed in **Table 2**. 0.61 L hydrogen gas was generated after 2 hours' treatment in current density of 10mA/cm^2 where the TOC removal efficiency reaches about 90%. In particular, no more than 1°ë of the volume is occupied in the confined space of 20m^3, where 18L hydrogen will be produced in case that 6kg water will be treated every day. It demonstrates that the volume of generated hydrogen is small in comparison to the size of the confined space. It can be collected and removed easily by other purification means in the space.

Table 2 Hydrogen and total gas generation under different current density with treating 200mL artificial humidity condensate in 2 hours

Current density (mA/cm^2)	Gas generation rate (mL/min)	Gas volume (L)	H_2 Volume (L)
5	4.3	0.51	0.34
10	7.7	0.92	0.61
15	12	1.44	0.96

The accelerated life test was carried out in 1mol/L sulfuric acid solution. The result shows that the working life of the BDD electrode can reach hundreds of hours, which indicates that the BDD electrode is stable and effective when long-term using. What's more, to ensure our daily water demand, we can use multiple sets of electrodes in combination in order to achieve the effective treatment in a short period of time. In conclusion, using the electro-catalytic oxidation technology to degrade artificial humidity condensate is an efficient and economic method.

References:

1. Samsonov, N.M., et al., Long-duration space mission regenerative life support. ***Acta Astronautica***, 2000. 47(2¨C9): p. 129-138.

2. Bobe, L., et al., Regenerative water supply for an interplanetary space station: The experience gained on the space stations "Salut", "Mir", ISS and development prospects. ***Acta Astronautica***, 2007. 61(1-6): p. 8-15.

3. Panizza, M. and G. Cerisola, Application of diamond electrodes to electrochemical processes. ***Electrochim. Acta***, 2005. 51(2): p. 191-199.

4. Peralta-Hern¨¢ndez, J.M., et al., A Brief Review on Environmental Application of Boron Doped Diamond Electrodes as a New Way for Electrochemical Incineration of Synthetic Dyes. ***International Journal of Electrochemistry***, 2012. 2012: p. 1-18.

5. Sun, J., et al., Electrochemical oxidation of aqueous phenol at low concentration using Ti/BDD electrode. ***Sep. Purif. Technol.***, 2012. 88: p. 116-120.

6. Guo, L. and G. Chen, Long-Term Stable Ti/BDD Electrode Fabricated with HFCVD Method Using Two-Stage Substrate Temperature. ***J. Electrochem. Soc.***, 2007. 154(12): p. D657.

7. Pierre, L.M., et al., Chemical analysis of potable water and humidity condensate: phase one final results and lessons learned. ***SAE Technical Paper***, 1999. 1999-01-2028.

(215t) Effluent Treatment Using Hydrodynamic Cavitation: Vortex Diode As a Cavitating Device

Hiremath, R. S.,
Bhandari, V. M., National Chemical Laboratory
Ranade, V. V., National Chemical Laboratory

Water pollution because of the discharge of industrial effluents in water bodies like rivers (& seas) has increased substantially in recent years. The specific contaminants leading to pollution in water include a wide spectrum of hazardous chemicals, heavy metal ions, pathogens, and physical changes such as elevated temperature, turbidity, discoloration, change in pH etc. The industrial wastewater effluents contain organic and inorganic substances in dissolved and suspended forms. Organic substances mainly comprise carbohydrates, fats, oils, grease, surfactants, proteins, pesticides and agricultural chemicals, volatile organic compounds, while inorganic matter includes salinity, hardness, pH, acidity, alkalinity, concentrations of ionized metals such as iron and manganese, and anionic entities such as chlorides, sulfates, sulfides, nitrates and phosphates. These result in high values of biological oxygen demand (BOD) and chemical oxygen demand (COD).

The dyestuff sector is one of the important segments of the chemical industry in India. The dyes are widely used in other industry sectors such as textiles, leather, paper, plastics, painting inks and foodstuffs. The dyestuff industries along with associated industry such as textiles are highly polluting as dyes and pigments are highly visible and easily detectable in effluent streams. It is required to minimize environmental pollution reducing COD apart from colour, even in cases where a small but visible release might be considered as toxicologically rather innocuous. Wastewater from textile processing plants is highly coloured and difficult to decolorize. In addition, the toxicity, mutagenicity and carcinogenicity of azo-dye degradation byproducts are of great environmental concern. Textile and dye wastewater exhibit low BOD to COD ratio (<0.1) indicating non-biodegradability of dyes, requiring specific designs/technologies for treatment.

The conventional technologies for effluent treatment can be divided into two main categories: biological (aerobic process, anaerobic process, anoxic process etc.) and physico-chemical (adsorption/ion exchange, coagulation, oxidation, membrane separation etc). Considering refractory nature of pollutants, most of these methodologies have limitations and no single method is generally suitable for effluent treatment for dye industry effluent.

In the present work, we have studied dye wastewater treatment on synthetic dye solution as well as on real industrial dye wastewater using hydrodynamic cavitation with Vortex Diode as a cavitating device. Hydrodynamic cavitation is relatively recent methodology that can be considered for treatment of effluents containing refractory pollutants. Design & development of new devices for newer applications is a promising area relatively unexplored so far. There are not many reports that indicate the application of hydrodynamic cavitation for wastewater treatment and most of the reported work is on conventional cavitating devices such as orifice and venturi with different configurations. Most of the studies have focused on degradation of Rhodamine B, Reactive Red, Orange G, Acid Red in low concentration ranges (upto75 mg/L) and with the use of oxidants. The results have not been satisfactory in most cases, especially under realistic wastewater conditions and at higher concentrations.

The present research work demonstrates hydrodynamic cavitation for effluent treatment using new cavitation reactor configuration-vortex diode and reports control of degradation by simple mechanical adjustment of the experimental parameters, to achieve significant decolourization/ mineralization of dyes that are difficult to degrade by other conventional means apart from appreciable COD reduction. Vortex diode relies on fluid vortex phenomena for its operation. The basic design of a vortex diode consists of a disc shaped chamber with cylindrical axial and tangential ports. The flow entering the device through the tangential port sets up a vortex, and establishes a large pressure drop across the device, generating cavities with subsequent collapse of cavities (implosion) with localized high temperature and pressure conditions, resulting into degradation of pollutants.

In the present work Auramine O was used as a model pollutant, not reported so far in the literature. Auramine O is a Basic Yellow dye containing amino and imino groups and also combined with hydrochloric acid. Basic dyes are cationic which have positive electrical charge and classified under the class of Diarylmethane. The effect of pressure drop, effect of feed concentration and effect of flow rate on degradation rate of Auramine O was studied over wide range of concentration from 80 mg/L to 1200 mg/L. The degradation rate was studied on the basis of COD and color reduction.

As the pressure drop increases, cavitation number reduces and hence more number of cavities are formed which results into higher cavitational yield, hence higher degradation was obtained at lower cavitation number. The highest value of degradation constant and regression coefficient was found to be 1.6×10^{-3} and 0.9943 respectively at 1.5 Kg/cm^2 pressure drop. The COD reduction increased from almost 15% to 26% and color reduction increased from 11% to 23% with an increase in pressure drop from 0.5 Kg/cm^2 to 1.5 Kg/cm^2.

With increase in the flow rate, i.e. an increase in the liquid velocity subsequently reduces the cavitation number which results into higher cavitational yield. Highest degradation rate constant was found to be 1.4×10^{-3} and regression coefficient was 0.9823 at flow rate of 715 LPH. The COD reduction increased from almost 8% to 21% and color reduction increased from 7% to 17% with an increase in flow rate from 390 LPH to 715 LPH. However, the COD reduction decreased to 16% and color reduction decreased to 14% as the flow rate was increased to 900 LPH. At increased flow rate cavities start coalescing to form a larger cavitational bubble (cavity cloud). These larger bubbles escape the liquid without collapsing or result into an incomplete and/or cushioned collapse, thus reducing the cavitational yield and thereby reducing the degradation rate after the optimum reached.

The degradation of Auramine O was investigated at different initial dye concentrations. It was observed that lower the dye concentration, shorter is the reaction period needed to degrade the dye. It was observed that the degradation rate decreases with increasing dye concentration. As the other parameters were not changed, the number of cavities and OH- radicals generated in solution approached a steady state. Considering this, the removal rate of apparent color could be decreased by increasing the initial dye concentration. Thus a reaction mixture containing more dilute solutions is expected to exhibit faster rates of degradation. Highest degradation rate was found to be 3.3×10^{-3} and regression coefficient was 0.9445 at 400 mg/L dye concentration. The COD reduction was observed to be 75%, 27%, 16% and 17% at initial feed concentrations of 80 mg/L, 400 mg/L, 800 mg/L and 1200 mg/L respectively.

To evaluate process performance on real industrial effluent, industrial wastewater with high COD in the range 16000 mg/L was used for the treatment studies. The sample was faint green in color with pH of 6.5. Two different cavitating devices viz. orifice plate and vortex diode were used and the degradation rate was studied in terms of COD reduction. The optimization was done in terms of fluid inlet pressure to the cavitating devices. The sample was treated for 150 minutes. Significant reduction in COD was observed in initial 60 minutes. The degradation rate was higher for vortex diode than orifice plate for same energy input. The effectiveness of cavitation process greatly depends on the geometry of the element generating the cavitation phenomenon and can be characterized by the cavitation number.

The results indicate that the COD reduction efficiency was significantly affected by parameters such as the initial concentration, bulk liquid temperature, pressure drop, and flow rate. Thus, the COD removal efficiencies can be improved by optimization of operating conditions. It has been found that there is substantial enhancement in the extent of degradation of the dye solution using hydrodynamic cavitation. Large-scale operation coupled with better energy efficiency makes this technique a viable alternative for conventional effluent treatment methods. Process Integration with Adsorption/Ion exchange/Oxidation can yield COD levels well within the prescribed limits of pollution apart from drastic colour reduction. Thus, the cavitation technology not only looks promising methodology on its own, but can also be integrated effectively with other conventional methods for achieving complete techno-economically feasible solution for wastewater treatment.

(215ae) Uniform Single Crystalline Sulphide-Based Quaternary Alloy Hexagonal Nanorings for Highly Efficient Photocatalytic Hydrogen Evolution

Tan, T. T. Y.,

Sulphide-based polynary alloy semiconductors are ideal photoabsorbers for solar energy applications because of their tunable electronic and optical properties. [1] Compared to binary materials, polynary alloy semiconductor nanomaterials are able to produce new properties while inheriting the properties from their parent binary materials. For example, the energy bandgap of polynary alloys could be precisely tuned by controlling the composition of alloyed nanomaterials without changing their size and shape, which leads to wider wavelength absorption over the visible to ultraviolet light range. [2] A common difficulty in the applications of these alloyed materials lies in achieving compositional homogeneity and morphological controllability due to the diverse starting materials having varying reactivity. Cation-exchange reactions can lead to the formation of $Zn_xM_{1-x}N$ (M=Cu, Cd etc, N=S, Se) polynary compounds with controlled composition and structure from their parent binary materials, which results in efficient H_2evolution in overall water splitting [3]. In principle, the solubility of involved materials is critical for the feasibility of ion-exchange reaction, and the products from ion-exchange reaction must be of a lower solubility than the initial ones. Most of the recent reports on ion-exchange reactions mainly involved binary compounds; their application in polynary alloy nanocrystals synthesis is still uncommon. In addition, the photocorrosion usually occurred in sulphide-based photocatalysts, which caused the degradation of their photocatalytic property.

In this work, we report an original ion-induced synthetic strategy to prepare single crystalline quaternary sulphide uniform nanorings ($Cu-MIn_xS_y$) with homogenous structure induced by metal ions. The Cu^{2+}-doped $NaInS_2$ nanorings were first synthesized via a copper ions induced Kirkendall effect, and the as-synthesized products could be further transformed to other quaternary sulphide nanorings by solubility induced cation-exchange strategy. Because of the higher solubility of sodium salts than that of other metals, this method presents a general strategy to fabricate quaternary sulphide alloys with defined morphology and structure (not limited to nanorings), and could be extended to the synthesis of other quaternary alloys with similar layer structure as the $NaInS_2$ nanocrystals. Furthermore, the as-synthesized quaternary sulphide nanorings exhibit efficient activity and good stability for photocatalytic H_2 evolution from water splitting. [4]

Reference:

[1] a) B. B. Kale, J. O. Baeg, S. M. Lee, H. Chang, S. J. Moon, C. W. Lee, *Adv. Funct. Mater.* **2006**, *16*,

1349-1354; b) J. Xu, C. Y. Luan, Y. B. Tang, X. Chen, J. A. Zapien, W. J. Zhang, H. L. Kwong, X. M. Meng, S. T. Lee, C. S. Lee, *ACS Nano* **2010**, *4*, 6064-6070.

[2] a) J. Akhtar, M. Afzaal, M. Banski, A. Podhorodecki, M. Syperek, J. Misiewicz, U. Bangert, S. J. Hardman, D. M. Graham, W. R. Flavell, D. J. Binks, S. Gardonio, P. O'Brien, *J. Am. Chem. Soc.* **2011**, *133*, 5602-5609; b) J. Xu, X. Yang, H. Wang, X. Chen, C. Luan, Z. Xu, Z. Lu, V. A. Roy, W. Zhang, C. S. Lee, *Nano Lett.* **2011**, *11*, 4138-4143.

[3] a) Y. Yu, J. Zhang, X. Wu, W. Zhao, B. Zhang, *Angew. Chem. Int. Ed.* **2012**, *51*, 897-900; b) J. Zhang, J. Yu, Y. Zhang, Q. Li, J. R. Gong, *Nano Lett.* **2011**, *11*, 4774-4779; c) X. Wu, Y. Yu, Y. Liu, Y. Xu, C. Liu, B. Zhang, Zhang, *Angew. Chem. Int. Ed.* **2012**, *51*, 3211-3215; d) J. Zhang, J. Yu, M. Jaroniec, J. R. Gong, *Nano Lett.* **2012**, *12*, 4584-4589.

[4] P. Hu, S. S. Pramana, S. Cao, C. K. Ngaw, J. Lin, S. C. Joachim Loo, T.T.Y. Tan, *Adv Mat*, **2013,** *25*, 2567-2572.

(215j) Graphene Membranes for Water Nanofiltration and Desalination

Yu, M.,

Graphene, a single layer of graphite, has been considered as an ideal membrane material because it is extremely thin and thus can provide high permeation flux. However, currently there are no technologies available to fabricate large area, ultrathin (<5 nm), defect-free graphene-based membranes that can highly selectively separate mixtures. A scalable technique was developed to prepare low-cost, ultrathin (down to 2 nm) and defect-free, graphene-based membranes with tunable pore sizes (sub-nanometer – 20 nm). Tunable pore sizes, from sub-nanometer to > 10 nm, and ultrathin thickness (< 5 nm), allow ultrathin, graphene-based membranes for desalination and nanofiltration for water purification. We found that water flux and rejection for nanometer-sized organic molecules can be adjusted by tuning proe sizes on graphene. This tunability of membrane pore sizes may allow water treatment from ultrafiltration, nanofiltration, to reverse osmosis (RO) desalination. This completely new type of membranes has the great potential to revolutionize separation using membrane technology, because they can provide both high selectivity and high flux.

(215z) Ionic Liquid Mediated Electrochemical Reduction of CO2 in a Microbial Electrolysis Cell

Li, Z.,
Wang, Q.,
Lei, L., Zhejiang Univeristy

Ionic liquid, 1-ethyl-3-methylimidazolium tetrafluoroborate (EMIM-BF4) has been proved to be a promising catalyst for CO_2 electrochemical reduction with a silver nanoparticle deposited electrode ADDIN EN.CITE <EndNote><Cite><Author>Rosen</Author><Year>2011</Year><RecNum>30</RecNum><DisplayText>[1]</DisplayText><record><rec-number>30</rec-number><foreign-keys><key app="EN" db-id="95dffxdd1zszaqeve5aprzxndxp0zr52awzv">30</key></foreign-keys><ref-type name="Journal Article">17</ref-type><contributors><authors><author>Rosen, Brian A.</author><author>Salehi-Khojin, Amin</author><author>Thorson, Michael R.</author><author>Zhu, Wei</author><author>Whipple, Devin T.</author><author>Kenis, Paul J. A.</author><author>Masel, Richard I.</author></authors></contributors><titles><title>Ionic Liquid-Mediated Selective Conversion of CO2 to CO at Low Overpotentials</title><secondary-title>Science</secondary-title></titles><periodical><full-title>Science</full-title></periodical><pages>643-644</pages><volume>334</volume><number>6056</number><dates><year>2011</year><pub-dates><date>Nov 4</date></pub-dates></dates><isbn>0036-8075</isbn><accession-num>WOS:000296494700047</accession-num><urls><related-urls><url><Go to ISI>://WOS:000296494700047</url><url>http://www.sciencemag.org/content/334/6056/643.full.pdf</url></related-urls></urls><electronic-resource-num>10.1126/science.1209786</electronic-resource-num></record></Cite></EndNote>[1]. As Brian et. al., reported, by applying 1.5 V cell voltage, efficient CO_2 reduction to CO can be achieved in a two-chamber electrolysis cell with EMIM-BF4 as the electrolyte ADDIN EN.CITE <EndNote><Cite><Author>Rosen</Author><Year>2012</Year><RecNum>29</RecNum><DisplayText>[2]</DisplayText><record><rec-number>29</rec-number><foreign-keys><key app="EN" db-id="95dffxdd1zszaqeve5aprzxndxp0zr52awzv">29</key></foreign-keys><ref-type name="Journal Article">17</ref-type><contributors><authors><author>Rosen, Brian A.</author><author>Haan, John L.</author><author>Mukherjee, Prabuddha</author><author>Braunschweig, Bjoern</author><author>Zhu, Wei</author><author>Salehi-Khojin, Amin</author><author>Dlott, Dana D.</author><author>Masel, Richard I.</author></authors></contributors><titles><title>In Situ Spectroscopic Examination of a Low Overpotential Pathway for Carbon Dioxide Conversion to Carbon Monoxide</title><secondary-title>Journal of Physical Chemistry C</secondary-title></titles><periodical><full-title>Journal of Physical Chemistry C</full-title></periodical><pages>15307-

15312</pages><volume>116</volume><number>29</number><dates><year>2012</year><pub-dates><date>Jul 26</date></pub-dates></dates><isbn>1932-7447</isbn><accession-num>WOS:000306725200014</accession-num><urls><related-urls><url><Go to ISI>://WOS:000306725200014</url><url>http://pubs.acs.org/doi/abs/10.1021/jp210542v</url></related-urls></urls><electronic-resource-num>10.1021/jp210542v</electronic-resource-num></record></Cite></EndNote>[2]. Here, in this work, we tried to evolve the electrolysis cell to a microbial electrolysis cell to decrease the cell voltage employed for CO_2 reduction process. Microbial electrolysis cell (MEC) is a type of bioelectrochemical system, in which oxidation reaction is catalyzed by electrochemical active bacteria with the solid anode as the electron acceptor in anode chamber and then the electrons pass through the out circuit power supply to cathode. Finally, the electrons drive the reduction reaction in cathode chamber. With the help of electrochemically active bacteria, the active energy for anodic oxidation reaction can be significantly reduced and thus results in a more negative anode potential. Because the cathodic reduction reaction has a lower redox potential than anodic oxidation reaction in MEC, more negative anode potential can decrease the cell potential that needed for driving the cathodic reduction reaction. Based on this principle, a typical electrochemically active bacteria shewanella oneidensis MR-1 was grown in the anode chamber using a carbon cloth electrode. 100 ml 18 mol% EMIM-BF4 was filled into the cathode chamber and the headspace is 20 ml. The cathode is silver nanoparticle deposited graphite block electrode. After sparging CO_2 overnight, different levels of cell voltage were applied and the gas composition in the headspace was analyzed by GC. To evaluate the effect of different electrode potential on CO_2 reduction, the two chamber reactor was firstly operated in abiotic three-electrode system for 12 hours. The counter electrode was a carbon cloth electrode and a Ag/AgCl reference electrode was placed in the anode chamber. The potential of the working electrode (a silver nanoparticle deposited graphite block electrode) was controlled at a range from 0 to -0.8 V (vs. Ag/AgCl) by a potentiostat. As shown in figure 1, when the working electrode potential was controlled at -0.4 V, the CO production rate was the highest. After 12 hours operation, the accumulated CO concentration was around 0.0035%. Surprisingly, decreasing the potential to a more negative level did not enhance the CO production but methane production was trigged. The methane production rate reached the maximal level at -0.5 V. Further decreasing the working electrode potential to -0.8 V decreases the methane production rate dramatically. As an alternative reduction reaction in cathode chamber, hydrogen evolution reaction became the predominant reaction at a potential more negative than -0.7 V (Fig. 2). These results are different from Brian et. al. reported. In their case, no methane production was detected in the gas phase and hydrogen production could be inhibited. The possible reason is our reactors were operated in batch mode for more than 12 hours for each test. During the long term operation, the pH, water content and other conditions might be varied. For example, the pH decreased from 5.69 to lower than 2.5 in 12 hours operation. These changes leads to methane and hydrogen production at different potential levels. Besides operating the reactor in the three electrode mode, the same experiments were conducted in a MEC mode. The anode potential during all the MEC tests was around -0.3 V and all experiments were operated for 24 hours. The cell voltage was varied from 0 to -0.5 V. In MEC mode tests, CO, methane, and hydrogen production rates follow a similar trend as that in three electrode mode. Especially for methane production, the final concentration in the gas phase in headspace was almost 100 times higher than that in the three electrode mode (Fig. 3). This is due to higher current density in MEC mode than in abiotic three electrode system with the help of electrochemically active bacteria catalyst. Also this high methane production rate indicated that methane production can be an important CO_2 reduction pathway besides CO_2 to CO reduction mechanism. Based on all these results, integrating ionic liquid mediated CO_2 electrochemical reduction with MEC can be a promising technology for CO_2 utilization and fixation.

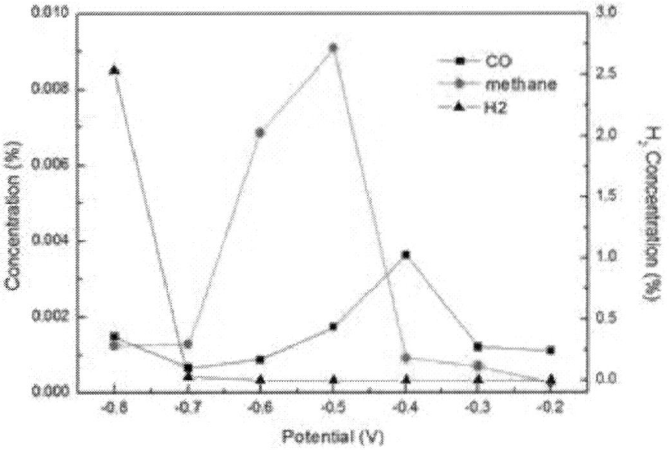

Figure 1. CO, methane and H_2 production in the three electrode system

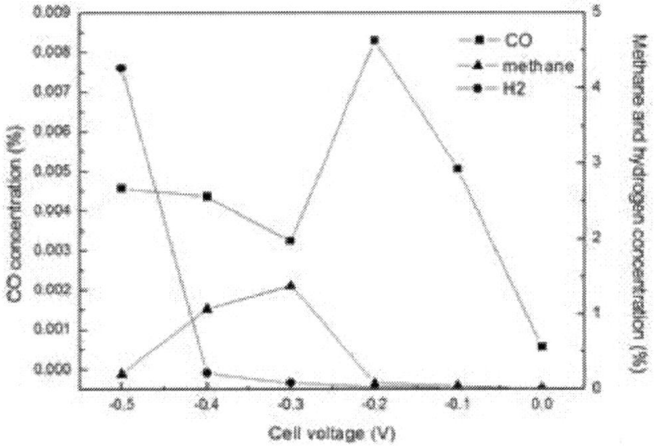

Figure 2. CO, methane and H_2 production in MEC

References

ADDIN EN.REFLIST [1] B.A. Rosen, A. Salehi-Khojin, M.R. Thorson, W. Zhu, D.T. Whipple, P.J.A. Kenis, R.I. Masel, Science 334 (2011) 643.

[2] B.A. Rosen, J.L. Haan, P. Mukherjee, B. Braunschweig, W. Zhu, A. Salehi-Khojin, D.D. Dlott, R.I. Masel, Journal of Physical Chemistry C 116 (2012) 15307.

(215aa) New Adsorbents and Material Modifications- An Approach for Ultraclean Fuels

Sorokhaibam, L. G., National Chemical Laboratory, Pune
Bhandari, V. M., National Chemical Laboratory
Ranade, V. V., National Chemical Laboratory

The present research analyses the performance and characterization studies of new class of modified solid adsorbents as an approach for production of ultraclean fuels. The conventional hydrodesulfurization (HDS) process has limitations in producing ultraclean fuel due to high operational cost involving the use of high pressure, temperature and H_2 in catalytic conversion of S compounds in liquid fuels to H_2S and corresponding hydrocarbon. The research is a continuation of our efforts in a) ultralow sulphur emission of transportation fuels as organosulphur compounds are responsible for SOx emission, a highly undesirable air pollutant and b) meeting increasingly stringent Environmental and Government regulations for sulphur content in transportation fuels. The study is also significant for future fuel cell applications which require practically zero sulphur content to avoid poisoning of catalysts.

There have been numerous studies on variety of adsorbents, single as well as modified, for the adsorptive desulfurization of fuels. The adsorbents studied in the literature include simple types such as activated carbons, zeolites, and also more complex types such as modified zeolites, metal oxides, adsorbents derived from biomass and π-complexation adsorbents. It was observed that the capacity for sulphur removal was far low for its commercial use in most cases and/ or performance is severely limited in real fuel desulfurization, mainly due to presence of refractory sulphur compounds and aromatics.

It is necessary to understand interactions of adsorbing components for effective removal of sulphur compounds with increased selectivity to refractory compounds. At present there are no theoretical guidelines in this regard and approach to selection and modification of adsorbent is purely empirical. The present study is directed to evaluate interactions of adsorbing sulphur compounds with the materials and material modifications in terms of nature of modifying metal species, percentage substitution of metal species, combined effect of different species on the adsorbent surfaces and overall improvement in deep desulfurization of fuels for the production of ultraclean fuels. Experimental studies have been carried out initially on single component adsorption of thiophene in iso-octane and benzothiophene in iso-octane using modified SHIRASAGI $GH2_X$ 4/6 and SRC_X 4/6 of Japan EnvironChemicals Ltd. Sulphur content in the feed model fuel was analysed using Total sulphur analyser. The efficiency in terms of sulphur removal/adsorption capacity was found to be substantially higher than most studies reported in literature for activated carbon, zeolites, silica based adsorbents, metal doped porous materials etc. Another main aspect of the experimental investigation is further modification of the adsorbent materials using transition metal salts especially of Ni, Cu etc and evaluation of the modified new material in adsorptive desulfurization. Effect of different parameters such as temperature, adsorbent dose and initial concentration on adsorptive desulfurization was also studied. Two widely used adsorption isotherms, Langmuir and Freundlich model were used to explain the nature of adsorptive desulfurization.

The main characterization studies include BET Surface analysis, FTIR, SEM-EDX and XRD. The adsorbents used in the present study have high specific area above 1000 $m^2.g^{-1}$ and total pore volume of 0.710 and 0.807 $cm^3.g^{-1}$ for SRCx and GH2x respectively. BJH pore size distribution analysis showed majority of the pores in microporous range with contributions from meso-microporous section in explaining the sorption behaviour of the modified adsorbents. Specific functional groups belonging to –OH, C-O and N-H group were identified by FTIR spectra which are likely to play crucial role in the desulfurization with specific interactions with sulphur moiety. The surface morphology as determined by SEM showed heterogeneous nature and irregular structure at the microscopic level. The surface analysis and composition was further elaborated by Energy Dispersive X-ray spectroscopy confirming the main compositions of the adsorbents as C and O with admixtures of Al and Si in case of GH2x and SRCx whilst Ni modified adsorbents showed the highest impregnation of ~4% in range of concentrations prepared with maximum strength 0.5 M Ni salt deposition. XRD patterns of modified adsorbents depicted two broad bands around $2\theta = 24$ and 43^0 indicating amorphous structure, characteristics of the carbonaceous adsorbent matrix. In XRD of GH2x and SRCx , no other peaks that can be assigned to oxides of Al and Si are located which further indicated that these components are well dispersed in the carbon matrix.

The desulfurization behaviour changes with the nature of sulphur compound which is attributed with the specific interaction of modified surfaces with the sulphur compounds. The interactions and their characterization is most critical for the production of ultraclean fuels as all refractory sulphur compounds need to be removed from the fuels for commercial applications. Role of aromatics is also very important in this regard as these are likely to compete in adsorption process thereby reducing removal capacity for sulphur compounds. Thus, there is always a further scope in developing newer material and modified materials that could selectively remove refractory sulphur compounds from diesel and other transportation fuels. The present research therefore addresses key environmental issue of sulphur removal with new and modified adsorbents materials offering promising approach towards environmental sustainability and pollution control. The methodology can also provide insights into production of ultra clean fuels when process intensified with HDS process.

(215v) Dynamic Modelling and Simulation of Anaerobic Digester For High Organic Strength Waste

Sharma, P., Indian Institute of Technology Roorkee
Ghosh, U. K., IIT Roorkee
Ray, A. K., Indian Institute of Technology Roorkee

High organic strength waste such as wastewater generated from various industrial activities and sludge generated from industrial and municipal wastewater treatment plant requires treatment prior to disposal in order to reduce environmental pollution load and also to comply with the disposal norms. Waste containing organicimpurities in nature must be treated biologically using life forms such as microorganisms, plants, etc. Anaerobic digestion is the most suitable treatment method for treating high organic strength waste as compared to other waste treatment methods such as aerobic and advanced treatment methods, because energy in the form of methane is recovered and also treated sludge has good soil conditioning values. Complete mix anaerobic digester is more suitable for treating high organic strength waste as compared to other because of proper heat transfer amd also proper mixing for maintaining uniformity throughout the reactor. Complete mix reactors are resistant to shock loading and also withstand high organic loading rates. Anaerobic digester suffers stability problem due to the accumulation of the volatile fatty acids and drop in pH. Once the failure of digester takes place it requires long time and efforts to restart it again. Dynamic modelling and simulation are useful tools for predicting the process stability by studying the behaviour under transient conditions and are also helpful in understanding the process operations.

In the present work an attempt has been made to develop simplified dynamic mathematical model for the anaerobic digestion of sludge.The simplified mathematical model for sludge digestion is based on the some important models described below. The inhibitory function developed by Andrews (1969) given below was an important investigation considering the inhibitory effect of unionized volatile fatty acids on methanogen bacteria. The unionized volatile fatty acids act as growth limiting substrate at lower concentration and cause inhibition at higher concentration for microorganisms utilizing them as substrate.

The inhibition model was given as

where,

$\mu = \hat{u} / (1+ (K_s/HS) +(HS/K_i))$

μ Specific growth rate , day^{-1}

$\hat{u}$ Maximum specific growth rate in the absence of inhibition , day^{-1}

HS Unionized substrate concentration, mg/L

K_s Saturation constant, mg/L

K_i Inhibition constant, mg/L

The model was restricted to fixed pH and only methanogenesis step was considered. Andrews *et al.* (1971) removed the limitation of fixed pH by considering the interaction between liquid, gas and biological phases and pH variation took place between 6-8. The buffering system was carbon dioxide-bicarbonate is assumed.
This model was also restricted to methanogenesis step. Hill *et al.* (1977) developed model for animal waste digestion based on two microbial culture acidogens and methanogens. They considered the inhibition caused by unionized volatile acids and unionized ammonia on growth kinetics of the methanogens.

The inhibition model was given as

Where,

$$\mu = \hat{u} / (1 + (K_s/VA) + (VA/K_{ia}) + (NH_3/K_{i2}))$$

μ Specific growth rate , day^{-1}

$\hat{u}$ Maximum specific growth rate in the absence of inhibition , day^{-1}

VA Unionized substrate concentration, mg/L

K_s Saturation constant, mg/L

K_{ia} Inhibition coefficient of acids, mg/L

K_{i2} Inhibition coefficient of ammonia, mg/L

NH_3 Concentration of unionized ammonia, mg/L

Havlik *et al.* (1986) developed model for digestion of complex organic substrates and also stated that the path of methane generation was either from acetate and via carbon dioxide reduction by hydrogen. The model developed by Moletta et al. (1986) was based on easily fermentable organics such as pea bleaching wastewater and synthetic substrate containing sucrose and organic acid. In the developed model biomass and metabolite production rates were described by distinct relations. No pH variation and no buffering system were assumed in this model. The dynamic Model developed by Mendoza *et al.*(1997) for anaerobic digestion of sewage sludge describes digester behaviour under non-ideal mixing conditions. He considered no buffering system in the developed model and the growth rate of microorganisms was assumed to be dependent on Monod Kinetics. Hydrolysis and death of microorganisms were described by first order reactions.

In 1997, an International anaerobic modelling task group was established in Japan as a common platform for the establishment of anaerobic digestion model for complex organic wastes.

The kinetic model developed in the present work describes the anaerobic digestion of sludges considering the hydrolysis of particulate material to soluble compounds by the extracellular enzymes produced by acidogens, fermentation / acidogenesis of soluble organics to volatile fatty acids by the acidformers and methanogenesis of volatile fatty acids to methane and carbon dioxide by methanogenic bacteria. In most of the cases first step in anaerobic digestion of waste depends on the nature of the organic waste. Sometimes fermentation may be the first step if most of the impurities are soluble in nature. The growth and decay of acidogens and methanogens are assumed to depend on Andrews's inhibitory function and also the inhibitory effect of unionized ammonia on methanogens. Hydrolysis of particulate organic compound is described by first order reaction. The microbial kinetic model expressions are linked to the complete mix anaerobic digester. Carbon dioxide and bicarbonate with ammonium ion is considered as the buffering system. Computer simulation of developed mathematical model helps to predict the dynamic behaviour of the developed model under batch mode, transient conditions, steady state conditions and also helpful in improving design.

The developed modified dynamic mathematical model describes the process operations in more quantitative way which is helpful in designing a better control system to enhance stability, improving performance of anaerobic digester and also optimizing process performance.

Keywords: anaerobic digester, organic waste, sludge, modelling and computer simulation

References:

1. Andrews, J.F., A dynamic model of the anaerobic digestion process. J. Sanitary Engineering Div., Proc. Am. Soc. Civil Eng. (1969) 95,SAI, 95.

2. Andrews, J.F. and Graef, S.P., Dynamic modelling and simulation of the anaerobic digestion process. Advances in Chemistry, American Chemical Society (1971),126-162.

3. Havlik, I., Votruba, J. and Sobotka, M., Mathematical modelling of the anaerobic digestion process: Application of dynamic mass-energy balance. Folia Microbiol (1986),31,56-68.

4. Hill, D.T. and Barth, C.L., A dynamic model for simulation of animal waste digestion. Journal of Water Pollution Control Fed.(1977),10,2129-2143.

5. Mendoza, R.B., and Sharratt, P.N., Modelling the effects of imperfect mixing on the performance of anaerobic reactors for sewage sludge treatment. J. Chemical Technol. Biotechnol.(1998),71,121-130.

6. Moletta, R., Verrier, D. and Albagnac, G., Dynamic modelling of anaerobic digestion. Water Resources (1986),20,427-434.

(215ab) Development of a Low Cost Carbon Materials and Its Functionalisation: for Waste Water Treatment

Paul, C. J., National University of Singapore
Mahanta, N., National University of Singapore

Carbon derivatives are interesting in current research due to their diverse nature with suitable surface functionality. Owing to that our research focused on the development of low cost hydrophilic carbon materials from waste resources. The physical and chemical property of the designed particles was studied by Raman spectroscopy and electron microscopic investigation. It was observed that the particles are amorphous in nature without any sorts of conjugation or aromatization. In order to improve the adsorption efficiency of such hydrophilic particles, the active functional group with suitable metal ion was chemically immobilized under basic conditions. As our designed material was targeted for environmental detoxification, so for safety view point the toxicity effect of the prepared nanoparticles was investigated using human cell line. From the study it was observed that the designed functionalized materials were nontoxic in nature. The developed functionalized materials were used for the removal of highly toxic elements, such as arsenic from water. In the adsorption experiment we studied both inorganic derivatives (arsenate) and organic derivatives, such as, dimethylarsenic acid (DMA) and monomethyl arsenic acid (MMA). From the adsorption study it was observed that, about 70 to 80% of its final adsorption efficiency for derivatives was achieved within the first 10 minutes time. The effect of pH and different ions on the adsorption behavior was investigated in detail. From this study, it can be seen that for all cases, at acidic range adsorption was maximum. The Langmuir equation well described the adsorption isotherm; the maximum adsorption capacity towards arsenate, MMA and DMA was found to be 110 mg/g, 79 mg/g and 39 mg/g, respectively at the optimal pH. This result revealed that, the trend for arsenic removal efficiency was followed a trend of as As (V) > MMA > DMA. The effect of humic acid and different coexisting anions during adsorption process was investigated in detail. Various characterization techniques, such as X-ray photoelectron spectroscopic study and Fourier transform infrared spectroscopy analysis was carried on the designed sorbent, before and after adsorption. Hence from the above results it can be concluded that, these novel metal ion immobilized carbon sorbents are efficient in removing arsenate from water and can be an attractive as low cost alternative for the treatment of arsenic contaminated waste water.

(215u) Removal of Disinfection By-Products (DPBs) By Electro-Coagulation Process in a Continuous Flow Reactor As a Localized Water Treatment Technology

He, C. C., State Key Laboratory of Chemical Engineering, Department of Chemical and Biological Engineering, Zhejiang University
Lo, S. L., Argonne National Laboratory

Nowadays, chlorine is commonly used as the disinfectants in water treatment process. However, it would react with nature organic matter (NOM) to product disinfection by-products (DBPs) such as trihalomethanes (THMs) and Haloacetic acids (HAAs). The above DBPs presented in drinking water have been considered as toxic materials because they would result in Bladder cancer, rectal cancer, colon cancer and brain cancer according to the laboratory animals. The standard regulations for THMs in drinking water are 0.1 mg/L in both European Union and Taiwan. On the other hand, the USEPA formulated the standard concentration for disinfectant-disinfection by-products from 100 ug/L to 80 ug/L and would be further restricted to 40 ug/L in the future.

Electrocoagulation (EC) is an electrochemical process for treating wastewater by using sacrificial anodes to produce coagulant precursors into solution. In that case, aluminum ions would be released from the sacrificial anode instead of traditional coagulant, e.g., aluminum salts, thereby increasing the treatment efficiency. In general, the removal efficiency of EC process for wastewater treatment is depended on the characteristic of wastewater quality and operation parameters. Therefore, the purposes of this investigation are to (1) evaluate the removal performance of THMs, biochemical oxygen demand (BOD) and suspended solid (SS) in wastewater by applying the continuous flow reactor with EC process, and (2) investigate the effects of different operating conditions including pH values, voltage and electrical conductivity on the removal efficiency of DBPs.

(215w) Biofiltration of Cu (II) Using Acclimated Mixed Culture

Gupta, S., Birla Institute of Technology & Science, Pilani
Raghuvanshi, S., Birla Institute of Technology & Science, Pilani
Gangadhar, G., Birla Institute of Technology & Science, Pilani
Majumder, S., Birla Institute of Technology & Science, Pilani

Heavy metal pollution by industrial activities and technological development is posing significant threats to the environment and public health because of its toxicity, non-biodegradability and bio-accumulation [1, 2]. The threat related to the disposal of heavy metal ions would increase significantly in the future years. Cupric compounds find many industrial applications such as in manufacturing of pesticides and fertilizers, mining, petroleum refineries, textile industries,synthetic rubber manufacturing ,preparation of catalysts etc. Copper, in lower concentration is not harmful to health (1mg/L). High dose of copper is harmful to all living organisms. It causes headache, vomiting, dizziness, nausea, damage to liver and kidneys and damage to marine organisms. A lot of physicochemical strategies, such as filtration, chemical precipitation, electrochemical treatment, oxidation/reduction, ion exchange, membrane technology, reverse osmosis, and evaporation recovery, have been developed for removing heavy metals from the polluted water [3]. However, most of them appear to be expensive, low efficient and labour intensive [4]. The main disadvantage associated with the above mentioned methods are the production of secondary pollutant obtained while separation of toxic metal compounds. In case of reverse osmosis the membrane has to be disposed when it is exhausted. So in all these methods the secondary pollutants produced have to be treated before dumping it to the environment. Considering these disadvantages, a new technique called bioremediation has been proposed for the removal of heavy metals from waste water. This can be solved by using biofiltration. Biofiltration uses a natural process of removing heavy metal ions from waste water using microorganisms. Microorganisms attach to the packing material in the form of biofilm. It acts as a catalyst and is able to utilize heavy metal compounds as a source of available carbon for further cell growth. This technique is environment friendly as it does not produce any secondary pollutants. Biofiltration does not require external energy and utilizes the capacity of microorganisms to transform the toxic heavy metals. The isolation of microorganisms that can sustain heavy metals and the understanding of the mechanisms can contribute to the development of improved detoxification processes.

The present study deals with the biofiltration of Cu(II) from aqueous solutions using an acclimated mixed culture developed from activated sludge. The biodegradation studies are conducted for 40 mg/L initial Cu(II) concentrations with different inlet concentrations such as 10, 20 and 30 mL/min. The final copper concentration is measured using Atomic Absorption Spectrophotomer. A maximum of 97% removal copper is achieved on the 23rd day of operation of the biofilter column for 30 mL/min of inlet flow rate of copper solution. The CFU analysis showed a decreasing trend of growth from top to down of the column.

References:

1. Perezmarin, A.B.; Ballester, A.; Gonzalez, F.; Blazquez, M.L.; Munoz, J.A.; Saez, J.; Meseguer Zapata, V., Study of cadmium, zinc and lead biosorption by orange wastes using the subsequent addition method, Bioresource. Technology, 99, 2008, 8101–8106.
2. Reddad, Z.; Gerente, C.; Andres, Y.; Thibault, J.F.; Le Cloirec, P., Cadmium and lead adsorption by a natural polysaccharide in MF membrane reactor: experimental analysis and modelling, Water Res., 37,2003, 3983–3991.
3. Xiao, X.; Luo, S.L.; Zeng, G.M.; Wei, W.Z.; Wan, Y.; Chen, L.; Guo, H.J.; Cao, Z.; Yang, L.X.; Chen, J.L.; Xi, Q., Biosorption of cadmium by endophytic fungus (EF) Microsphaeropsis sp. LSE10 isolated from cadmium hyperaccumulator Solanum nigrum, Bioresource Technology, 101,2010, 1668–1674.
4. Chen, G.Q.; Zeng, G.M.; Tang, L.; Du, C.Y.; Jiang, X.Y.; Huang, G.H.; Liu, H.L.; Shen, G.L., Cadmium removal from simulated wastewater to biomass byproduct of Lentinus edodes, Bioresource Technology, 99, 2008, 7034–7040.

(215k) Removing Microorganisms From Wastewater in a Painting Industry By Advanced Oxidation Processes

Chang, W., MATGAS Research Center and Carburos MetÃ¡licos, Air Products Group
Vega, L. F., MATGAS Research Center and Carburos MetÃ¡licos, Air Products Group
Guri, S., Carburos MetÃ¡licos, Air Products Group

Advanced Oxidation Processes (AOP) are defined as those which involve the generation of hydroxyl radicals in sufficient quantity to affect water purification. Ozonation, in particular, is a well known technique to remove several pollutants from water and an alternative to bleach for general disinfection processes; in addition, ozonation combined with hydrogen peroxide treatment has been widely studied as a most efficient alternative, decreasing reaction time and increasing pollutant removal efficiency.

The goal of this work was test the efficiency of AOP to remove microorganisms found in wastewater effluent from a painting industry. The treated water can then be reused in the paint production process. The presentation will summarize the main results obtained in the assays carried out in a 2L capacity Halia® AOP bench test reactor with the aim of eliminating those microorganisms as a tertiary treatment. Two treatments were tested: ozonation and ozonation with the addition of hydrogen peroxide (O_3, O_3+H_2O_2). For the ozonation treatment a concentration of 2 ppm of ozone was applied and water sample was circulating through the reactor by 30 minutes in order to warranty complete reaction. In the other treatment a 2 ppm of O_3 and a 1 ppm of H_2O_2were combined and the reaction time, in this test, was 15 minutes. The characterization of water samples was done before and after the treatments, analyzing: DQO, turbidity, colour and microbiological counts.

For both treatments, the results showed that the microorganisms were completely removed, DQO values decreased up to around 50%, as well as turbidity and colour, validating the use of this technology for this application.

(215n) Life Cycle Assessment of the Gasplasma Process: An Innovative Waste Management Option

Tagliaferri, C.,
Evangelisti, S.,
Lettieri, P., University College London
Chapman, C., Advanced Plasma Power

A growing environmental concern of waste management drives new research toward the identification of low impact waste treatment and disposal options. Life Cycle Assessment (LCA) can be used as a tool to provide information about the environmental impact of these waste disposal methods. We performed an attributional Life Cycle Assessment with system expansion using GaBi 6 LCA software (PE international) for an advanced thermal conversion technology called Gasplasma. Refuse Derive Fuel (RDF) is fed to the innovative core of the Gasplasma process: a plasma reactor follows the gasification of waste in order to obtain an efficient cracking of problematic tars and highly polluting compounds of the raw gas. The high purity syngas is suitable, after cleaning, for electrical power generation in a gas engine. Electricity, together with the stable vitrified plasma slag and the steam from heat recovery, constitute the valuable products of the Gasplasma process.

Advanced Plasma Power developed a pilot plant of the Gasplasma process and supplied process simulation data of a commercial scale plant. We collected, organized and grouped those data for the life cycle inventory.

We performed the hot spot analysis of the Gasplasma process analysing the environmental impact of each unit of the plant. This study takes into consideration the impact categories as defined by CML (Guineè, 2002). The hot spot analysis shows that each unit contributes differently to the environmental impact depending on the type of indicator considered and also depending on the emission and the indirect burden.

The two units of the Gasplasma process that show the most marked impact are the flue gas to stack and the quench that is the cooling unit of the gas cleaning section. The flue gas to stack contributes to the Global Warming Potential for 85% of the total plant burden whereas the effluent emitted from the quench increases the fresh water toxicity, the eutrophication potential and the acidification potential by 60%, 44% and 12%, respectively.

The burden of the gasifier and the plasma is due to the high amount of energy required to produce oxygen and electricity supplied to these two units, respectively. Sodium hypochlorite and urea productions, supplied to the syngas and flue gas cleaning sections are the main contributor for the ozone layer depletion (95% of the total burden).

We also tracked the energy input and output of all activities. This analysis reveals the high primary energy requirement for the production of pure oxygen supplied to gasifier and for the production of electricity supplied to the plasma. The two major energy outputs from the plant are the electricity produced in the gas engine but also the sensible heat of the hot flue gas released to the atmosphere at high temperature.

We performed a sensitivity analysis, considering the marginal electricity production technology. The marginal technology was assumed to be a Combined Cycle Gas Turbine plant with natural gas. The GWP of the Gasplasma process is 20% lower in this case, compared to the scenario with the electricity grid mix (electricity production weighted on the base of the UK specific percentage from the different sources).

References

J. B. Guineè, Handbook on Life Cycle Assessment, Operational Guide to the ISO Standards, (Ed.) 2002, Kluwer Academic Publishers, Dordrecht.

(215l) Kinetics Analysis for Development of a Rate Constant Estimation Model of Ultrasonic Degradation Reaction

Kobayashi, D., Tokyo University of Science
Honma, C.,
Matsumoto, H.,
Takahashi, T., Tokyo University of Science
Shimada, Y., Tokyo University of Science
Kuroda, C.,
Otake, K., Tokyo University of Science
Shono, A., Tokyo University of Science

Ultrasound is used in various applications such as medical imaging, non-destructive testing of materials, underwater ranging, welding of thermoplastics, and chemical reactions. Especially, ultrasound has been found to be an attractive advanced technology for the degradation of hazardous organic compounds in water. The ultrasonic degradation of dyes has been investigated by many researches. Sonochemical degradation of organic compounds in aqueous solution occurs by pyrolysis and/or reaction with hydroxyl radicals. Moreover, kinetics of sonochemical degradation has been proposed. However, the effects of ultrasonic frequency on degradation rate were not investigated quantitatively. In our previous study, we proposed a simple model for estimating the apparent degradation rate constant of methylene blue based on ultrasonic power and the SE_{KI} value in the range of frequency between 20 kHz and 500 kHz. However, we have not investigated the ultrasonic degradation of methylene blue in high frequency region around 1 MHz. The effects of initial concentration of methylene blue and volume of sample solution have not been investigated yet, either. In this study, the degradation process using methylene blue as model hazardous organic compounds by ultrasonic irradiation was investigated. The ultrasonic frequency was operated in the range from 20 kHz to 1.6 MHz, the initial concentration of methylene blue was in the range from 0.005 to 0.04 mM, and the sample solution volume was in the range from 0.1 to 0.3 L. Our proposed model can apply to this study which extended the frequency range. The effects of initial concentration and sample solution volume can also be estimated using this model in the range from 0.01 to 0.04 mM and from 0.1 to 0.3 L, respectively. It is also postulated that threshold power is required for degradation to progress, and frequency has less influence on threshold power during degradation. In addition, this model can apply to ultrasonic degradation of other organic hazardous compounds such as phenol.

(226a) New Developments on Membranes for CO2 Separation and Capture

Chen, Y.,
Ramasubramanian, K., The Ohio State University
Severance, M. A., The Ohio State University
Tong, Z.,
Vakharia, V., The Ohio State University
Zhao, L., The Ohio State University
Zhao, Y., The Ohio State University
Dutta, P. K., The Ohio State University
Ho, W. S. W., The Ohio State University
Originally presented on: 11/5/2013 8:30:00 - 9:00:00

This presentation covers new developments on carbon dioxide-selective membranes for: (1) hydrogen purification for fuel cells and (2) carbon capture from flue gas. We have synthesized CO_2-selective membranes by incorporating amino groups into polymer networks. The membranes have shown high CO_2 permeability and selectivity vs. hydrogen, carbon monoxide and nitrogen up to 180°C. We have elucidated the effect of amine steric hindrance in the solid membrane, showing significant enhancement for CO_2 transport. Hydrogen sulfide permeates through the membrane much faster than CO_2, allowing H_2S removal in the treated synthesis gas before water-gas-shift (WGS) reaction. Our initial experiments have shown a nearly complete removal of H_2S from 50 ppm in synthesis gas to about 10 ppb in the hydrogen product. Using the membrane, we have obtained <10 ppm carbon monoxide in the hydrogen product in WGS membrane reactor experiments via CO_2 removal. The data have been in good agreement with our modeling predictions. The membrane is being scaled up for commercialization. For CO_2 capture from flue gas in coal-fired power plants for sequestration, we will discuss a new concept of inorganic/polymer composite membranes along with initial membrane transport results. We will also discuss the process modeling along with cost sensitivity for the CO_2 capture to achieve the US DOE target of £35% increase on the cost of electricity (COE) and £$40/tonne CO_2 captured for 2025. Cost calculations show that this target can be met with an air-sweep process to achieve 90% CO_2 recovery and 95% purity of the CO_2 product. This process operates at slightly above atmospheric feed pressure (about 1.1 atm) with feed compression for cost optimization and requires a low membrane module price of \$2.5/ft^2 (\$27/m^2), a CO_2/N_2 selectivity of about 150 combined with a high CO_2 permeance of 3000 GPU.

(226b) Organic Solvent Nanofiltration Molecular Separations In Organic Liquids

Livingston, A. G., Imperial College London

Originally presented on: 11/5/2013 9:00:00 - 9:30:00

Organic liquids are ubiquitous in chemical science based industries, which range in scale from refining to pharmaceutical production. It is generally accepted that 40-70% of capital and operating costs in these industries are dedicated to separations; and a substantial fraction of this cost is related to processing of organic liquids, both as product streams and solvents. Organic Solvent Nanofiltration (OSN) technology has the potential to provide game changing alternatives to conventional concentration and purification technologies such as distillation, liquid extraction, adsorption and chromatography, reducing energy requirements and so improving sustainability.

In order to achieve this potential, membrane modules must meet several challenges. They must be stable in organic solvents, offer attractive fluxes and rejections for systems of interest, and give reliable and predictable service lifetime and performance. The obvious benefits of membrane processing have been apparent for many years, and have attracted research, development and commercialisation efforts from academic groups, end users, and membrane suppliers. In the last few years these efforts have resulted in a growing range of commercially available membranes, and an increasing number of industrial applications.

This presentation will firstly briefly describe recent developments leading to a new class of OSN membranes and modules. The DuraMem™ series OSN membranes are made through post-formation processing including crosslinking of polymer membranes, which makes them stable in polar solvents such as acetone, DMF and THF. These membranes are commercially available from Evonik MET as spiral elements. Some of the key challenges overcome in developing a membrane from lab bench discovery to a commercially available product will be described.

The presentation will then describe recent research in OSN membrane formation, and OSN applications. New membranes include thin film composites, based on interfacial polymerisation (TFC-IP), having good flux and rejection in solvents including DMF and THF. There are a growing number of applications of OSN to industrial separation problems, and some will be outlined, together with the advantages OSN brings to commercial systems. Prediction of process performance for organic liquid systems will be discussed. Finally, the attributes of "ideal" membranes, and likely limitations on system performance will be outlined.

(226c) Sustainability Issues in Membrane Systems for Bioprocessing

Zydney, A. L., The Pennsylvania State University

Originally presented on: 11/5/2013 9:30:00 - 10:00:00

Although product yield and purity tend to be the dominant issues in the development of membrane systems for bioprocessing applications, membranes can also play a major role in the development of more sustainable bioprocesses. This talk will highlight several examples of sustainability issues and opportunities in membrane technology for bioprocessing applications. Ultrafiltration is currently used in the final concentration and formulation of nearly all recombinant protein products. In addition to giving very high product yield and quality, ultrafiltration also provides significant energy savings relative to alternative (e.g., thermal) processes for product concentration. Membrane chromatography can be used to significantly reduce the energy costs compared to traditional packed columns due to the significant reduction in the required pressure drop in the highly porous membranes. There are also exciting new opportunities for using membrane systems to develop continuous bioprocesses with dramatically smaller environmental footprints. These continuous processes are based on perfusion bioreactors, e.g., hollow fiber membrane bioreactors with the cells immobilized in the shell region of a hollow fiber module or more traditional bioreactors in which the cells are continuously recycled by a microfiltration unit. New research on continuous countercurrent tangential chromatography has demonstrated that this technology can provide continuous product capture using high performance resins that are processed as a flowing slurry through a series of static mixers and hollow fiber membrane modules. These continuous systems have productivities that are 10-100 times greater than conventional batch processes, dramatically reducing capital costs and environmental impact. Membrane systems should continue to play a critical role in the development of sustainable bioprocesses.

(226d) Membrane Separations As "Lever" Points in Biomass-Based Processes
Pellegrino, J., University of Colorado
Originally presented on: 11/5/2013 10:00:00 - 10:30:00

From a technical perspective, sustainability is a second law of thermodynamics analysis that has become synonymous with development of non-fossil fuel energy carriers. In that broad milieu, harvesting photons into biomass and converting these into different chemical forms has been a "grand challenge" for many decades. The challenge is having a "process make sense" from both an economic and life-cycle (environmental) perspective (the latter being where the second law usually shows up). I will illustrate how membrane-based separations can be "enabling" for potentially sustainable processes within both the algal and lignocellulosic biomass to fuels and chemicals framework. Useful membrane materials already exist for some applications, while optimum material properties and module designs are needed for others. Examples of techno-economic analysis of algal biomass processes will illustrate how membrane separations became a "lever" point for a different idea (for the appropriate species).

(155e) Progresses in Membrane Integrated Systems for Water, Energy and Raw Materials Production

Drioli, E., Institute on Membrane Technology of the National Research Council - (ITM-CNR)
Chung, T. S., National University of Singapore
Macedonio, F., Institute on Membrane Technology of the National Research Council - (ITM-CNR)
Originally presented on: 11/5/2013 10:30:00 - 11:00:00

In Nature complex molecular separation processes are controlling strategic aspects of life and evolution. Multifunctional membrane operations are often the core of the systems. Science and Engineering have been realized important progresses in the understanding of these phenomena and in their design for reproducing their potentialities artificially at industrial level or in medicine.

Artificial polymeric and inorganic membranes have been already studied and realised with important impacts on strategic aspects of our life. Dominant role has today membrane engineering for solving water stress by desalination and water reuse. Significant contribution to new form of energy systems might be related in the next future to the utilization of salinity gradient, via PRO or RED. The design of integrated membrane operations where fresh water production, energy production, and metals and various salts recovery will be realized simultaneously is becoming an interesting possibility.

New membrane reactors and membrane contactors will contribute to the realization of these objectives.

(225a) Fundamental Developments and Economic Feasibility of AOPs Involving Ultrasound for Environmental Remediation

Adewuyi, Y. G., North Carolina A&T State University
Peters, D. R. W., University of Alabama at Birmingham
Originally presented on: 11/5/2013 8:30:00 - 8:45:00

Recently, advanced oxidation processes (AOPs) have found world-wide applications for water treatment and environmental remediation and control of air toxics. AOPs involving sonochemistry (i.e., ultrasound), combinative sonophotochemical, sonophotocatalytic and sonoelectrochemical processes, hydrodynamic cavitation, persulfate and/or activated persulfate and other advanced oxidation processes are of particular emerging interests in the research community. These processes offer the potential for shorter reaction cycles, cheaper reagents, and less extreme physical conditions, leading to less expensive and perhaps smaller plants. AOPs have promise to destroy many emerging organic contaminants in water and are being considered in potable water treatment, wastewater treatment, site remediation, and industrial applications. Numerous aqueous emerging organic micropollutants such as perfluorinated surfactants and perfluroalkyl acids (PFAAs), pharmaceuticals, hormones, steroids, antibiotics, personal care products, disinfection by-products and other emerging pollutants; have been studied. This paper will provide a state-of-the-art and unified fundamental chemistry and reaction kinetics involved in the innovative advanced oxidation processes via cavitation, activated aqueous persulfate and other oxygenated oxidants; and will evaluate the performance and economics of AOPs involving ultrasound and hydrodynamic cavitation against conventional water treatment technologies. It is intended to advance our understanding of these emerging technologies and outline directions for future research.

References

Mahamuni, N.N.; **Adewuyi, Y.G.** Advanced Oxidation Processes (AOPs) Involving Ultrasound for Waste Water Treatment: A Review with Emphasis on Cost Estimation. Ultrason. Sonochem. **2010**, 17, 990-1003 http://dx.doi.org/10.1016/j.ultsonch.2009.09.005

Adewuyi, Y.G. Sonochemistry in Environmental Remediation. 2. Heterogeneous Sonophotocatalytic Oxidation Processes for the Treatment of Pollutants in Water. Environ. Sci. Technol. **2005**, 8557-8570. http://dx.doi.org/10.1021/es0509127

Adewuyi, Y.G. Sonochemistry in Environmental Remediation. 1. Combinative and Hybrid Sonophotochemical Oxidation Processes for the Treatment of Pollutants in Water. Environmental Science & Technology. **2005**, 39, 3409-3420 http://dx.doi.org/10.1021/es049138y

Adewuyi, Y.G. Sonochemistry: Environmental Science and Engineering Applications. Industrial & Engineering Chemistry Research. **2001**, 40, 4681-4715 http://dx.doi.org/10.1021/ie010096l

(225b) Advanced Oxidation of Estrone in Water and Wastewater

Sarkar, S.,
Nakhla, G., University of Western Ontario
Ray, M. B., The University of Western Ontario
Originally presented on: 11/5/2013 8:45:00 - 9:00:00

Natural estrogen, estrone (E1) is one of the most potent endocrine disrupting compounds (EDC) which is of primary concern due to its widespread presence in the aqueous environment and also due to its adverse impact both on aquatic life and potential risk to human health. Conventional waste water treatment plants are not designed to remove it, and most of it remains adsorbed on the activated sludge, while part of it is being discharged with secondary effluent, requiring tertiary treatment for water recycle, surface water protection, as well as surface water augmentation. Advanced oxidation processes (AOPs) are seen as promising technologies for removal of estrogens from environment. In this work, a comprehensive research was conducted to determine the relative performance of various advanced oxidation processes such as O_3, H_2O_2, UV, and combinations of UV/ O_3, UV/H_2O_2, ultrasonication, and Fenton agent for the removal of E1 from pure water and secondary effluent. In addition to the parent compound (E1) removal, performance of the AOPs was characterized using removal of total organic carbon (TOC) and estrogenicity in the effluent. Intermediate analysis indicates that UV photolysis produces intermediates, which absorb radiation at wavelength lower than 254 nm making them recalcitrant causing slow removal of TOC. Energy order calculations were performed and cost analysis indicates that ozonation to be the best oxidation process for the removal of E1, TOC and estrogenicity. About 0.5 mg/L of E1 degradation was complete within 30 minutes of continuous purging of water using 1 mg/L of ozone in water. Although, the initial rate of E1 degradation in secondary effluent was decreased by 50% compared to that of pure water, the final removal of E1 was not significantly affected in secondary effluent.

(225c) Analysis of Ozone Absorption and Reaction Kinetics in the Oxidation of Organics in Effluents in a Stirred Tank Reactor

Mathews, A. P., Kansas State University
Panda, K. K., Kansas State University
Originally presented on: 11/5/2013 9:00:00 - 9:15:00

The presence of high molecular weight organic compounds such as dyes, pharmaceuticals, and endocrine disruptors in receiving waters is of concern due to their toxicity and carcinogenicity to humans and aquatic organisms. Ozone is a powerful oxidant that can be used to oxidize such compounds in effluents. However, the use of ozone in these applications is somewhat limited at present due to the relatively high cost of ozone and ozonation systems. This is in part due to mass transfer limitations, and the lack of understanding of fundamental factors that contribute to poor utilization of applied ozone. Previous studies have concluded that mass transfer of ozone form the gas to the liquid phase is the main rate limiting step in ozonation. However, with high molecular weight compounds, the diffusion and transport of the compound to the reaction plane, the production of byproducts and secondary reactions, can also influence the overall oxidation kinetics. In this study, the mass transfer and reaction kinetics regimes are modeled based on fundamental principles and applied to analyze the oxidation of a high molecular weight dye in a stirred tank reactor with a bubble diffuser. Ozonation reactions are mediated primarily by ozone molecules at low pH values (pH ~3), and by hydroxyl radicals at higher pH values. The oxidation kinetics model for the dye is based on the consideration of two parallel reactions in the reactor. The first step involves a fast decomposition reaction that results in the production of primary products. In the second step, the primary products react further with ozone at a slower rate to produce secondary end products. Depending on the relative rates of reaction and mass transfer, the gas-liquid reactions can occur in the liquid film, the bulk liquid, or a combination of both. The decolorization reaction is moderately fast, and is assumed to occur in the film and the bulk liquid. The second reaction is slow, and is assumed to occur in the bulk liquid only. This paper will present experimental studies conducted at pH ~7 for the oxidation of high molecular weight dyes along with model description, verrification, and sensitivity analyses to optimize the performance of stirred tank ozonation systems.

(225d) Electrochemical Wastewater Treatment: Applications and Scale-Up

Woisetschlager, D., Graz University of Technology
Siebenhofer, M., Graz University of Technology
Originally presented on: 11/5/2013 9:15:00 - 9:30:00

In recent years great progress has been made in the research of electrochemical wastewater treatment (EAOP) with the aim to develop a universal method for the treatment of persistent water constituents. Especially anodic oxidation of organic substances with boron-doped diamond (BDD) electrodes has demonstrated its high potential for the degradation of persistent pollutants.

In this project the principles of electrochemical oxidation in wastewater treatment were developed and the applicability of this technology was verified. The degradation trends were characterised for synthetic and industrial wastewater effluents. The influence of the water matrix and the operating parameters were monitored and analysed. Boron-doped diamond electrodes prove to be a highly appropriate electrode in terms of oxidation power, application and stability. With this electrode configuration the purification of a variety of industrial wastewaters was successful. Side reactions and the formation of side products due to chloride, carbonate or ammonium load were also investigated. In the experiments the formation and prevention of chlorinated organic reaction products (AOX) was evaluated as well. The results of the research activities confirm that electrochemical wastewater treatment can establish as highly reliable technology.

(225e) Case Study - In-Situ Oxidation of Ground Water Contaminated By Petroleum Hydrocarbons, Acetone and Chlorinated Hydrocarbons At a Former Industrial Facility in Northern California

Qadir, R. M., Enpro Solutions, Inc.

Originally presented on: 11/5/2013 9:30:00 - 9:45:00

In-situ treatment of dissolved petroleum hydrocarbons and solvents presents many unique challenges due to variability in the subsurface hydrogeology, pH, and chemistry of the soil and ground water matrix. In addition, the petroleum hydrocarbon contaminants present a suite of chemicals in a variety of carbon chains, alkane lengths, and aromatic structures. Proposed in-situ treatments must be carefully tailored to the subsurface characteristics, land use, site-specific cleanup standards, and time constraints. This presentation describes the design and implementation of in-situ ground water remediation by injecting a mixture of Advanced Oxygen Release Compound (ORC-Advanced®) and Regenox® (a solid alkaline oxidant) to ground water contaminated by petroleum hydrocarbons, acetone, and chlorinated hydrocarbons at a former acetylene gas manufacturing site located in Northern California.

Nine ground water monitoring wells were installed and used to track water quality parameters and ground water contaminant concentrations within the treatment zone during remediation; and up-gradient and down-gradient to the treatment zone. Baseline ground water samples were collected and analyzed to estimate the chemical dosage and design the amendment mixture. The amendment consisted of a mixture of Regenox® for rapid, direct oxidation of contaminants and ORC Advanced®, for slow controlled release of oxidants for long-term bio-stimulation. A total of 10,023 gallons of amendment containing over five tons of oxidant was injected at 45 locations using multiple direct-push drill rigs. Eight quarters of post-treatment ground water monitoring showed that concentrations of all dissolved contaminants in ground water were successfully reduced to below cleanup standards. For example, acetone was reduced from a pre-remediation maximum concentration of 54,000 micrograms per liter (μg/L) to 1,600 μg /L.

Challenges encountered during implementation included surfacing of the injected material that reacted vigorously with naturally occurring carbon and anthropogenic contaminants. Subsurface conditions varied widely across the treatments area, including the presence of lime sludge and subsurface obstructions such as buried railroad ties, pipelines and concrete slabs. In addition, since site access constraints required that all work be performed on weekends, multiple drill rigs and support vehicles operated within a fairly small area, increasing logistical and safety concerns.

(225f) Engineering Challenges of Implementing the Peroxone Process for in-Situ Oxidation of Ground Water Contaminated By Acetone and Petroleum Hydrocarbons At a Former Industrial Facility

Qadir, R. M., Enpro Solutions, Inc.

Originally presented on: 11/5/2013 9:45:00 - 10:00:00

In-situ treatment of dissolved chemicals of concern (COC) presents unique challenges due to variability in the subsurface hydrogeology, pH, and chemistry of the soil and ground water matrix. At this former acetylene manufacturing facility, very high concentrations – currently up to 15 x 10^6 micrograms per liter (µg/L) of acetone are present in the ground water! Other COCs include gasoline range petroleum hydrocarbons and benzene, toluene, ethylbenzene and xylenes (BTEX). The Peroxone process – simultaneous ozone (O_3) and hydrogen peroxide (H_2O_2) injection - was selected for in-situ chemical oxidation of dissolved species in the ground water. This advanced oxidation process produces highly reactive hydroxyl (OH^-) free radicals during the spontaneous decomposition in the ground water of O_3 to oxygen (O_2). Adding H_2O_2 increases the OH^-concentration and the overall oxidation rate. In addition, both O_3 and H_2O_2 break down into O_2 which leads to increased bio-stimulation.

In 2010 six dual-completion (two well casings in the same well bore) injection wells were installed to deliver the oxidants to the subsurface. Six ground water monitoring wells were used to track water quality parameters and ground water COC concentrations within the treatment zone during remediation. In addition, one well is placed up-gradient to the treatment zone and another off-site sentinel well is placed down-gradient to the site. Baseline ground water samples were collected and analyzed to estimate the mass-in-place of COCs dissolved in the ground water and adsorbed onto subsurface soils. The natural chemical oxygen demand (COD) based on naturally occurring organic materials and minerals present in the soil and ground water was estimated. The total COD is many times greater than that required to react with the COCs alone. The optimal oxidant dosage and flow rates required to overcome natural COD and consume the COCs over a period of several years were estimated. A packaged remediation system of appropriate capacity was installed to produce ozone from compressed air and deliver a mix of ozone, enriched O_2 and H_2O_2 to the subsurface through injection wells. The Peroxone process control system software allowed individually varying process variables such that injection flow rate, injection time and other parameters could be customized for each individual well. A complete injection cycle consisted of delivering oxidant to all six wells in turn for the injection times determined by the individual set point at each well. On site monitoring well data show that source area COC concentrations are being reduced. The off-site sentinel well data show that COC concentrations down-gradient to the source area have been reduced by as much as 90% since the Peroxone system was installed.

Challenges encountered during system installation included the presence of subsurface obstructions such as buried utilities and concrete slabs; and logistical and safety concerns due to limited site access and restricted work hours. A major operating challenge was process alarms and shutdowns caused by apparent leaks that turned out to be migration of the injected ozone from the sub-surface through secondary containment piping back to the process equipment trailer. Additional challenges include monitoring remediation progress and optimizing process variables and injection cycles over time.

(225g) Heterogeneous Fenton Oxidation of Synthetic Dye in Water With Surface-Modified Iron-Amended Activated Carbon

Kim, J. R., University of Hawaii at Manoa
Santiano, B., University of Hawaii at Manoa
Kim, H., University of Hawaii
Kan, E., University of Hawaii at Manoa
Originally presented on: 11/5/2013 10:00:00 - 10:15:00

The present study aims to develop effective adsorption and oxidation of synthetic dye in wastewater by using newly synthesized nano iron catalyst-deposited activated carbon. Recently synthetic dye-containing wastewater has gained more attention due to its mass discharge, high toxicity and low biodegradation. For enhancing adsorption of dye and oxidative regeneration of dye-exhausted activated carbon, the novel synthesis of nano iron-deposited granular activated carbon (GAC) was developed. It was to amend ferrous iron onto the acid-pretreated GAC followed by Fe amendment onto the acid-pretreated GAC when pH of iron solution was higher than the pH at point of zero charge ($pH_{, pzc}$) of the GAC. Methylene blue (MB) in water was adsorbed onto the iron-amended GAC (Fe-GAC) followed by single or multiple applications of H_2O_2. Batch experiments were carried out to study the adsorption isotherm and kinetics at at 20°C and pH 6 with the initial concentration of 20-200 mg/L and the solution pH of 6.0. It showed that the adsorption isotherm for MB onto Fe-GAC followed Langmuir isotherm, yielding maximum monolayer adsorption capacity (q_m) of 238.1 ± 0.78 mg/g which is higher than the virgin GAC with q_m of 175.4 ± 13.6 mg/g at 20°C and pH 6. The kinetics of adsorption results indicated that pseudo-second order kinetics represented adsorption of MB. The heterogeneous Fenton oxidation of MB in the Fe-GAC showed increasing the H_2O_2 concentration from 7 to 140 mmol H_2O_2/mmol MB led to enhancing the oxidation efficiency of MB in the GAC from 62.6% to 100%, respectively. The half-life of the H_2O_2 showed the range between 1.4 to 39.4 h depending on the H_2O_2 concentrations at a fixed iron loading of 6 mg Fe/g GAC. The oxidation of MB in the GAC increased at the higher initial concentration of H_2O_2 and Fe^{2+} because of the increased generation of hydroxyl radicals. Further enhancement of oxidation of MB in the Fe-GAC was made by the multiple injection of H_2O_2 while minimizing OH radical scavenging often occurring at high concentration of H_2O_2. The re-adsorptive capacity of the Fe-GAC showed 87-92% over the repeated multiple cycle of adsorption and oxidation of MB indicating effective treatment of MB in wastewater. Further works will include development of diffusion-reaction model and the optimization of the operating conditions over the repeated cycles of adsorption and oxidation.

(225h) Reaction Kinetics and Mechanistic Studies of Nitric Oxide Removal By Combined Aqueous Persulfate and Ferrous-EDTA Systems

Adewuyi, Y. G., North Carolina A&T State University
Khan, M. A., North Carolina A&T State Uuniversity
Originally presented on: 11/5/2013 10:15:00 - 10:30:00

Nitrogen oxides (NO_x – mainly NO_2 and NO) and sulfur oxides (SO_x – mainly SO_2) are the most prominent acid gases emitted from the burning of fossil fuel especially from the coal fired power plants, and are responsible for widespread problems of air pollution, health hazards, acid rains etc. Of these oxides NO_2 and SO_2 are readily soluble in water and can be scrubbed easily from the exhaust stream but nitric oxide (NO) is very sparingly soluble in water and cannot be separated easily. It is well known that the commonly practiced methods of removing NO such as selective catalytic and non-catalytic processes have high capital costs and undesirable problems of high temperatures and handling of harmful chemicals, and alternative cost-effective and environmentally friendlier processes are of ardent interest. It is also well known that suitable oxidizing agents can increase the solubility of NO significantly in water and our group's previous studies involved detailed experimental and mechanistic studies of the removal of NO by aqueous oxidizing agents, of which, peroxydisulfate or persulfate ($S_2O_8^{2-}$) activated by temperature showed significant removal (up to 90% at 90^0C). Later work involving the simultaneous activation by thermal and ferrous ion showed further 10% increment in the NO removal at all temperatures. But the energy requirement for high temperature process to obtain higher removal rates could be one of the main stumbling blocks for further developments. A number of reports have suggested NO removal by iron chelating agent especially ferrous ethylenediamine tetraacetic acid (Fe^{II}-EDTA) where NO is combined with Fe^{II}-EDTA and separated from the solution. The main drawbacks this process are removal of Fe^{II}-EDTA(NO) complex and fast oxidation of Fe^{II}-EDTA to inert Fe^{III}-EDTA and high cost of EDTA. The problem was somewhat overcome in BiodNO_x process where Fe^{II}-EDTA(NO) is broken down by denitrifying bacteria and Fe^{III}-EDTA is reduced by iron reducing bacteria and the resultant Fe^{II}-EDTA is recycled to the reactor again. But the process was impeded by the slow action of the biological process and uncertainty involving in bacteria culture and the additional cost of another unit. It is widely reported in the literature that Fe^{III}-EDTA can also be reduced by persulfate ion ($S_2O_8^{2-}$) and sulfate radical ($SO_4^{\bullet-}$) and the resulting Fe^{II}-EDTA used for the remediation of other chemical pollutants. However, to the best of our knowledge the simultaneous and synergistic application of peroxydisulfate ($S_2O_8^{2-}$) and Fe^{II}-EDTA for the removal of NO has never been studied and is the main focus of this work.

NO removal by combined aqueous persulfate and ferrous EDTA was studied in a bubble column reactor in semibatch mode. Firstly, the optimum Fe^{II}:EDTA ratio was determined for different temperatures and the

1:1 ratio was found to give the highest NO removal rates. The experiments were carried out at different temperatures and the results were compared to the thermal- and iron-activated persulfate alone removal experiments. It was shown that the combined persulfate-ferrous-EDTA system resulted in significantly higher NO removal at low temperatures, almost 30% higher than the thermal- and iron-activated systems. Also, the pH for optimal NO removal was found to be around neutral (pH 7.0). The concentrations of the ferrous, ferric ion and iron-EDTA complex species were measured by spectrophotometric methods and total iron material balance verified. Finally, kinetics and mechanistic reaction pathways for the complex process were investigated using a physicochemical mathematical model involving the use of pseudo-steady-state-approximation technique and film theory of mass transfer. The model was solved numerically using fourth order RK method in Matlab to find the theoretical concentration profile and results used to validate the experimental data. We are currently investigating the effects SO_2 and the various process parameters using design of experiments (DOE) methodologies.

References

Adewuyi, Y.G., Khan, N.E. Modeling the Ultrasonic Cavitation-Enhanced Removal of Nitrogen Oxide (NO) in a Bubble Column Reactor. AIChE J. **2012**, 58, 2397 - 2411. http://onlinelibrary.wiley.com/doi/10.1002/aic.12751/pdf

Khan, N.E.; **Adewuyi, Y.G.** Absorption and Oxidation of Nitric Oxide (NO) by Aqueous Solutions of Sodium Persulfate in a Bubble Column Reactor. Ind. Eng. Chem. Res. **2010**, 49 8749-8760. http://pubs.acs.org/doi/abs/10.1021/ie100607u

Adewuyi, Y.G., Owusu, S.O. Ultrasound-induced Aqueous Removal of Nitric Oxide from Flue Gases. Effects of Sulfur Dioxide, Chloride and Chemical Oxidant. J. Phys. Chem. A. **2006**. 110, 11098-11107. http://dx.doi.org/10.1021/jp0631634

Owusu, S.O; **Adewuyi, Y.G.** Sonochemical Removal of Nitric Oxide from Flue Gases. Ind. Eng. Chem. Res. **2006**, 45, 4475-4485. http://dx.doi.org/10.1021/ie0509692

Adewuyi, Y.G.; Owusu, S.O. Aqueous Absorption and Oxidation of Nitric Oxide with Oxone for the Treatment of Tail Gases: Process Feasibility, Stoichiometry, Reaction Pathways and Absorption Rate. Ind. Eng. Chem. Res. **2003,** 42, 4084-4100. http://dx.doi.org/10.1021/ie020709+

Adewuyi, Y.G.; He, X.; Shaw, H.;Lolertpihop, W. Simultaneous Absorption and Oxidation of NO and SO_2 by Aqueous Solutions of Sodium Chlorite. Chem. Eng. Commun., **1999**, 174, 21-51. http://dx.doi.org/10.1080/00986449908912788

(225i) Sonication Treatment of Sump Water Containing Diesel Fuel

Peters, D. R. W., University of Alabama at Birmingham
Bessette, R., Sound Environmental Systems
Originally presented on: 11/5/2013 10:30:00 - 10:45:00

Pervasive use of diesel fuel around rail yards impacts stormwater, by increasing petroleum releases to surface waters. On-site treatment of diesel sump water generally involves gravity oil-water separation. However, oil-water separators cannot meet National Pollution Discharge Elimination System (NPDES) stormwater discharge standards for oil and grease because they best remove "free" oil, not emulsified oil. Oil-water gravity separation treatment is least effective during stormwater runoff events. An on-site, cost-effective treatment technology treatment technology of diesel sump water is lacking.

This research had as its objective to determine sonication's potential to treat diesel laden stormwater. Sonication effectively oxidizes and pyrolyses a wide range of hydrocarbons in aqueous solution. A mass balance, using gravimetry and chromatography, determined the physical-chemical fate of diesel range organics (DRO). The rate of DRO decrease as a function of sonication time was investigated. Bench-scale sonication results from an initial diesel fuel concentration of 203 mg/L achieved the NPDES discharge level of 15 mg/L within 10 minutes treatment time. Sonication of diesel fuel produced carbon monoxide, short-chain hydrocarbon gases, short-chain hydrocarbons in aqueous solution, solid hydrocarbon precipitates, and residual diesel range organics. After 20 minutes treatment of No.2 diesel fuel (198 mg/L, 203 mg/L, and 2033 mg/L) all produced carbon monoxide and short-chain carbon compounds (< C6) in the gas phase. This indicated that lysis and oxidation of the diesel fuel hydrocarbons occurred. Evidence of diesel fuel hydrocarbon lysis was found in the sequence of changing chromatograms from each initial concentration's subjugation to sonication over a period of 20 minutes. The chromatograms showed a decrease of hydrocarbons in the diesel fuel range and a simultaneous shift of carbon compounds appearing the shorter (< C10) region of the chromatogram and "shouldering" into the solvent peak. There was a shift of the longer carbon chains to shorter carbon chains as sonication time increased from 0 to 10 minutes.

Sonication was more energy intensive per unit mass treated than oil water separation. However, sonication's [potential to eliminate peripheral costs required by oil-water gravity separation may prove cost effective.

(225j) UV-Based Treatment of Fluoroquinolone Antibiotics in Wastewater

Snowberger, S., University of Maryland, Baltimore County
He, K., University of Maryland, Baltimore County
Blaney, L., Lehigh University
Originally presented on: 11/5/2013 10:45:00 - 11:00:00

Over the past few decades, pharmaceutical use has become significantly more prevalent due to advances in technology and medicine; however, this rise in pharmaceutical consumption has been correlated with increases in the concentrations of pharmaceuticals detected in municipal wastewater. Several reports have expressed concerns over the presence of antibiotics in wastewater due to the potential for inhibition of select microorganisms and the development of antibiotic resistance. Furthermore, low concentrations of novel biologically active compounds may pose threats to aquatic ecosystems. Trace concentrations of pharmaceuticals persist through wastewater treatment plants and are discharged to surface water, compounding the issue.

This research examines transformation of pharmaceutical antibiotics using UV irradiation and the UV-H_2O_2 advanced oxidation process. Five fluoroquinolone antibiotics, namely ciprofloxacin, enrofloxacin, levofloxacin, norfloxacin, and sarafloxacin, were investigated. Transformation occurred through direct photolysis (UV), as well as through reaction with hydroxyl radicals (UV-H_2O_2). The reaction kinetics were examined by quantifying the quantum yield and the fluence-based rate constants. For each experiment, 10^{-7} M of the antibiotics were dosed into 2 L of 10 mM phosphate buffer (for pH control). In the case of UV-H_2O_2 testing, 10^{-5}-10^{-7} M hydrogen peroxide was added to the experimental solution. The solution was then run through a batch-recycle reactor containing a low-pressure UV lamp that emits 2.55×10^{-4} Einstein/min at 254nm. Experiments generally lasted ten minutes for direct photolysis and five minutes for UV-H_2O_2; samples were withdrawn at appropriate time intervals. Fluoroquinolone concentrations in experimental samples were analyzed using high performance liquid chromatography with fluorescence detection (HPLC-FLD), which allowed for measurement of all five fluoroquinolones in one sample run.

To examine the effect of complex water matrices on transformation, combinations of multiple fluoroquinolones (total concentration of 10^{-7} M) were simultaneously dosed into the reactor at varying molar ratios. Data resulting from these experiments verified the quantum yield and fluence-based rate constants for all five fluoroquinolones. In addition, the impact of real wastewater matrices was examined by spiking fluoroquinolones into influent and tertiary effluent from a regional wastewater treatment plant. The resulting transformation efficiencies were used to quantify the impact of wastewater-derived organic matter on treatment of fluoroquinolone antibiotics. The impact of dissolved organic matter on UV photolysis and UV-H_2O_2advanced oxidation was determined and provides important insight into transformation of pharmaceuticals by direct photolysis and hydroxyl radicals.

The results of this research include the following: fluence-based rate constants for the five fluoroquinolones at wastewater treatment-relevant pH; the impact of complex wastewater matrices on transformation kinetics (*i.e.*, multiple fluoroquinolones and wastewater-derived organic matter); and the improvement in transformation kinetics as a function of H_2O_2 concentration. As more wastewater treatment plants utilize UV disinfection processes, the results of this research will prove useful in developing effective solutions to the emerging threat from pharmaceuticals.

(242a) Simultaneous Adsorption and Heterogeneous Oxidation of Endocrine Disrupting Compounds in Wastewater Using Nano Metal Catalyst-Deposited Carbon Nanotubes

Cleveland, V., University of Hawaii at Manoa
Bingham, J. P.,
Kan, E., University of Hawaii at Manoa
Originally presented on: 11/5/2013 8:30:00 - 8:51:00

A process for the removal of bisphenol A (BPA), one of the endocrine disrupting compounds (EDC) from wastewater was developed using nano metal catalyst-deposited carbon nanotubes (CNT). For the past decade BPA have gained more attention as emerging contaminants, and have been linked to multiple cancers below the concentration of ppb. This process relies on simultaneous adsorption and heterogeneous *in-situ* Fenton oxidation for degradation of BPA. The present study investigated how acidic surface modifications of CNTs and the novel amendment of iron onto the CNTs affect the adsorption and oxidation of BPA. The virgin CNTs were treated with acid to increase acidic surface oxide groups while lowering the pH at the point of zero charge (pH, $_{PZC}$) of the CNT, and improve electrostatic interactions with the iron catalyst for enhancing dispersion and deposition of iron onto the CNT. The novel iron amendment method encourages a uniform distribution of iron nanoparticles onto the surface of the CNTs. The SEM and TEM image analysis confirmed the uniform distribution of the iron over the CNT. Adsorption and desorption of BPA onto the CNTs was found to reach equilibrium in less than 5 min. The iron-amended acid-treated CNTs and the iron-amended untreated CNTs followed the Langmuir model of adsorption, indicating a single layer of adsorption. The iron-amended CNTs showed slightly lower maximum adsorption capacity for BPA in water compared to the virgin CNTs. The oxidation of BPA with the iron-amended acid-treated CNTs showed greater than 95% removal of BPA and a H_2O_2 half-life ranging from 36-1386 min at 4.5-45 mmol H_2O_2/mmol BPA at pH 3.5 over the repeated cycle of treatment of BPA. The heterogeneous Fenton oxidation of BPA in water revealed quick adsorption, effective degradation of BPA, and high regeneration of adsorptive capacity with no iron sludge production compared to homogeneous Fenton oxidation. Since the limitation of the process is the oxidation rate of BPA, further investigation will focus on enhancing the reaction kinetics of BPA oxidation by optimizing the ratio of BPA:H2O2:Fe and operating conditions such as temperature, pH and mixing.

(242b) Study of Carbon-Hydrogen Bond On the Microstructure of Hydrogenated Graphite

Zhang, Y., University of Birmingham; Waseda University

Originally presented on: 11/5/2013 8:51:00 - 9:12:00

There is a need to understand the hydrogen sorption mechanism(s). In this work, hydrogenated graphite samples were characterized by TGA-Mass Spectrometry, Raman spectros-copy and high resolution TEM. It was found that graphite milled for various times (0-40 h) had different d-spacings and CHx ($x \leq 3$) ratios. Graphite milled up to 10 h that was subsequently heated (under argon to 990 °C) desorbed only hydrogen, while above 10 h the amount of desorbed hydrogen decreased and an increasing amount of methane was also evolved. The 10 h sample had no measurable sp3 bonding, and desorbed 5.5 wt% hydrogen when heated. While the 40 h sample had a d-spacing of 0.48 nm, both sp2 and sp3 bonding, and desorbed 5.1 wt% pure hydrogen. The mechanism of graphite hydrogenation is discussed in terms of hydrogen atoms being predominately trapped at the graphite edges, rather than between graphene layers.

References

[1]. S. Orimo, G. Majer, T. Fukunaga, A. Zuttel, L. Schlapbach and H. Fujii, *Applied Physics Letters*, 1999, **75**, 3093-3095.

[2]. T. Ichikawa, H. Fujii, S. Isobe and K. Nabeta, *Applied Physics Letters*, 2005, **86**.

[3]. H. Miyaoka, T. Ichikawa, T. Fujii, W. Ishida, S. Isobe, H. Fuji and Y. Kojima, *J Alloy Compd*, 2010, **507**, 547-550.

(242c) Efficient Removal of Both Arsenic and Fluoride By Lanthanum Modified Carbon Nanoparticles

Li, G., National University of Singapore
Yu, Y., National University of Singapore
Chen, J. P., National University of Singapore
Ma, Y., National University of Singapore
Originally presented on: 11/5/2013 9:12:00 - 9:33:00

In this study, a novel lanthanum modified carbon nanoparticle was prepared and applied in the removal of both arsenic and fluoride. Electron microscopic investigation showed that the prepared carbon particles were around 50-70 nm in diameter. The adsorbent showed good adsorption efficiency of both arsenic and fluoride in a wide range of solution pH. In the kinetic study, it was observed that the adsorption process was fast and around 90% of its final adsorption was achieved within the first 10 min. The Langmuir equation well described the adsorption isotherm. At the optimal pH, adsorption capacities of arsenic and fluoride were around 102mg/g and 122mg/g, respectively. The ionic strength had a slight influence on the arsenic and fluoride removal. The effect of common existing anions such as PO_4^{3-} HCO_3^{-} SO_4^{2-} and SiO_3^{2-} and humic acid on the adsorption process was investigated in detail. From this investigation it was concluded that the La-immobilized carbon nanoparticle showed the fast removal of toxic arsenate and fluoride from water with very high adsorption efficiency which is beneficial to its application in the water treatment.

(242d) Efficient Removal of Fluoride By Lanthanum Modified Seaweeds Sargassum S.P

Li, G., National University of Singapore
Yu, Y., National University of Singapore
Chen, J. P., National University of Singapore
Originally presented on: 11/5/2013 9:33:00 - 9:54:00

In this study, a novel lanthanum modified seaweeds (Sargassum S.P.) was prepared and applied in the fluoride removal from the water. The adsorbent showed good adsorption efficiency on the fluoride removal in a wide range of solution pH and the maximum adsorption capacity reached 45mg/g at pH 4. In the kinetic study, it was observed that the adsorption process was very fast and around 75% of its final adsorption was achieved within the first 10 min. The Frenudlich isotherm equation well described the adsorption process. At the optimal pH, adsorption capacities of fluoride were around 125mg/g. The fluoride removal efficiency increased with the increase of the ionic strength in the solution. The effect of common existing anions such as PO_4^{3-}, HCO_3^-, SO_4^{2-}, $SiO_3^{2-}$and AsO_4^{3-} and humic acid on the adsorption process was also investigated in detail. The presence of SO_4^{2-} did not show a significant impact on the fluoride removal while the adsorption efficiency was decreased with the increase of the concentration of other ions and humic acid. From this investigation it was concluded that the La-immobilized seaweed showed the fast fluoride removal from water with very high adsorption efficiency which is beneficial to its application in the water treatment.

(242e) Size-Dependent Removal of Arsenite From Aqueous Solution By Akaganeite (β-FeOOH)

Wu, S. H., School of Chemical Engineering and Technology
Sun, Y. L. Sr., Department of Chemical Engineering, School of Chemical Engineering and Technology, Tianjin University,
Jia, S. Y. Sr., Department of Chemical Engineering, School of Chemical Engineering and Technology, Tianjin University,
Zan, T. T. Jr., Department of Chemical Engineering, School of Chemical Engineering and Technology, Tianjin University,
Wang, J. Sr.,
Originally presented on: 11/5/2013 9:54:00 - 10:15:00

Crystal size and morphology had a significant effect on its adsorption behaviors. Therefore, in the present study, we have synthesized three different sizes (from <10 nm="" to="">100 nm) and morphologies of akaganeite nanometer materials to investigate their different performance of the adsorption of the arsenite. The kinetics data of these three adsorbents showed little difference and were best described by the pseudo-first-order and pseudo-second-order equation which assumed that the adsorption rate depended on both physical and chemical adsorption. The adsorption equilibrium isotherm fitted a Langmuir isotherm model well and the maximum loading capacity could reach 790 mg of arsenite/g of adsorbent. The value was comparatively higher than previously reported. This was due to the high specific surface and the unique combination between akaganeite and arsenite. The influence of the pH on the adsorption rate was also discussed. The results showed that the adsorption rate of the size smaller than 100 nm had little change with the increasing of pH. While the adsorption rate of the size larger than 100 nm presented apparent difference as pH shifted from 7.3 to 10. A mechanism that chloride ion was replaced by arsenite had been proposed. The phosphate and silicate showed significant competitiveness with arsenite, while sulfate showed little influence. Besides, there was no change on the crystalline structure of the adsorbents after adsorption. This work offered a novel understanding on the adsorption mechanism of the arsenite by the akaganeite.

(242f) Removal of Ammonia From Water Using Nanoscale Zero Valent Irons Modified Fly Ash-Derived Zeolite Adsorbents

Li'an, H.,
Minmin, L.,
Originally presented on: 11/5/2013 10:15:00 - 10:36:00

Fly ash-derived zeolite A was impregnated with different weigh ratios of nanoscale zero valent irons to produce new adsorbents (x) Fe-Z. Adsorbents were tested to remove ammonia from water solutions. Simultaneous removals of ammonia were examined via batch adsorption experiments. The new adsorbents were characterized by the X-ray diffraction (XRD), Brunauer-Emmett-Teller-N2 surface areas, pore size distributions, Transmission Electron Microscope (TEM) and Hysteresis loops in order to probe effects of the weigh percentages of NZVI impregnated. Virgin and exhausted adsorbents were examined via X-ray photoelectron spectroscopy (XPS) analyses to elucidate adsorption mechanisms. Maximum adsorption capacity of the ammonia for (50) Fe-Z adsorbent was 50.2mg/g. The maximum adsorption capacity of (50) Fe-Z is more than that of other adsorbents for ammonia removal from water. The performance was affected by the pH of aqueous solution, the initial concentration and reaction temperature. The adsorption mechanisms of ammonia on (x) Fe-Z are attributed to the ion exchange and electrostatic adsorption. Some parts of NZVI were oxidized and substituted the aluminum in the framework of zeolite A in the synthesis process of (x) Fe-Z. It made more sodium ions be exchanged with positive ions after NZVI impregnated into zeolite A. In addition, the negative surface charge of (x) Fe-Z attracted the positive ammonia ions. Furthermore, the dispersion of NZVI on the surface of (x) Fe-Z was benefit for (x) Fe-Z to be separated magnetically. Moreover, the presence of iron hydroxide promoted adsorption of ammonia on the surface of (x) Fe-Z.

(242g) Drinking Water Disinfection Using Silver Nanoparticle Impregnated Activated Carbon Hybrid

Biswas, P., Indian Institute of Technology Bombay
Bandyopadhyaya, R., IIT Bombay
Originally presented on: 11/5/2013 10:36:00 - 10:57:00

In most household drinking water purification systems, activated carbon (AC) is extensively used as a filtration medium. However, prolonged use results in biofouling, which can be eliminated by utilizing the biocidal effect of silver (Ag). In the present work, Ag nanoparticle has been impregnated in AC, for disinfection and purification of drinking water.

Ag nanoparticles were synthesized by three different methods, namely, UV reduction (A), UV reduction in presence of ammonia (B) and thermal reduction (C). Silver nitrate was used as the precursor, while tri-sodium citrate acts as both a reducing as also a stabilizing agent, to synthesize a stable Ag nanoparticle dispersion in water. Methods A, B and C showed mean primary nanoparticle diameter of 28.1 nm, 44.8 nm and 26.7 nm respectively, with standard deviation of 12.6 nm, 19.4 nm and 11.7 nm respectively. So, method B was not desirable due to the larger particle size. Between the other two, method C does not work at higher precursor concentration and results in aggregation and sedimentation of nanoparticles. So method A was selected, as usage of higher initial precursor concentration was possible, due to the slow rate of reduction, resulting in higher quantity of Ag particles being formed. The AC used in the present work is hydrophobic in nature, with a contact angel with water of 112°. To improve wettability of an aqueous Ag dispersion during the subsequent wet-impregnation step, the AC granules were therefore first plasma treated under a pure oxygen atmosphere. Consequent to this, FTIR spectroscopy showed a definitive increase in polar functional groups like, carboxyl, ketones on the outer surface of AC, which hence resulted in Ag impregnation mostly on the outer surface of AC, instead of the pore-interiors of AC, during the wet-impregnation step. FEG-SEM images confirmed the presence of Ag nanoparticles on the outer surface of the Ag-AC hybrid, with a loading of 0.78 wt. % of Ag in the Ag-AC hybrid.

Disinfection studies were performed using these hybrid granules, in both batch and continuous mode, using a wild type model microorganism, namely, *E. coli* K12 (MTCC 1302), starting with an initial cell concentration of 10^4 CFU/ml, typical of contaminated drinking water. Cells were counted using the plate-count method and the antibacterial activity of Ag-AC was confirmed by the zone of inhibition method. For a typical loading of 8 mg of Ag-AC hybrid per ml of cell suspension, batch experiments showed a first order rate constant of 0.21 min^{-1} with respect to *E. coli* concentration. Subsequently, residence time required for continuous column-mode disinfection experiments were calculated based on batch kinetic data. To this end, a small column with a diameter of 1 cm and a height of 25 cm was used. Inlet water with a flow rate of 0.4 ml/min and having 10^4 CFU/ml *E. coli* was successfully treated by Ag-AC, so that no cells in the outlet water stream was detected.

In general from such hybrids, there is a slow loss of Ag due to formation of Ag-ions in presence of water. Our batch study however showed only 0.31 wt% of Ag release out of the total Ag loaded in the Ag-AC hybrid, even after a 108 h of contact period with water. The overall release rate was thus significantly low and confirms the potential of a long period of usage for antibacterial activities of Ag-AC. This was possible as Ag was impregnated with a good adhesion on the plasma treated functionalized surface of the hybrid material. Thus, the high biocidal activity coupled with a low rate of Ag release establishes the Ag-AC hybrid of the present study as a good material for drinking water disinfection. Hence, water purification, disinfection and biofouling prevention can be potentially achieved in a single water-filter column, suitable for potable quality water. Further ongoing work is aimed towards this goal.

(245a) Implementation of Tower Gardens On An Urban University Campus

Peters, D. R. W., University of Alabama at Birmingham
Barbour, N.,
Gaston, S.,
Burns, M.,
Originally presented on: 11/5/2013 8:30:00 - 8:55:00

As society faces the problems of food deserts, diseases of civilization due to malnourishment, and countless environmental harm such as pollution or droughts, creating sustainable communities may serve as a solution to cope with these problems. Implementing green spaces in a form of vertical gardens will have positive environmental effects and offer improved access and quality of fresh produce in the urban as well as rural areas. Growing produce in tower gardens and consuming more unprocessed foods, can potentially reduce rates of undernourishment and many illnesses of longevity.

The Tower Garden® utilizes aeroponics to recycle 100% of its nutrients and water. Furthermore, the system reduces the need for use of pesticides, insecticides, or herbicides. Vertical gardens also use less than 10% of both water and land that are used in traditional farming. They have an undeniable benefit of high yield fresh, nutritious, produce.

Several Tower Gardens are being installed on building rooftops on the University of Alabama at Birmingham (UAB) campus. Raised soil bedded roof gardens are already in-place on several buildings. Knowledge is being gained as to what is most successful in a roof garden in terms of soils and plant materials. Also under study is the growth of plant materials under varying degrees of water and nutrients, and differing locations with respect to sunlight. The use of the Tower Garden will provide the researchers with another system to test and compare with the success of the raised soil bedded gardens already in place. The rooftop gardens promote a sustainable environment as they enhance cooling on rooftops and provide increase insulation for the host building rooftop. This serves to reduce energy use and lowers the building operating costs. The Tower Gardens add two additional benefits to the positive nature of vegetated roofs: (1) the plants grow in an air/water/nutrient mix, with most of the water returned to the atmosphere via evaporation, and (2) the main plant sources are edible produce. For this research project, the Tower Gardens make use of building collected rainwater and equipment condensate as the water source, thereby reducing the storm water runoff from rooftops.

All the benefits of vertical farming contribute to a more sustainable planet.

(245b) New Developments in Implementing Green Engineering in the Chemical Engineering Curriculum

Shonnard, D. R., Michigan Technological University
Allen, D. T., The University of Texas at Austin
Nguyen, N., U.S. Environmental Protection Agency
Austin, S., Chemical Engineering Branch, US EPA
Originally presented on: 11/5/2013 8:55:00 - 9:20:00

The goal of this presentation is to outline updated green engineering and sustainable engineering content for undergraduate and graduate chemical engineering education. The work of chemical engineers is central to the development of sustainable chemical and energy production systems and to meet global human needs. Incorporating environmental issues and sustainability into the design of engineered products, processes, and systems generally means evaluating environmental, economic, and social performance. This presentation will summarize two approaches to incorporate green and sustainable engineering into undergraduate chemical engineering education; **i.** a stand-alone elective course, and **ii.** modules for required core chemical engineering courses. The modular approach, one or more chapters in a forthcoming 2nd edition of "Green Engineering: Environmentally-Conscious Design of Chemical Processes", Pearson-Prentice Hall, can serve as stand-alone supplementary materials for core chemical engineering courses. For example, in a thermodynamics course, environmental equilibrium partitioning and fate properties of chemicals could be a supplemental topic and these properties can be predicted with group contribution or other structure-activity relationships. In a mass balance course, these environmental partitioning and fate properties can be used in conservation of mass problems involving accidental releases of pollutants to various environmental media; air, water, soils, and sediments. These and other examples of incorporating green engineering concepts, methods, and tools in core chemical engineering courses will be presented.

(245c) Incorporating Process Integration Methodologies Into Senior Capstone Design Courses

Dunn, R. F., Polymer and Chemical Technologies, LLC
Debelak, K. A., Vanderbilt University
Originally presented on: 11/5/2013 9:20:00 - 9:45:00

Significant efforts have been devoted to the development of process integration design methodologies over the past several decades. Some of the main developments in process systems analysis using process integration design tools are the identification of heat exchange networks, mass exchange networks and water allocation networks. These systems-based tools are widely applicable and are now essential elements of undergraduate process design instruction. Furthermore, numerous opportunities exist to incorporate these tools in the analysis of senior capstone design projects. In addition, senior design projects based primarily on process integration design methodologies have recently been developed. Experience and lessons learned in applying process integration in senior design projects in the undergraduate curriculum will be presented.

(245d) Assembling Engineering for Sustainable Development Programs

Jensen, C. D., Colorado School of Mines

Originally presented on: 11/5/2013 9:45:00 - 10:10:00

This talk will present four years of action research aimed at developing an Engineering for Sustainable Development (ESD) degree program at a Higher Education Institution known for expertise in mineral extraction, geosciences, materials, and science-technology-engineering-mathematics topics. The action research presented will highlight six stages of research that when incorported into the education process illustrate the breadth and depth that ESD programs can take. The action research presented continues to evolve and recent outcomes (e.g. a collaborative international summer curriculum) will be presented. This research is a testament that traditional engineering programs can be progressive and make use of holistic concepts and not only reductionist education philosophy or principles.

(337a) Ambient-Air Plasmas for Frugal and Sustainable Water Disinfection in the Developing World

Pavlovich, M. J., University of California - Berkeley
Galleher, C., Univeristy of California - Berkeley
Ono, T., Tokyo Institute of Technology
Machala, Z., University of California - Berkeley
Clark, D. S., University of California
Graves, D. B., University of California - Berkeley
Originally presented on: 11/5/2013 15:15:00 - 15:37:00

Ambient-condition air plasmas have been shown to decontaminate water and aqueous solutions through the production of reactive oxygen and nitrogen species (RONS) that are toxic to bacteria and other microorganisms. One method of creating air plasma at ambient conditions is the "dielectric barrier discharge" (DBD) technique, which creates a small, confined plasma region near a pair of electrodes separated by a dielectric layer such as glass or ceramic; RONS diffuse away from the plasma region and into a contaminated aqueous volume. We have previously shown that DBD-based devices can disinfect water in two distinct "modes": a high-power, nitrogen oxides-rich discharge creates a persistent antimicrobial effect in water that can last at least one week [1], while a low-power, ozone-rich mode can rapidly decontaminate aqueous solution [2].

We describe a design and configuration of air plasma reactors for the novel application of water disinfection in low-resource settings, including the developing world. Ambient-air plasma is a promising method for low-resource sanitation because it requires only electricity and air to create a disinfectant. The plasma reactors used in this study are assembled from inexpensive materials including ceramic tubes and steel rods, and comparatively low-cost energy sources and power supplies like solar-energy rechargeable lead-acid batteries and neon-sign transformers are used to ignite the plasma. In this study, we compare the gas- and liquid-phase RONS chemistry created by these "frugal" devices to the chemistry observed by using more expensive and precisely tunable laboratory devices. In addition, we show how the disinfection efficacy, using *E. coli* as a model bacterium, depends on transformer voltage, dissipated power, air flow rate through the reactors, and reactor configuration. Our preliminary results indicate that the concentrations of some reactive species can differ dramatically based on the reactor geometry and material and the power source. However, the chemistry measured here agrees reasonably well with the established laboratory-scale plasma chemistry in some important aspects, including the ozone-nitrogen oxide "mode transition". Therefore, air plasma treatment remains a promising strategy for inexpensive and effective water disinfection.

References:

[1] M. J. Traylor, M. J. Pavlovich, S. Karim, P. Hait, Y. Sakiyama, D. S. Clark, D. B. Graves, *Journal of Physics D: Applied Physics* **2011**, *44*, 472001.

[2] M. J. Pavlovich, H.-W. Chang, Y. Sakiyama, D. S. Clark, D. B. Graves, *Journal of Physics D: Applied Physics* **2013**, *46*, 145202.

(337c) Chemical Degradation of Xanthate By Pulse Corona Discharges

Cerda, K. V., University Catolica del Norte
Arce, P. E., Tennessee Technological University
Originally presented on: 11/5/2013 15:59:00 - 16:21:00

This work is focused on the degradation of xanthates waste used in flotation process from copper mining companies, which are organosulfur compound useful in the extraction of certain ores, in this case sulfide ore. These wastes are often deposited on tailings - pile in areas surrounding the mine site, and sometimes they drain-down. For this reason, it will be useful to degrade them instead of being deposited. A very promising high oxidation technique based on electrical discharges will be used as the key method for the degradation of xanthate in this research project. This presentation will overview both oxidation technique as well as the plan to conduct preliminary studies of feasibility for the degradation of xanthate in aqueous solution. Preliminary results of the investigation will be discussed.

(337d) Pursuing a Viable Plasma-Based Water Treatment Process: Identifying Transport Limitations and Investigating the Effect of Reactor Design On the Degradation of Bisphenol A

Mededovic, S., Clarkson University
Bellona, C., Clarkson University
Dai, F., Clarkson University
Franclemont, J., Clarkson University
Holsen, T., Clarkson University
Stratton, G., Clarkson University
Zhang, C., Clarkson University

Originally presented on: 11/5/2013 16:21:00 - 16:43:00

To improve the feasibility of plasma-based water treatment technology, a study was conducted to identify and characterize the design parameters that influence treatment effectiveness and the physical phenomena that determine treatment efficiency. A compound's octanol-water partition coefficient (i.e., hydrophobicity) has been determined as the major factor that influences the compounds' relative ability to diffuse into the plasma channel. Experiments conducted with nine different solutes have shown that hydrophilic compounds tend to stay in the bulk liquid whereas hydrophobic compounds such as Bisphenol A (BPA) readily diffuse into the plasma channel where they react with OH and other radicals. Thus the electrical discharge as an advanced oxidation technology appears to be the most effective for the degradation of compounds with limited solubility in water.

The effects of several reactor design parameters were investigated, including electrode geometry, electrode spacing, reactor geometry, and the ground electrode plate area, on the degradation of BPA. The hybrid-series reactor in which the high voltage (plasma) electrode is in water and the grounded electrode is above the water's surface was about four times as effective as the reference configuration with both electrodes in water. The reactor geometry, particularly the distance between the high voltage electrode and the reactor wall, was observed to have a major impact on the characteristics of the electrical discharge, which in turn influenced the BPA removal rate. Tripling the reactor radius yielded a threefold increase in the peak current and thus the power deposited into plasma. Similarly, the area of the grounded electrode plate was found to affect the discharge characteristics with a twofold increase in the plate area corresponding to an increase of about 40% in the peak current and input power. The distance between the electrodes was identified as the key parameter. It was found that the optimum distance depends on the input voltage and solution conductivity, but is generally near the minimum distance at which no sparking between the electrodes occurs. At the same input power of 50 W, the optimization increased the extent of removal of a 5 mg/L BPA solution from 32% to 94% in less than one hour.

(337e) Enhanced Photocatalytic Degradation of Malachite Green By Visible Light Sensitive ZnO/ZrO2 Doped TiO2 Nanoparticles

Shende, A., South Dakota School of Mines and Technology
Shende, R. V., South Dakota School of Mines and Technology
Originally presented on: 11/5/2013 16:43:00 - 17:05:00

The large energy band gap of TiO_2 permits activation only in the UV portion of the solar radiation. To obtain visible light sensitive material ZnO/ZrO_2 doped TiO_2 nanoparticles were synthesized by the sol-gel method using Pluronic 123 surfactant. The structure and surface morphology of the photocatalyst was studied using X-ray diffraction, HRTEM, BET and diffuse reflectance spectroscopy. Photodegradation rates of malachite green (MG) in solar radiation using the $ZnO/ZrO_2/TiO_2$ and P-25 TiO_2 were compared. Effect of pH, catalyst load and dye concentration on the photoconversion efficiency were determined. The degradation products were identified by GC-MS. The toxicity of the degradation products were tested using E.coli.

(337f) Endogenous Hormones Degradation Using Nanotubular Oxide Layer Grown On Tiw Alloy By Electrochemical Anodization

Rodrigues, C. A., Universidade Federal de Sao Paulo
Oliveira, M. E.,
Barroso, B. L.,
Gomes, M. L. M.,
Bertazzoli, R., Universidade Estadual de Campinas
Moraes, M. D. L. L.,
Originally presented on: 11/5/2013 17:05:00 - 17:27:00

Advanced Oxidative Processes (AOP) are a significant green technology for application in water purification and organic pollutant degradation. The main mechanism of AOPs is the generation of highly reactive free radicals, such as the hydroxyl radical (•OH). Among different available AOPs, heterogeneous photocatalysis seems to be one of the most popular technologies for wastewater treatment. This process uses semiconductor materials as catalysts and UV light. The efficiency of the electrochemical and photoelectrochemical treatment is strongly dependent on the choice of the anode material and its morphology. The manufacture of Ti and Ti alloys oxide nanotube arrays by electrochemical methods has attracted a great deal of attention for applications in environmental photocatalysis due to its photoinduced reactivity[1]. In this work it is proposed the degradation of endogenous hormones, estrone (E1) and 17α-ethinylestradiol (EE2) by photoelectrocatalysis process using self-organized oxide semiconductor nanotubes on a titanium/tungsten alloy applying an anodization process. The growth of self-organized oxide semiconductor nanotubes on a Ti-x%W alloy (x= 0.5, 2.5 and 5.0% wt) was investigated as a function of the applied potential, electrolytic solution (aqueous or organic) and anodization time. The titanium alloy was prepared in an arc melting furnace using a non-consumable tungsten electrode and water-cooled copper crucible in a high purity argon atmosphere. The samples were solidified in an ingot form that was submitted to a rolling process to manufacture the alloy sheet X-ray fluorescence was used to determine the chemical composition of the alloy. The anodization experiments were carried out with a DC power supply and an acrylic electrochemical cell, with a platinum cathode and the Ti alloy anode separated by 2.5 cm. An aqueous solution of 0.2 M HF and two types of ethylene glycol solutions (0.2 M NH_4F and 0.2 M HF) were used as electrolyte solution. All the anodization experiments were carried out at 120 V, with applied time that varied between 5 to 120 minutes. SEM-FEG was used to characterize the nanotubes and X-ray fluorescence results confirmed that the composition of the alloys was as required. The best results of anodization were obtained when 0.2 M HF with ethylene glycol solution was used with the Ti-0.5w%W electrode, with an anodization time of 60 minutes at 120 V. Under these conditions the oxide nanotubular layer was homogeneous and continuous through the surface. After anodization, the nanotubular oxides on Ti0.5%W were annealed at temperature greater than 450°C in air atmosphere for 2 hours and then, cooled back to room temperature in a box furnace to obtain a crystalline structure. X-ray diffraction was used to characterize the crystallization behavior of oxide layer. The degradation of endogenous hormones was carried out in a photoelectrochemical cell. A Ni screen was used as a cathode, and the self-organized oxide nanotube on titanium alloy produced were used as anode. The efficiency of these treatment processes was monitored by following the TOC (Total Organic Carbon) and the E1 and EE2 concentration decay by HPLC technique (High Performance Liquid Chromatography). UV lamp was positioned to illuminate the anode electrode in order to induce the photodegradation of the organic compounds. The percentage of degradation and total organic carbon removal, and the energy consumed per volume of effluent treated were investigated during electrolysis in photoelectrochemical cell.

Acknowledgments:

CNPq (Proc. 483285/2011-0), Fapesp (Proc. 2011/51226-3), LNLS and UNICAMP

References:

[1] Nah, Y.C.; Ghicov, A.; Kim, D.; Berger, S.; Schmuki, P. *Electrocatalysis* 12 (2008) 16154 – 16155

(337g) Flame Aerosol Synthesized Cr Incorporated TiO2 for Visible Light Photodegradation of Acetonitrile

Inturi, S. N. R., University of Cincinnati
Boningari, T., University of Cincinnati
Suidan, M., University of Cincinnati
Smirniotis, P., University of Cincinnati
Originally presented on: 11/5/2013 17:27:00 - 17:49:00

The photocatalytic activity of flame aerosol made TiO_2 loaded with various amounts of Cr was studied as a catalyst for both liquid and gas phase photodegradation of acetonitrile (ACN) under visible light (400–800 nm). It was found that the optimal concentrations of Crand TiO_2 exist (Ti/Cr atomic ratio = 40) for the efficient oxidation of ACN in the gas phase.XRD patterns showed a decrease in the anatase phase with increase in the amount of Cr loading. Temperature programmed reduction (TPR) and UV–vis spectra profiles clearly indicated that the Cr^{6+} phase is dominant in the samples with the atomic ratio of Ti/Cr equal to 50, 40, and 30. The rest of the catalysts with Ti/Cr ratios of 100, 75, 20, and 10 were found to possess Cr^{3+} species along with Cr^{6+} species. Our XPS results illustrated that the relative atomic percentage value of Ti^{3+}/Ti^{4+} characterized by XPS was significantly high for the Cr/TiO_2 nanoparticles with Ti/Cr = 40 atomic ratio (Ti^{3+}/Ti^{4+} = 1.14, 42.1%), whereas, other samples demonstrated low atomic percentage value of Ti^{3+}/Ti^{4+} (Ti^{3+}/Ti^{4+} = 0.18-1.05). Moreover, Cr interacts with TiO_2 nanostructure in the interface of flame-made nanoparticles, bulk Cr oxide exists over the surface of TiO_2 nanostructure.The photodegradation ability of TiO_2/Crcatalyst with Ti/Cr atomic ratio of 40 was highly related to the existence of Cr^{6+}species which strongly interacted with TiO_2.Among all the catalysts tested, the system with Ti/Cr atomic ratio 40 demonstrated a superior catalytic performance with the rate constant of 0.812 $m^3g^{-1}mol^{-1}$ under visible light irradiation.The proposed route of the catalytic activity of the above material in visible light involves the reaction of dopant level electrons withsurface Cr^{6+}, which makes available valence band holes to performoxidation reactions.The photocatalytic performance of flame aerosol synthesized Cr incorporated TiO_2 material in both liquid and gas phase will be presented in the persentataion.

(359a) Nanoparticle Emission Characterization and Management in Nanocoating Material Spray Operation

Uttarwar, R., Wayne State University
Huang, Y., Wayne State University
Originally presented on: 11/5/2013 15:15:00 - 15:36:00

Nanoparticle emission occurs during nanomaterial manufacturing, application, and the use phase of the end product. Nanopaint is a very promising type of nanocoating material, in which nanoparticles are added into the resin of the paint. During paint spray, a certain fraction of paint unable to land on the surface of substrate will enter the environment. The nanoparticles contained in tiny paint droplets are thus emitted, which will fly in the air for a certain period of time, and then either land on the wall of the spray booth or follow the ventilation air flow to a drainage water system. To minimize the adverse environmental and health impact and to improve nanopaint transfer efficiency, the nanoparticle emission must be minimized and managed.

In this paper, we introduce a CFD-based modeling and simulation approach to characterize nanoparticle transport dynamics and quantify nanoparticle emission. The models are incorporated into a comprehensive system model set which includes a process energy efficiency model and a product quality model that characterizes the formation of nanocoating film on substrates. The model set is then used for the optimization of the entire nanopaint spray operation, where coating quality, material and energy use efficiency, and emission minimization are all optimized. Case studies will demonstrate the efficacy of the methodology with comparison of literature data.

(359b) Innovative Room-Temperature Preparation of One-Dimensional NiO Nanowires and Their Energy and Environmental Applications

Meng, X., Wayne State University
Ng, K. Y. S., Wayne State University
Deng, D., Wayne State University
Originally presented on: 11/5/2013 15:36:00 - 15:57:00

Synthesis of one-dimensioned (1-D) nanostructures has recently attracted much attention for energy and environmental applications. Generally, high temperature (e.g., CVD) and high pressure (e.g., hydrothermal) methods are employed to prepare 1-D nanostructures and usually the production is only at milligram scale. Alternative methods for room temperature and ambient pressure-based synthesis of 1-D nanostructures are always highly desirable to drive down the cost and scale up to gram-scale production. Herein, we will report a novel method for the fabrication of multifunctional NiO nanowires at room temperature and ambient pressure. The formation mechanism at room temperature is revealed by extensive characterization by FESEM, TEM and XRD for samples prepared under different controlled conditions. This presentation will also cover the preliminary results that demonstrate excellent potential for energy and environmental applications of those 1-D nanowires produced at gram scale.

(359c) The Direct Activation of Targeted Partial Oxidation Reactions in Liquid Phase Via Enhanced Corona Discharge Techniques From Nanofibrous Emitter Electrodes in Microreactor Channels

Yokochi, A. F. T., Oregon State University
Pommerenck, J., Oregon State University
Pommerenck, J., Oregon State University
Kreider, P., Oregon State University
Xiangru, F.,
Alanazi, Y., Oregon State University
Originally presented on: 11/5/2013 15:57:00 - 16:18:00

In this study, the application of carbon nanotubes, grown with randomly dispersed diameters ranging from 10 to 50 nanometers, has been extended to the microscale for liquid phase corona discharge. Metal nanowires and nanotubes have also been examined. While the length scale of the diameter of the nanostructures are hardly on the atomic scale as at the tips of most of the atomically-sharp metal emitters, the material properties of the carbon nanotube emitters is a relativistic playground for electrons and have displayed durability up to 100 h in corona discharge process. A surface covered with a forest of the nanofibrous emitter electrodes activates a larger portion of the fluid than did the single atomically fine metal tips. The application of these discharge processes to the microscale invokes the typical improvements in heat and mass transfer to cool the anode (where the majority of the heat dissipation occurs), the ability to activate these reactions at extremely low electric potentials, and effectively transfer reactive species into the fluid. Flow through systems use forced convection to cool the anode emitter which allows nearer separations. The carbon nanotube allows the turn-on electric potential to be reduced. These enhancements have grand implications for phenomena such as frothy electric discharge which occurs in conducting aqueous fluids and for dc corona discharge in gas phase. The energy requirements of which have been reduced by several orders of magnitude (typical reports list 150 mA at 900 V and hour scale treatment times depending on whether the study is examining glow discharge electrolysis or arc discharge-this study reports 6 mA at almost any desired electric potential (depends on the distance of the carbon nanotube covered wire to the pin and material constraints) and minute scale residence times for similar degradation ratios). This is the first attempt in literature to take nanostructured emitters down to microscale dimensions for liquid phase corona discharge reaction engineering to the author's knowledge.

The target reaction is the oxidative destruction of organics in groundwater with phenol being the example organic. The advantages of the corona discharges at low turn-on electric field strengths will be discussed with respect to the sanitization of groundwater. The ability to operate at 50 V rather than 10 kV at the same current density as at the macroscale is powerfully important both when considering the simplicity of the required pulsing technology, effectiveness of the corona process and reactor design. This improvement is made viable by understanding the corona discharge mechanism, microreactor technology and enhanced nanofibrous electrodes. The application of microscale reaction engineering technology and the nanofibrous emitter electrode corona discharge process to aqueous systems will have remarkable implications in every area of the chemical industry. Purifications, reaction chemistry and separations can be accomplished using the enabling techniques developed during this study to achieve energy savings two orders of magnitude greater than currently available world-wide for electric discharge technologies.

(359d) Magnetic Nanocomposite Microparticles for On/Off Binding of Persistent Organic Pollutants

Gutierrez, A. M., Universidad de los Andes
Newsome, B. J., University of Kentucky
Dziubla, T. D., University of Kentucky
Hilt, J. Z., University of Kentucky
Originally presented on: 11/5/2013 16:18:00 - 16:39:00

Magnetic nanocomposite microparticles (MNMs) were produced using iron oxide nanoparticles incorporated into a polyphenolic-based polymer matrix with high affinity for organic pollutants. This platform allows for the specific binding of chlorinated organics, the rapid magnetic separation of bound organics from contaminated water sources, and the thermal destabilization of the polymer matrix for contaminant release and material regeneration. Quercetin multiacrylate (QMA), an acrylated form of the nutrient polyphenol with known affinity for chlorinated organics, was crosslinked with polyethylene glycol (PEG) diacrylate using a free radical polymerization in the presence of magnetic iron nanoparticles, and the magnetic nanocomposite was subsequently cryomilled to form the MNMs. Particles were characterized using transmission electron microscopy (TEM), dynamic light scattering (DLS), Fourier transform infrared spectroscopy (FTIR), and thermal gravimetric analysis (TGA). Pollutant binding studies were performed using model chlorinated organic pollutants, polychlorinated biphenyls (PCBs, specifically PCB 126), to determine binding affinity and capacity, as well as optimal binding kinetics, and this was quantified using LC-MS/MS. It was demonstrated that the MNMs effectively bound PCBs with the addition of QMA resulting in greater affinity. An alternating magnetic field (AMF) was used to heat and destabilize the binding in the polymer matrix leading to PCB release from the particles, and the percentage of uptake and release was determined. Repeated PCB binding/release was performed to determine MNM stability and reusability.

(359e) Carbon Nanocomposite Fabrics Toward Cr(VI) Removal From Polluted Water

Zhu, J., Lamar University
Wei, S., Lamar University
Guo, Z., Lamar University
Originally presented on: 11/5/2013 16:39:00 - 17:00:00

Cotton fabrics can be converted to carbon fabrics by conventional annealing in inert atmosphere. Magnetic carbon nanocomposite fabrics decorated with uniformly distributed iron oxide nanoparticles will be fabricated via microwave-assisted and conventional annealing processes. Both annealing method and purging gas species on the hybrid structure morphology and nanoparticle crystalline structure will be studied and their corresponding Cr(VI) removal performance from polluted water will be comparatively investigated. Cotton fabrics, carbon fabrics and carbon nanocomposite fabrics will be comparatively studied for their Cr(VI) removal performance. And also, the Cr(VI) concentration, adsorbent loading and adsorption time will be systematically investigated to determine the adsorption capacity and rate. The open structure of the nanocomposite fabrics are projected to remain high treatment capacity. The decorated nanoparticles are expected to enlarge the Cr(VI) removal efficiency and capability. The Cr(VI) removal mechanisms will be exploited by studying the magnetization change before and after adsorption treatment.

(359f) Pilot Scale Evaluation of Nano-Size Adsorbent for Arsenic Contaminated Water Treatment

Yu, Y., National University of Singapore
Chen, J. P., National University of Singapore
Li, G., National University of Singapore
Mahanta, N., National University of Singapore
Originally presented on: 11/5/2013 17:00:00 - 17:21:00

In order to investigate the arsenic removal by nano-size adsorbent in the pilot scale, a water treatment system with the capacity of 10 tons/d was designed in our study. The nature reservoir water contaminated by arsenic was sequentially treated by adsorption, coagulation, sedimentation and membrane filtration. A novel nanostructured zircomium-manganese binary oxide was added in the adsorption process, and $Al_2(SO_4)_3$ was chose as the coagulant. When the initial arsenic concentration was 100μg/L, only 2.5 mg/L adsorbent was needed to ensure the effluent safety. This system showed a good stability and the final concentration of arsenic and turbidity meet the WTO drinking water standard. The turbidity increased after the addition of adsorbent and coagulation, while through the sedimentation and membrane filtration the turbidity was decreased to below 1 NTU. Only a slight influence on the removal of As(V) was found in the study with the presence of the co-existing ions such as Cu^{2+}, Pb^{2+}, HCO_3^- and F^-. The adsorbent dosage needed to be increased when a certain amount of As(III) existed in the influent. The toxicity of the effluent was analyzed by human cell lines. The results found the samples were cytocompatible without any toxic nature.

(359g) Composition and Shape Controlled Syntheses of ZnxCd1-Xs Photocatalysts for Reducing CO2 to Methyl Formate in Methanol

Jiang, W., Tianjin University of Technology
Yin, X., Tianjin University of Technology
Xin, F., Tianjin University
Sui, D., Tianjin University of Technology
Originally presented on: 11/5/2013 17:21:00 - 17:42:00

Band structure-controlled $Zn_xCd_{1-x}S$ photocatalysts were prepared at 160 °C by the hydrothermal synthesis method and used for reducing CO_2 to methyl formate (MF) in methanol via ultraviolet irradiation. The structures, morphologies and optical properties of the obtained samples had been characterized by X-ray diffraction (XRD), energy dispersive X-ray spectra (EDS), scanning electron microscopy (SEM), transmission electron microscopy (TEM), high-resolution transmission electron microscopy (HRTEM) and UV-vis absorption spectra (UV-vis). The band gap of such a solid solution of the photocatalysts could be tuned between 2.59 and 3.18 eV by varying composition *x*. In XRD analysis, the peaks systematically shifted to larger angles as the Zn content increased. The shape of the $Zn_xCd_{1-x}S$ photocatalysts changed from microsphere to inflorescence and then to rod with a decrease of Zn content in the particles. Under ultraviolet light (λ=365 nm) irradiation, the photocatalysts showed composition-dependent photocatalytic efficiencies for reducing CO_2 to MF. The samples prepared at the condition of composition x=0.42, 160 °C, 12 h possessed the best activity with the MF formation rate of 85.7 mmol/h/g•cat. The catalytic performance was due to an optimal balance among the crystalline structure, the morphology and the light absorption properties.

(361a) First Principles Investigation of SOx (x = 0 - 4) Chemistry On Pt (111) and Pd (111)

Sharma, H., University of Connecticut
Sharma, V.,
Mhadeshwar, A. B., University of Connecticut
Ramprasad, R., University of Connecticut
Originally presented on: 11/5/2013 15:15:00 - 15:35:00

First principles investigation of SO_x ($x = 0$ - 4) chemistry on Pt (111) and Pd (111)

-

Hom N. Sharma[1], V. Sharma[2], Ashish B. Mhadeshwar[1,3] and R. Ramprasad[1, 2,*]

[1]Department of Chemical and Biomolecular Engineering, University of Connecticut, Storrs, CT

[2]Material Science and Engineering, University of Connecticut, Storrs, CT

[3]Currently at ExxonMobil Research & Engineering, Annandale, NJ

*rampi@ims.uconn.edu

Sulfur oxides (SO_x ($x = 0 - 4$)) in the diesel engine exhaust are one of the major factors contributing toward deactivation of the noble metals (Pt/Pd) based emissions aftertreatment catalysts [1, 2]. Interactions of SO_x with the catalyst metals and supports are associated with changes in various structural, morphological, and electronic properties, which can lead to sulfate formation [3, 4]. Â Despite many experimental and computational studies, SO_x interactions with metal surfaces and their impact on catalytic activity are not clearly understood. In the present study, we explore the SO_x ($x = 0 - 4$) chemistry in terms of interactions and oxidation pathways on Pt(111) and Pd(111) surfaces using first principles density functional theory (DFT) [5]. We have identified various stable configurations (figure 1), which reveal the charge redistribution patterns, and molecular orbital interactions of SO_x on both surfaces during adsorption. The possible reaction pathways and estimated activation barriers for SO_x oxidation on the metal surfaces ? identified using climbing image nudged elastic band (CI-NEB) method [6] ? are found to be in good agreement with the experimental observations. This comprehensive and comparative study provides important insights about

SO_x chemistry on Pt and Pd surfaces, which can be useful to design and improve sulfur resistance catalyst materials.

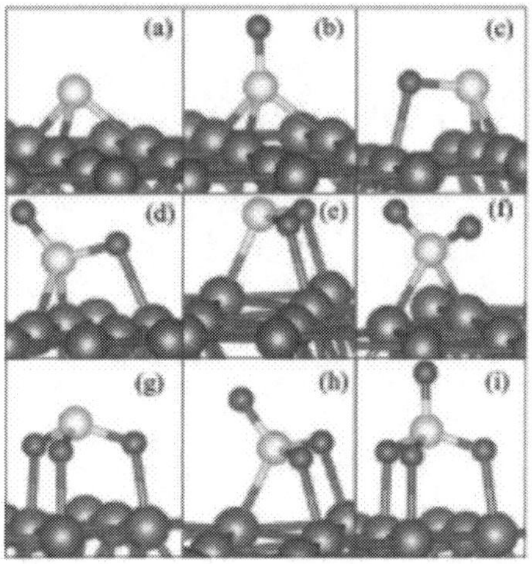

Figure 1: Representative stable configurations of SO_x molecules on Pt(111): (a) S (fcc) η^1 S_f, (b) SO (fcc) η^1-S_f, (c) SO (fcc) η^2 S_bO_a, (d) SO_2 (fcc) η^2 S_bO_a, (e) SO_2 (fcc) η^3 $S_aO_aO_a$, (f) SO_2 (bridge) η^1 $S^{\perp}$, (g) SO_3 (fcc) η^3 $O_aO_aO_a$, (h) SO_3 (fcc) η^3 $S_aO_aO_a$, and (i) SO_4 (fcc) η^3 $O_aO_aO_a$. The number super-scripted to η represents the number of atoms in a molecule coordinated to the metal surface; and the subscripts *a, b,* and *f* represent the atop, bridge, and fcc hollow sites, respectively.

References

[1] Â T. Kolli, M. Huuhtanen, A. Hallikainen, K. Kallinen, R. Keiski, Catal. Lett. 127 (2009) 49-54.

[2] Â A. Russell, W.S. Epling, Catal. Rev. Sci. Eng. 53 (2011) 337-423.

[3] Â J.A. Rodriguez, J. Hrbek, Acc. Chem. Res. 32 (1999) 719-728.

[4] Â H.N. Sharma, S.L. Suib, A.B. Mhadeshwar, in: Novel Materials for Catalysis and Fuels Processing, ACS, 2013, In press.

[5] Â G. Kresse, J. FurthmÃ¼ller, Phys. Rev. B. 54 (1996) 11169.

[6] Â G. Henkelman, B.P. Uberuaga, H. Jonsson, J. Chem. Phys. 113 (2000) 9901-9904.

(361b) In-Situ Drifts Studies of Selective Catalytic Reduction of NO With C3H6 On Cu-Chabazite Catalyst

Raj, R., University of Houston

Harold, M. P., University of Houston

Balakotaiah, V., University of Houston

Originally presented on: 11/5/2013 15:35:00 - 15:55:00

In-situ DRIFTS Studies of Selective Catalytic Reduction of NO with C_3H_6 on Cu-chabazite Catalyst

Richa Raj, Michael P. Harold[1], and Vemuri Balakotaiah[2]

Department of Chemical & Biomolecular Engineering, University of Houston, Houston, TX, United States.

Abstract

In-situ DRIFTS studies were carried out to elucidate the underlying mechanism of NO_x reduction with C_3H_6. *In-situ* DRIFTS was performed using a Nicolet Nexus 470 spectrometer, equipped with a small collector with a dual sample environmental chamber (Thermo Fisher Scientific). The sample of catalyst powder was prepared by scraping the washcoat off the monolith channels. The feed gas, at a total flow rate of 50 cm^3/min at STP, was mixed using Bronkhorst mass flow controllers; and its composition, with all gases present, was 500 ppm C_3H_6, 500 ppm NO, 1% or 5% O_2, 0 or 2.5% H_2O balanced by He. Before each experiment, the catalyst sample of 60 mg was treated with 5% O_2 in He for 1 hour at 600°C.

Fig. 1 shows the evolution of surface species with time for $C_3H_6 + O_2$ reaction. The growth of the bands around 1550 cm^{-1} and 2900 cm^{-1} correspond to stretching of C-C and C-H bonds of adsorbed hydrocarbon species, respectively. The broad band near 3060 cm^{-1} corresponds to vibration of C-H bond of condensed aromatic rings. The growth of these bands with time suggest the formation and accumulation of coke like species on the catalyst surface formed via oligomerization of adsorbed C_3H_6 derived hydrocarbon species. The appearance of band at 1660 cm^{-1} corresponds to C=O stretching of acrolein like species. Similarly, the peak at 1450 cm^{-1} corresponds to adsorbed carboxylate species. Hence, oxygenates are the intermediates in C_3H_6 oxidation. Fig. 2 shows the spectra obtained for different times on introduction of NO in feed after exposure to $C_3H_6 + O_2$. On exposure to NO a new peak at 2250 cm^{-1} is observed which is attributed to ?NCO species. There is increase in the peak intensity corresponding to ?NCO species with time suggesting its accumulation on the catalyst surface. It clearly indicates that N-containing hydrocarbon species act as intermediate in NO_x reduction with C_3H_6.

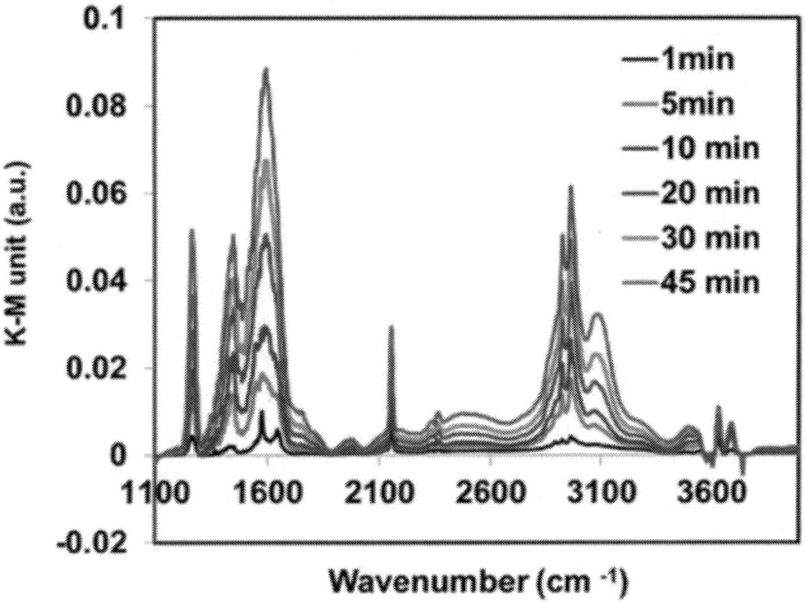

Figure 1 DRIFTS spectra obtained at different times upon Cu-SSZ13 exposure to 500 ppm C_3H_6 at 380 °C (500 ppm C_3H_6, 1% O_2, balanced by He)

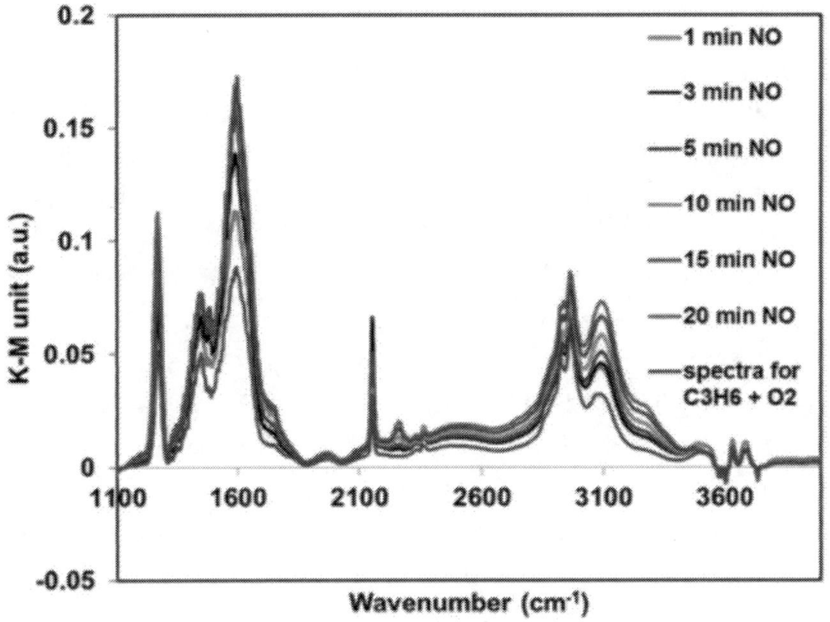

Figure 2 Emergence of new peak upon exposure to 500 ppm NO after 45 minutes of exposure to 500 ppm C_3H_6 at 380 °C (500 ppm C_3H_6, 500 ppm NO, 1% O_2, balanced by He)

[1]Corresponding author: Tel.: +1 713 743 4322; Fax: +1 713 743 4323; Email: mharold@uh.edu (M. P. Harold)

[2]Corresponding author: Tel.: +1 713 743 4318; Fax: +1 713 743 4323; Email: bala@uh.edu (V. Balakotaiah)

(361d) Development of Low-Temperature SCR Catalysts for the Cold-Start Applications of NOx Reduction

Boningari, T., University of Cincinnati
Ettireddy, P. R., Tenneco Inc
Kotrba, A., Tenneco Inc.
Spinks, T.,
Smirniotis, P., University of Cincinnati
Originally presented on: 11/5/2013 16:15:00 - 16:35:00

Forthcoming low emissions standards for vehicles are forcing automobile and catalyst manufacturers to focus on reducing cold start NO_x emissions since a large portion of NO_x emissions occur during the cold start period. Consequently, control of emissions during the cold start operation of a vehicle containing an internal combustion engine is highly desirable. In pursue of our research interest in selective catalytic reduction (SCR) of NO_x, we prepared a series of TiO_2, TiO_2-SiO_2 (1:1 mole ratio) supported catalysts by adopting incipient wetness technique and investigated for the catalytic reduction of NO_x with ammonia. In order to simulate the realistic diesel exhaust conditions, we have evaluated the potential catalytic performance of all the prepared catalysts under industrially relevant conditions with excess (10 vol.%) oxygen in the feed. To our surprise, Cu/TiO_2 catalyst showed highly impressive low temperature deNO_x performance in the 150-200°C temperature regimes. The V/TiO_2, Ce/TiO_2 formulations achieved better stability in the temperature range of 200-350 °C, which will add several advantages over high temperature operations. Therefore, the effect of dopants, water vapors and SO_2 poisoning will be investigated. The chemical environment of these catalysts were characterized by using different spectroscopic and non-spectroscopic methods such as N_2 physisorption, temperature programmed reduction (TPR), X-ray diffraction (XRD), and X-ray photoelectron spectroscopy (XPS). More details about the reaction, durability of the catalysts and characterizations will be discussed in the presentation.

(361e) Effect of Mg/Al Ratios in HT Supports On NOx Storage-Reduction Processes Over Pt-BaO/Hydrotalcite(HT)-Derived Support

Jeong, S., Seoul Nat Univ
Youn, S.,
Kim, D. H., Pacific Northwest National Laboratory
Originally presented on: 11/5/2013 16:35:00 - 16:55:00

Efficient NOx abatement catalyst operated under lean condition must be developed to promote the use of more fuel-efficient diesel and/or lean-burn engines since conventional three-way catalyst does not work properly. For this purpose, selective catalytic reduction (SCR) catalyst or NOx storage-reduction (NSR) catalyst were commercially being used. NSR technology, which is more appropriate for the small-sized engine, primarily consisted of noble metals such as Pt, alkali earth metal (e.g. Ba) and support (e.g. Al_2O_3 or $MgAl_2O_4$). It was demonstrated that $MgAl_2O_4$ support provided the improved NOx storage activity at higher temperature compared with Al_2O_3 support, implying that support material can play a crucial role in affecting the stability of barium nitrate species.

Hydrotalcite (HT) material, which can be precursors to $MgAl_2O_4$ spinel structure, has received more attention due to its various advantages, such as large surface area, controllable acid-base properties, and the improved thermal stability. In this work, HT materials with different Mg/Al ratios were prepared, resulting in the different amount of $MgAl_2O_4$ and Al_2O_3 in the support after calcination. Then, Pt-BaO active components were loaded on these HT-derived supports, and the catalysts were extensively investigated by using NOx storage measurement and various analytical techniques.

Combination of activity measurement and characterization results indicates that HT-derived supports give rise to the higher thermal stability of barium nitrates and Pt species as well as superior activity than one supported on Al_2O_3, moreover, can be optimized further for the high temperature NSR application by changing Mg/Al ratios in HT support. This work provides fundamental understanding for optimizing HT-derived $MgAl_2O_4$ support, which would be a strong candidate for the high temperature NSR application.

(361f) NH3-SCR Over Cu/SAPO-34 - Zeolite Acidity and Active Site Changes As a Function of Cu Loading

Wang, D., University of Houston
Zhang, L., University of Houston
Kamasamudram, K., Cummins Inc.
Epling, W., University of Houston
Originally presented on: 11/5/2013 16:55:00 - 17:15:00

NH_3 Selective Catalytic Reduction (NH_3-SCR) is a widely applied technology to efficiently treat NO_x emissions from heavy duty diesel vehicles. Copper-exchanged chabazite (CHA) catalysts exhibit both superior activity and hydrothermal stability for this reaction in comparison with Cu-BEA and Cu-ZSM-5, and have therefore been recently adapted for on-vehicle application. In this study, a series of SAPO-34 catalyst with various Cu loadings (ranging 0.7-4.0 wt%) were prepared by the solid state ion exchange method. The acid properties were characterized by XRD, NH_3-TPD, in-situ DRIFTS, H_2-TPR and UV-vis techniques and active sites for NH_3-SCR were also probed. The data show that both Bronsted acid sites and Lewis acid sites were present in the as-prepared samples. The Lewis acid sites were created by Cu ions and increased monotonically with increasing the Cu loading. Furthermore, four different Cu species were discovered in the samples: isolated Cu^{2+} ions, Cu^{+} ions, large CuO particles and nano-sized CuO. However, the proportion of these species on each sample was found to be different, resulting in various SCR behaviors. For example, the amount of isolated Cu^{2+} ions was proved to increase monotonically with increasing Cu loading. However, the total number of Cu^{+} ions increased as the Cu loading increased up to 2.5 wt% then decreased with higher Cu loading. Isolated Cu^{2+} species were found to be the active sites over Cu-SAPO-34 catalysts at low temperature (130-300°C). However, the NH_3-SCR activity at high temperature (300-550°C) could also be attributed to the nano-sized CuO species that are probably located in the zeolite pores. In addition, the particle size of the CuO species has various impacts on NH_3 oxidation activity and selectivity. The data show that large CuO particles could promote the competitive NH_3 oxidation reaction and therefore decrease the NO conversion of standard SCR at high temperature. On the other hand, the nano-sized CuO particles actually increased the selectivity of NH_3 oxidation to NO. Moreover, Cu^{+} ions likely do not affect the SCR activity.

(361g) In-Situ Generated Hydrogen Driven Reductive Photocatalysis

Gullapalli, S., Rice University
Velazquez, J. C., Rice University
Pretzer, L. A., Rice University
Wong, M. S., Rice University
Originally presented on: 11/5/2013 17:15:00 - 17:35:00

Photocatalytic removal of organic contaminants is of great interest since UV light is a readily available energy source. However most photocatalytic reactions are oxidative in nature and result in toxic by-products. We develop a bifunctional semiconductor-metal composite material that functions as a reduction photocatalyst using *in-situ* generated hydrogen (H_2). The Pd-on-Au/TiO_2 catalyst (0.1wt% Pd, 1wt% Au) was synthesized by reducing dihydrogen tetrachloropalladate (II) onto a commercially available Au/TiO_2 catalyst. As the test reaction, the hydrodechlorination of trichloroethene (TCE, a common groundwater contaminant) was observed to occur in batch reactor studies, under both aerated and deaerated conditions upon illumination with a 400 W xenon light source. Analysis of the reactor headspace confirmed that reductive hydrodechlorination products are formed. No photocatalytic TCE oxidation is observed. Control experiments without TCE reveal that *in-situ* H_2 is generated by photocatalytic water splitting, irrespective of the presence of air. Other catalysts such as commercially available Au/TiO_2 and Pd/TiO_2 were also investigated. We therefore infer that Pd-on-Au/TiO_2 simultaneously photocatalyzed water splitting and catalyzed reductive TCE hydrodechlorination in the same batch reactor. Extension of this study to other bifunctional semiconductor-metal systems and groundwater contaminants could significantly simplify future remediation efforts.

(364a) Energy-Optimal Control of An Integrated UF-RO Seawater Desalination Plant

Gao, L., University of California, Los Angeles
Rahardianto, A., University of California, Los Angeles
Gu, H., University of California, Los Angeles
Christofides, P. D., University of California, Los Angeles
Cohen, Y., University of California, Los Angeles
Originally presented on: 11/5/2013 15:15:00 - 15:40:00

In recent years, reverse osmosis (RO) desalination has emerged as a leading method for desalting seawater, inland brackish water, and water reuse applications due to the versatility of this process relative to other water desalination technologies. Large-scale RO desalination plants are common in many parts of the world today. Despite this, as popularity of the process grows, there is a growing demand for small-scale distributed systems. Minimizing energy use is always a concern when optimizing RO plant operation but energy recovery devices are impractical for small-scale systems low efficiencies.

In order to carry out energy-optimal operation of a small-scale RO desalination plant, it is important that the control system be able to maintain freshwater production and energy-optimal operating conditions simultaneously. In order achieve all this, a novel model-based RO controller was developed at UCLA. This controller was then implemented into a novel RO system integrated with its ultrafiltration (UF) pre-treatment system also built at UCLA and field tested at Port Hueneme US Naval Base.

Previous research has been done by Zhu et al. [1] that demonstrate the use of the specific energy consumption (SEC) with respect to RO water recovery as a framework to determine energy-minimal RO operating conditions. Further analysis is done in this framework in order to develop a simple algorithm to calculate the energy-minimal operating point. The analysis shows that the energy-optimal operating condition always occurs at the maximum physically possible RO water recovery.

This algorithm is combined with a simple RO model which is derived from the classical RO flux equation. This model calculates the RO feed flow rate and the RO feed pressure required to achieve the RO water recovery and RO permeate flow rate set-points. These set-points are then communicated to local controllers for actuation.

Experiments were carried out to observe the performance of this novel model-based controller. It was demonstrated that the controller was successful in accurately achieving the RO permeate flow rate and RO feed pressure set-points. It was also shown that the controller was able to independently control the two set-points even though the control variables are dynamically coupled. Disturbances were induced through unsteady-state operation of the UF pre-treatment system and it was demonstrated that the system successfully rejected the disturbance by maintaining its set-points despite fluctuations in the feed water conditions. It was concluded after the study that this novel model-based controller enabled robust control of an integrated UF-RO system under energy-optimal conditions.

1. Zhu, A., P. D. Christofides and Y. Cohen, "Effect of Thermodynamic Restriction on Energy Cost Optimization of RO Membrane Water Desalination," Industrial and Engineering Chemistry Research, 48, 6010-6021 (2009).

(364b) Iron Electro-Coagulation for Simultaneous Arsenic Removal and Microbial Attenuation: Mechanisms of E.Coli Removal and Inactivation

Delaire, C., UC Berkeley
Amrose, S., UC Berkeley
Van Genuchten, C., University of California
Gadgil, A., University of California
Originally presented on: 11/5/2013 15:40:00 - 16:05:00

Around 60 million people in South Asia drink groundwater from arsenic contaminated shallow aquifers. Research over the last two decades has focused on arsenic removal alone to mitigate this problem, largely ignoring possible microbial contamination of shallow groundwater. Recently, fecal indicators and pathogens were detected in shallow tubewells in Bangladesh, and diarrheal diseases are still prevalent in the region. Therefore comprehensive treatment technologies addressing both microbial and arsenic contamination of drinking water are needed, and will potentially increase health impacts and social acceptability.

Iron electro-coagulation (EC) is a simple technology relying on the quick dissolution of a sacrificial Fe(0) anode to produce iron precipitates that serve as an adsorbent for arsenic and as a coagulant for microbes. In the process, strong oxidants generated by Fenton-like reactions convert As(III) into As(V), which is more amenable to adsorption. The mechanisms of arsenic removal with iron EC have already been described. In this work, we demonstrate that iron EC can simultaneously remove arsenic and the model organism *E.coli* in synthetic groundwater similar to contaminated aquifers in South Asia. The iron dosage (in mg-Fe/L) required to achieve 4-log attenuation of *E.coli* is very sensitive to pH and varies from 25 mg-Fe/L at pH 6.6 to 140 mg-Fe/L at pH 7.5. In this pH range, iron precipitates generated in synthetic groundwater have a negative surface charge, whose variation cannot entirely explain the sensitivity of *E.coli* attenuation to pH. We find that inactivation as well as coagulation with iron precipitates both contribute to *E.coli* attenuation. We propose to analyze these two mechanisms.

Fe(II) can cause oxidative stress on bacterial membranes and inside cells due to the production of strong oxidants by Fenton-like reactions. We examine the effect of pH and strong oxidant scavengers on *E.coli* inactivation during iron EC. In order to probe oxidative stress on *E.coli* membranes, we quantify membrane permeability changes and lipid peroxidation. We discuss the possible role of anodically produced Fe(II) as a bactericidal agent in our system.

Covalent bonding between bacterial terminal phosphate groups and iron oxides has been previously reported. We use Fourier Transform Infrared Spectroscopy to analyze how *E.coli* binds to a range of EC-precipitates (amorphous ferrihydrite, magnetite, carbonate green rust, etc) that can be generated by varying EC operating conditions.

A better understanding of the mechanisms of *E.coli* removal and inactivation by iron EC allows us to discuss the impact of water quality parameters, like pH and natural organic matter. Finally, we discuss the implications of our results on the potential for iron EC to provide a viable solution to contaminated groundwater in South Asia.

(364d) Comparison of Efficiency for Impellers Commonly Used in Anoxic Denitrification

Grenville, R. K., Philadelphia Mixing Solutions Ltd.
Giacomelli, J. J., Philadelphia Mixing Solutions Ltd.
Higbee, R. W., Philadelphia Mixing Solutions Ltd.
Hutchinson, T. M., Philadelphia Mixing Solutions Ltd
Knecht, C. J., Philadelphia Mixing Solutions Ltd.
Roszczenko, J., Philadelphia Mixing Solutions Ltd.
Wyczalkowski, W., Philadelphia Mixing Solutions Ltd.
Barnes, J.,

Originally presented on: 11/5/2013 16:05:00 - 16:30:00

The hydraulic efficiency of impellers operating in stirred vessels can be defined in the same way as a pump - the ratio of the flowing fluid's kinetic energy to the mechanical energy input by the mixing device.

In this presentation the hydraulic efficiencies of several impeller types commonly used in the process industries are compared to the efficiencies of impellers used for mixing in anoxic denitrification.Â Values for the impellers' efficiencies are compared with values previously reported in the literature.

Results of a full-scale mixing trial in which the solids distribution was measured for three impellers will also be presented and results discussed in the light of the impellers' relative efficiencies.Â The local solids concentration was measured at several locations within the vessel and the power input per volume determined when each local measurement was found to be within Â± 5 % of the average.Â Hydrofoil impellers were found to be 3 ? 5 times more energy efficient than a Hyperboloid impeller in achieving this target.

(364c) Design of Anaerobic Fluidized Bed Reactor Technology for Microbial Sulfate Reduction in Surface and Ground Water Sources

Samee, M., University of Southern California
Ravindran, V., University of Southern California
Pirbazari, M., University of Southern California
Originally presented on: 11/5/2013 16:30:00 - 16:55:00

Sulfate contamination in surface waters and ground waters is a significant problem in the United States. The major sources of sulfate contamination are rock weathering, volcanoes, and biochemical processes. Human activities such as mining, waste discharge and fossil fuel combustion were also important sources. Sulfate is often used as a means to kill those noxious algal blooms due to the presence of excess nutrients, nitrogen and phosphorus.

The wide use of sulfate in agricultural areas has created serious environmental problems in rivers and lakes. Similar problems have been caused across the nation due to mining of sulfide and subsequent oxidation f sulfide to sulfate The Upper Potomac River in the Maryland-District of Columbia region was identified as a major area affected by sulfate contamination due to agricultural runoff of sulfate fertilizers. In the lakes of Minnesota, sulfate contamination arises from mining activities associated with sulfide ores. In the Florida Everglades covering the neighborhoods of Miami and Fort Lauderdale, the sulfate added to control algal blooms in wetland marshes has led to sulfate contamination. The reactions of mercury (from rainfall) and sulfate results in a "cocktail" of Methyl-mercury, a highly toxic substance causing severe environmental and ecological problems. Similar problems have been experienced in the Missouri and Mississippi regions due to agricultural and industrial sulfate discharges.

Microbial sulfate reduction is recognized as an effective technology for removing sulfate from sulfate contaminated water sources. The sulfate reducing bacteria have adaptability and can acclimate themselves to different environmental conditions such as pH, temperature and toxic metals. Microbial sulfate reduction can be economically and efficiently accomplished using anaerobic fluidized bed reactors with recycling. The sulfate contamination of surface water sources in the United States was in the range of 400-1500 mg/L.

The present research is directed at evaluation and design of anaerobic fluidized bed reactor (FBR) process for the removal of sulfate present in surface waters and potable water sources. The FBR process employs granular activated carbon (GAC) or sand as the packing medium for microbial support. The GAC was chosen owing to its ability to promote microbial growth and to resist shock to biologically toxic and inhibitive constituents. However, sand is sometimes preferred as a medium for biofilm support owing to its low cost and easy availability. One important aspect of the technology was the choice of appropriate electron donor as several electron donors are currently available including hydrogen, acetate, lactate, methanol and ethanol. The process was based on a preliminary thermodynamic and kinetic approach for feasibility evaluation of each electron donor. Initially, completely mixed batch reactor (CMBR) studies were conducted to investigate the process feasibility with each electron donor and to determine the biokinetic parameters for sulfate reduction. These included the Monod maximum substrate utilization and half-saturation coefficients, the decay coefficient, and the yield coefficient. The effects of various biological process variables including pH, temperature, and carbon-to-sulfur ratio on the extents and rates of overall sulfate conversion were evaluated. After a preliminary screening, lactate and acetate were tested as potential electron donors. The estimates of biokinetic coefficients and other biological parameters such as biomass concentrations, and sulfate loading provided sufficient information for modeling and designing FBR systems.

The FBR model that we have developed was tested for performance prediction and forecasting of the microbial sulfate reduction process. The important parameters in the model were as follows: (1) bioreactor size, (2) biomass concentration and biofilm thickness, (3) bed expansion due to hydraulic flow and biofilm growth, (4) expansion index and expanded bed height, (5) type, size and density of support media, and (6) superficial fluid velocity in the reactor. The most important FBR design parameters were substrate per unit of biomass, and biomass per unit mass of support medium. Model calibration was based on biological and transport parameters determined from independent laboratory experiments and/or correlation techniques. The model was tested under two different scenarios: (1) employing GAC as an adsorbing medium, and (2) using sand as a non-adsorbing medium for biofilm formation. Chemostat studies were employed in an earlier investigation to determine the Monod coefficients and other biological parameters. They also provided useful information for FBR design such as biomass concentrations, biomass loading, sulfate loading and electron donor levels for the model.

The FBR studies were conducted under different experimental and operating conditions to determine the process efficiency as a function of time. The FBR model was tested and validated for different process operating conditions. Process design and upscaling strategies were developed, and non-dimensional groups relevant to process dynamics were identified. The FBR model successfully predicted the process dynamics with reference to sulfate removal and carbon source utilization, and was validated by experimental results.

The results of this study demonstrated that the FBR system represents a reliable, efficient, and cost effective technology for removing sulfate from potable water supplies. It was found that the FBR system using GAC was significantly more efficient than the FBR using sand in sulfate reduction. However, the FBR using sand had certain advantages in so far as requiring lower hydraulic retention time, and therefore entailing smaller reactor sizes and lower energy costs. The FBR model successfully predicted the process dynamics with reference to sulfate removal and carbon source utilization. Furthermore, it was found useful in the performance prediction of laboratory-scale FBR systems and provided the means for process upscaling to pilot-scale systems using dimensional analysis and similitude techniques.

(374a) Simulation Tools for Nanoparticle-Based Composite Processing and Property Prediction

Bolintineanu, D. S., University of Minnesota
Lechman, J. B., Sandia National Laboratories
Schunk, P. R., Sandia National Laboratories
Originally presented on: 11/5/2013 15:15:00 - 15:40:00

We present a set of tools for the modeling of various aspects of solution-based nanomanufacturing processes. First, we discuss several discrete element modeling (DEM) techniques for the simulation of nanoparticle/colloidal suspensions, including efficient algorithms for the inclusion of hydrodynamic interactions, capillary forces, long-range colloidal interparticle forces, granular contact forces, solvent evaporation and highly aspherical particles. We apply these techniques to the simulation of a battery electrode manufacturing process, and show that the predicted microstructures provide relevant insights for the processing parameters. In the second half of the talk, we discuss methods for the prediction of the bulk properties of disordered, particle-based composites. We present several approaches that rely on a finite element model (FEM) representation of a sample of the material, which can be extracted from a DEM simulation or taken from a tomographic image of the real composite. Overall, this work presents an extensive, multiscale simulation toolset for connecting particle-based nanomanufacturing processing conditions to resulting microstructures and ultimately to material properties.

(374b) Tin Oxide Nanowires and Their Hybrid Architectures for Kinetically Fast Redox Couples in Dye-Sensitized Solar Cells

Vendra, V. K., University of Louisville
Nguyen, T., University of Houston
Druffel, T., Conn Center for Renewable Energy Research
Sunkara, M., University of Louisville
Amos, D. A., University of Louisville
Originally presented on: 11/5/2013 15:40:00 - 16:05:00

The redox electrolyte is an important component of the dye-sensitized solar cells. Iodide/tri-iodide electrolyte used in high efficiency cell is not suitable for practical application due to its corrosive nature, toxicity, competition between dye and tri-iodide for light aborption. One of the major drawback of using alternate redox couples is their fast recombination kinetics and mass transport limitations.

In this study, we show that tin oxide nanowires show a substantial improvement in the photovoltaic performance when compared to titania nanoparticles, when electrolytes having fast recombination kinetics are employed in dye-sensitized solar cells (DSCs). State of the art methods in literature employ titania nanoparticles use passivating layers of Al_2O_3 or poly(methyl siloxane) to lower the electron recombination rate. There have been reports investigating the performance of nanowire in conjunction with alternate redox couples. The single crystalline nature allows for fast electron transport in tin oxide nanowires. Further, prior work from our group has also shown that tin oxide nanowires exhibit two orders of magnitude higher electron lifetimes when compared to titania nanoparticles when iodide/tri-iodide is used as an electrolyte. In this study we employed tin oxide nanowires as photoanodes for DSCs and investigated their performance when kinetically fast electrolytes such as ferrocene/ferrocenium,tempo/tempo,ferricyanide /ferrocyanide are used as redox electrolytes.

Surface modifications such as $TiCl_4$ treatment and atomic layer deposition of titania and alumina were performed to further lower electron recombination in tin oxide nanowires. Fundamental properties such as electron transport time scales and electron recombination time scales were determined for tin oxide nanowires and hybrid architectures where the surface of tin oxide nanowires is coated with titania and alumina.

(374c) Multiscale Sustainability Assessment in Nanocoating Material Design and Manufacturing

Song, H., Wayne State University
Uttarwar, R., Wayne State University
Huang, Y., Wayne State University
Originally presented on: 11/5/2013 16:05:00 - 16:30:00

Nanopaint has drawn great attention in the coatings industry, as it may offer surface coatings with significantly improved performance or new functionalities and properties, such as surface texture alteration control, self-healing, toxic chemical/acid/corrosive agent repelling, glass opaqueness control, building exterior's infrared reflecting/absorbing properties control, and even pathogen growth control. Nevertheless, it is under scrutiny for the environmental and health risks which may pose in nanopaint manufacturing and nanocoating realization. Therefore, it is crucial to analyze the sustainability performance of nanocoating materials and prepare a roadmap for their successful development and application in the future.

In this paper, we introduce a systematic sustainability assessment methodology by resorting to multiscale science and sustainability science. Different from known methodologies focusing on the assessment of manufacturing process and manufactured product, this methodology initiates the assessment starting at the nanomaterial formulation design stage. A variety of material design parameters, such as those about polymer system, nanomaterial type, shape, size, and orientation, solvent and additives, are analyzed using a number of sustainability indicators, which will be incorporated into the analysis of coating manufacturing stages. Multiscale models are developed to predict nanoparticle emission severity in nanopaint spray and curing, nanocoating properties, product realization process performance on quality assurance, waste reduction, and operational safety. The efficacy of the multiscale sustainability assessment methodology will be demonstrated by the case studies on nanocoatings with super-scratch resistance performance and barrier property. The introduced methodology is generic and can be applied to study the sustainability performance of a wide range of nanocoating materials in their early design stage.

(374d) Continuous Nanoparticle Sizing and Characterization Via Microfluidic Interfacial Fluorescent Complexation

Meng, F., Texas A&M University
Ugaz, V. M., Texas A&M University
Originally presented on: 11/5/2013 16:30:00 - 16:55:00

A variety of characterization tools are available to sensitively analyze collected nanoparticles (e.g., condensation particle counters, differential mobility analyzers, scanning mobility particle sizers (SMPS), ion chromatography, mass spectrometry, and electron microscopy). But these methods often require a dedicated laboratory infrastructure to operate optimally, and are therefore not straightforward to adapt for online use. We have developed a new microfluidic-based approach that overcomes many of these limitations by harnessing the inherently steep chemical gradient established at the interface between co-flowing streams containing a nanoparticle-laden suspension and a fluorescent dye. This sharp mismatch acts to localize adsorptive dye-nanoparticle complexation interactions within a narrow interfacial zone, instantaneously producing an intense and easily detectable fluorescence signature. These interactions are inherently dominated by phenomena occurring at the nanoparticle surface, introducing the exciting possibility to extract information about particle size and morphology from the fluorescence profile. We demonstrate this capability experimentally using ZnO and TiO_2 nanoparticles, and introduce a reaction model that enables interfacial fluorescence to be predicted so that kinetic parameters associated with the underlying surface complexation can be quantified. Our approach offers potential for continuous operation that is particularly attractive in manufacturing settings where robust on-line characterization tools are needed.

(374e) Safer Formulation Concept for Flame-Generated Engineered Nanomaterials

Sotiriou, G. A., ETH Zurich
Gass, S.,
Cohen, J., Harvard University, School of Public Health
Pyrgiotakis, G., University of Florida
Pratsinis, S. E., ETH Zurich
Demokritou, P., Harvard School of Public Health
Originally presented on: 11/5/2013 16:55:00 - 17:20:00

The likely success or failure of the nanotechnology industry depends on the environmental health and safety of engineered nanomaterials (ENMs). While efforts toward engineering safer ENMs are sparse, such efforts are considered crucial to the sustainability of the nanotech industry. A promising approach in this regard is to coat potentially toxic nanomaterials with a biologically inert layer of amorphous SiO_2. Core-shell particles exhibit the surface properties of their amorphous SiO_2 shell while maintaining specific functional properties of their core material. A major challenge in the development of functional core-shell particles is the design of scalable high-yield processes that can meet large-scale industrial demand. Here, we present a safer formulation concept for flame-generated ENMs based on a one-step, in flight SiO_2 encapsulation process, which was recently introduced by the authors as a means for a scalable manufacturing of SiO_2 coated ENMs. Firstly, the versatility of the SiO_2-coating process is demonstrated by applying it to four ENMs (CeO_2, ZnO, Fe_2O_3, Ag) marked by their prevalence in consumer products as well as their range in toxicity. The ENM-dependent coating fundamentals are assessed and process parameters are optimized for each ENM investigated. The effects of the SiO_2-coating on core material structure, composition and morphology, as well as the coating efficiency on each nanostructured material, are evaluated using state-of-the-art analytical methods (XRD, N_2 adsorption, TEM, XPS, isopropanol chemisorption). Finally, the biological interactions of SiO_2-coated vs. uncoated ENMs are evaluated using cellular and animal bioassays, providing valuable evidence for reduced toxicity for the SiO_2-coated ENMs. Results indicate that the proposed 'safer by design' concept bears great promise for scaled-up application in industry in order to reduce the toxicological profile of ENMs for certain applications.

References

[1] Sotiriou G.A., Sannomiya T., Teleki A., Krumeich F., Vörös J. and Pratsinis S.E. Non-toxic dry-coated nanosilver for plasmonic biosensors. *Adv. Funct. Mater.* **20**, 4250-4257, (2010).

[2] Sotiriou G.A., Hirt A.M., Lozach P.Y., Teleki A., Krumeich,F. and Pratsinis S.E. Hybrid, silica-coated, Janus-like plasmonic-magnetic nanoparticles. *Chem. Mater.* **23**, 1985-1992, (2011).

[3] Gass S., Cohen J.M., Pyrgiotakis G., Sotiriou G.A., Pratsinis S.E. and Demokritou P. Safer Formulation Concept for Flame-Generated Engineered Nanomaterials. *ACS Sustainable Chem. Eng.* **in press**, DOI: 10.1021/sc300152f, (2013).

[4] Demokritou P., Gass S., Pyrgiotakis G., Cohen J.M., Goldsmith W., McKinney W., Frazer D, Ma J, Schwegler-Berry D, Brain J, Castranova V., An in vivo and in vitro toxicological characterisation of realistic nanoscale CeO_2 inhalation exposures *Nanotoxicology* **in press**, DOI: 10.3109/17435390.2012.739665 (2013).

(374f) Green Pathways for Development of Nanostructured Aerogel Photocatalysts Effective At Both UV and Visible Range

Li, H., University of South Florida
Sunol, S. G., University of South Florida
Sunol, A. K., University of South Florida
Originally presented on: 11/5/2013 17:20:00 - 17:45:00

Nano-structured highly porous $TiO_2/WO_3/Fe^{3+}$ aerogel photocatalyst are prepared, characterized, and tested for model photo-catalytic reactions. The catalyst structure is tailored to capture environmental pollutants and destruct them under both u-v and visible light through oxidation.

A novel and green method is utilized to prepare the unique surfactant templated aerogel photo-catalyst that has highly porous nanostructure with high surface area and tailored pore size distribution. Sol-gel process is combined with supercritical extraction, deposition and drying. Supercritical drying with heat treatment results in titanium dioxide anatase crystal form. Templates used further enable retention and tuning of the nano-pore structure and surface properties.

The catalyst was characterized using AFM, SEM, FIB, XRD, XPS, BET and porosimetry prior to using in reactions. The band gap of the catalyst particles were determined using diffuse reflectance.

Resulting aerogel $TiO_2/WO_3/Fe^{3+}$ has similar photo-catalytic capability compared to highly optimized commercial Degussa P25 under UV exposure and offer much superior photo-catalytic capability under visible light exposure. The model reaction utilized employed Methylene Blue photo oxidation under Visible and UV light.

(387a) Findings On Life Cycle Aquatic Ecotoxicity Impacts of Ionic Liquids

Mehrkesh, A., University of Colorado Denver
Karunanithi, A. T., University of Colorado Denver
Originally presented on: 11/5/2013 15:15:00 - 15:34:00

Ionic liquids (ILs) are a class of chemicals that are being developed as a replacement for volatile organic solvents for a range of industrial applications (e.g. CO_2 capture, aliphatic/aromatic separations, battery electrolytes, cellulose dissolution). Although ILs can potentially decrease health impacts associated with air emissions due to their nonvolatile nature, many of them are highly soluble in water and can become persistent aquatic pollutants. Environmental impacts of tailored ionic liquids arise both from their direct release into the environment during or after use phase as well as indirect emissions during their synthesis. In this work, we use the USEtox model to quantify aquatic ecotoxicity impacts over the life cycle of an Ionic liquid (IL), 1-butyl-3-methylimidazolium bromide ($BMIM^+ Br^-$), from cradle to grave. The aquatic ecotoxicity impacts from synthesis of this IL as well as its release into the enviorment have been taken into account. Life cycle impact assessment methods define environmental impacts in terms of characterization factors. In this work, for the first time we have developed these characterization factors for ILs. The USEtox, a multimedia fate and transport model, was used to derive fate factors and exposure factors needed for this calculation while a large number of literature data related to toxicity of ILs in a variety of species was used to derive the effect factor. Finally we compare the ecotoxicity production impacts with that of direct exposure.

(387b) Comparative Life Cycle Assessment of Vaccine Cold Chain Infrastructure in Developing Countries

Mousavi, B., University of Colorado Denver
Karunanithi, A. T., University of Colorado Denver
Originally presented on: 11/5/2013 15:34:00 - 15:53:00

An important component of vaccine programs in low-and middle- income countries is the cold chain, which is a collection of cold rooms, refrigerators, freezers and cold boxes that keep vaccines at an appropriate temperature. Vaccines require cold storage (between 2° C and 8° C) in order to remain stable and be potent when administered. Heat sensitive vaccines lose potency after exposure to temperatures above the recommended range and freeze sensitive vaccines lose potency when exposed to freezing temperatures. In the next few years there is an urgent need to streamline and expand efficient and reliable cold chain infrastructure in these countries. In this context, the importance of designing and building new cold chain infrastructure that minimizes the environmental impact of energy, materials and processes both within the country and globally cannot be overstated.

In this study a complete life cycle assessment (LCA) analysis for 5 different types of vaccine refrigerators is done for the case of ghana. All of these refrigerators work in regions where accessibility to grid electricity is very minimal. Firstly, the cradle to grave life cycle inventory for each case and secondly the complete analysis of Global Warming Potential (GWP), Human Health (HH), acidification, ozone depletion and smog impacts are presented. After this LCA analysis, policy makers in developing countries can decide to choose the best option (eco-friendly) based on any environmental impacts.

The 5 different types of vaccine refrigerators selected for this study are:

- solar powered battery supplemented refrigerator (SPBS)
- Solar Powered battery-free refrigerator (SPBF)
- Fully Kerosene refrigerator (FK)
- Hybrid Kerosene- electricity refrigerator (HKE) (biomass cogeneration generates electricity)
- Fully LPG Refrigerator (FL)

The SPBS and SPBF systems work in Vapor-Compression cycle refrigeration system and the other types work in absorption refrigeration cycle. All of these types are specific products from famous companies that produce vaccine refrigerators for developing countries all around the world.

The functional unit of this study was set as providing 1 liter cold storage for vaccines during one year. The life span time for SPBS and SPBF was 15 years and for FK,HK and FL is 5 years.

The cradle to grave unit processes of this study include: refrigerant production, insulation foam production, other materials production, photovoltaic modules production, battery production, assembly of all parts of a refrigerator, transportation, LCI of kerosene and LPG from extraction to combustion and discharge of refrigerant and insulation foam to air at the disposal phase.

The Impact assessment results show that FK and SPBF vaccine refrigerators have respectively the highest and lowest GWP and HH impacts. On the other hand, if the HKE used instead of FK all of the environmental impacts decrease dramatically.

(387c) Hybrid Life Cycle Assessment Model of Silicon Photovoltaics

Yao, Y., Northwestern University
Chang, Y., Northwestern University
Masanet, E., Northwestern University
Originally presented on: 11/5/2013 15:53:00 - 16:12:00

Research on sustainable energy resources is receiving increasing attention because of growing concern about severe pollution associated with fossil-fuel power plants. Solar energy is one of the most promising alternatives. Photovoltaic (PV) technology is gradually becoming competitive in power generation market for its abundant resources and affordable electricity[1]. Among all solar technology, silicon PV technology has been successfully commercialized and has the greatest market share[2]. However, any energy production will also consume energy, generate pollutants and result in some environmental impacts [3]. In order to assess the energy sustainability and environmental impacts of silicon PVs, a lot of work has been done on applying Life Cycle Assessment (LCA) tools to solar energy. Currently, the majority of assessments adopt process-based LCA[1] which is a bottom-up method that is limited by system boundary variations, difficulty for replicating results, and inability to provide assessments for future products due to frequent use of proprietary data. Constrained by incomplete system boundary, results of process-based LCA studies often fail to evaluate the entirety of energy and environmental impacts of solar technology on the whole society and ecosystem. Another tool for LCA studies is Economic Input-output LCA (EIO-LCA) model which has a more comprehensive system boundary, allows for comparisons in systems-level and capacity of assessing future product developments[4]. However, EIO-LCA has time lag for current practices, highly depends on government statistical data, and suffers from sectorial aggregations[5].

In this paper, a hybrid LCA model for silicon PVs, which takes advantage of both process-based method and EIO-LCA, was developed. This hybrid model expanded the system boundary to all supply chain materials of silicon PV by matrix operation (e.g. usage of petroleum coke to produce silicon carbide for wire sawing, electricity generation and usage of all upstream materials). Another strength of this hybrid LCA model is that the geographical shift of PV manufacturing, such as expanding Chinese silicon PV production, can be easily estimated by using publicly accessible input-output data of that country. Since silicon PV has been commercialized for several decades, the variation of technology parameters caused by geographic shift is not significant[1]. However, the difference of sustainability performance of PV made in different countries can still be considerable because of diverse economic structure and energy mix. This difference can be captured and assessed by this hybrid LCA model.

For model demonstration, a case study of U.S. multicrystalline PV production was conducted. Results show that the GHG emissions and energy intensity are 41 g CO_2-eq/kWh (based on 1,800 kWh/m^2/year, 14% module efficiency, 0.75 performance ratio and 30 years lifetime) and 5811 MJ/m^2 (balance of system is included), respectively. The Energy Payback Time (EPBT) is calculated to be 2.54 year. Results of this study are 30 – 60% higher than previous studies using process-based LCA model after harmonization[1]. This is caused by the comprehensive boundary of hybrid model, covering more upstream materials in the supply chain and thus more GHG emissions and energy considerations in research system. Sensitivity analysis was conducted to filter dominant factors influencing the sustainability of multicrystalline PV made in the U.S. from a more comprehensive view, which can provide useful recommendations for future research. Given that this model also shows specific energy consumption and GHG emissions of all sectors in U.S. economy caused by multicrystalline PV production and operation, study findings provide policy maker and general public with a more holistic view of the impacts of PV technology over the whole society and ecosystem. More case studies for silicon PV made in other countries, such as China, will be shown in this paper to give a comparison and show the fluctuation of the sustainability performance of silicon PV caused by geographic shift.

References

[1] D. D. Hsu, P. O'Donoughue, V. Fthenakis, G. A. Heath, H. C. Kim, P. Sawyer, *et al.*, "Life Cycle Greenhouse Gas Emissions of Crystalline Silicon Photovoltaic Electricity Generation," *Journal of Industrial Ecology,* vol. 16, pp. S122-S135, 2012.

[2] "Solar Generation 6," European Photovoltaic Industry Association (EPIA)2011.

[3] EPA, "Annual Energy Outlook 2013," U.S. Energy Information Administration2013.

[4] C. T. Hendrickson, L. B. Lave, and H. S. Matthews, *Environmental Life Cycle Assessment of Goods and Services: An Input-Output Approach*: RFF Press, 2006.

[5] Y. Chang, R. J. Ries, and Y. Wang, "The quantification of the embodied impacts of construction projects on energy, environment, and society based on I–O LCA," *Energy Policy,* vol. 39, pp. 6321-6330, 2011.

(387d) Root-Cause Analysis in the Design of Chemical Processes

Gangadharan, P., Lamar University
Chen, L., Lamar University
Lou, H. H., Lamar University
Originally presented on: 11/5/2013 16:12:00 - 16:31:00

Root-Cause Analysis techniques have found widespread application in quality control, management, failure analysis and process safety. A sustainability root-cause analysis is a powerful tool that will enable designers to identify and display the fundamental causes, related causes and root causes that impact the sustainability of a process. This facilitates the understanding of potential interrelationships between causes which lead to the sustainability issue. A base case in the design of chemical/energy systems generally has high potential for improvement. The challenge is to identify the bottleneck issues that can have a significant impact on overall sustainability.

The Pareto chart and the Fish-bone diagram, in conjunction with the sustainability metrics, form the framework of the sustainability root-cause analysis. The sustainability assessment incorporates economic, environmental, societal, and efficiency criteria. Because of the multi-dimensional feature of sustainability, accounting for the impacts of various design factors and the cause-and-effect relationships can be very difficult. After selecting a process from a group of alternative designs, it is further examined in a root-cause analysis, to identify the major causes that affect the parameters considered in the sustainability assessment, and then eliminate them with appropriate corrective action.

To illustrate the use of the sustainability root-cause analysis tools, several case studies including biodiesel production, polygeneration and natural gas reforming are employed. The Pareto analysis and Fish-bone diagram are invaluable aids to visualize complex relationships and screen out significant factors and correlations for sustainability improvement. This methodology can provide critical guidance to design for sustainability.

(387e) Development of Product-Based Energy Intensity Metrics for the California Food Industry

Walker, M. E., Illinois Institute of Technology
Hutchins, M., UC-Berkeley
Lettieri, D., UC-Berkeley
Romanin, V., UC-Berkeley
Blinde, P.,
Masanet, E., Northwestern University
Originally presented on: 11/5/2013 16:31:00 - 16:50:00

Product-based energy intensity metrics describe the energy footprint of manufactured products. These metrics are valuable for several reasons. From an industry standpoint, determining these metrics are an important first step to the development of an internal energy benchmarking strategy. From a marketing point of view, industrial and home consumers alike are putting an increased focus on green manufacturing and carbon footprints; companies that can highlight their own energy and manufacturing efficiencies may benefit in the marketplace. Finally, in terms of emissions regulation, the development of product-based energy metrics and a product-based benchmarking strategy is the preferred approach.

This presentation focuses on the development of product energy intensity metrics in the California food industry for integration with the state's Cap & Trade program. The development of such metrics for food processing presents a number of challenges, stemming primarily from the fact these plants manufacture multiple products and may lack detailed process monitoring capabilities. This presentation highlights the issues facing the determination of these metrics and discusses the strategies and methodologies applied to develop these metrics in the CA food processing industry.

(387f) Estimations of Direct and Indirect Emissions From Mobile Air Conditioner Units in Automobile Transportation: The Cool Mile Functional Unit

Blowers, P., The University of Arizona
Alayedi, I.,
Dougless, A., U of Arizona
Gerwe, B., U of Arizona
Mesta, A., U of Arizona
Rugel, C., U of Arizona
Wang, Y. Y.,
Originally presented on: 11/5/2013 16:50:00 - 17:09:00

Hydrofluorocarbons (HFCs) may become regulated in the near future by the US Environmental Protection Agency because of their contributions to global climate change. HFCs are currently used as refrigerants and solvents because they are relatively inexpensive and inert. This work focuses on the use of HFCs in automobile air conditioning applications because of the large number of vehicles in use in the US, and the rapid rise of the application of refrigeration in mobile air conditioning systems (MACs) worldwide, with R134a being the predominantly used refrigerant. R-134a (1,1,1,2-tetrafluoroethane,) has emissions of about 38.6 x 10^6 kg/year per year from all sources, with approximately 47% of those due to emissions from vehicles. If R-134a is banned, replacements need to be evaluated in order to determine if greenhouse gas (GHG) emissions are indeed reduced when both direct emissions, in the form of leaks, and indirect ones from changes in fuel mileage, are accounted for.

Papasavva, et al, hafe already analyzed emissions patterns for a new replacement compound, HFc-1234hf, which is a hydrofluoro-olefin from fleet-wide use of the refrigerant in different locations around the world under ambient weather conditions, but did not consider indirect emissions impacts. This refrigerant has also been technically evaluated for use in passenger cars and it appears to be a suitable drop-in replacement. Other refrigerant choices are possible and are investigated in this work, using the "number of cool-miles" as the functional unit in order to capture the changes in emissions due to air conditioning use during drawdown (initially cooling of a vehicle from a hot start) and/or stable maintenance of temperature on longer drives so that the importance of different variables can be explored.

This work shows that some refrigerants have reduced direct emissions due to the refrigerant having a lower global warming potential (GWP), but that indirect emissions due to running the refrigerant through the system sometimes outweigh those direct emissions reductions. Future work should explore how to capitalize on opportunities to minimize the sum of the direct and indirect emissions.

(387g) Regional Air Quality Improvement By Scheduling of Multiple Chemical Plant Start-Ups

Cai, T., Lamar University
Xu, Q., Lamar University
Originally presented on: 11/5/2013 17:09:00 - 17:28:00

Chemical plant concentrated regions may suffer localized and transient air pollution events that violate national ambient air quality standards (NAAQS). Flaring emissions, especially intensive start-up flaring emissions from chemical plants, have potentially significant impacts on local air quality. Thus, when multiple plants in an industrial zone plan to start-up within a same time period, their start-up plans should be evaluated and optimally controlled so as to avoid unexpected air-quality violations in any air-quality concern regions (AQCRs). In this paper, a general systematic methodology for multi-plant start-up emission evaluation and control has been developed. The methodology starts with collecting regional meteorological information such as wind speed and temperature; geographical information of all of the involved chemical plants and AQCRs; as well as plant operation data such as the start-up time window, start-up duration, and estimated emission profile. Next, a regional air-quality evaluation based on Gaussian dispersion model will be conducted. If any air quality violation is predicted to an AQCR, a multi-objective scheduling problem will be generated and solved to optimize the start-up sequence and start-up beginning time for all chemical plants. The scheduling model minimizes the overall air quality impacts to all of AQCRs as well as minimize the total start-up time mismatch of all plants, subject to the principles of atmospheric pollutant dispersion. This study may provide valuable quantitative decision supports for multiple stake holders, including government environmental agency, regional chemical plants, and local communities.

(405a) Lean NOx Reduction By H2/CO/C3H6 Using NSR and SCR Catalysts

Kota, A. S., University of Houston
Luss, D., University of Houston
Balakotaiah, V., University of Houston
Originally presented on: 11/6/2013 8:30:00 - 8:50:00

Transient NO_x reduction performance of NSR and SCR catalysts is studied under adiabatic conditions using 1+1 dimensional two phase model. In the NSR catalyst, NO_x is stored under lean conditions on the model Pt/BaO/Al_2O_3 catalyst and subsequently regenerated by a rich mixture of H_2, CO and C_3H_6. A micro-kinetic reaction scheme is developed based on literature reported steady state and cyclic experimental studies of both NO_x Storage and Reduction (NSR), and Selective Catalytic Reduction (SCR) processes. The kinetic parameters are initially estimated for each reductant separately, and are then modified based on comparison with experimental results. Mechanistic analysis of the reaction scheme provides clear insights into the dependence of NO_x reduction efficiency on feed temperature and molar ratio of reducing agents. The model is able to reasonably accurately predict the NO_x conversion and selectivities observed during the experimental studies. The kinetic scheme is further used to investigate the behavior of coupled NSR-SCR catalysts and study the effects of various parameters like cycle time, rich feed ratio, feed temperature, precious metal loading and catalyst architecture on NO_x conversion under both aerobic and anaerobic conditions. Spatio-temporal concentration and temperature profiles are analyzed and used to determine the optimum operating conditions for the coupled catalysts, especially at low temperatures.

(405b) Fast Lean-Rich Cycling for Enhanced NOx Conversion

Perng, C., University of Houston

Harold, M. P., University of Houston

Originally presented on: 11/6/2013 8:50:00 - 9:10:00

Abstract

Conventional NO_x storage and reduction (NSR) has emerged as an important emission aftertreatment technology for reducing NO_x (NO and NO_2) to N_2 under lean conditions typical of diesel or lean-burn gasoline vehicles [1]. Typical cycle timing is 1-2 minutes for the lean phase and 3-20 seconds for the rich phase. In the past year, Toyota Inc. has presented the ?Di-Air? technology which involves much faster cycling and the use of hydrocarbon reductant. Published data show a large expansion in the temperature operating window for achieving high NO_x conversion [2]. In this study we have carried out NSR over a wide range of cycle timing to examine the generality of rapid cycling and to understand causative mechanisms for conversion enhancement.

NO_x reduction experiments have been carried out using the rapid injection of C_3H_6 over a NSR catalyst containing $Pt/Rh/BaO/Ce/Al_2O_3$ combined with a fast switching valve in close proximity to the catalyst. The injection frequency was varied for a fixed cycle-averaged propylene concentration. In contrast to traditional NSR cycling systems, NO_x conversion was observed to increase as injection frequency increases, up to an injection frequency of 0.5 Hz. The shorter cycle times result in a steadier rate of exothermic catalytic oxidation of propylene, characterized by nearly pseudo steady state values for the catalyst temperature and O_2 concentration. This also allows a higher conversion of propylene to be obtained throughout the length of the cycle. In addition, the effect of the lean phase length and thermal durability were studied. High conversions were obtained using less reductant (>90% conversion for a cycle-averaged stoichiometric number of 6) and over a wider temperature range (>90% conversion for a catalyst temperature of 250-550°C) than previous literature values. Figure 1 shows typical results. The extension of the high conversion temperature window implies that the decomposition of barium nitrate is not the limiting factor, likely due to the diminished role of NO_x storage. Furthermore, the inhibitory effect of Ba carbonate formation on NO_x conversion was determined to be less significant at high injection frequencies. This can be attributed to low utilization of dynamic storage sites, which reduces the impact of carbonate formation. These results are consistent with hydrocarbon intermediates on the catalyst surface providing competition for the propylene oxidation reaction and reducing NO_x.

References

[1] Harold, M.P., ?NOx Storage and Reduction in Lean Burn Vehicle Emission Control: A Catalytic Engineer's Playground,? Current Opinion in Chemical Engineering, **1**, 303-311 (2012).

[2] Bisaiji, Y., Yoshida, K., Inoue, M., Umemoto, K., Fukuma, T., "Development of Di-Air - A New Diesel deNOx System by Adsorbed Intermediate Reductants," *SAE Int. J. Fuels Lubr.* 5(1):380-388, 2012, doi:10.4271/2011-01-2089.

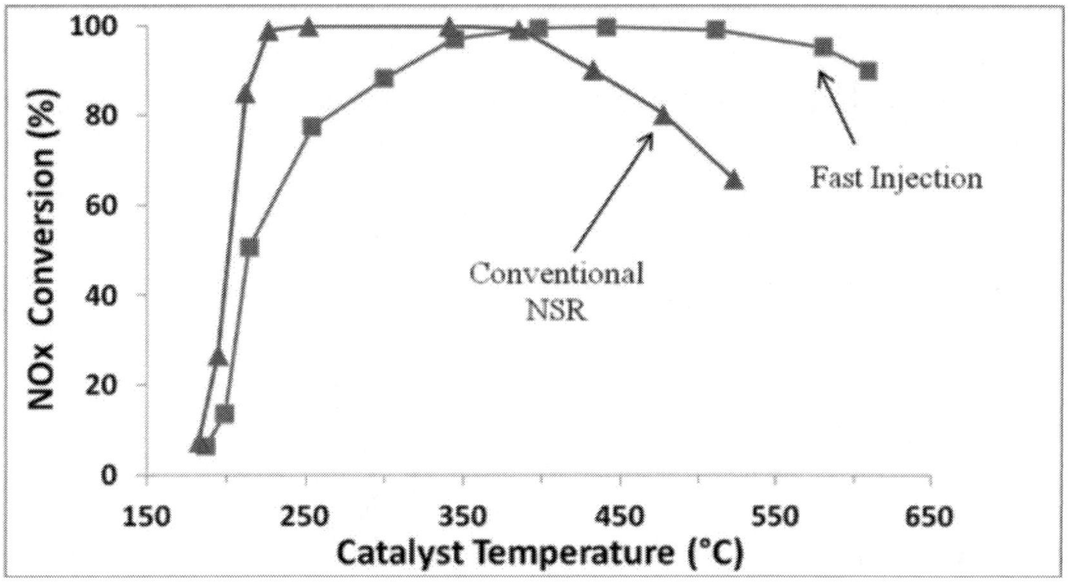

Figure 1. Comparison of temperature effect on NO_x conversion using conventional NSR and fast injection method.

(405c) Experimental and Kinetic Study of the SO2 Oxidation Reaction Over a Pt/Al2O3 Catalyst

Hamzehlouyan, T.,
Sampara, C.,
Epling, W., University of Houston
Li, J., Cummins Inc.
Kumar, A., University of Houston

Originally presented on: 11/6/2013 9:10:00 - 9:30:00

Sulfur oxides in diesel engine exhaust interact with the after treatment catalysts and deactivate them through sulfur poisoning. The diesel oxidation catalyst (DOC) is typically the most upstream device and in order to understand this deactivation behavior, as well as the formation of S species that can poison downstream catalysts systems, i.e. selective catalytic reduction (SCR) and lean NOx trap (LNT) catalysts, the SO_2 oxidation kinetics as well as its effect on the oxidation activity of the DOC need to be understood. In this study, an SO_2 oxidation experimental study was performed and a kinetic model was developed in order to properly describe the SO_2oxidation reaction. The kinetic model was validated using a set of experimental data which was not included in the fitting procedure.

A monolith supported Pt/Al_2O_3 (50 g/ft^3 Pt) catalyst was used in the experiments. A separate reactor upstream was used to generate SO_3 when SO_3 was required in the feed for the kinetic and deactivation studies. For all experiments, the outlet gas concentrations were measured using a MKS MultiGas MG-2000 FT-IR analyzer. Experiments were run in order to determine the SO_2 oxidation rate dependency on the SO_2, O_2 and SO_3 concentrations. The kinetic experiments were performed in a differential reactor regime to minimize effect of temperature gradients and to avoid mass transfer limitations due to the concentration gradients. A long time SO_2 exposure was used as a pretreatment method where the catalyst was exposed to 50 ppm SO_2 in N_2 at 240°C overnight. Reproducibility within ±9% was achieved for the data points with SO_2conversions higher than 5%.

Apparent activation energy of SO_2 oxidation over Pt/Al_2O_3 was determined under different concentrations of SO_2, O_2 and SO_3. Five sets of feed composition were selected and the apparent activation energy was measured by randomly varying the temperature between 240 and 308°C while keeping the feed composition constant. The effect of SO_3 on the activation energy was studied doing another set of experiments with no SO_3 in the feed. In order to generate the experimental data to be used in the model development, a set of experiments were designed where the inlet concentrations of SO_2, O_2 and SO_3 as well as temperature were randomly varied over specific ranges. Prior to experiments, the catalyst was pretreated with 50 ppm SO_2 in N_2 at 240°C overnight. In the steady state SO_2 oxidation experiments, the concentrations varied over the ranges 50-150 ppm for SO_2, 0-105 ppm for SO_3 and 5–13% for O_2over the temperature range of 240-330°C. Each data point was recorded after the steady state was reached.

The reaction order experiments were performed in the presence of SO_3 in the feed and reaction orders were estimated using a log linear least square analysis. The reaction orders of 0.88 and -0.24 were obtained for SO_2 and O_2, respectively, and the SO_3 reaction order was found to be -0.45, demonstrating its inhibition effect. An average activation energy of 101 kJ/mol was obtained for SO_2 oxidation when SO_3 was present in the feed. In the case of no SO_3 in the feed, an apparent activation energy of 59 kJ/mol was estimated, verifying the SO_3inhibition effect.

There is no proven mechanism for SO_2 oxidation over Pt catalyst in the literature. However, some researchers proposed Langmuir-Hinshelwood type mechanisms for the SO_2 oxidation reaction over Pt-based catalysts [1,2]. In the present work, a microkinetic model based on a Langmuir-Hinshelwood mechanism was proposed for the catalytic oxidation of SO_2 on Pt/Al_2O_3 and a one dimensional steady state model was developed for a single channel of a monolith. A plug flow reactor model was assumed and the set of algebraic differential equations was solved at various temperatures to predict the SO_2conversion as a function of temperature. A set of kinetic parameters, taken from the literature [2], was used in solving the reactor model where some parameters were fit to match the experimental data. External mass transfer was also implemented in the reactor model to be able to perform some parametric studies under extreme reactor conditions. A separate set of experimental data which was not included in the parameter optimization was used to validate the kinetic model.

The relative importance of each step in the reaction mechanism was studied at different temperatures to identify the rate determining step. It was found that the reversible surface reaction between the adsorbed SO_2 and adsorbed oxygen is the rate determining step over the entire temperature range studied. Dissociative adsorption of oxygen on Pt was also found to be important at lower temperatures, ie below 350°C, where its rate is slower than those of SO_2 adsorption and SO_3 desorption. The modeling results revealed that at low temperatures, ie 200°C, the adsorbed SO_2 had the highest coverage on the Pt surface followed by oxygen and SO_3. However, as temperature increases, the adsorbed oxygen becomes the most abundant surface intermediate which is consistent with the results observed in the literature [1]. Ultimately, a parametric study was performed using the developed model to predict the mass transfer controlled regime as well as the catalyst activity under different operating conditions.

References

1. Benzinger W., Wenka A. and Dittmeyer R., *Appl. Catal. A: General* 397, 209 (2011).
2. Dawody J., Skoglundh M., Olsson L., Fridell E., *Applied Catalysis B: Environmental,* Vol 70 (2007).

(405d) Ozone Removal Using New Structure of Matter for High Performance Heterogeneous Catalysis At Short Contact Time

Gu, Q., Auburn University
Tatarchuk, B., Auburn University
Originally presented on: 11/6/2013 9:30:00 - 9:50:00

Pleated Microfibrous Entrapped Catalyst (MFEC) manufactured from 8 μm diameter nickel fibers were tested for ozone decomposition at turbine bleed air conditions to demonstrate the improved catalytic performance over conventional monolith reactors.

In this research, ozone has been used as model compound. MFEC reactors were investigated under aircraft bleed air conditions of high temperature (100-200℃) and high face velocity (10-30 m/s) resulting an interlayer contact time of 67-200μsec. Precious metal (Pd, Ag) and transition metal (Mn) catalysts were impregnated on entrapped particles (e.g. γ-Al_2O_3) using incipient wetness method. Ozone test concentration was set at a high-demanding 1.5 ppmv. Results showed that a high level of ozone decomposition was achieved with a significant reduction of catalyst consumption. Compared with conventional aircraft filters, this can be a huge advantage in terms of material cost and labor. Because the entrapped small particles (150-250μ m) improve the contact efficiency significantly, this kind of reactor has shown high catalytic activity even though the contact time for the reactant has been shortened to 100μ s range due to the high face velocity. The ozone reaction kinetics has been analyzed using a first order plug flow model, which showed the reaction was external mass transfer limited. Mass transfer rate, effective reaction rate and overall efficiency factor were compared among packed bed, monolith and MFEC reactors. Also catalyst aging tests were carried out according to the frequency of commercial aircraft ozone abater replacement, which evaluated long-term performance of MFEC. XPS, TPR and chemisorption tests have been used to characterize catalyst deactivation mechanisms. Besides, several other aspects including humidity, system pressure been investigated for affection on the reaction.

(405e) Thermal Analysis of Active Catalytic Diesel Particulate Filter Regeneration

Su, C., University of Notre Dame
Yuan, Y.,
Zokoe, J.,
Henrichsen, M.,
McGinn, P., University of Notre Dame
Originally presented on: 11/6/2013 9:50:00 - 10:10:00

A novel K-containing glass catalyst [1] was applied to ceramic filters for diesel soot combustion. The results show that the catalytic coating on the ceramic substrate can reduce the temperature where half of the engine soot is oxidized (T_{50}) to as low as 450°C under loose contact conditions, and offers catalytic stability for long term combustion cycling. A major technological challenge in operation is that a rapid shift to idle may create local hot region with a temperature much higher than under stationary feed conditions. This excessive transient temperature rise may cause local melting or cracking of the ceramic filter. For this reason, a one-dimensional computational fluid dynamics (CFD) model has been developed to simulate the exhaust flow and chemical reactions in the Diesel Particulate Filter (DPF) under different active flow schemes, e.g. drop-to-idle conditions. The regeneration strategy was thus tailored for the application of the novel low soot oxidation temperature catalyst to avoid overheating. Implications of the suggested regeneration strategy to fuel economy were also investigated.

Reference:

[1] C. Su, P.J. McGinn, Appl. Catal.B-Environ.,138 (2013) 70.

(405f) Kinetics of Passive Soot Oxidation On a Model Un-Catalyzed and Commercial Catalyzed Diesel Particulate Filter

Kumar, A., University of Houston
Stafford, R., Cummins Inc
Kamasamudram, K., Cummins Inc.
Currier, N., Cummins Inc
Yezerets, A., Cummins Inc.
Originally presented on: 11/6/2013 10:10:00 - 10:30:00

Diesel Particulate Filters (DPFs) are widely employed in advanced aftertreatment systems to meet tailpipe particulate matter (PM) emission requirements. DPF captures soot, generated by engine, via filtration through porous cordierite substrate. The soot accumulation causes an increase in back pressure resulting in higher fuel consumption. Therefore, periodic DPF regeneration is engaged to oxidize soot accumulated in the filter to decrease backpressure. The DPF regeneration can be achieved by active regeneration which requires temperatures in excess of 550°C and causes fuel penalty and hydrothermal aging of other after-treatment components like DOC and SCR. Another way to regenerate the DPF is by utilizing more potent oxidizing agent NO_2 in the exhaust, called passive soot oxidation. Passive soot oxidation can be achieved below 400°C that will decrease fuel consumption and minimize aftertreatment system degradation. Previous laboratory studies have reported passive soot oxidation using packed bed reactor with soot and model catalyst mixtures. In this work, we developed kinetic model of passive soot oxidation on a model un-catalyzed and commercial catalyzed diesel particulate filter.

We developed a procedure to deposit soot in 1" diameter x 3" long DPF cores, using a prototype soot generator by rich combustion of propane. Varying levels of soot loaded DPF cores (1 to 5 g/L) were prepared for passive soot oxidation kinetics analysis. Multiple square pulses (1 min long) of NO_2 were introduced over the soot loaded DPF, maintained at isothermal temperature, to achieve differential conversion of NO_2 and soot. The developed pulse technique allows the estimation of the soot oxidation rates as a function of soot consumption. Temperature and NO_2 concentration was varied over a range to develop the model with NO_2 concentration, soot loading, and temperature as key parameters.

A comprehensive passive soot oxidation model that was developed and validated using commercial catalyzed DPFs will be presented.

(405g) Visible Light Induced Degradation and Disinfection Using Multifunctional Ag/AgCl - Activated Carbon Composite Photocatalysts

Gamage McEvoy, J.,
Cui, W., Hebei United University
Zhang, Z., University of Ottawa
Originally presented on: 11/6/2013 10:30:00 - 10:50:00

In this study, novel Ag/AgCl – activated carbon (AC) composite photocatalysts were prepared and their activity tested for visible light induced photocatalysis. These composites combine the high visible light activity of Ag/AgCl, which arises due to its surface plasmon resonance and charge separating effects, with the improved mass transfer and adsorption characteristics of activated carbon. The composites were characterized by X-ray diffraction, scanning electron microscopy, X-ray photoelectron spectroscopy, N_2 adsorption-desorption porosimetry and UV-visible light diffuse reflectance spectroscopy, and the properties of the composites and the effect of the incorporated catalyst on the AC matrix were elucidated. The composites were found to possess good adsorption capabilities, and the sorption kinetics, isotherms, and mechanism were investigated using methyl orange (MO) as a model organic pollutant. The photocatalytic degradation of MO was studied under visible light, and was modelled using Langmuir-Hinshelwood kinetics. The effects of solution concentration, pH, and radical quenchers on the overall reaction rate were also studied, and a mechanism of photocatalysis proposed where the silver acts as the electron/hole generator, and polarization of the photoinduced charge relative to the AgCl facilitates electron-hole separation, while the AC works to concentrate the MO around the active sites. The prepared Ag/AgCl – AC was also investigated for use as an antibacterial and photocatalytic inactivation agent against a model microorganism, *Escherichia coli* K-12. The antibacterial and photocatalytic effects were confirmed, and their relative contributions to inactivation were compared via zone of inhibition, multiple-blot, and standard plate count assays. The photocatalyst was found to inactivate up to 99.5% of bacteria under irradiation for sixty minutes in a 5 g/L slurry. The composite also possessed some antibacterial action due to the incorporated nanosilver; however, the action of the photo-induced radical species on the bacteria was thought to dominate the process in the presence of irradiation. The mechanism for photocatalytic cell death was probed using selective plate assays, measurements of potassium ion leakage from the cells during treatment, and Fourier Transform Infrared spectroscopy (FTIR) to investigate structural changes which occurred. From these studies, inactivation was thought to be initiated by the action of photo-induced radical species on the cell wall. Due to its activity for visible light induced degradation and disinfection, the novel Ag/AgCl - AC prepared provides interesting potential for use as visible light active photocatalysts in applications such as solar photocatalytic water treatment.

(421a) Multicomponent Ppb Competitive Adsorption in the Presence of Varying Relative Humidity

Cheng, P., Auburn University
Tatarchuk, B., Auburn University
Originally presented on: 11/6/2013 8:30:00 - 8:50:00

Contaminant gas removal is a critical process for current and future power supply systems including fuel cells, gas turbines (GT) or a combination on ships. Jet fuels such as JP-5 are widely used in marine power supply systems, resulting in emission of volatile organic compounds (VOCs). The contaminants can affect the fuel cell performance of both anode and cathode, even exposed to an ultra-low concentration (<1 ppm). In this work, toluene and thiophene were introduced as gas phase contaminants at various initial concentrations (0.1 ppm, 0.2 ppm and 0.4 ppm). A packed bed consisting of activated carbon was used as an adsorbent operated at a face velocity of 0.9 m/s. Concentration of contaminants upstream and downstream was determined by GC/FID. The experiments for adsorption behavior (e.g. breakthrough time, saturation time and separation time) were carried out on different types of activated carbon with pretreatment. BET surface area, average pore volume and pore-size distribution were measured for each type of activated carbon in order to characterize their adsorption capacity. Furthermore, the effect of relative humidity (RH) was studied at such trace concentrations.

(421b) Removal of Low Vapor Pressure Perfluoroalkanes By Activated Carbon

Schindler, B. J., U.S. Army Research, Development and Engineering Command, Edgewood Chemical Biological Center
Buchanan, J. H., U.S. Army Research, Development and Engineering Command, Edgewood Chemical Biological Center
Mahle, J. J., U.S. Army Research, Development and Engineering Command, Edgewood Chemical Biological Center
Peterson, G. W., U.S. Army Research, Development and Engineering Command, Edgewood Chemical Biological Center
Glover, T. G., SAIC
Originally presented on: 11/6/2013 8:50:00 - 9:10:00

With increased use of highly fluorinated compounds has come an increased need for knowledge of the fundamental properties of these chemicals, such as vapor pressure, to determine how they will behave in the environment and how to remove them. A modified vapor saturation method was used to measure the ambient temperature vapor pressure of 3-(perfluorobutyl)propanol, (perfluorooctyl)ethylene , perfluorohexanoic acid, ethyl perfluorooctanoate, and perfluoro-3,6-dioxaheptanoic acid between 254.15 K and 309.15 K. The vapor pressure values are then used to generate a feed concentration at 139 PPM. The adsorption capacity of three of the chemicals on BPL activated carbon. At low relative humidity 3-(perfluorobutyl)propanol has a greater capacity than ethyl perfluorooctanoate, which has a greater capacity than (perfluorooctyl)ethylene. At high relative humidity the trend remains the same but the capacities differ.

(421c) The Design and Analysis of Activated Biochar Material for the Control of Elemental Mercury Emission From Industrial Facilities Into the Global Atmospheric Pool

Shewchuk, S. R.,
Dalai, A. K., University of Saskatchewan
Smith, D.,
De, M.,
Azargohar, R.,
Originally presented on: 11/6/2013 9:10:00 - 9:30:00

Mercury in the environment is a global environmental issue. One of the main sources of elemental mercury into the atmospheric environment is coal burning electricity generating power plants. Mercury is a trace contaminant in the burning of coal. Essentially, this anthropogenic elemental mercury is emitted into the earth's atmospheric environment. It then mixes globally. Once it deposits to the earth's surface, it is the aquatic systems of planet earth that are the most sensitive to the elemental mercury. The deposited mercury methylates the aquatic systems of the planet with the accumulation of this material occurring in selective fish tissue and then subsequently into the human being with the consumption of fish tissue.

Biological carbon or biochar has been prepared and used by the University of Saskatchewan Chemical Engineering Department, the Saskatchewan Research Council (SRC) and SaskPower in a research program over the years, to test its effectiveness as a capture agent for elemental mercury gas in Saskatchewan's coal-fired power plant flue gas streams.

The source of the biological carbon products for these programs has come from Advanced BioRefinery, which is a commercial firm that acquires its biological carbon from Canada's forest industry and other similar venues.

The biochar material is then activated at the University of Saskatchewan and prepared into an appropriate particle size range at SRC for injection into a coal-fired power plant's flue gas stream within the province of Saskatchewan, Canada. After activation, this biochar material is ground into a series of small aerosol particles in the size range of approximately 50 microns in diameter for direct in-stack injection into an operational coal-fired power plant's flue gas stream. Once in the flue gas stream, these biochar aerosols capture elemental mercury atoms which are subsequently sequestered from emission into the environment.

In this presentation we will discuss the need for the activation process of the raw biochar product. It has been discovered in this research that it is the surface aerosol microporosity that is the primary agent responsible for the elemental stack gas mercury capture efficiency. Various procedures have been employed to enhance the biochar elemental mercury capture effectiveness within an operational setting.

We have prepared two levels of aerosol microporosity described as an optimum level at 287 m^2/g and a reference level at 207 m^2/g. Once these various levels of biochar aerosol microporosity were injected into a coal burning power plan's flue gas stream, it was discovered that the optimum level of biochar aerosol microporosity reduced the elemental mercury emission rate into the environment by approximately 88 per cent. The reference level of aerosol microporosity reduced the elemental mercury emission rate in to the environment by approximately 60 per cent.

The primary conclusion of this research work with biochar is that it is the aerosol surface microporosity that is the main capture agent for in-stack elemental mercury atoms. Biochar aerosols can be prepared and sold into the international market as a sorbent-based emission control agent for in-stack elemental mercury emissions control for the coal burning power stations of the world.

(421d) Mercury Interaction With the Fine Fraction of Coal Fly Ash in a Simulated Coal Power Plant Flue Gas Stream

Jew, A. D., Stanford University
Rupp, E. C., National Energy Technology Laboratory
Jung, J. E., Stanford University
Farfan, G.,
Brown, G. E. Jr.,
Wilcox, J., Stanford University
Originally presented on: 11/6/2013 9:30:00 - 9:50:00

Mercury is a potent neurotoxin and a significant contaminant released through the burning of coal in coal-fired power plants. Besides Hg, coal-fired power plants produce large quantities of fly ash that interact with Hg contained in flue gas emissions. Because fly ash has the potential to be a sink for Hg and is used in products such as cements, it is necessary to understand how Hg is associated with the ash. This work focuses on both bulk and < 100 nm sized fly ash particles reacted with a simulated flue gas stream containing Hg vapor. Fly ash samples were analyzed by a variety of laboratory- and synchrotron-based techniques to determine the size fraction of fly ash responsible for the majority of Hg uptake, determine any correlations between Hg and fly ash components (both organic and inorganic constituents), and to determine the Hg phases present after fly ash reaction in the simulated flue gas stream.

The fine fraction of fly ash from a coal-fired power plant was separated to a particle size ≤ 100 nm. This fine fraction was reacted in a simulated flue gas containing CO_2, H_2O, O_2, NO_x, SO_2, HCl, and Hg. These gases were initially passed through a flame at ~1,100°C to simulate gaseous reactions found in a coal-fired power plant boiler, and then the gas stream was reacted with the fly ash in a packed bed reactor at ~140°C. Reacted and unreacted samples were analyzed by a large suite of techniques. Characterization of the ash was done using scanning electron microscopy (SEM), transmission Fourier transform infrared (FTIR) spectroscopy, x-ray diffraction (XRD), and electron microprobe. To understand how Hg correlates with different constituents of the fly ash, samples were analyzed by both electron microprobe and synchrotron-based x-ray fluorescence (s-XRF) mapping. X-ray absorption spectroscopy (XAS) was used to identify the Hg-bearing phases present in the samples. Micro x-ray absorption near edge structure (μ-XANES) spectroscopy was coupled with s-XRF to identify the form of Hg found in Hg hot spots identified during s-XRF mapping. Bulk extended x-ray absorption fine structure (EXAFS) spectra of the reacted fly ash were collected by slow cooling to 77K to identify the Hg-phases present in the sample, including elemental Hg if present.

Synchrotron XRF mapping of both the bulk and fine fraction reacted ash samples indicates that the fine fraction (≤ 100 nm) dominates the uptake of Hg in fly ash. XRF mapping also showed that Hg is present in two major regions: those with high Fe concentrations and Hg hot spots not associated with Fe. XRF mapping shows a correlation between Fe and Cl in Fe-rich regions, whereas Hg hot spots have associated S and Br as well as minor amounts of As and Se. Mercury was not found to be associated with Ti, alkali metals, alkaline earth metals, or the light transition metals in these samples. Electron microprobe analyses showed that some regions of the samples show a correlation between Hg and C, whereas in other regions there is no association of Hg with C. μ-XANES spectroscopy of the Hg hot spots indicates the presence of cinnabar (α-HgS) with no indication of the higher temperature polymorph, metacinnabar (β-HgS), being present. Bulk EXAFS analysis indicates that Hg phases present include Hg-Br, Hg-Cl, Hg bound to organics, Hg bound to Fe-oxides (presumably hematite), Hg-O (montroydite-like phase), and cinnabar. We did not detect elemental Hg in the fly ash, suggesting that Hg uptake by fly ash involves the oxidation of Hg(0) to Hg(II). FTIR analysis of the fly ash samples before and after reaction shows a significant intensity decrease of the carboxylic acid functional group absorption bands. This decrease is interpreted as suggesting that the carboxylic groups bind to Hg in the unburned carbon portions of the fly ash. Overall, these results show a complex interaction between Hg and coal fly ash. By understanding the speciation of Hg in fly ash, more refined model experiments using synthesized forms of phases present in the sample can be done to determine the stability of the Hg in the fly ash that is either disposed of in landfills or used in cement products.

(421e) Mercury Sorption Technology in Coal-Fired Power Plants

Meeks, N. D., Southern Company Services, Inc.

Originally presented on: 11/6/2013 9:50:00 - 10:10:00

Development of mercury sorption technology has recently emerged as an environmental challenge, most recently in response to the US EPA Mercury and Air Toxics Standards Rule, finalized in 2012. This talk wiill cover recent full-scale sorption trials of activated carbon, non-carbon sorbents, as well as pilot-scale studies of fixed structure mercury sorbents, including mixed matrix membrane technology. In particular, the talk will focus on overcoming challenges associated with conventional sorbents including Hg/SO3 selectivity. **(Most recent data will be included in the abstract closer to the meeting)

(421f) Gas-Phase Mercury Removal Using Sulfur Doped Porous Carbon

Reddy, K. S. K., The Petroleum Institute
Al Shoaibi, A.,
Srinivasakannan, C., The Petroleum Institute
Originally presented on: 11/6/2013 10:10:00 - 10:30:00

Gas phase mercury removal is a vital unit operation in gas processing industries. The present work attempts to prepare a sulfur impregnated carbon at optimized experimental conditions and compares its elemental adsorption capacity with the number of commercially available carbon based adsorbents. The effect of adsorption temperature on mercury adsorption capacity has been estimated for the prepared sulfur impregnated carbon. The adsorption capacity was found to increase with increase in adsorption temperature owing to the chemisorption nature of the adsorption. The adsorption isotherms were generated at three different temperatures and were found to close adhere to the Langmuir Isotherm model. The adsorption capacity was found to increase until 140°C, while decrease beyond, which was attributed to the softening and agglomeration of sulfur. The maximum adsorption capacity of 4325 μg/g was observed at a temperature of 140°C. A comparison of the relative adsorption capacity of various adsorbent at 30°C, reveled the adsorption capacity of the sulfur impregnated carbon prepared in the present work much higher than the commercially available carbons. The high adsorption capacities with simple preparation techniques favor the commercial mercury adoption process.

(421g) Study of H2S Removal At High Temperature With 20%Fe2O3/Al2O3 Sorbent Applying Factorial Experimental Design Method

Huang, C. Y., Institute of Nuclear Energy Research
Chyou, Y. P., Institute of Nuclear Energy Research
Originally presented on: 11/6/2013 10:30:00 - 10:50:00

In gasification-based energy systems, e.g., an IGCC (Integrated Gasification Combined-Cycle) system, the coal gasification process will generate large amount of high-temperature syngas, while release H_2S, COS, $CS_2$and other harmful gaseous pollutants. Many developed countries in the world were devoted to research and development of high-temperature dry desulfurization technology, which can improve the thermal efficiency as well as reduce the cost of power generation and environmental protection.

In this study, iron nitrate is impregnated on porous alumina and calcinated at 973K for 6 hours to prepare the desulfurization sorbent. The sorbent is used as a high-temperature desulfurization agent for the removal of hydrogen sulfide in a fixed-bed reactor. To improve the sulfur capacity of desulfurization sorbent, this paper applies the 2^5 factorial experiment designs that not only apply the ANOVA table and regression analysis to analyze the experimental data, but also verify the data completely by experiment step design before studying the important operational factors. Finally, the impact factors of desulfurization reaction of the main effects and the interaction effects can be identified. The 2^5 sets experimental data via conversion reduce the impact of interaction number. From analysis obtained that the main effect is the reaction temperature (positive) and CO_2 concentration (negative). The interaction effect is the CO_2 concentration & reaction temperature.

(422a) Engineering Protein Delivery Systems to Protect Pyrethroids (λ-Cyhalothrin) Against Photodegradation

Lee, H., University of Illinois at Urbana-Champaign
Thirumalai, P.,
Feng, H., University of Illinois at Urbana-Champaign
Originally presented on: 11/6/2013 8:30:00 - 8:55:00

Engineering Protein Delivery Systems to Protect Pyrethroids (λ-Cyhalothrin) against Photodegradation

Hyoungill Lee[a], Pranav P. Thirumalai[a], Hao Feng[a*]

[a]- Department of Food Science and Human Nutrition, University of Illinois at Urbana-Champaign, Urbana, IL, USA-61801

[*] Corresponding author and Associate Professor, Department of Food Science and Human Nutrition, University of Illinois at Urbana-Champaign, 382F-AESB, 1304 West Pennsylvania Avenue, Urbana, IL 61801, USA.

Phone: 01-217-244-2571; Fax: 01-217-333-9329; Email: haofeng@uiuc.edu

Abstract:

In recent years, Liu et al. have shown that pyrethroid pesticides have been widely used due to their long lasting stability and high insecticidal activity. Several studies have shown the adverse effects of chemical and/ or photodegradation of pesticides in food and water systems. A review by Legrini et al. shows Advanced Oxidation Processes (AOPs) as a potential in this area. In this work we have used soy protein stabilized nanoemulsions and nanocomplexes with pyrethroids, such as λ-Cyhalothrin, to protect the pesticides against UV exposure. Soluble Soy Protein Isolate (SPI) was prepared using pH shifting (pH 8, 9, 10, 11 and 12) and a combination of pH shifting and ultrasound (100% amplitude for 1, 2, 3, 4 and 5 minutes). The combination of pH shifting and ultrasound showed significantly greater recovery and surface hydrophobicity of soluble SPI, irrespective of the pH and ultrasound treatment time. The turbidity and the particle size (nm) of nanoemulsions increased from 0.6783 to 3.9445 and 85.1 nm to 241.9 nm respectively, with an increase in oil:protein (w:w) ratio from 0.25:1 to 2:1. Soy protein stabilized (pH 12 + sonication) nanocomplexes with λ-Cyhalothrin and nanoemulsions showed at least 40% more protection than λ-Cyhalothrin in distilled water, when exposed to UV radiation for 6 hours. Thus, both the nanocomplexes and nanoemulsions prepared from 90% soluble SPI with a combination of pH shifting and ultrasound showed greater protection of pesticide against photodegradation.

Keywords: Photodegradation, pyrethroids, ultrasound, nanotechnology, nanoemulsion, nanocomplex

(422b) Toxicity Effects Associated With Exposure of Lung Epithelial Cells to Polymer Nanocomposites and Nano-Platelets of Silicates

Agarwal, S., West Virginia University
Eldawud, R., West Virginia University
Dinu, C. Z., West Virginia University
Gupta, R. K., West Virginia University
Originally presented on: 11/6/2013 8:55:00 - 9:20:00

Montmorillonite, a layered silicate in the form of platelets having a thickness of about 1 nm, an aspect ratio exceeding 200 and a surface area of about 650 m^2/g, is commercially compounded into polymers such as nylons and polyolefins. Even at low filler loading levels (about 5 wt%), the resulting polymer nanocomposites (PNCs) have improved properties, including reduced flammability when compared to the matrix polymer. Such enhanced property is observed when the nanofiller is deagglomerated and uniformly distributed throughout the plastic, with the nanofillers contributing to the formation of char that acts as a heat and mass transfer barrier. However, when such polymer nanocomposites are burnt, inhalation exposure to the nanoparticles contained within the PNCs can occur. Here we started to investigate the toxicological effects of associated with exposure of human epithelial cells to silicate nanoplatelets to determine the any harmful effects. For this, rectangular strips of PNCs were first ignited in air using a methane flame, and a Combustion™ DMS500 Differential Mobility Spectrometer was used to measure the number concentration of particles emanating from the burning sample. Measurements indicated the presence of nanoparticles in the 10 to 100 nm range and a concentration that was as much as $2.7 \times 10^9/cm^3/s$; these numbers are comparable to those encountered in the exhaust from heavy-duty diesel engines. Next, human lung epithelial cells were exposed to various concentrations of nanoplatelets and the effects of the nanoplatelets on the cells morphology, structure and functions were analyzed. Our studies allowed the characterization of the cytotoxicity associated with the combustion products of nanoplatelets and promises to help develop a tool for the prediction of large-scale burning behavior of polymer nanocomposites.

(422c) The Age of Carbon Nanotubes and Biomolecular Convergence for Cancer Therapeutics

Dinu, C. Z., West Virginia University

Originally presented on: 11/6/2013 9:20:00 - 9:45:00

Carbon nanotube (CNT) is the new "it" in nanotechnology revolution. CNT's ease of production as well as their exceptional chemical, mechanical and physical properties, and the ability to be functionalized with molecules of interest based on their high stability and high aspect-ratios, have recently prompted CNTs use both for delivery of drugs as well as for effective killing of tumor cells. In particular, interest was given to improving the engineering technologies that allow CNT synthesis and functionalization for safe and highly efficient delivery of the therapeutic agents. Nevertheless, given the nascent state of nanotechnology, much remains to be learned about the properties, characteristics and effects of carbon nanomaterials when exposed to biological systems, and the complexity of nano-bio-interface-based reactions when considering developing carbon nanomaterial-based therapeutics. In particular, the assessment of CNTs exposure risks and complementary, the benefits of using carbon nanomaterials in the biomedical fields, requires information and vision. This presentation reviews the current status in toxicological and pharmacological profiles of CNTs, the current efforts in my group in understanding nanotube toxicity in relationship to nanotube structure, properties and aspect ratio and proposes a paradigm shift for controlling immunomodulatory activity of nanotubes for providing the next generation of targeted therapeutics that require convergence to minimize host toxicity, while maximizing cancer cell genotoxicity.

(422d) Monitoring Ligand Exchange On the Surface of Gold Nanoparticles Using Isothermal Titration Calorimetry

Kitchens, C. L., Clemson University
Hart, A. E., Clemson University
Mefford, O. T., Clemson University
Powell, B. A., Clemson University
D'Unger, D., Clemson University
Originally presented on: 11/6/2013 9:45:00 - 10:10:00

Ligand exchange is an important factor in the function and end fate of nanomaterials. A better understanding of binding strengths of various functional groups to the surface of gold nanoparticles, GNPs, is needed in order to better predict how they will behave in natural and biological environments. Fluorescent Resonance Energy Transfer (FRET) was used to develop a hierarchy of competitive binding strengths of 5 fluorescently tagged functional groups using fluorescence spectroscopy. The functional groups investigated were; COOH, NH_2, OH, SH, and SO_3.GNPs bound with fluorescently tagged ligands were titrated into a known concentration of non-fluorescently tagged unbound ligands and the fluorescence intensity was monitored until ligand exchanged reached equilibrium. The study resulted in the following hierarchy with the lowest observed binding strength listed first; OH $<$ NH_2 $<$ COOH $<$ SO3 $<$ SH. Kinetics of the FRET data was used to calculate competitive binding constants. Isothermal Titration Calorimetry, ITC, was performed with the same ligands used in the FRET experiments in order to confirm the binding energies obtained from the FRET results.

(422e) Regulatory Association of Cell Responses Induced By Metal and Metal Oxide Nanoparticles

Liu, R., University of California, Los Angeles
France, B., University of California, Los Angeles
George, S., University of California, Los Angeles
Zhang, H., University of California, Los Angeles
Xia, T., University of California, Los Angeles
Nel, A. E., University of California, Los Angeles
Bradley, K., University of California, Los Angeles
Rallo, R., Universitat Rovira Virgili
Cohen, Y., University of California, Los Angeles
Originally presented on: 11/6/2013 10:10:00 - 10:35:00

A data mining approach was developed and applied to discover relationships among fourteen different types of cell responses (including ten signaling pathway activities and four cytotoxicity effects) for two different cell lines in response to exposure to six different metal and metal oxide nanoparticles. Association rules were identified for a High-Throughput Screening (HTS) data developed for murine macrophage (RAW264.7) and transformed bronchial epithelial (BEAS-2B) cells. The dataset consisted of a total of 55,296 data readings for nanoparticle-regulated pathway activities over an exposure period up to 24 h and exposure concentration range of 0.39-200 μg/mL. HTS data were processed for outlier removal and standardized via inter/intra-plate normalization with subsequent hit-identification (for significantly regulated pathways), which provided a reliable multivariate (multi-pathways) toxicity profiles for the nanoparticles. The association rules identified from the multivariate toxicity profiles were then pruned to remove redundant rules. The final set of non-redundant association rules revealed pathways that are co-regulated such that "significant regulation" of one or more pathways implied the existence of other (associated) specific pathways that are also significantly regulated. In order to explore the validity of data-driven hypothesis generation, specific pathway-activity studies were carried out with ZnO nanoparticles confirming the association rule identified for the p53 and Mitosox pathways. This pathway relationship revealed that blocking of the transcriptional activity of p53 lowered the Mitosox signal (i.e. p53 leads to mitochondrial dysfunction). The present approach to data-driven hypothesis generation (i.e., association rules discovery) has important implications for streamlining multi-parameter HTS assays, support for improved understanding of toxicity mechanisms, and selection of endpoints for the development of nano-Structure Activity Relationships (nano-SAR).

(422f) Investigations of the Oral Uptake of Titanium Dioxide Nanoparticles Via the Buccal Mucosa

Teubl, B.,
Leitinger, G., Medical University of Graz
Frohlich, E.,
Schneider, M.,
Lehr, C. M.,
Zimmer, A., University of Graz
Roblegg, E., University of Graz
Originally presented on: 11/6/2013 10:35:00 - 11:00:00

The development of engineered nanomaterials (ENM) and their commercialization for application in (consumer) products, medical and diagnostic devices and/or pharmaceutical drug delivery vehicles presents an enormous challenge for the scientific, regulatory, industrial and public field. Titanium dioxide (TiO_2) nanoparticles, for example, are manufactured worldwide in large quantities and TiO_2 pigments (< 2.5 µm) account for 70% of the total production volume [1]. However, decrease of particle size (from the microscale to the nanoscale) has been identified as a key parameter for an increased toxicity behavior. Regarding the respiratory tract it has been clearly demonstrated that TiO_2, which is one of the priority materials (concerning risk assessment) of the Organisation for Economic Co-operation, causes respiratory tract cancer [2]. However, studies dealing with the oro-gastrointestinal route are still lacking and the basic understanding of how nanoparticles interact with oral biological tissues is still missing. Previously, we demonstrated that nanoparticles are small enough to overcome the barriers of the oral cavity, likely entering systemic circulation and hence, translocate in the human body [3,4]. Thus, in this study we aimed at investigating interactions of three distinct TiO_2 nanoparticles (NM 100, pigment, anatase; NM 101, 7 nm, anatase; NM 105, 22 nm, anatase/rutile) with the oral cavity, in particular the buccal mucosa (lining the cheek area). To correlate particles´ properties to their biological interactions, characterization was carried out with photon correlation spectroscopy (PCS) in physiological media and with electron microscopy. Furthermore, the surface hydrophobicity was quantified with the Rose Bengal adsorption method. Ex-vivo penetration/permeation studies were investigated with static diffusion cells using porcine buccal mucosa. The penetration depth of TiO_2 nanoparticles was visualized by multiphoton microscopy (MPM). Furthermore, intracellular localization was examined via transmission electron microscopy (TEM) and the detected particles were verified with energy filtered TEM (EFTEM) combined with electron energy loss spectroscopy (EELS). Additionally, potential cytotoxicity effects were evaluated in-vitro with human oral TR146 cells and production of reactive oxygen species (ROS) was studied.

The data revealed that the investigated TiO_2 nanoparticles were hydrophilic and exhibited a high aggregation tendency in aqueous and physiological media. The particles penetrated/permeated the buccal epithelium in a size-dependent manner: NM 100 (two primary size distributions: 148 nm and 34 nm) and NM 105 (primary particle size: 22 nm) permeated the upper parts and penetrated into the deepest parts of the mucosa (i.e., basal lamina and connective tissue). However, NM 101 (7 nm particles) only penetrated the upper epithelium. Regarding intracellular localization, most of the particles were found freely distributed in the cytoplasm and not in endocytotic vesicles. These findings suggest that further entering mechanisms, apart from endocytosis, exist in the buccal mucosa. The cell viability/integrity was not compromised and no acute toxicity was observed after 4 and 24 h exposure time. However, NM 105 caused a significant increased metabolic activity ($P<0.001$) after 4 h, indicating a stress situation of cells. This was due to the production of reactive oxygen species (ROS) after short-time incubation.

It can be concluded that all investigated TiO_2 nanoparticles are able to rapidly penetrate the buccal mucosa. However, the penetration depth is mainly dependent on the particle size. Larger particles (i.e., NM 100 and NM 105) penetrate the upper and the lower parts of the buccal mucosa, suggesting particles′ distribution in the systemic blood circulation and furthermore, translocation in different organs. Smaller particles (i.e., NM 101) fail to penetrate lower parts and are only detected in the superficial layer of the oral epithelium. However, additional studies are necessary to evaluate possible interactions of freely distributed particles (in the cytoplasm) with intracellular proteins, organelles and cell nuclei. Additionally, ROS production of NM 105 may result in inflammation, mitochondria-mediated cell death and/or DNA damage. These data strongly emphasize that the oral cavity should be considered in further risk assessment, since it acts as the first line in defense, not only for oral but also for inhaled nanoparticles.

[1] Baan R., Straif K., Grosse Y., Secretan B., Ghissassi F., Cogliano V., *Carcinogenicity of Carbon Black, Titanium Dioxide and Talc*, Lancet Oncol 2006; 7: 295-296

[2] Trouiller B., Reliene R., Westbrook A., Solaimani P., Schiestl R., *Titanium Dioxide Nanoparticles Induce DNA Damage and Genetic Instability in vivo in Mice*, Cancer Res 2009; 69: (22) 8784-878

[3] Roblegg E., Fröhlich E., Samberger C., Zaversky M., Teubl B., Zimmer A., *Evaluation of a Physiological In-Vitro System to Study the Transport of Nanoparticles through the Buccal Mucosa*, Nanotoxicology 2012; 6: 399-413.

[4] Teubl B., Meindl C., Eitzelmayr A., Zimmer A., Fröhlich E., Roblegg E., *In-vitro Permeability Studies of Neutral Polystyrene Particles Through the Buccal Mucosa*, Small, 2013; 9(3): 457-66. doi: 10.1002/smll.201201789

(423a) Incorporating Inherently Safer Designs to Improve Sustainability of Industrial Facilities
Reed, M., MMI Engineering
Peterson, E., Scandpower Inc., a member of the Lloyd's Register Group
Originally presented on: 11/6/2013 8:30:00 - 8:46:00

Incorporating Inherently Safer Designs to Improve Sustainability of Industrial Facilities

Morgan Reed

MMI Engineering

11490 Westheimer Road, Suite 150 Houston, Texas 77077

MReed@MMIEngineering.com

Eric Peterson

MMI Engineering

11490 Westheimer Road, Suite 150 Houston, Texas 77077

EPeterson@MMIEngineering.com

Keywords: Inherent Safety, Sustainability, Process Safety, Process Hazard Analysis, and CFD

Abstract

An increasing number of facilities across the globe are reaching a significant age of maturity. Those facilities were constructed under different regulations and guidelines than those used today. Renovations, upgrades and retrofits are abundant in the oil, gas, and petrochemical facilities and these new constructions must follow the current regulations. The new guidelines will enhance the facility to become more sustainable and inherently safer based on newer design and construction practices. This paper discusses the coming challenges to facilities and the need to incorporate inherently safer designs in the upgrades, modifications to the facilities.

(423b) Designing Greener Plasticizers: Influence of Geometry of Central Group and Side Chains

Erythropel, H. C., McGill University
Shipley, S., McGill University
Brown, T.,
Maric, M., McGill University
Leask, R. L., McGill University
Cooper, D. G., McGill University
Originally presented on: 11/6/2013 8:46:00 - 9:02:00

Plasticizers are important additives to brittle polymers such as poly(vinyl chloride) (PVC). They are mixed with the resin to lower the glass transition temperature (Tg) and their blends result in soft and flexible materials. Yet plasticizers are not bound to the polymer matrix and will slowly leach out of the blend into the environment. Many studies have shown di (2-ethylhexyl phthalate) (DEHP) and some of its stable metabolites such as its monoester mono (2-ethylhexyl) phthalate (MEHP), 2-ethyl hexanol and 2-ethyl hexanoic acid to be toxic. MEHP is under further scrutiny as it is believed to be an endocrine disruptor. The above-mentioned compounds are generally considered ubiquitous environmental contaminants due to their slow breakdown by soil bacteria. This has led to efforts to design new, "green" plasticizers.

Small diester molecules were prepared synthetically by Dean-Stark esterification due to their resemblance to DEHP. These molecules were based on maleic acid (*cis*-isomer), fumaric acid (*trans*-isomer) and succinic acid (saturated analogue), and were esterified with alcohols ranging from ethanol (C2) to n-octanol (C8) as wells as 2-ethyl hexanol. Using a twin screw-extruder, these compounds were incorporated into unplasticized PVC at various concentrations (16% - 33% by weight) and subsequently tested for their plasticizer properties. The Tg reduction was measured using differential scanning calorimetry (DSC). Tensile test bars according to ASTM standards were prepared using a hot press, which were then tested for elongation at break and the secant modulus at a constant strain rate of 5 mm/s. Biodegradation tests were carried out with the pure compounds using the common soil bacterium *Rhodococcus rhodocrous* in shake flasks at 30°C. The medium consisted of minimum mineral salt medium, yeast extract, an auxiliary carbon source hexadecane and the plasticizer. The flasks were inoculated with 1ml of pre-grown culture, and whole shake-flask extractions were performed on a regular basis. The flasks were extracted using chloroform with added internal standard for further analysis by gas chromatography (GC). Using calibration curves for known compounds, plasticizer and metabolite concentrations were calculated. Unknown metabolites such as the respective monoesters were identified by analyzing synthesized standards.

The influence of the central structure was evaluated by comparing 2-ethylhexyl diesters of the proposed molecules to DEHP and another commercial plasticizer: di (2-ethylhexyl) adipate (DEHA). From Tg and tensile strength data, the fumarate compound was eliminated from the list of alternative "green" plasticizers due to poor Tg reduction and poor tensile strength data compared to DEHP. Both di (2-ethylhexyl) maleate and succinate did perform better than DEHP in terms of Tg reduction, and not statistically different in terms of tensile strength.

In the biodegradation experiments a strong influence of the central structure was found: the maleate compound showed almost no hydrolysis of either of the two ester bonds of the molecule over the course of 30 days, as observed similarly with DEHP. In both DEHP and di (2-ethylhexyl) maleate the esters are situated in *cis*-position towards another. The fumarate, in which the two ester bonds are in *trans*-position to one another, an increase in hydrolysis rate was observed, and assuming first-order kinetics, a half-life of 13 days was calculated. For both saturated molecules, the half-life dropped further, to about 6 days in the case of DEHA, and about 2.5 days in the case of the di (2-ethylhexyl) succinate. As expected, there was a buildup of 2-ethanol and 2-ethyl hexanoic acid, however the corresponding monoester was not observed for any of the compounds. In the case of the slow-degrading maleate this is explained by the fact that little of the diester was degraded. It was expected to see some monoester produced by biodegradation of the fumarate and succinate, however, this was not the case. It was concluded that the central structure played a crucial role in degradability of these compounds, and that a saturated molecule that is free to rotate about the central bond is more easily hydrolysed.

Using these results, the succinate compounds were further studied. Instead of the 2-ethylhexyl side chain, even-numbered unbranched alcohols were used to create the ester bond, from ethanol to n-octanol. All four compounds were better at lowering the Tg than DEHP at 29 weight-% of plasticizer in the blend. There was a trend to a lower Tg with increased side chain length, which in turn also meant increased overall molecule length. When the results for dihexyl succinate were compared to the branched di (2-ethylhexyl) succinate, it was found that their Tg at 29 weight-% in PVC were similar, suggesting that the ethyl branch did not play a role in Tg lowering.

Similar to the di (2-ethylhexyl) biodegradation experiments, a rough first-order kinetics approach was chosen to compare relative hydrolysis rates. A trend towards larger half-lives with increasing side chains (less than 1 day for diethyl succinate and about 3 days for di-n-octyl succinate) was observed. When comparing dihexyl succinate to di (2-ethylhexyl) succinate, a difference of about 1 day in half-life was observed, indicating a small steric effect of the ethyl-branch on hydrolysis, however this could also result from a lower solubility of the compound in the aqueous biodegradation broth. None of the succinate diesters with unbranched side chains showed a buildup of any kind of metabolite; in fact it could be shown that the bacteria preferentially grew on the plasticizer or respectively on its metabolites, such as the liberated alcohols. This was shown by running parallel experiments without auxiliary carbon source.

We have demonstrated the potential of small diester molecules to replace DEHP for plasticizing PVC. Given these results, a compromise will have to be found between longer side chains needed to obtain good plasticizer properties, and short side chains for better biodegradability. More mechanical testing and toxicity testing are needed to fully characterize these green plasticizers.

Erythropel, H.C., M. Maric, and D.G. Cooper, *Designing green plasticizers: influence of molecular geometry on biodegradation and plasticization properties.* Chemosphere, 2012. **86**(8): p. 759-66.

Erythropel, H.C., et al., *Designing green plasticizers: Influence of alkyl chain length on biodegradation and plasticization properties of succinate based plasticizers.* Chemosphere, 2013. **91**(3): p. 358-365.

(423c) A Program for Predicting Air Pollution in Chemical, Biological, Radiological, Nuclear and Explosives (CBRNE) Disasters

Lee, K., Yonsei University
Park, J., Yonsei University
Yoon, Y., Inje University
Her, S.,
Moon, I., Yonsei University
Woo, S.,
Yoon, H.,
Originally presented on: 11/6/2013 9:02:00 - 9:18:00

A Program for Predicting Air Pollution in Chemical, Biological, Radiological, Nuclear and Explosives (CBRNE) Disasters

Kwanghee Lee[a], Jaedeuk Park[a], Yi Yoon[b], Sungyun Her[c], Il Moon[a,*]

[a]Department of Chemical and Biomolecular Engineering, Yonsei University, 50 Yonsei-ro, Seodaemun-gu, Seoul 120-749, Korea

[b]Research Institute of Disaster and Emergency Management, Inje University, 197 Inje-ro, Jung-gu, Gyeongsangnam-do 621-749, Korea

[c]National Emergency Management Agency, 209 Sejongdae-ro, Jongro-gu, Seoul 110-760, Korea

Chemical, biological, radiological, nuclear and explosives (CBRNE) disasters involve industrial accidents at chemical and nuclear power plants, incidents from the use of weapons at terrors or wars, and accidents at transportation of hazardous materials. Those disasters incur release of severely hazardous materials into atmosphere, and are featured to be very difficult and expensive to forecast as well as highly lethal once an incident occurs. Internationally representative CBRNE accidents may include those of nuclear power plant at Chernobyl, Ukraine and chemical plant at Bhopal, India, bringing out massive damages to both the accident nation and its neighbors. In recent years, many chemical and radiological accidents have taken place in South Korea, for example, the accident of Hydrogen Fluoride (HF) release at Gu-mi in 2012 killed 5 people and made all Gu-mi residents impossible to sustain their daily business and life. A majority of countries which operate chemical and nuclear power plants have their own emergency response systems. Among them, National Atmospheric Release Advisory Center (NARAC), USA can be thought as being the most advanced worldwide. South Korea has made enormous efforts to develop a set of systems for responsively and effectively managing CBRNE disasters; Chemical Accident Response Information System (CARIS) and Atom-CARE as a system for managing combined chemical and radiological accidents. However, one of the main problems is that the two systems are operated separately in different government ministries, leading to inefficiency in dealing with CBRNE disasters. This study is concerned with the development of a new prototype package for predicting air pollution of combined chemical and radiological accidents, as shown in Figure 1. One of main features of the program is an addition of a new zone taking into account damage on sensitive civilians, namely, "sensitive zone" to existing three zones classified in terms of levels of dispersions of hazardous materials. The prototype package can potentially be expected to be of great help for the development of a program for managing CBRNE disasters in South Korea.

Fig. SEQ Figure * ARABIC 1. Scheme of integrated program

REFERENCES

[1] Michael M, Bradley. NARAC: an emergency response resource for predicting the atmospheric dispersion and assessing the consequences of airborne radionuclides. Journal of Environmental Radioactivity 2007;96:116-21.

[2] Han MH, Kim EH, Suh WT, Hwang WT, Jeong HJ. Development of radiological dose assessment systems to support a radiological emergency preparedness in Korea. Annals of Nuclear Energy 2012;43:187-91.

(423d) Experimental Study and Theoretical Analysis On Decomposition Mechanism of Benzoyl Peroxide

Lv, J., Nanjing University of Science and Technology
Chen, W., Nanjing University of Science and Technology
Chen, L., Nanjing University of Science and Technology
Originally presented on: 11/6/2013 9:18:00 - 9:34:00

Benzoyl peroxide (BPO), a source of radical used as initiator, is an important member in the organic peroxide family. To master its thermal characters, experimental study and theoretical analysis were employed on studying its decomposition mechanism. The autocatalytic process, which was deduced from dynamic experimental data by differential scanning calorimeter (DSC), was testified by isothermal DSC tests. Thermal scanning unit (TSU) was taken to predict pressure hazard of BPO decomposition and accelerating rate calorimeter (ARC) tests were conducted to investigate the thermal behavior of BPO in adiabatic condition. The high pressure and fast pressure release rate indicated a high possibility of accompanied pressure hazard once the runaway decomposition was trigged. Combined with experimental results, quantum chemistry method was used to locate the most possible reaction path and calculate thermodynamic energies of BPO molecule in gas phase. The calculated total released heat during the whole decomposition was in good accordance with the values tested by thermal analysis instruments. And bond dissociation enthalpy (BDE) obtained at different theory levels provided a comparable value in runaway possibility analysis.

(423e) Determination of Ignition Probabilities for LNG Spills

Gopalaswami, N., Mary Kay O' Connor Process Safety Center, Texas A & M University

Originally presented on: 11/6/2013 9:34:00 - 9:50:00

An accidental release of liquefied natural gas (LNG) can lead to the formation of a liquid pool & vapor cloud, which can ignite in the presence of ignition sources. The determination of ignition probability for LNG spills is an important part of quantitative risk analysis to understand the risk people may be exposed to. This paper deals with a review of the models that are currently applied in industry & provides a novel methodology to determine ignition probability. The new ignition probability model takes into account meteorological factors like wind direction, wind class & congestion level, in addition to factors affecting ignition probability which include mass flow rate of flammable materials, its reactivity & the number of ignition sources. An illustrative case study of an LNG setup present at the Brayton Fire Training Field in College Station, Texas is used for calculation of ignition probabilities. A sensitivity analysis is also performed on the new parameters that are involved in the ignition probability calculation. The estimation of ignition probability involves both qualitative identification of sources and its quantification, the results of which can be applied to risk analysis of LNG facilities.

(423f) Zero Valent Silver-Based Electrode for 2, 4-Dinitrotoluene Detection

Mbah, J. C., Tuskegee University

Moorer, K., Tuskegee University

Vahdat, N., Tuskegee University

Hernandez-Rivera, S., University of Puerto Rico

Originally presented on: 11/6/2013 9:50:00 - 10:06:00

Zero Valent Silver-based Electrode for 2, 4-Dinitrotoluene Detection

Mbah J. C.-Tuskegee University

Moorer K. - Tuskegee University

Vahdat N. - Tuskegee University

Hernandez-Rivera S.-University of Puerto Rico-Mayaguez

The prevalent use of nitro-aromatic materials including, 2, 4, 6-trinitrotoluene (TNT) for explosive devices has made their detection crucial in the fight against act of terrorism and detection of military explosives. But rather than attempting to detect TNT, researchers focus on detecting 2, 4-dinitrotoluene (DNT) which is a common impurity in TNT ?based explosive. DNT has a higher vapor pressure than TNT, allowing for more sensitivity and easy detection. In the production of munitions and explosives, DNT is used as a gelatinizing and waterproofing agent. Furthermore, it can be found in landmines, contaminated water, waste sites, and a variety of other locations making it readily accessible to terrorists. High sensitivity and selectivity, combined with the ability to lower the deployment cost of sensors using mass production, is essential in winning the war on explosives-based terrorism.

In this study, the donor based electrode for the detection of DNT was synthesized by modifying the surface of carbon fiber material with zero valent silver (Ag) via chemical deposition. DNT detection was conducted electrochemically in aqueous media of pH 6.0. The cyclic voltammetric oxidation of DNT is diffusion controlled with detection sensitivity increasing with Ag loading of the electrode. The only oxidation product is benzoic acid, which suggests that the difference in performance of Ag-C electrode compared to other available materials is directly ascribable to the electrode fabricated in this study. Detection limit of DNT was 5μM with the electrode showing no signs of degradation after several cycles, and supported by surface behavior of characterized Ag-C electrode. Kinetic study was also investigated. In another study, the fabricated Ag nanoparticle on carbon fiber substrate was used as an efficient surface enhanced Raman spectra (SERS) substrate to detect the adsorbed DNT molecules with a detection limit of 50 ppm. In advent, our developed SERS substrates could have great potential in detecting other nitro-aromatic based-explosive materials, such as TNT molecules.

The objective of this research is to engineer an efficient surface for rapid detection of trace quantities of nitro-aromatic based explosives with particular emphasis on DNT with the following specific aims:

1. Deposit zero valent silver nanoparticles on carbon fiber surface which provide higher surface area-to-volume ratio that can lead to enhanced reactivity with environmental contaminants that degrade through adsorption mechanism.

2. Detect and splits DNT with high selectivity, linearity, and reversibility to a more benign material by electroanalytical means.

3. Modify material surface for surface enhanced Raman spectra with excellent response time and dynamic range.

(423g) Environmental Implications of Phenol Gasification in Supercritical Water

Huelsman, C. M., University of Michigan
Savage, P. E., University of Michigan
Originally presented on: 11/6/2013 10:06:00 - 10:22:00

Supercritical water gasification (SCWG) has been proposed as both a wastewater treatment technology and a biomass conversion technology. Phenol is an important candidate for study in either case, because it is both a regulated EPA priority pollutant commonly found in industrial wastewater effluents and a good model compound for the phenolic structures found in biomass. Additionally, several investigations of SCWG using whole biomass feedstocks identified phenol as a persistent byproduct. Regardless of whether phenol is considered a wastewater pollutant or a byproduct from biomass gasification, our own experiments show that phenol leads to the formation of little gas and a swath of heavier molecular weight compounds when exposed to supercritical water. Most of the non-gaseous products identified in these experiments are polycyclic aromatic hydrocarbons (PAHs), EPA priority pollutants, or have known human health and environmental effects. We experimentally quantified as many of these compounds as possible under various SCWG processing conditions. We then used this experimental data in conjunction with the UNEP/SETAC environmental toxicity ("USEtox") model and other models to characterize the human and ecotoxic impacts due to hypothetical emissions of phenol SCWG byproduct streams into freshwater.

(423h) Modeling the Transport of Mercury in Indoor Spaces Following a Release of Mercury Droplets

Gaglione, A., Eastern Research Group, Inc
Hou, A., Eastern Research Group, Inc
Rindfusz, K., Eastern Research Group, Inc.
Originally presented on: 11/6/2013 10:22:00 - 10:38:00

A mathematical model for predicting potential exposure concentrations of mercury resulting from the breakage of mercury-containing products has been developed. This modeling work estimates airborne mercury concentrations in indoor spaces following the breakage and subsequent release of mercury from mercury-containing products, such as switches, thermostats, and hygrometers. This model was intended to estimate exposure on a per-breakage-event basis. Additional activity factors, such as the frequency of product breakages or the exposure duration to a broken product within these occupational settings, are not included. Two occupational settings are modeled: office and industrial.

This model, named the Mercury Evaporation in Indoor Spaces Model (MEISM), applies engineering principles of liquid-gas phase mass transfer to an industrial hygiene scenario. MEISM models the evaporation of mercury using boundary layer theory and estimates airborne concentrations as a function of time using a near-field – far-field approach to account for imperfect air mixing. The near-field – far-field approach divides the indoor space into two, well-mixed zones and calculates the airborne concentrations in each zone using a mass balance similar to that of a CSTR.

Multiple literature contributions have described the oxidation of mercury in a humid air environment. These studies have noted the exponential decline of airborne mercury concentrations after the initial release. This decline is likely due to the formation of a mercury oxide layer on the surface of the evaporating mercury, which provides a resistance to mercury evaporation. To account for this important phenomenon, MEISM used multiple mercury exposure studies from the literature to regress a first-order rate constant of the oxidation of mercury in humid air. The first-order rate of mercury oxidation is modeled as an exponential decay of mercury surface area available for evaporation. An Arrhenius relationship was used to describe the first-order rate constant as a function of temperature. These regressions resulted in an Arrhenius activation energy of 41.5 kJ/mol and a pre-exponential factor of 578,388 hr^{-1}. This model predicts a first-order rate constant of 0.0312 hr^{-1} at 25 °C.

For the purpose of modeling, MEISM assumes the spilled mercury forms several hemispherical droplets on the floor, and the entire surface area of each hemisphere is available for evaporation. To capture uncertainties of the various model inputs, MEISM uses probabilistic modeling to estimate airborne mercury concentrations following a mercury release. Probability distributions were assigned to the model inputs of temperature, mercury mass within the product, number of mercury droplets formed, indoor air speed, and the oxidation first-order rate constant. For the breakage of a thermostat, switch, and hygrometer in both model office and industrial settings, MEISM predicted the following 95th percentile peak and eight-hour time weighted average (TWA) concentrations in the near field (the zone closest to the mercury source):

		Offices		Industrial	
Product Type	**Mercury Mass Range (g)**	**Near Field 95th Percentile Peak Concentration (μg/m³)**	**Near Field 95th Percentile 8-hr TWA Concentration (μg/m³)**	**Near Field 95th Percentile Peak Concentration (μg/m³)**	**Near Field 95th Percentile 8-hr TWA Concentration (μg/m³)**
Thermostats	2.8 – 16.8	7.43	6.55	0.46	0.422
Switches	0.05 – 70	37.2	32.8	2.3	2.13
Hygrometers	3 – 7	4.17	3.67	0.26	0.236

(423i) The Influence of Solid Inertants On the Explosion Behavior of Dusts From Mechanical Waste Treatment Facilities

Held, K., Montanuniversität Leoben

Raupenstrauch, H., Montanuniversität Leoben

Originally presented on: 11/6/2013 10:38:00 - 10:54:00

Conservation of natural resources is a main target in today's lives wherefore the reusing and recycling of waste is very important. To recycle waste material it is necessary to treat the mass flows in special ways. In most instances the waste material passes through mechanical treatments. Especially for the production of substitute fuel mechanical treatment is a common treatment operation. During this mechanical treatment the fractions get shredded, separated and prepared for further recycling. Nevertheless, besides all those positive effects, the mechanical treatment of waste material also generates a fraction of fine particles, which gets dispersed in air. These occurring dusts either have to be separated with suction cleaning or they get deposit unregulated on plant components. Anyway all occurring dusts are fraught with the risk of dust explosions if they are burnable and dispersed in air.

Dusts from different waste treatment facilities were investigated in order to identify the dust explosion potential and its influencing factors. Dusts deposited on plant components as well as filter dusts were sampled. The reason for taking samples from deposit dusts is that they can cause dust explosions if they get dispersed in air. Filter dusts are relevant materials because they are dispersed in air when they get suctioned and might form hazardous explosive atmospheres. Burnable dust dispersed in air can cause dust explosions if the dust cloud is getting in contact with an effective ignition source.

The explosibility characteristics for the investigated dusts were determined by Siwek 20-Liter testing. Experiments were conducted to identify the influence of solid inertants, as well as the influence of their particle size and particle shape. These investigations are done from the intention to predict the explosion behavior of dusts based on the amount, size and shape of their inert fraction.

Because of batch wise delivery of the recycling material and similarly input there is a variation of composition of the treated material over time. That is why it was necessary to check if the occurring dusts respective the explosion behavior of the occurring dusts is altering with time. Beside of short-term fluctuations caused by varying input material also seasonal changes were considered. The comparison of samples taken from the same plant within different periods of time showed that despite of changes in charge material, the explosion behavior of occurring dusts within one operation unit in a plant is reproducible.

Solid Inertants

A known way of reducing the risk of dust explosions is to mix the burnable dust with a solid inertant [1]. This method is either used to generate a resulting mixture, that is rendered non-explosible or the admixture is done in purpose of mitigation of the consequence of a dust explosion [2, 3]. Although there is good knowledge of solid inerting, it is not widely used in process industries. In most applications mixing with solid inertants can lead to unacceptable product contamination so this procedure is only in rare cases practical from a technical perspective.

In the case of dusts from mechanical waste treatment facilities risk reduction due to solid inertants is an interesting approach, because of the fact, that the dusts itself contain a huge amount of inert fraction. The inert fraction is on average at least 40 mass percent.

To identify the influence of inert fraction, dust samples from a mechanical sorting plant were investigated. The ash content of the samples was increased and an inhabitation as shown in Figure 1 could be seen. On this basis it could be shown that the explosion behavior of dust samples from different operation units in the same plant are just differing due to the different inert fractions. So the inert fraction is within one plant a valuable parameter for the prediction of the explosion characteristics of the occurring dusts.

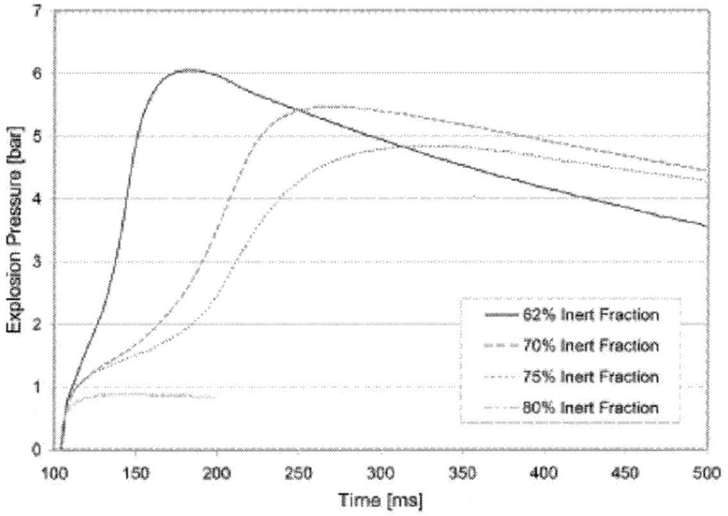

Figure 1: Pressure-time trace for the reaction of dust sample from mechanical sorting with different inert fractions

Particle
Size and Particle Shape

Beside the high amount of inert particles, the fraction of coarse particles in the occurring dusts is very high. Most of the dusts show very wide particle size distributions. For the investigations of the explosion behavior in the 20-Liter Siwek Apparatus as well as in the modified Hartmann Apparatus only particles beyond 500 µm were mentioned. However because of the fact that the particle class of 500 ? 1000 µm is of the same magnitude as the particle class smaller 500 µm, the influence of coarse particles in the mixture was investigated.

An increased mean particle diameter of an explosible dust caused by an increase of particle size over the whole size distribution can significantly decrease the inerting level [1]. Larger particles are more difficult to disperse than smaller particles, which is another relevant aspect of the characterization of dust explosions in mechanical waste treatment plants. Especially, in due consideration of the high LEL (lower explosion limit) of the outlined dust samples a restraint in dispersion properties is considerable.

In Figure 2 the notable results are given. Already at a coarse fraction content of 5 % the explosion characteristics switches to the characteristics of the coarse fraction. Therefore, regardless of separation phenomena, an inhibition of the explosion intensity can be expected because of the presence of coarse particles in the dust samples.

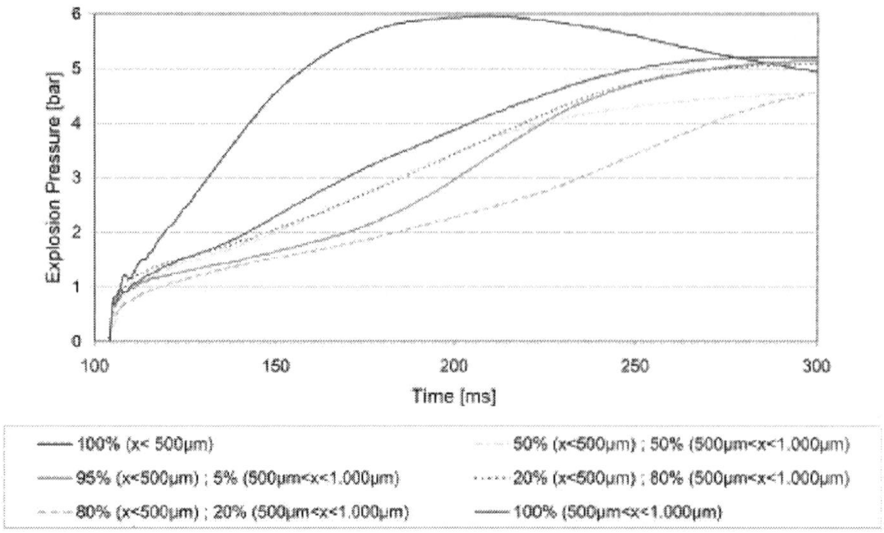

Figure 2: Effect of coarse particles on the explosion characteristics of dust samples from mechanical waste treatment facilities

Particle Size and Particle Shape of Solid Inertants

According to the influence of inert fraction and particle size investigations concerning the particle size and particle shape of solid inertants in the dust samples were done. Therefore a sample from a document destruction plant with an ash content of 42 % was prepared by admixture of glass beads and glass fibers respectively. Due to this the inert fraction of the sample was raised by 5 %.

With an addition of 5 % it could be seen that the effect of beads is nearly negligible while the admixture of fibers has considerable effect. The added materials are of the same equivalent diameter, which is comparable with the particle size distribution of the original sample. Although the explosion pressure of the prepared samples is equal the explosion intensity in terms of rate of pressure rise is strongly weakened (cf. Figure 3). This shows the importance of the particle shape of solid materials that act as inhibitors.

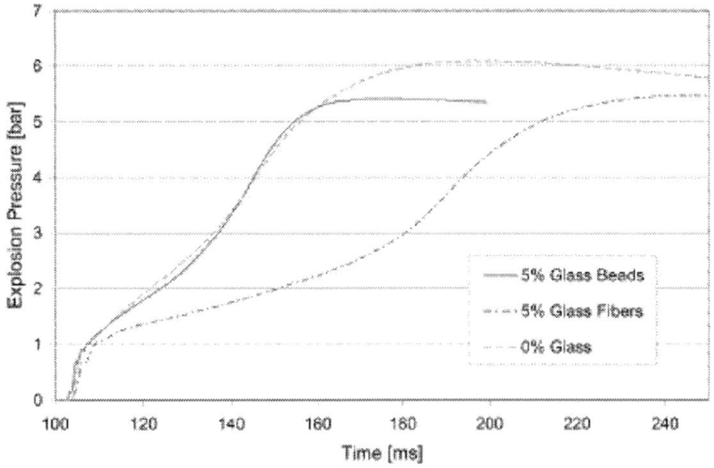

Figure 3: Effect of the admixture of inert material of different particle shape on the pressure rise over time

As seen in Figures 4 and 5 the Siwek 20-Liter testing of the document destruction dust sample with the admixture of 5 % glass beads are compared to those of the same dust sample mixed with 5 % glass fibers. The explosion hazard is therefore significantly magnified by the particle shape of the present inert material.

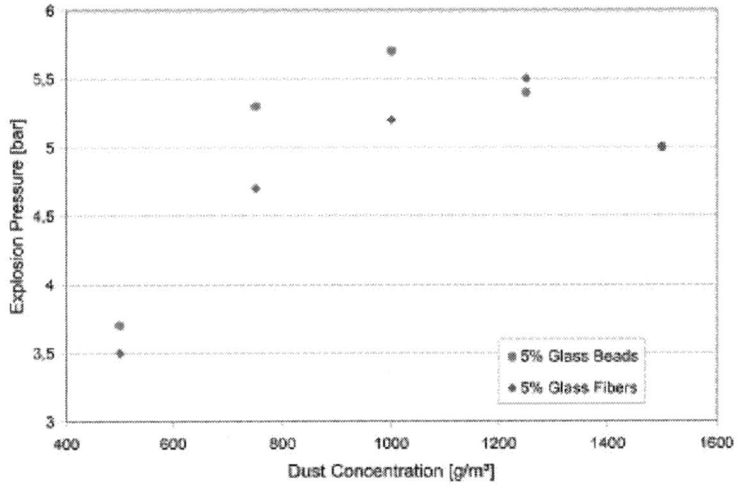

Figure 4: Explosion pressure as a function of dust concentration for dust from document destruction mixed with 5 % glass beads and accordingly 5 % glass fibers

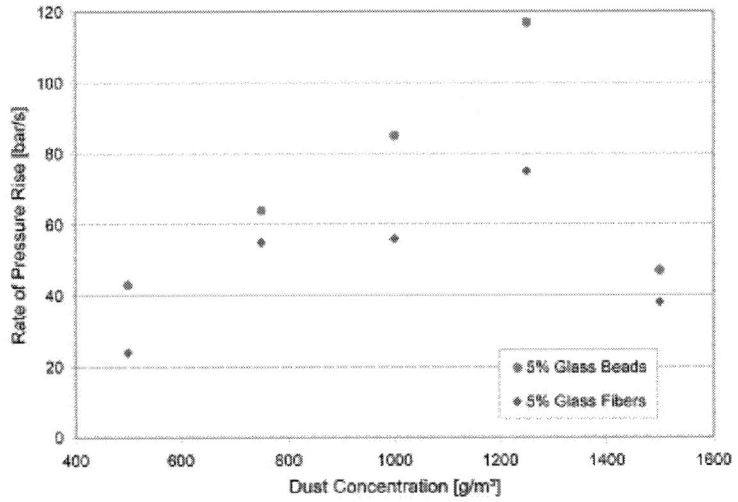

Figure 5: Rate of pressure rise as a function of dust concentration for dust from document destruction mixed with 5 % glass beads and accordingly 5 % glass fibers

These investigations show that the explosion behavior of dusts from mechanical waste treatment facilities is strongly influenced by the presence of inert material and its particle shape as well as its particle size. Previously analyses of different dusts of mechanical waste treatment facilities and the effect of their influencing factors on the explosion reactions were done. This is the basis for further investigations on the effect of particle size and shape of the ash content, which should permit the estimation of the explosion behavior.

Further analyses with variation of the particle size distribution and particle shape as well as the material of the solid inertants will be done. The investigations will be done with dusts from document destruction as well as dusts from substitute fuel production and other mechanical waste treatment facilities. The comparison of the influencing effects provides important information on the characterization of these dust explosions.

References

[1] Amyotte, P., Pegg, M., et al. (2007). Moderation of dust explosions. Loss Prevention in the process industries, 20. 675-687

[2] Chatrathi, K. & Going, J. (2000). Dust deflagration extinction. Process Safety Progress, 19. 146-153

[3] Dastidar, A. & Amyotte, P. (2002). Determination of minimum inerting concentrations for combustible dusts in a laboratory scale chamber. Process Safety and Environmental Protection, 80. 287-297

(427a) Metal Recovery From Spent Catalyst By Means of Biotechnological Strategies to Scale-Up Studies

Bisht, V. Jr.,
Singh, B. Sr., Teri
Sarma, P. M., the energy resources Institute
Mandal, A. K., The Energy Resources Institute
Lal, B., The Energy Resources Institute
Originally presented on: 11/6/2013 8:30:00 - 8:48:00

A bioleaching study aimed at recovering metals from hazardous spent hydro processing catalyst was carried out. The exhaust catalyst was rich in nickel (25%) vanadium (2.3%) and molybdenum (30%). Involved microorganisms were iron/sulfur oxidizing bacteria. Investigated factors were elemental sulphur addition, ferrous iron addition. Ferrous iron resulted to be essential for metal extraction. A mixed culture of two strains of Fe/S oxidizing bacteria (*A. Ferrooxidans,A.thiooxidans)*This culture was cultivated under acidic conditions (pH 2), in the ideal liquid growth media 9K : Two solution were prepared separately and blended only after their sterilization by autoclave. The first solution was obtained dissolving the following reagents in 1000 mL $(NH_4)SO_4$ 2.og, KCl 0.1g,H_2PO_4 0.5g, $MgSO_4$.7H2O 0.5g and Ca (NO_3) 0.01g. and Sulfur as energy source.The second one was an energetic solution of 49.1g $FeSO_4$, dissolved in 300ml of distilled water. Sulfuric acid 4M was used to a adjust pH to 2, before serlization of the solutions. scale-up (from 50 to 500ml) of mixed cultures of *Acidithiobacillus thiooxidans* and *Acidithiobacillus ferrooxidans* was done for bioleaching. The experiments were carried out in Erlenmeyer flasks with media containing both sulfur powder and ferrous sulfate as substrates, at pH 2.0. The pH decreased to 0.93 in 4 days for 50ml , 0.66 in 7 days for 100ml and 0.98 in 18 days 500ml batch respectively. The media was centrifuged to get the filtrate and filtrate was used for leaching experiments, Nickel (92%), V (95%), Mo (4%) was achieved by the 0.9pH supernatant from the spent catalyst procured from petroleum refinery. Pellets were used for further scale-up studies. The biomass was checked using Folin-Lowry protein estimation method and Barium chloride method was used for sulfate ion estimation.

(427b) Synthesis of P25 Doped CoFe2O4/Graphene Magnetic Catalyst and Its High-Performance Photocatalytic Activities Under Visible Light Irradiation

Sun, J.,

Originally presented on: 11/6/2013 8:48:00 - 9:06:00

With a facile one step hydrothermal method, a novel magnetically separable P25/CoFe2O4/Graphene photocatalysts have been fabricated in a mild condition. A systematic investigations on improving the photocatalytic activity of the catalysts was made via a strategy of locating P25 on CoFe2O4/Graphene suface. Transmission Electron Microscopy (TEM) indicated that the P25 and CoFe2O4 nanoparticles are well-dispersed and successfully anchored on the graphene sheet surface. X-ray diffraction (XRD), Raman spectroscopy and X-ray photoelectron spectroscopy (XPS) further suggested the similar results. After adsorption equilibrium, the photocatalytic activities were evaluated in terms of the photodegradation among various dyes, such as methylene blue(MB), methyl orange(MO) and neutral dark yellow(GL) in aqueous solution under visible-light irradiation. Amazingly, the as-prepared P25/CoFe2O4/Graphene nanocomposites exhibited much higher photocatalytic activities than the CoFe2O4/Graphene composites. The miraculous improvement of such photocatalytic properties might be attributed to the remarkable synergistic effects and the superior characteristics of graphene. Furthermore, the P25/CoFe2O4/Graphene nanocomposites provided a magnetically separable ability, which could be recovered in a suspension system by an applied magnetic field, solving the difficulty of recyclable use.

(427c) Removal of Cr(VI) From Aqueous Solutions By Self-Flocculating Yeast

Chen, L., Dalian University of Technology
Bai, F., Dalian University of Technology
Originally presented on: 11/6/2013 9:06:00 - 9:24:00

In this study, the ability of self-flocculating yeast to remove Cr(VI) from aqueous solutions was investigated. Effect of pH, contact time and temperature on the biosorption was examined, and it was found that pH was a significant factor, affecting not only the removal ability of Cr(VI) by the self-flocculating yeast but also the contact time required for the biosorption to reach equilibrium, while the biosorption was not significantly affected by temperature at the range from 30 to 40 °C suitable for the self-flocculating yeast to perform. The fitness of the experimental data with the Freundlich and Langmuir isotherms was further evaluated, and good correlation was observed, indicating both monolayer biosorption and the presence of heterogeneous surfaces. Since no Cr(VI) was directly detected in the solution when the self-flocculating yeast after adsorption was treated by HNO_3 under high temperature, the removal mechanism of Cr(VI) was deduced to be its reduction under acidic conditions on the surfaces of the self-flocculating yeast cells. The first order dependence of the biosorption rate on each of Cr(VI) and biomass concentrations was experimentally validated, and a pseudo second-order kinetic model was established for the biosorption process.

(427d) Speciation of Chromium Adsorbed On Activated Carbon and Stabilization of Lime Precipitated Chromium Residue With Activated Carbon

Mao, L., Wuhan University of Science and Technology
Tong, S., Wuhan University of Science and Technology
Fan, M., Wuhan University of Science and Technology
Jia, C. Q., Department of Chemical Engineering and Applied Chemistry, University of Toronto

Originally presented on: 11/6/2013 9:42:00 - 9:57:00

Speciation of Chromium adsorbed on Activated Carbon and Stabilization of lime precipitated chromium residue with Activated Carbon

Chromium-containing waste residue is a hazardous industrial solid waste due to the migration of toxic chromium into the environment. It seems urgent for the further treatment of chromium-containing waste residue in China at present. Compared with other methods for treatment of chromium-containing waste residue, stabilization/immobilization technique is characterized by cost-effective, easier operation and in-situ application available by addition of additives. In this work, the modified activated carbon was employed as an additive to reduce the leaching of chromium so as to achieve the stabilization of the lime precipitated chromium residue(LPCR). Also, the leaching potential of Cr adsorbed on modified activated carbon was evaluated by a five-step sequential extraction procedure. The information will be helpful for the establishment of a simple, economical and feasible disposal method for the chromium-containing waste residue.

The LPCR sample was collected from Wuhan Iron and Steel (Group) Corporation. The contents of total Cr and hexavalent chromium in LPCR were 24000mg/kg and 1856mg/kg. Activated carbon sample (AC) was modified from commercial activated carbon with 20wt% HNO_3 at 20 °C. The Cr-laden activated carbons for the sequential extraction were prepared by static adsorption of Cr(III) ions and Cr(VI) ions respectively from standard solution with modified activated carbon..

The extraction conditions of the five-step sequential extraction procedure employed to investigate the fractions of chromium in LPCR and those of trivalent chromium(Cr(III)) and hexavalent chromium(Cr(VI)) adsorbed on activated carbon were listed in table 1. The contents of total chromium and Cr(VI) in the leachate were determined by spectrophotometry with diphenylcarbazide as chromogenic agent at a wavelength of 540nm according to GB/T 15555.5-1995 and GB/T 15555.4-1995.

The distribution of Cr in LPCR and Cr(III) and Cr(VI) adsorbed on AC were listed in table 2. It is considered that heavy metal present as the residual of the five fractions can be expected to be the most stable. The results of the sequential extraction showed that the residual fraction of total Cr and Cr(VI) in

LPCR are 74.89% and 49.32%, which indicated that the unstable fractionations of chromium especially Cr(VI) in LPCR have some environmental mobility. Nearly 94% of adsorbed Cr(III) on AC was present as the most stable residual fraction while the fractions of exchangeable and bound to carbonates only accounted for 0.44%, suggesting the possible chelation of Cr(III) with the surface oxygen groups on the modified activated carbon. Adsorbed Cr(VI) on AC seemed less stable than Cr(III) with 71.35% residual fraction. This maybe result from the selective adsorption of the surface binding sites to the different charged hydrolyzed Cr(III) and Cr(VI). Therefore, there seems a need to study the chemical interaction between the surface functional groups of activated carbon and heavy metals. The enhancement of adsorption via chemical interaction is expected to enhance the stability of heavy metals.

Batch extraction and continuous column leaching experiments were carried out for mixtures of LPCR with activated carbon to investigate the leaching behavior of Cr and the effectiveness of activated carbon as an additive. The batch extraction was performed in an over-end-over rotator at 30rpm. The leaching experiments were conducted using a series of LPCR/LPCR-AC packed columns (30mm-ID, 350mm-height). The leaching solutions with different pH values flowed through the columns. A significant decrease (54% for total Cr, 52% for Cr(VI)) on Cr was observed through activated carbon addition at 1:10 additive: LPCR ratio. The results confirmed the positive effect of activated carbon on the stabilization of Cr in LPCR.

Table 1 Extraction conditions of the fiveÂ¨Cstep sequential extraction procedure

Fraction	Extraction conditions and reagents
F1.Exchangeable	3.0g LPCR, 30mL 1 mol/ L $MgCl_2$, pH = 7.0,30 °C , 1 h
F2.Bound to Carbonates	30mL 1 mol/ L NaAc , pH = 5.0,30 °C , 5 h
F3.Bound to Iron and Manganese Oxides	30mL 1 mol/ L 0.04 mol/ L NH_2OHÂ°Â¤HCl , 85 °C , 5 h
F4.Bound to Organic Matter	5 mL 0.02 mol/ L HNO_3, 8.3 mL 30 % H_2O_2, pH = 2.0, 85 °C , 2 h; 8.3 mL 30 % H_2O_2, pH = 2.0, 85 °C , 2 h; 8 mL 3.2mol/L NH_4 Ac in 20 % HNO_3, 30 °C , 0.5 h;
F5.Residual	0.2g F4 step residue, 2 mL H_2O, 10mL HNO_3 heating, cooling, 5mL HNO_3, heating for 3 times; 2mL H_2O, 3mL H_2O_2 for 3 times

Table 2 Distribution of Cr in LPCR and AC by five-step sequential extraction (%)

(427e) Blue Preserve - Recovery Of Oil Spilled In Sea A Review

Maheshwari, U., Birla Institute of Technology & Science (BITS)
M, B., Birla Institute of Technology & Science (BITS)
Originally presented on: 11/6/2013 9:57:00 - 10:12:00

In our day-to-day life, we sometimes come across in newspapers that an oil tanker collides in mid sea and that large quantity of oil is spilled away, often accompanied by gory pictures of sea/ marine life. This is usually followed by protests by environmentalists against the company demanding compensation to bring back the beauty of the sea. This is a normal show which takes place and we the people are not aware of the problems going to be faced in the future both economic and the environmental damages. In addition to the cost incurred in cleaning of the sea, precious fuel is lost; resulting in the oil price hike. This paper aims at suggesting the means to recover the spilt oil thereby saving valuable fuel oil as well as guarding the beauty and safety of the environment. This paper deals with the sea which is polluted by oil, what are the causes of oil spills, how oil spill occur, what are the effects of marine-Eco system, what are the methods are available for cleanup of oil spill and latest advanced methods of recovery.

(427f) An Eco-Friendly Technology for Metal Dissolution Using Biologically Produced Acid From Chelated Exhaust Catalyst Using Acidithiobacillus Thiooxidans and Its Kinetic Aspect

Singh, B. Sr., Teri
Sarma, P. M., the energy resources Institute
Mandal, A. K., The Energy Resources Institute
Lal, B., The Energy Resources Institute
Originally presented on: 11/6/2013 10:12:00 - 10:27:00

Bioleaching of metals from hazardous spent hydro-processing catalyst was attempted in the second stage after growing the bacteria with sulfur in the first stage. The first stage involved transformation of elemental sulfur particles to sulfuric acid through an oxidation process by acidophilic bacteria (*Acidithiobacillus thiooxidans)*. In the second stage, the acidic medium was utilized for the leaching process. Nickel and Molybdenum containing spent catalyst were chelated with EDTA (0.8M) and leached from the solid materials to liquid medium by the action of sulfuric acid that was produced by acidophilic bacteria and compared with commercial acid. Involved microorganisms were sulfur oxidizing bacteria. Experiments were conducted varying the reaction time, amount of spent catalysts, amount of elemental sulfur and temperature. At 15 g/L spent catalyst concentration and 10 g/L elemental sulfur, 94.3% Ni, 46.3% Mo, were recovered after 7 hrs of heating at 100^0C. Chemical leaching with commercial sulfuric acid of the similar pH that was produced by bacteria was compared. Thermodynamic parameters were calculated and the nature of reaction was found to be exothermic. Leaching kinetics of the metals was represented by different reaction kinetic equations, however, only diffusion controlled model showed the best correlation here. During the whole process Mo showed low dissolution because of substantiate precipitation with leach residues as MoO3. Bioleach residues were characterized by EDX and SEM.

(427h) Detoxification of Halogenated Aromatic Compounds By Anaerobic Cultures

Ding, C.,
Wang, S.,
He, J., National Univeristy of Singapore
Originally presented on: 11/6/2013 10:27:00 - 10:42:00

PolyBrominated diphenyl ethers (PBDEs) and Polychlorinated biphenyls (PCBs) are widespread global contaminants due to their widespread industrial usage and improper disposal. They are classified as persistent organic pollutants (POPs) and are known for their endocrine disrupting behavior, neurotoxicity, and possible human carcinogenicity. At the contaminated sites, PBDEs and PCBs frequently coexist with other halogenated compounds such as chloroethanes, and chloroethenes. The presence of multiple halogenated compounds usually poses toxicity to dehalogenating microbes, because few dehalogenating cultures are capable of detoxifying a broad spectrum of halogenated compounds. So far, little is known about microorganisms that carry out dehalogenation of these congeners under anaerobic conditions. Here we report an anaerobic culture capable of reductively debrominating congener 47 (100% removal), congener 99 (88% removal), and congener 100 (92% removal) to the non-brominated diphenyl ether within 2-4 weeks. We also obtained a sediment-free culture which is able to sequentially remove halogens from PCBs and other halogenated compounds. For example, this culture dechlorinates the commercial PCB mixture - Aroclor 1260 - mainly by removing flanked para- and doubly flanked meta- chlorines, debrominates octa-brominated diphenyl ether (octa-BDE) mixture predominantly to tetra-BDEs, and dechlorinates tetrachloroethene (PCE)/1,2-dichloroethane (1,2-DCA) completely to ethene. The 16S rRNA gene-based techniques revealed that Dehalococcoides are the dechlorinators in both cultures. Quantitative PCR results showed that the growth of Dehalococcoides linked tightly with PBDE/PCBs debromination. Given the ubiquity and recalcitrance of PBDEs/PCBs, the anaerobic cultures are promising for bioremediation applications in the restoration of halogenated compounds contaminated environments.

(427i) Screening for Efficient Hydrocarbon Degradation By Bacterial Isolates From Crude Oil Contaminated Sites of Assam, India

Ghoshal, A. K., Indian Institute of Technology Guwahati

Originally presented on: 11/6/2013 10:42:00 - 10:57:00

Screening for efficient hydrocarbon degradation by bacterial isolates from crude oil contaminated sites of Assam, India

Abhijit Sarma Roy[1], Sounak Bera [2], A.K.Ghoshal*[1, 2]

1Centre for Energy, Indian Institute of Technology Guwahati, Assam-781039, India

[2] Department of Chemical Engineering, Indian Institute of Technology Guwahati, Assam-781039, India

***Corresponding author** (E-mail:aloke@iitg.ernet.in)

India is one of the fastest growing developing nations in the world and consequently the quest for energy reserves is very high. Mineral oil is one of the chief energy resources being utilized in India. A significant portion to the countries oil demand is met from foreign imports and thus crude oil prices has a major say in the economy of the country. India has been steadily improving its crude oil production and thus drilling, storage, transport activities etc. are on a high. Crude oil contamination of land and water bodies in and around drill sites is an emerging problem in different region of the country and Assam being the pioneering state in crude oil drilling activities in India is also facing similar problems. Different crude oil components on contact with soil, alters the soil intrinsic characteristics and renders the soil unfit for plant growth. Beside the microbial flora associated with the soil undergo a drastic change due to crude oil contamination. Besides this different chemical compounds are being increasingly used to aid different activities during drilling to increase the oil yield, many of these chemical compounds are also very much toxic to the soil. Different research groups around the world have been actively participating in research related to remediation of such crude oil contaminated sites, biological remediation using hydrocarbon degrading bacterial specie are being reported by researchers across the globe. Biological remediation of crude oil contaminated sites is very effective because it ensures complete degradation of the toxic compound as well as it is very cost effective. In the present study the main focus was laid towards isolation of efficient bacterial sp. from different crude oil contaminated sites of Assam and accordingly four sites Guwahati Refinery, Guwahati, Digboi Refinery and drill sites, Digboi, Duliajan and Jorajan were selected. 23 bacterial species belonging to different genera were isolated from fifteen soil/ sediment/ sludge and crude oil samples collected from the four contaminated sites. Isolation was done in Minimal salt medium (MSM) by enrichment culture technique. The Physicochemical and hydrocarbon contaminant characterization of all the fifteen samples collected from the four sites were also done. The pH of the samples ranged from 4.96 to 10.3, conductivity of the samples was between 0.6 to 28.6Âµsimencm^{-1}. Highest TPH (Total Petroleum Hydrocarbons) was found (494.6g/kg) and lowest was (5.26g/kg). Highest aliphatic compounds was (121011.3mg/kg, 12.1 %) and lowest recorded was (5.26mg/kg). Highest moisture content of the samples was (52.66%) and lowest moisture content was found to be (4.52%). Heavy metals- Arsenic, mercury and chromium were present in trace quantities while. Sodium, potassium, iron (Fe III), chloride, sulfate, nitrate and phosphates were also found to be present in the samples. Among the 23 bacterial isolates screened for degradation ability 7 isolates were found to be gram's negative while 16 isolates were found to be Gram's positive. Few biochemical characters of the isolates were also being done following standard protocols. In gelatin hydrolysis test only 4 isolates showed positive results while in catalase test only one of the isolates showed negative results. In starch hydrolysis test only 5 isolates showed positive results while the rest all of the strains were found to be negative. Emulsification activity is an important criteria for selection of hydrocarbon degrading bacteria, in the E_{24} assay for assessing the emulsification assay of the bacterial isolates 10 strains showed positive result while the others showed no or insignificant activity. All the strain showed growth in MSM media amended with 100 ppm phenol as the sole source of carbon. One of the strain designated as DJ2 was able to grow at a phenol concentration of 2000 ppm. 16s rRNA sequencing of the bacterial isolates were also completed for accurate identification of the bacterial species and the sequence obtained were deposited in NCBI GenbankÂ Â and the accession number KC992288 to KC992310 were generated for the 23 strains. *In vitro* hydrocarbon degradation ability of the isolates with BTEX (Benzene, Tolune, Ethylbenzene and Xylene) is being continued now. Our preliminary findings suggests that some of the bacterial isolates have good potential of biodegradation of different crude oil fractions and these strains might be useful in developing an efficient consortia for actual field scale bioremediation of crude oil contaminated sites of Assam.

Keywords: Crude oil contamination, BTEX, Bioremediation, TPH

(430a) Impact of Acids On the Structure of Linde Type A Zeolites for Use in Reverse Osmosis Membranes for Wastewater Recovery

Jamieson, H., Arizona State University
Waller, A., Arizona State University
Yin, H., Arizona State University
Lind, M. L., Arizona State University

Originally presented on: 11/6/2013 8:30:00 - 8:55:00

Abstract:

Membrane separations and osmotic processes are attractive options for wastewater reclamation, and with growing concerns over a depleting global water supply such technologies are rapidly becoming more valuable. Previously, hydrophilic Linde Type-A (LTA) zeolites have been incorporated into polyamide reverse osmosis membranes, creating high flux membranes called Thin Film Nanocomposite membranes[1], [2]. In our ongoing work, we are exploring the use of zeolite nanocomposite membranes for the recovery of potable water from wastewaters. We are specifically interested in the wastewaters generated by NASA on the International Space Station (ISS). Currently, only 70% of the wastewater generated on the ISS is recovered; the remaining 30% of the water is stored for later disposal in the form of brine[3]. With the retirement of the NASA space shuttle fleet, the need to minimize use of ground-based resources is a high priority; as a result, NASA desires to maximize water recovery[4]. Our research focuses on using LTA zeolite nanocomposite membranes to recover water from the urine and urine brine wastewaters produced on the ISS. The small pore size of the LTA zeolite should make it ideally suited to allow passage of water, but rejection of small neutral organic molecules such as urea, a main component of urine[5].

In this talk, we will present our results evaluating the resistance of LTA zeolites to the components NASA uses to acidify the urine solutions. NASA treats the urine wastewaters with a multi-component acidic solution to inhibit bacterial growth. We have created synthetic urine brines in our lab and found that these have adverse effects on the crystal structure of the LTA zeolites. We produced solutions of potassium phosphate at pH 4.2 and 5, and sulfuric acid, and chromium trioxide solutions at pH values varying from 0.5 to 5 to identify the most damaging acid component. We exposed zeolites to the synthetic urine, and acid-specific solutions for 1 to 18 days to test the effects of pH and acid type over time. We characterized the samples with X-ray diffraction, scanning electron microscopy and Fourier transform infrared spectroscopy to evaluate any structural changes in the zeolites. Significant degradation was observed over a period of 9 to 18 days at pH values below 2 for all acids, and immediate dissolution occurred at or below pH 1. Potassium phosphate showed completely amorphous structure at pH 4.2, likely due to formation of aluminum phosphate. We will discuss the potential mechanisms for zeolite degradation, including acid hydrolysis at low pH. The results of this experimentation suggest that LTA zeolites are inappropriate for use in low pH environments.

References:

[1] Jeong, B. -H. *et al.* Interfacial polymerization of thin film nanocomposites: A new concept for reverse osmosis membranes. *Journal of Membrane Science*. **294**, 1-7 (2007).

[2] Lind, M. L. *et al.* Influence of zeolite crystal size on zeolite-polyamide thin film nanocomposite membranes. *Langmuir: the ACS Journal of Surfaces and Colloids*. **25**, 10139-10145 (2009).

[3] Carter, L., Marshall, N. & Flight, S. Status of the regenerative ECLS water recovery system. *AIAA International Conference on Environmental Systems; 11-16 Jul. 2010; Barcelona; Spain*1-11 (2010).

[4] Link, D. E., Carter, D. L. & Higbie, S. Development of an advanced recycle filter tank assembly for the ISS Urine Processor Assembly. *American Institute of Aeronautics and Astronautics*1-8.

[5] Database of Zeolite Structures. International Zeolite Association. http://www.iza-structure.org/databases/, IZA.

(430b) Arsenic Rejection By Nanofiltration Membranes: Experimental Results and Model Prediction

Deng, B., University of Missouri
Fang, J., Delon Hampton and Associates
Originally presented on: 11/6/2013 8:55:00 - 9:20:00

Nanofiltration (NF) membranes, NF270 and TFC-SR-2, were selected for arsenic rejection in dead-end stirred cell filtration equipment. The membranes were characterized using Atomic Force Microscopy, Attenuate total reflection-Fourier transform infrared spectroscopy, potentiometric titration, and contact angle measurements. The results showed that NF270 was rougher than TFC-SR-2, and both membranes contained typical carboxylic and amide groups. NF270 had a contact angle of 30.7^{o}, and TFC-SR-2 was more hydrophilic with a contact angle at 20°. Potentiometric titration results showed that both membranes were negatively charged while pH>3.0, and carried more negative charges with increasing pH. The pore sizes of NF270 and TFC-SR2 were 0.49 and 0.46 nm, respectively, which were determined by fitting rejection data of uncharged solutes (glucose and maltose) into Spiegler-Kedem-Steric Hindrance Pore Model.

Arsenic rejection experiments showed that rejection of As(V) by NF270 and TFC-SR-2 membranes increased with increasing initial concentrations due to dilution effect and the rejection rate could be higher than 90%. The rejection of As(V) increased with increasing pH. As(III) rejection was not affected by pH while less than 9.2 because of its existence in neutral form. Arsenic rejection increased with increasing operating pressure and stirring speed. Ionic strength has influences on arsenic rejection due to shielding phenomenon. Co-occurring ions, such as sulfate, phosphate, and calcium, could reduce the arsenic rejection. As(V) rejection rate decreased with increasing concentration of co-occurring ions in the range of 0.0001M to 0.01M. Rejection rate of As(III) also reduced with the presence of co-occurring ions but was not affected by their concentration.

The As(III) rejection by NF270 and TFC-SR-2 membranes was fitted into Spiegler-Kedem-SHP model. The model calculated rejection was much higher than the experimental data. The poor fitting was possibly due to the affinity interaction between arsenite molecules and membranes. The formed hydrogen bond rendered the easy entry of arsenite into membrane structure, consequently, lowered the rejection of arsenite. The rejections of As(V) as a function of permeate flux by NF270 and TFC-270 were modeled using Donnan Steric Pore Model (DSPM) model combined with concentration polarization theory. The extended Nernst-Planck equation was solved numerically using a fourth order Runge-Kutta method. Membrane charge density and pore size are two important factors which have a major impact on membrane performance. The calculated rejections were in good agreement with the experimental results.

(430c) Fouling-Resistant Membranes for the Treatment of Flowback Water From Hydraulic Shale Fracturing: A Pilot Study

Miller, D. J., University of Texas at Austin
Huang, X.,
Li, H., Advanced Hydro, Inc.
Kasemset, S., University of Texas at Austin
Lee, A., University of Texas at Austin
Agnihotri, D.,
Hayes, T.,
Paul, D. R., The University of Texas at Austin
Freeman, B. D., The University of Texas at Austin
Originally presented on: 11/6/2013 9:20:00 - 9:45:00

Polyacrylonitrile hollow fiber ultrafiltration (UF) and polyamide spiral wound reverse osmosis (RO) membrane modules were surface-modified by contact with an aqueous solution containing dopamine to deposit polydopamine on the membrane surfaces and other wetted parts inside the modules. UF modules were further modified by grafting poly(ethylene glycol) (PEG) brushes to the polydopamine coating. Polydopamine and polydopamine-*g*-PEG coatings increase hydrophilicity of the membrane surfaces and have previously been shown to improve fouling resistance towards model oil/water emulsions in laboratory studies. In a pilot-scale test treating hydraulic fracturing flowback water from the Barnett Shale region of Texas, the fouling performance of modified UF and RO membrane modules was compared to that of unmodified analogs. UF modules were used to remove most of the highly fouling organic matter in the feed before desalination by a train of RO elements. Polydopamine-modified UF modules maintained higher flux, lower transmembrane pressure difference, and improved cleaning efficiency relative to unmodified modules. The polydopamine coating did not appear to improve RO fouling behavior, presumably because most of the organic foulants had been removed by UF pretreatment of the feedwater. However, higher and more stable salt rejection was observed in modified RO modules than in unmodified modules.

(430d) Microbial Adhesion to Advanced Water Treatment Membrane Materials: Implications for Biofouling

Kirschling, T. L., National Institute of Standards and Technology
Greenlee, L. F., National Institute of Standards & Technology
Originally presented on: 11/6/2013 9:45:00 - 10:10:00

Membrane based water purification will be essential for providing potable water from saltwater and wastewater as global water scarcity becomes more pressing. One of the biggest challenges to membrane based water treatment processes is biofouling which causes significant increases in transmembrane pressure drop and decreases in flux. This performance decline increases both energy requirements and the number of required cleaning cycles leading to significant cost increases. Novel process improvements and advanced materials have the potential to decrease biofouling, but an understanding of how these factors impact membrane associated microbial communities and their interactions with the underlying material is lacking. This study characterizes the microbial foulants recovered from a reverse osmosis (RO) pilot plant undergoing a six month study to assess the use of ozone pretreatment prior to the RO train. Pyrosequencing of the 16S rDNA gene was used to determine which microbial strains were prevalent in the biofoulant material. Sequencing data was compared with isolated bacterial strains from the membrane surfaces. Species of the genus *Mycobacterium* were found to be dominant in both ozonated and control trains. The adhesive properties of relevant isolates were studied along with their adhesion to various polymer films using atomic force microscopy (AFM) and quartz crystal microbalance (QCM) methods. The results of these studies with a *Mycobacterium sp.* isolate will be discussed.

(430e) Real-Time Monitoring Approach and System for Detection and Characterization of RO Membrane Fouling

Thompson, J. F., University of California, Los Angeles
Rahardianto, A., University of California, Los Angeles
Cohen, Y., University of California, Los Angeles
Originally presented on: 11/6/2013 10:10:00 - 10:35:00

Mineral scaling and membrane fouling remain significant impediments in reverse osmosis (RO) membrane desalination of various water sources, including inland brackish water, seawater and wastewater effluent for water reuse. In order to mitigate membrane fouling and mineral scaling it is imperative to be able to monitor the onset of the above in real time, particularly for water sources of high mineral scaling propensity. Accordingly, a novel membrane monitoring system was developed (for RO plant monitoring) consisting of a high-pressure transparent membrane cell and an automated detection system for high resolution optical scanning of the membrane surface. System pressure and feed flow rate are self-adjusted using a custom control system so as to set the desired level of concentration polarization in the membrane channel, at the desired level relative to that which is expected in the RO plant. Online image analysis software, which is integrated into the control system, analyzes the extent of mineral scaling to quantify the extent of surface coverage by particulate fouling or mineral scale and the number density of mineral crystals. Equipped with fluorescent marker injection and an inline spectrometer for monitoring, the integrated monitoring system also provides the capability for characterization of the feed-side mass transfer coefficient and marker rejection, thereby enabling estimation of the level of concentration polarization in the RO plant. The overall approach has been tested for operational scenarios in which calcium sulfate (gypsum), calcium carbonate (calcite), and silica are the mineral scalants of concern.

(430f) A Molecular Marker Approach for Characterization of RO Membranes and Solute Transport in Spiral-Wound RO Membrane Systems

Surawanvijit, S., University of California, Los Angeles
Thompson, J., University of California, Los Angeles
Rahardianto, A., University of California, Los Angeles
Cohen, Y., University of California, Los Angeles
Frenkel, V., ARCADIS U.S. Inc.
Originally presented on: 11/6/2013 10:35:00 - 11:00:00

Over the last few decades, there has been a marked growth in the application and integration of reverse osmosis (RO) with conventional processes for water production. RO processes are now used not only for seawater and brackish water desalination, but are also finding use in agricultural drainage water desalting, nitrate and color removal from impaired groundwater, as well as in water reuse applications. Even though RO is effective in rejecting solutes as small as monovalent ions, there are factors that may impact membrane performance (e.g. flux and rejection). These include membrane fouling and mineral scaling due to concentration polarization, inadequate hydrodynamic mixing in membrane modules, as well as loss of membrane integrity (e.g., due to prolonged exposure to disinfectants such as chlorine, as well as physical damage caused by manufacturing defects, faulty installation and maintenance, failure of components, and stress and strain on membrane surface from operating condition). In order to ensure robust process performance, it is paramount to obtain real-time information on solute transport parameters, hydrodynamics conditions, and membrane integrity. Accordingly, in the present study, an online fluorescent marker-based approach was developed for spiral-wound RO membrane systems for real-time monitoring of: (a) membrane mass transfer coefficient on the feed-side channel which enables assessment of concentration polarization conditions; (b) solute residence time distribution which provides information regarding hydrodynamic conditions in membrane modules and thus inference regarding potential structural changes within membrane elements, as well as membrane fouling and mineral scaling; and (c) transport parameters (e.g., solute solution-diffusion transport parameter and reflection coefficient) that enable characterization of membrane integrity. The approach involved a controlled introduction of a fluorescent marker dose to the feed stream and online monitoring of the marker concentration in the permeate and concentrate streams. Experimental studies were carried out, with an especially-designed online fluorescent marker detection system, using both intact and compromised (by exposure to chlorine and via mechanically induced damage). Results from the present study demonstrated the technical feasibility for real-time online assessment of concentration polarization, marker transport parameters, and highly sensitive detection of membrane integrity in spiral-wound RO membrane systems.

(436a) Hydrogen Production From Water

McDaniel, M. D.,
Ekerdt, J. G., University of Texas at Austin
Originally presented on: 11/6/2013 8:30:00 - 9:00:00

Silicon-based photocathodes show promising potential for photoelectrochemical (PEC) water splitting for renewable production of hydrogen. Trade-offs between stability and efficiency have limited the success of Si-based photoelectrodes. To overcome this issue, we utilize a thin layer (≈ 2 nm) of strontium titanate, $SrTiO_3$ (STO), which is grown epitaxially on p-doped Si (001) by molecular beam epitaxy. The STO layer greatly enhances the stability of the Si-based photocathode. In addition, the high quality STO layer and clean, low defect interface between STO(001) and Si(001) provides minimal trap states for charge carriers. Moreover, there is no conduction band offset between STO and Si. This allows the photo-generated electrons to easily move from the Si substrate to the STO surface, where hydrogen ions are reduced to form gaseous hydrogen. Based on our electrochemical measurements, we find that the STO-Si photocathode is both stable and exhibits high conversion efficiency. Studies with continuous Ti/Pt films for H_2 evolution yield a current density above 15 mA/cm^2; patterned metal layers are expected to give higher current densities. The results indicate that Si-based photocathodes for PEC water splitting can be improved using a thin STO layer, providing a viable method for clean hydrogen production.

(436b) Core-Shell Photoelectrochemical Electrodes for Water Splitting

Peng, Q., Duke University
Kalanyan, B., North Carolina State University
Hoertz, P.,
Miller, A.,
Kim, D. H., North Carolina State University
Hanson, K., UNC-Chapel Hill
Alibabaei, L., UNC-Chapel Hill
Liu, J., Duke University
Meyer, T., UNC-Chapel Hill
Parsons, G. N., North Carolina State University
Glass, J. T., Duke University
Originally presented on: 11/6/2013 9:00:00 - 9:18:00

Photoelectrochemical (PEC) processes offer one of most promising ways to convert solar energy into fuels (e.g. H_2, and CH_3OH), especially H_2. In order to synthesize high efficient and stable PEC water splitting electrodes, one must optimize the light absorption, charge separation, and charge transfer processes of the electrode simultaneously. Core-shell nanostructures with transparent conductive cores and semiconductor shells show great promise for optimizing PEC performance of photoactive materials, such as TiO_2. However, limited research has been performed in this field.

We will present the result of using transparent conductive network consisting of low cost, solution-processed antimony doped tin oxide nanoparticles (nanoATO) and tin doped indium oxide nanoparticles (nanoITO) for enabling core-shell TiO_2 PEC electrodes. In detail, we will present the synthesis of TiO_2 PEC electrodes by coating solution-processed nanoATO on FTO glass with TiO_2 through atomic layer deposition (ALD). The conductive, porous nanoATO film-supported TiO_2 electrodes, yielded a highest photocurrent density of 0.58 mA/cm^2 under AM 1.5G simulated sunlight of 100 mW/cm^2. This is approximately 3× the maximum photocurrent density of planar TiO_2 PEC electrodes on FTO glass. Transient photocurrent measurements showed that nanoATO films reduce charge recombination by accelerating transport of photoelectrons through the less defined conductive porous nanoATO network. Further, we will show that the effect of interfacial reaction during the TiO_2 ($TiCl_4/H_2O$) ALD on transparent nanoATO and nanoITO on their PEC performance. The scalable solution-processed porous transparent nanoATO and nanoITO films are promising as a framework to replace other conductive scaffolds (Si nanowire forest, CNTs, and $TiSi_2$ nanonets) for PEC electrodes. The results are not only valuable for PEC, but also informative for the related DSSC research.

(436c) Design of Photocatalysts Active in Infrared and Visible Range

Lee, D. C., Korea Advanced Institute of Science and Technology (KAIST)

Originally presented on: 11/6/2013 9:18:00 - 9:36:00

Infrared- and visible-active semiconductors are promising candidates for light-harvesting materials. When the semiconductor nanocrystals are within quantum confinement regime, they exhibit size tunable bandgap, enabling light-harvesting even more controllable. Furthermore, the dynamics of excited states can also be controlled through a design of heterostructure semiconductor nanocrystals. We synthesized PbSe/CdSe/CdS core/shell/shell nanocrystals with varying structural design, e.g., spheres, pyramid and tetrapod. These nanocrystals show varying electron-hole exchange interactions and radiative recombination lifetimes depending on their dimensions. We also prepared semiconductor/metal nanojunction and used the nanomaterials as photocatalysts for reduction of various dyes by growing metal tips at the corners of semiconductor nanocrystals. Interestingly, the morphology change leads to the difference in the tip growth pattern. In addition, the growth of metal particles affects the Fermi level of the entire hybrid composites, leading to the change in their photocatalytic activity. In the presentation, the implications of the heterostructure quantum dots and their hybrid composites with metal nanoparticles will be discussed in the context of energy applications.

(436d) Efficient Photoelectrochemical Water Splitting With Si-Based Metal-Insulator-Semiconductor Photoelectrodes

Esposito, D. V., University of Delaware
Lee, Y., National Instittue of Standards and Technology
Talin, A. A., Sandia National Laboratory
Moffat, T. P., National Institute of Standards and Technology
Originally presented on: 11/6/2013 9:36:00 - 9:54:00

Photoelectrochemical (PEC) water splitting is an attractive pathway for renewable production of hydrogen, but the efficiency and stability of semiconducting photoelectrodes must be improved. One promising approach to achieve both high efficiency and good electrochemical stability is the metal-insulator-semiconductor (MIS) photoelectrode architecture.[1,2] The MIS design consists of catalytic metal structures, or collectors, deposited on a 1-3 nm thick oxide-covered semiconductor. Of great importance to this design is the thin insulating oxide layer, which must protect the semiconductor from the potentially corrosive electrolyte while simultaneously mediating minority carrier tunneling between the semiconductor and collectors. In recently published work,[2] we have demonstrated significant improvement in the performance of Si-based MIS photocathodes through interfacial engineering. These planar photocathodes exhibit excellent stability and appreciable photoelectrode conversion efficiencies, up to 2.9%, but further improvement in efficiency must be achieved in order to make this process economically viable.

In this work, nano- and micro-structured Si-based MIS photoelectrodes fabricated by reactive ion etching and vapor-liquid-solid chemical vapor deposition are explored as a means to further increase the efficiency of Si-based photoelectrodes. The use of 3D-structured electrodes offer several potential advantages for achieving higher efficiency, including increased surface area for catalysis, improved collection of photo-generated carriers, and decreased optical reflection losses. However, the use of 3D photoelectrodes also introduces detrimental effects that create trade-offs between optical, catalytic, and electronic properties. In this talk, design principles for optimizing the performance of 3D-structured photoelectrodes are described and a comparison of the performance of planar and 3D-structured MIS photoelectrodes is presented. The incorporation of MIS photoelectrodes into stand-alone PEC water splitting devices will also be discussed. Calculations based on the method of detailed balance show that PEC devices utilizing Si-based MIS photoelectrodes are capable of achieving solar-to-hydrogen conversion efficiencies greater than the 2015 efficiency target of 15% set by the Department of Energy.[3]

References

[1.] H.J. Lewerenz, et al., *Electrochem. Acta*, **2011**, 56, 10726.

[2.] D.V. Esposito, I. Levin, T.P. Moffat, A.A. Talin, *Nature Materials*, **2013**, (in press).

[3.] Fuel Cells Technologies Office Multi-Year Research, Development and Demonstration Plan. US Department of Energy Efficiency and Renewable Energy. Available from: http://www1.eere.energy.gov/hydrogenandfuelcells/mypp/ (2012)

(436e) Graphene Oxide Photocatalysts for Water Splitting and Its Upconverted Photoluminescence

Yeh, T. F.,
Teng, H.,
Originally presented on: 11/6/2013 9:54:00 - 10:12:00

Graphene-Oxide Photocatalysts for Water Splitting and Its Upconverted Photoluminescence

Te-Fu Yeh[1] and Hsisheng Teng[1,2,*],

[1]*Department of Chemical Engineering and Research Center for Energy Technology and Strategy, National Cheng Kung University, Tainan 70101, Taiwan*

[2]*Center for Micro/Nano Science and Technology, National Cheng Kung University, Tainan 70101, Taiwan*

**E-mail:* hteng@mail.ncku.edu.tw

Abstract

Graphene oxide (GO) is a p-type semiconductor with a suitable band positions for photocatalytic watersplitting. GO sheets have a hydrophilic 2D carbon structure that allows for extensive chemical modification. This study presents strategies of tuning the conductivity type of GO from p-type to n-type for simultaneous evolutions of H_2 and O_2 under visible light irradiation. Further converting the n-type graphene oxide sheets to quantum dots results in stoichiometric evolution of H_2 and O_2 at 2-to-1 ratio. The quantum dots also serve as an energy-converting medium with upconverted photoluminescence to promote the activity of metal-oxide photocatalyts.

Introduction

Polymeric semiconducting molecules, which have a large contact area with water, attract significant attention as alternatives to metal-containing photocatalysts. Graphene oxide (GO) derived from graphite oxidation is a molecule-like semiconductor [1]. GO can be extensively dispersed in water to molecular scale and has a large exposed area. Graphene is a zero gap semiconductor whereas the GO gap increases with the oxidation level [2,3]. The bonding with electron-withdrawing oxygen functionalities results in the p-type conductivity of GO. N-type conductivity appears when graphene covalently bonds to electron-donating nitrogen functionalities [4-6].

In addition to chemical doping, we exploit the quantum-confinement effect to modulate electronic properties. Unlike two-dimensional graphene, the zero-dimensional graphene quantum dots (GQDs) have a finite band gap depending on their size [7]. The confined p-electron in sp^2 domain gives the quantized discrete levels [8]. With these features, GQDs exhibit upconvered photoluminescence (PL) [9].

This study presents the potential of GO as a photocatalyst for water splitting. P-type GO can catalyze H_2 evolution with the presence of electron donors. Ammonia-treated GO (NGO) exhibits n-type conductivity and promising photocatalytic activity for generating H_2 and O_2 simultaneously, but with the ratio below the stoichiometric value (that is, 2). Miniaturized N-doped graphene oxide quantum dots (NGO QDs) function as photocatalysts for stoichiometric H_2 and O_2 generation from water cleavage under visible light irradiation. We also demonstrate the upconversion ability of NGO QDs.

Experimental

GO was prepared using a natural graphite powder through a modified Hummers' method. NGO was obtained through nitridation of the as-prepared GO performed at room temperature using a flow of NH_3 gas. NGO QDs were prepared from oxidation of nitrogen-doped graphene. Photocatalytic reactions were conducted at approximately 25 Â¢XC in a gas-enclosed inner irradiation system or a side irradiation system.

Result and Discussion

We demonstrated the photocatalytic activity of GO in the gas-enclosed circulation system with mercury-lamp irradiation. Continuous H_2 evolution over metal-free GO was observed for reactions conducted in pure water and an aqueous methanol solution [1]. The total evolution of H_2 from aqueous methanol solution far exceeds the amount of H_2 obtained from pure water. GO cannot catalyze O_2 generation even with the presence of the electron scavenger Ag^+ ion.

We modified the as-synthesized GO photocatalyst by ammonia treatment to form NGO, which exhibited n-type conductivity. The change of conductivity type was contributed by the electron-donating property of the introduced amino groups. NGO effectively catalyzes O_2 evolution under mercury-lamp irradiation [6]. The n-type conductivity of NGO attracts hole migration toward the interface region for O_2 evolution. NGO gave rise to simultaneous evolutions of H_2 and O_2 from the side irradiated system under visible light irradiation [6]. The molar ratio of the evolved H_2 and O_2 was smaller than the stoichiometric ratio 2-to-1.

Quantum confinement effect elevates the conduction band edge, promoting the electron injection for H_2 evolution from water. We found stoichiometric evolutions of H_2 and O_2 from pure water over NGO QDs with visible-light irradiation. Figure 1 shows the PL spectra of NGO QDs excited by long-wavelength light (660-900 nm), with the upconverted emission located in the range of 450-530 nm. The upconverted PL emission of NGO QDs excited $NaTaO_3$ to catalyze H_2 evolution from an aqueous methanol solution under visible light irradiation.

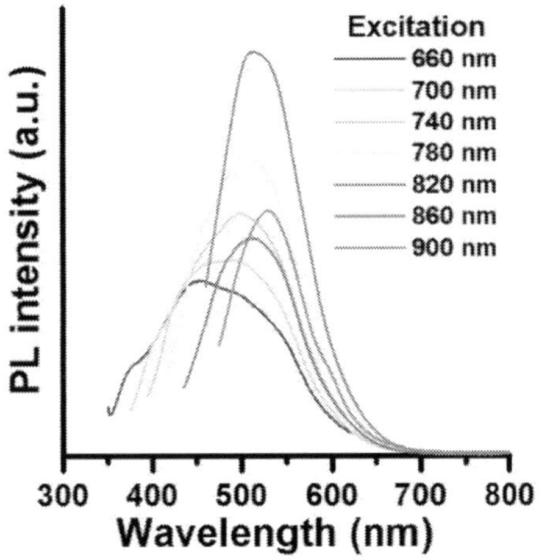

Figure 1. Upconverted PL from NGO QDs. The excitation wavelength ranges from 660 nm to 900 nm. The upconverted emission located in the range of 450-530 nm.

Conclusions

The p-type GO catalyzed H_2 evolution from an aqueous methanol solution under mercury-lamp illumination. The introduction of nitrogen functionalities converted the p-type GO to an n-type NGO, which was capable of catalyzing simultaneous H_2 and O_2 evolutions from water under visible light. The NGO QD catalyst achieved overall water splitting ($H_2/O_2 = 2$) under visible light irradiation. The upconversion ability of NGO QDs made wide-gap photocatalysts active under visible-light illumination.

Acknowledgements

This research is supported by the National Science Council of Taiwan (101-2221-E-006-243-MY3 and 101-2221-E-006-225-MY3) and the "Aim for the Top-Tier University and Elite Research Center Development Plan" of National Cheng Kung University.

References

[1] T.F. Yeh, J.M. Syu, C. Cheng, T.H. Chang, H. Teng, *Adv. Funct. Mater.* **2010**, 20, 2255.

[2] T.F. Yeh, F.F. Chan, C.T. Hsieh, H. Teng, *J. Phys. Chem. C* **2011**, 115, 22587.

[3] T.F. Yeh, J. Cihlář, C.Y. Chang, C. Cheng, H. Teng, *Materials Today* **2013**, 16, 78.

[4] X. Wang, X. Li, L. Zhang, Y. Yoon, P. K. Weber, H. Wang, J. Guo, H. Dai, *Science* **2009**, 324, 768.

[5] C. C. Hu, H. Teng, *J. Phys. Chem. C* **2010**, 114, 20100.

[6] T.F. Yeh, S.J. Chen, C.S. Yeh, H. Teng, *J. Phys. Chem. C* **2013**, 117, 6516.

[7] L. A. Ponomarenko, F. Schedin, M. I. Katsnelson, R. Yang, E. W. Hill, K. S. Novoselov, A. K. Geim, *Science* **2008**, 320, 356.

[8] G. Eda, Y. Y. Lin, C. Mattevi, H. Yamaguchi, H. A. Chen, I. S. Chen, C. W. Chen, M. Chhowalla, *Adv. Mater.* **2010**, 22, 505.

[9] H. Li, X. He, Z. Kang, H. Huang, Y. Liu, J. Liu, S. Lian, C. H. A. Tsang, X. Yang, S. T. Lee, *Angew. Chem. Int. Ed.* **2010**, 49, 4430.

(436f) The Study of Cu2ZnSnS4 Nanocrystal/TiO2 Nanorod Arrays Heterojuction Photoelectrochemical Cell for Hydrogen Generation

Ho, T. Y.,
Chen, L. Y.,
Originally presented on: 11/6/2013 10:12:00 - 10:30:00

In this study, we reported TiO_2 nanorods arrays (TiO_2 NRAs) synthesized by hydrothermal method with controllable densities, diameters and lengths for photoelectrochemical hydrogen generation. The densities and diameters of the TiO_2 NRAs could be easily controlled by the concentration of titanium precursor (titanium isopropoxide, TTIP) and the amount of the HCl. With different reaction time, the lengths of TiO_2 NRAs could be increased from 2 μm to 9 μm. The photocurrent density and conversion efficient were increased with the lengths of TiO_2 NRAs. The maxima current density and conversion efficiency are 2.5mA/cm^2 at 1.23V vs RNHE and 0.95% with 4-5 μm in length under 100mA/cm^2 Xenon lamp illumination with AM1.5 filter, respectively. To improve the conversion efficiencies, various sensitizers, such as dye or quantum dot, are usually used to cover the absorption in the visible light range. Herein, Cu_2ZnSnS_4 (CZTS), which owns direct band gap semiconductor with 1.5eV and suitable band positions, was employed to extend the absorption range. In this study, CZTS nanoparticles were direct decorated on TiO_2 NRAs by solvothermal method with metal chloride and thiourea as precursor and ethanol as solvent to improve the conversion efficiency for hydrogen generation.

(436g) Development of New Generation of Sub-Nanometer Catalysts for Sustainable Energy Applications

Wu, Q., Stony Brook University/BNL
Shen, P., Stony Brook University / BNL
Zhao, S., Stony Brook University
Ramakrishnan, G.,
Su, D., Brookhaven National Laboratory
Li, Y.,
Orlov, A., Stony Brook University / BNL
Originally presented on: 11/6/2013 10:30:00 - 10:48:00

Developing novel heterostructured nanomaterials for energy applications as water splitting under visible light has enormous potential for energy production. Currently the existing materials have very low quantum efficiency, which indicate that new approaches are required in catalysts design. In this work we have developed and characterized sub nanometer Au and Pt nanoparticles, which greatly enhanced rate of hydrogen production of CdS. These heterostructured materials were characterized by TEM, STEM, XRD, UV-Vis spectroscopy and DRIFTS. We have also conducted comprehensive studies of these catalysts in photocatalytic reactors, where we observed a substantial hydrogen production under visible light.

(436h) ZrO2 Incorporated ZnO/TiO2 Mesoporous Photocatalyst for H2 Generation

Hussein, A., South Dakota Shool of Mines and Technology
Shende, R., South Dakota School of Mines and Technology
Originally presented on: 11/6/2013 10:48:00 - 11:06:00

The present research deals with the synthesis of ZrO_2 incorporated coupled ZnO/TiO_2 photocatalyst and its effectiveness on H_2 generation from aqueous methanol solution. Ordered mesoporous ZrO_2 supported coupled TiO_2/ZnO materials were synthesized using surfactant as structure directing agent, titanium isopropoxide, zinc acetate dihydrate and zirconium n-propoxide precursors. The ZrO_2 incorporated coupled ZnO/TiO_2composites were characterized by XRD, UV-Vis spectra, SEM/TEM, XRF, Raman and N_2 adsorption-desorption isotherm. Photocatalytic hydrogen evolution was carried out in a 7 mL quartz reactor at ambient temperature. The focused intensity and area on the quartz cuvette were 0.60 W and 2 cm^2, respectively. In a typical experiment, 2.0 mg of $ZrO_2/ZnO\text{-}TiO_2$ photocatalyst was dispersed in 3.5 mL of H_2O containing 0.5 mL methanol or 10 mg biomass as a sacrificial agent. A 300 W xenon lamp (Oriel) equipped with a UV-filter (240 nm - 400 nm) was employed as the light source. Prior to irradiation, the suspension of the catalyst was degassed with ultrapure nitrogen for 30 min to completely remove the dissolved oxygen and to ensure the reactor in an anaerobic condition. From 3 mL reactor overhead space, 0.1 mL of gaseous samples were withdrawn periodically and analyzed using the gas chromatograph (GC), from which photocatalytic effectiveness towards H_2 generation was evaluated and compared with coupled TiO_2/ZnO material.

(444a) Novel Catalysts and Reactor Concepts for Sustainable Processes: the Use of LCA Data in Process Design

Yaseneva, P.,

Lapkin, A., University of Cambridge

Originally presented on: 11/6/2013 8:30:00 - 8:55:00

Novel catalysts and reactor concepts for sustainable processes: the use of LCA data in process design

Polina Yaseneva and Alexei Lapkin

Department of Chemical Engineering and Biotechnology, University of Cambridge, Cambridge, CB2 3RA, United Kingdom

Abstract

We present a multi-step assessment framework, which was developed to guide decision making process in a process development context. Reaction mass intensity metric, gate-to-gate flowsheet analysis and LCA are used at appropriate stage gates. Specifically results of LCA assessment will be discussed for the case studies of a novel catalyst for water purification process, an intensive process of a pharmaceutical API synthesis and a novel organometallic catalytic process for intermediates synthesis.

Introduction

Assessment of sustainability of novel processes for decision-making purposes is a complex task, as frequently uncertainty of the data precludes the use of full LCA methodology, whereas proxy methods such as mass or energy metrics provide incomplete information. At the same time, information about sustainability of novel technologies is now routinely requested even within research projects. Thus, most EU projects in the area of chemical technologies are accompanied by LCA studies.

Within a large integrated project SYNFLOW (www.synflow.eu) we have developed a three-step approach, using reaction mass efficiency metric for selection of chemical routes, gate-to-gate mass and energy metrics for assessment on flow sheet options and finally LCA for evaluation of the demonstration case studies. These steps coincide with stage-gating within a project: confidence in chemistry to be taken up for process development, evaluation of process options with a focus on clean technologies (novel solvents, process intensification) and finally the detailed analysis of complete process at the stage of demonstration with LCA, targeting detailed flowsheet optimisation.

Here we present results of LCA studies for three specific examples. We show LCA of a novel carbon nanotube-based catalyst for water purification, focusing on comparison of the manufacture of the catalyst, a study of a novel flow-process of stoichiometric reduction of artemisinin, leading to manufacture of an important pharmaceutical API (focusing on comparison of batch vs continuous process) and finally a study of a novel catalytic Buchwald-Hartwig amination under overall flow conditions.

Results and Discussion

Carbon nanotube catalyst for water purification

A highly active catalyst was developed for reduction of an emergent aqueous pollutant, bromate, based on 0.3 %wt Pd deposited onto carbon nanotubes, grown on sintered metal fibre. The TOF of the catalyst was found to be an order of magnitude higher than that of a conventional catalyst based on alumina support (Pd/Al_2O_3). However, there remained the question whether high-temperature CVD process used in the manufacture of carbon nanotubes resulted in a cleaner overall process. The LCA study focused on the manufacture of the two catalysts. In this specific case all environmental impacts for the new catalysts were lower. Detailed analysis of contributions of the individual manufacturing steps to various impacts, e.g. analysis of contributions to cumulative energy demand shown in Figure 1, allows to reveal further options for optimisation of the novel catalyst manufacture [1].

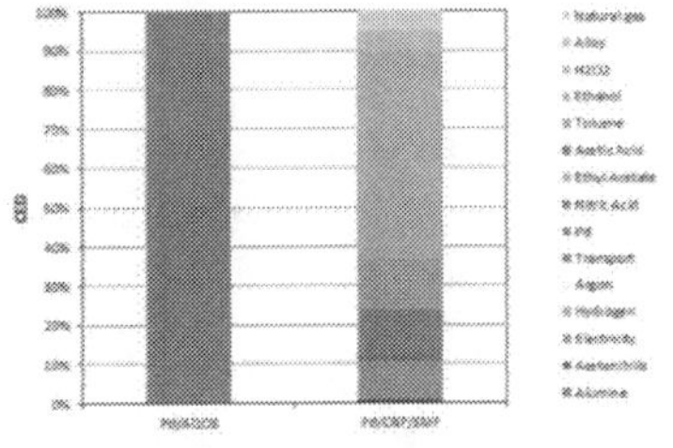

Figure 1. Individual contributions to cumulative energy demand (CED) of the manufacture of the two catalysts.

Artemisinin reduction under intensive flow conditions

We have developed a flow process for reduction of artemisinin to overcome the limitations of a conventional batch process [2]. In the current study we extended the reaction to the final step in the manufacture of the antimalarial API, but also performed an LCA study to compare the novel flow process with the conventional batch reaction [3]. Analysis of LCA results revealed that the solvent used in the manufacture and storage of the reducing agent required for the flow process is having a massive negative impact, see Figure 2. It is becoming critically important to replace the solvent used in a particular stage of the overall process. In this case LCA is instrumental in identifying the critical stage of the overall process that requires optimisation.

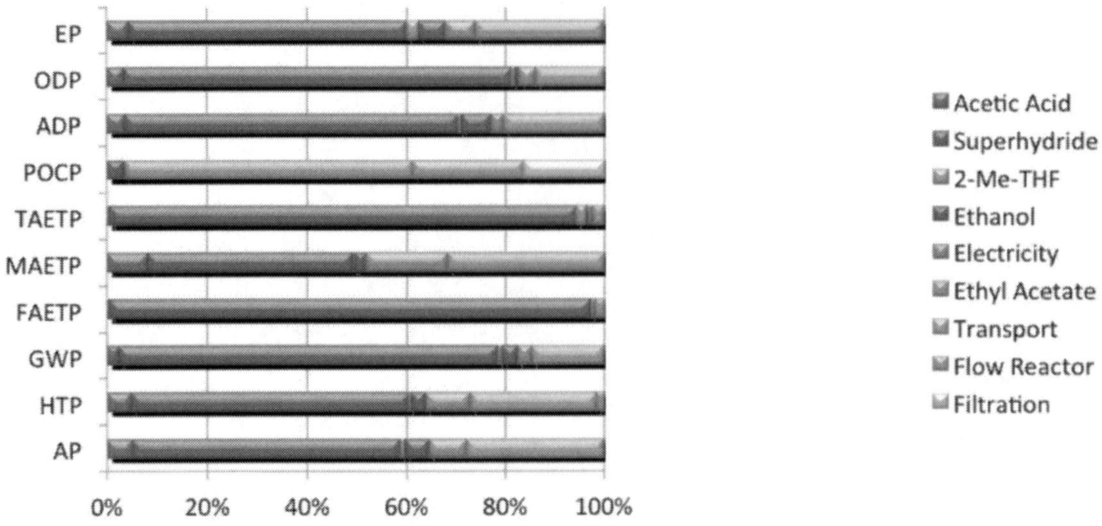

Figure 2. Individual contributions to environmental impacts in artemisinin reduction under flow conditions.

Buchwald-Hartwig amination in flow

The three-step decision making process is illustrated on the example of Buchwald-Hartwig amination. Reaction mass efficiency, process efficiency and LCA are evaluated for this reaction at the different stages in the process design. This approach complements the parallel chemistry-process design methodology being developed within Synflow project, ultimately aiming at a significant reduction of time required for commercialisation of novel technologies, which are also sustainable. Thus, for the initial reaction options we have used RME metric to evaluate the importance of solvent and catalyst recycle. The new process is currently being scaled-up to a demonstrator at an Invite facility of Bayer. We will show the results of stage-gating metrics and initial LCA data.

[1] P. Yaseneva et al., Reduction of bromates on carbon nanofibre supported structured Pd catalysts: experimental and life cycle assessment study, Submitted.

[2] X. Fan, V. Sans, P. Yaseneva, D. Plaza, J.M.J. Williams, A. Lapkin, Facile stoichiometric reductions in flow:

example of artemisinin, Org Proc Res Dev 16 (2012) 1039-1042.

[3] P. Yaseneva et al., Optimisation of the flow process of manufacture of artemether and its LCA study, Submitted.

Acknowledgement: This work was supported by the European Community's Seventh Framework Programme under project "SYNFLOW", NMP2-LA-2010-246461.

(444b) Techno-Economic and Environmental Assessment of Ionic Liquid-Based Biorefineries

Giarola, S., Imperial College London
Shah, N., Imperial College London
Originally presented on: 11/6/2013 8:55:00 - 9:20:00

Although the peak oil is still a debated issue, petroleum supply is expected to become ever more vulnerable of interruptions and current economic system reliance on fossils is exposing all economic sectors to unsustainable price fluctuations [1]. Concerns of maintaining an adequate supply of resources, such as fuels, food, raw materials for a growing population, are also raising a pressing interest for a change in our way to conceive energy and resource provision system [2]. A paradigm shift in the industrial sector from today's large fossils exploitation towards more sustainable practices is also advocated to face current trend of carbon dioxide emissions production, which may soon be approaching greenhouse gas (GHG) levels dangerously close to tipping points of an accelerated global warming [3]. Thus, the future challenge to address the growing world energy demand may be met as long as the supply of higher levels of resources and energy is achieved in a sustainable way promoting low-carbon technologies and renewables [4]. In this respect, the spread of biomass exploitation, as a unique source that can supply both energy and carbon for the production of the commodity chemicals and products, has the potentials to lead to the settlement of a new biomass-based economy [5-6].

Biomass is an abundant, renewable and carbon neutral resource, which is currently significantly exploited for biofuels production from edible components of food crops. However, the major challenge for a future bio-based economy would very likely rely on lignocellulosic feedstock exploitation, being more abundant and less expensive than starchy crops as well as avoiding any direct conflicts with the food chain [7]. Industrial biorefineries have been identified as the most promising route to the creation of a new domestic biobased industry, where biomass conversion processes into fuels, power, and chemicals are integrated through an efficient combination of technologies, to minimise the environmental footprints and ensure the sustainability of all products generated. By producing multiple products, a biorefinery can take advantage of the differences in biomass components and intermediates and maximise the value derived from the biomass feedstock [8]. However, cost-effective technologies for processing lignocellulosic materials into fuels and chemicals are not commercially available. In this context, a central role in determining the overall system profitability is played by the pretreatment unit, where biomass is deconstructed and the three main components (i.e. cellulose, hemicellulose and lignin) are made accessible to subsequent processes [9].

Since the study of Wilkes and Zaworotko [10], ionic liquids have become increasingly popular as reaction and extraction media. The property of ionic liquids which has led to them being classed as green solvents is their very low volatility, thus they do not emit volatile organic compounds and are suitable for reactions at atmospheric pressure. Recently, ionic liquids have begun to be investigated for biomass related processes, yet there are many technological applications and large commercial interests in the field [2]. The use of ionic liquids as lignocellulose deconstruction solvents is also extremely promising [11]. Among the potential applications, ionic liquid can be used to decrystallise cellulose and disrupt the lignin-hemicellulose network. The possibility of removing lignin with ionic liquids and recovering it as precursor for commercial products is also an interesting option [11]. However, deconstruction with ionic liquid will only be viable if some drawbacks are overcome, in particular the cost of ionic liquids. Process design and optimisation is required in order to have high solvent recovery and recycle, but also consumables use reduction, energy integration and waste minimisation. Economic aspects should also be coupled with the concept of ecological sustainability during the process optimisation strategy [12]. Process system engineering techniques play a major role in achieving all the above targets.

This work proposes an investigation of techno-economic performance of a biorefinery based on ionic liquid-pretreatment. The analysis, based on mathematical modelling and process design techniques, aims at identifying the key physical phenomena, the main cost centres affecting the process economics and proposing strategies to develop more efficient biochemical conversion routes for the biomass. This is performed along with the system environmental performance appraisal based on Life Cycle Assessment (LCA) techniques [13]. The final purpose is to provide a mathematical tool for an in-depth economic and environmental analysis of the ionic liquids fractionation, identifying the process enhancements alternatives and outlining potential synthetic routes to biocommodities.

References:

1. Lutz, C., U. Lehr, K. S. Wiebe (2012). Economic effects of peak oil. Energy Policy, 48, 829-834.
2. Tan, S. S. Y., D. R. MacFarlane (2009). Ionic liquids in Biomass Processing. Topics in current chemistry, 290, 311-340.
3. Foley, J. A. (2005). Tipping Points in the Tundra. Science, 310, 627-628.
4. International Energy Agency (2012). World Energy outlook. IEA, Paris (FR).
5. Clark, J. H., F. E. I. Deswarte, editors (2008). Introduction to Chemicals from Biomass Padstow: John Wiley & Sons, Ltd.
6. Ragauskas, A. J., C. K. Williams, B. H. Davison, G. Britovsek, J. Cairney, C. A. Eckert, et al. (2006). The Path Forward for Biofuels and Biomaterials. Science, 311, 484-489.
7. The Royal Society (2008). Sustainable biofuels: prospects and challenges. The Royal Society, London (UK). www.royalsociety.org
8. National Renewable Energy Laboratories (2009). Biomass Program. U.S. Department of Energy, Golden (CO).
9. Aden, A., M. Ruth, K. Ibsen, J. Jechura, K. Neeves, J. Sheehan, et al. (2002). Lignocellulosic biomass to ethanol process design and economics utilizing co-current dilute acid prehydrolysis and enzymatic hydrolysis for corn stover. Technical Report. National Renewable Energy Laboratory (NREL), Golden (CO).
10. Wilkes, J. S., M. J. Zaworotko (1992). Air and water stable 1-ethyl-3methylimidazolium based ionic liquids. Journal of the Chemical Society, Chemical Communications, 965-967.
11. Brandt, A., J. Gräsvik, J. P. Hallett, T. Welton (2013). Deconstruction of lignocellulosic biomass with ionic liquids. Green Chemistry, 15, 550-583.
12. Kralisch, D., D. Reinhardt, G. Kreisel (2007). Implementing objectives of sustainability into ionic liquids research development. Green Chemistry, 9, 1308-1318.
13. International Organisation for Standardisation (1997). ISO 14040: Environmental Management - Life Cycle Assessment - Principles and Framework. International Organisation for Standardisation: (ISO), Geneva.

(444c) Optimal Simultaneous Production of Bio-Ibutene and Bioethanol From Switchgrass

Martin, M., University of Salamanca
Grossmann, I. E., Carnegie Mellon University
Originally presented on: 11/6/2013 9:20:00 - 9:45:00

Ibutene is an important intermediate that has been typically produced from the cracking of the C4 fraction of crude oil. It is the basis for the production of a common additive to gasoline in search for a cleaner burning fuel, the methyl-tert-butyl ether (MTBE), as well as a monomer for the polymerization such as butyl rubber. Its importance can be somehow reflected in its price, around $2/kg. However, the dependency on the crude oil and the already limited availability due to the large number of applications has increased the need for new sources of this chemical. Ibutene has recently been used as intermediate for the production of oxygenated diesel substitutes from glycerol (Chen et al 2011, Vlad et al 2010, Martín & Grossmann 2013). In this application the main drawbacks are, on the one hand, the production of diesel substitutes using a fossil fuel based intermediate and, on the other hand, the economics since from the energy and water consumption standpoint its use for the enhanced production of diesel substitutives is competitive with the process that directly sells the glycerol as byproduct.

In this work, we propose the optimal flowsheet for the production of ibutene from switchgrass. A superstructure embedding a number of alternatives is proposed. Two technologies are considered for switchgrass pretreatment, dilute acid and ammonia fibre explosion (AFEX) so that the structure of the grass is broken down. Surface response models are developed for predicting the effect of the operating conditions on the yield of each pretreatment. Next, enzymatic hydrolysis follows any of the pretreaments to obtain fermentable sugars, mainly xylose and glucose. Ibutene is obtained by fermentation of the sugars. However xylose cannot be easily converted, and thus we also evaluate the possibility of using it to produce bioethanol. Next the ibutene has to be separated mainly from CO_2 for which, PSA or membrane separation are considered while in case ethanol is produced, a multi-effect distillation column followed by molecular sieves are used to dehydrate the ethanol. The problem is formulated as an MINLP with simultaneous optimization and heat integration based on Duran and Grossmann's model (1986). Finally, an economic evaluation is performed.

The most promising process involves the use of dilute acid pretreatment and membrane purification of the ibutene. The production of ibutene using renewable sources is feasible, and may be competitive for its use to generate diesel substitutes for glycerol etherification. However, the decision related to the production of ibutene alone or the simultaneous production of ibutene and ethanol depends on the prices for bioethanol and for switchgrass. For low and current biomass cost the production of ethanol ad ibutene is interesting for bioethanol process above $1/gal.

References:

Chen, P., Min, M., Chen, Y., Wang, L., Li, Y., Chen, Q., Wang, C., Wan, Y., Wang, X., Cheng, Y., Deng, S., Hennessy, K., Lin, X., Liu, Y., Wang, Y., Martinez, B., Ruan, R., 2009. Review of the biological and engineering aspects of algae to fuels approach. Int. J. Agric. Biol. Eng. 2, 1–30.

Duran, M.A.; Grossmann, I.E. (1986) Simultaneous optimization and heat integration of chemical processes. *AIChE, J.,* 32, 123-138

Martín, M., Grossmann, I.E. (2013) Simultaneous dynamic optimization and heat integration for the co-production of diesel substitutes: Biodiesel (FAME & FAEE) and glycerol ethers from algae oil. To be Submitted

Vlad, E., Bildea, C.S. and Bozga, G., 2010, Integrated design and control of glycerol etherification processes, Bull. Inst. Pol. Iasi, LVI (LX), 4, 139-148.

(444d) Synthesis and Design of Biorefinery Processing Networks With Uncertainty and Sustainability Analysis

Cheali, P., Technical University of Denmark
Gernaey, K. V., Technical University of Denmark (DTU)
Sin, G., Technical University of Denmark
Originally presented on: 11/6/2013 9:45:00 - 10:10:00

Synthesis and Design of Biorefinery Processing Networks with Uncertainty and Sustainability analysis

Peam Cheali, Krist V. Gernaey and Gürkan Sin

Department of Chemical and Biochemical Engineering, Søltofts Plads, Building 229, Technical University of Denmark, DK-2800 Lyngby, Denmark

Chemical industries usually rely on fossil based feedstock, which is a limited resource. In view of increasing energy demands and the negative environmental and climate effects related to the use of fossil based fuels, this motivates the development of new and more sustainable technologies for processing renewable feedstocks, with the aim of bridging the gap for fuel, chemical and material production. This project is focusing on biorefinery network design, in particular for early stage design and development studies.

Optimal biorefinery design is a challenging problem. It is a multi-objective decision-making problem not only with respect to technical and economic feasibility but also with respect to environmental impacts, sustainability constraints and limited availability & uncertainties of input data at the early design stage. It is therefore useful to develop a systematic methodology to support the evaluation of processes and the generation of sustainable alternatives for identifying the optimal processing routes. One particular challenge here is to include proactively sustainability analysis as part of the synthesis of biorefinery networks. Another challenge is the handling of several sources of uncertainties such as availability and composition of renewable feedstock, the technical performance of alternative processing technologies and the availability of future markets for biorefinery products, among others.

As part of earlier work in our research group, a systematic methodology to identify/generate optimal biorefineries was developed using the superstructure-based approach, and was implemented in a computer-aided framework. The methodology consists of tools and methods including databases, models, superstructure, and solution strategies to represent, describe and evaluate various combinations of processing networks. The optimization of the network is formulated as a mixed integer nonlinear programming type of problem and solved in GAMS. The methodology was applied for designing optimal biorefinery networks considering biochemical routes. Furthermore, the methodology has also been applied to soybean oil processing networks and industrial wastewater treatment networks. In the work presented here, the methodology for designing optimal biorefinery networks was expanded with inclusion of thermochemical routes, which means the scope and the size of the biorefinery network problem was extended at the level of the database, the models and the superstructure relevant for thermo-chemical conversion routes (the conversion of biomass feedstocks (corn stover, poplar wood) to fuels and chemicals (FT-gasoline, FT-diesel, bioethanol and higher alcohols) via thermal decomposition processes like pyrolysis and gasification).,

In this study, we extend the methodology, models and database by incorporating uncertainty and sustainability analysis as well. Appropriate ranges for uncertain parameters are identified with their correlation/covariance structure and latin hypercube sampling (LHS) is used to sample parameters from their respective domain of uncertainty. The parameter samples are then used as input for solving a deterministic and stochastic optimization problem. The sustainability analysis was performed in two ways: First, it was performed retrospectively to the optimal biorefinery network solution obtained after the MINLP by using an in-house software (SustainPRO) that employs ICHEME sustainability metrics. Secondly, the sustainability analysis was included proactively as part of the MINLP optimization problem that is performed to find the optimal processing path with respect to multi-criteria assessment including technical, economics and sustainability.

The expanded database and superstructure with uncertainty and sustainability analysis form a powerful process synthesis toolbox to be used in design of future biorefineries with multi-criteria evaluation (technical and economic feasibility, environmental impact and sustainability).

(444e) Ecosystems As Unit Operations for Innovative and Sustainable Process Design

Bakshi, B. R., The Ohio State University
Urban, R. A., The Ohio State University
Ziv, G.,

Originally presented on: 11/6/2013 10:10:00 - 10:35:00

Since their development in the 1920s, the concept of unit operations has played an important role in the design of chemical processes through the development of general models and software tools. With increasing importance of reducing the environmental impact of chemical processes, the concept of unit operations expanded to include processes for reducing pollutants in the streams leaving the process boundary. Systems engineering methods based on pinch analysis and optimization have also played an important role in enhancing the efficiency of manufacturing systems. For most chemical processes, it is not practical or economically feasible to completely eliminate material emissions or waste, and the chemical industry continues to be a major producer of hazardous waste and air pollutants, including carbon dioxide.

Ecosystems have the natural ability to treat a large variety of pollutants, and provide goods and services to manufacturing systems. For example, wetlands can provide fresh water by removing contamination by chemicals such as pesticides, oils, phenolic compounds, nitrates, etc. Trees provide water, biomass and oxygen, along with the ability to remove air pollutants such as carbon monoxide, particulate matter, sulfur oxides, carbon dioxide, etc. In fact, an unsustainable system may be defined as a system that produces waste in excess of the natural cleaning capacity of ecosystems. Conversely, sustainability requires that engineering be done within ecological constraints. Unfortunately, most current sustainable engineering methods, including life cycle assessment are not able to satisfy this requirement.

In this presentation, we will "rediscover" the bounty of nature and its importance in sustaining chemical processes. This will involve expanding the concept of unit operations by including ecological processes, to enable engineering design within ecological constraints. Process flowsheets will be designed as networks of technological and ecological systems. Monetary and environmental objectives will be considered, and results for objectives such as maximum profit, zero water, and zero carbon dioxide will be compared. Application to a biodiesel manufacturing process will show that the techno-ecological network is superior to a conventional technology-only process in terms of economic and environmental objectives. Case studies of other chemical processes will demonstrate the need to modify the chemistry or processing methods to permit full benefit of ecosystems. This can lead to new innovation that may not be discovered without including ecosystems among the design options. This approach of techno-ecological networks may be extended to larger spatial scales to result in unique ways of designing and operating supply chains and entire life cycles so that they stay within ecological constraints.

(444f) Process Analysis of Continuous Catalytic Gasification As a Waste to Energy Alternative

Tillie, C. F., Cleveland State University
Riedthaler, M. J., Cleveland State University
Adhikari, S., Cleveland State University
Reeves, S. A., Cleveland State University
Gatica, J. E., Cleveland State University
Originally presented on: 11/6/2013 10:35:00 - 11:00:00

Presently, space exploration waste is either carried onboard for the duration of the mission or discarded into a trash module which is sent back into Earth's atmosphere for destruction. On the other hand, with an estimated cost of $10,000 per pound, the cost of delivering mass to locations beyond low earth orbit (LEO) locations is prohibitive.

Since approximately 15 pounds of waste are produced daily by a crew of four during a mission, the intrinsic value of this waste can be greatly enhanced by waste to energy technologies. Moreover, conceptualizing waste gasification as an in-situ alternative for processing of trash is also an efficient route to control waste, while maintaining a healthy habitable environment, during long-duration missions (Santiago-Maldonado et al., 2010).

The gasification of long-chain polymers is hypothesized to follow a mechanism of four combined reactions —two parallel oxidation reactions which are kinetically controlled, accompanied by two series reaction where oxidation products (carbon dioxide and carbon monoxide) are converted into methane and hydrogen. Integrating this process into spaceflight missions would greatly reduce the onboard mass requirements, as these products can be used as propellant fuel. Further, under certain conditions the process could be also be adapted to produce oxygen and water, two essential life support species for space missions.

This paper explores scale-up alternatives to implement low-temperature continuous gasification processes as potential waste management processes. A kinetic model formulated from batch gasification experiments is integrated into a process simulator along with a detailed thermodynamic description.

Emphasis is placed on the preparatory treatment of the waste slurries, the reactor module, and equilibrium stages for recycling reactants and recovering the catalyst. Detailed characterization of low and mid fidelity simulants is presented.

(477a) Remediation of Diesel-Contaminated Soil With a Biostimulation and Bioaugmentation Joint Process

Li, X.,

Lu, D., Tsinghua University

Zheng, L., Tsinghua University

Originally presented on: 11/6/2013 12:30:00 - 12:44:00

Remediation of diesel-contaminated soil with a biostimulation and bioaugmentation joint process

Xiaosen Li, Diannan Lu and Zheng Liu*

Department of Chemical Engineering, Tsinghua University, Beijing 100084, China

lixs09@mails.tsinghua.edu.cnludiannan@mail.tsinghua.edu.cn

*Corresponding author: Phone: 86-10-6277-9876

Email: liuzheng@mail.tsinghua.edu.cn

A new remediation process that integrates biostimulation and bioaugmentation was proposed for diesel-contaminated soil. In addition to the determination of the residue diesel in soil, the total dehydrogenase activity, the microbial community changes in terms of composition and quantity of both microbial community and functional genes were probed by molecular biological techniques. In this research, five bacterial strains capable of degrading diesel were isolated from contaminated sites, in which Rem-E gave the highest degradation efficiency compared to the others and thus used as the additive strain for the bioaugmentation. As for biostimulation, monopotassium phosphate and ammonium sulfate were added to achieve an optimal C:N:P being determined as 100:1.25:1. After 10 day biosimulation, the content of diesel reduced from 2.7% to 1.6%. The joint process by biostimulation and bioaugmentation, however, gave a reduction of disel content from 2.7% to 0.6% in 40 days. Pyrosequencing for the total DNA extracted from the soil showed the evolution of the microbial community and qPCR revealed the change of the functional genes during the remediation process. The result revealed that coupling degrading strain and source of phosphorus and nitrogen led to the highest dehydrogenase activity and degradation rate, which also recover almost all the microbial community in the soil after the remediation. This work showed the great potential of the coupling technology composed of bioaugmentation and biostimulation for the remediation of the oil-contaminated soil. It also offers a viable route for the application of molecular biological techniques as enabling tools in the monitor, design, implementation and assessment of bioremediation practice.

(477b) Electrically Enhanced Contaminant Degradation in Sediment Caps

Zhang, R., Tianjin University of Technology

Originally presented on: 11/6/2013 12:44:00 - 12:58:00

In-situ capping is known as an attractive and cost-effective method for remediation of contaminated sediments. Sediment caps with bioremediation will ultimately significantly improve it performance, but reports about methods to enhance contaminant degradation in sediment caps still very little. Electrically enhanced bioactive barrier □BIOLANCE□is a new process for in situ remediation reducible compounds in soil or sediments. Chlorinated benzenes are significant contaminants to environment.

Many chlorinated organic compounds are persistent organic pollutants due to their resistance to biodegradation and chemical degradation. The objective of this study was to determine if electrodes placed in sediment and poised at several volts could create a redox gradient and provide electron acceptor/donor to stimulate chlorinated contaminant degradation, which will prepare for sediment caps application. The results demonstrated that choose graphite as electrodes can lead to sustainable evolution of hydrogen. Hydrogen evolution displayed zero-order kinetics with the time in the initial stages in different voltage. Even high voltage can produce more hydrogen, but it can not keep longer time because the overpotentials on electrode surfaces will prevent its function. In our experiments, 4V is the best choice. At this condition, the redox and pH change were detected. It is possible that reductive dechlorination could have been achieved with lower redox potentials in the presence of enough hydrogen electron donors through electrodes. Just while sulfate reducing or the other favorable electron acceptor may have reducing priority than chlorinated compounds. The experiments of T-cell capping biodechlorination use electrodes will be further studied.

(477c) pH-Dependent Transport of Energy By-Products in the Subsurface

Prigiobbe, V., University of Texas at Austin

Hesse, M., University of Texas at Austin

Bryant, S., University of Texas at Austin

Originally presented on: 11/6/2013 12:58:00 - 13:12:00

pH-dependent transport of energy by-products in the subsurface

Valentina Prigiobbe[a], Marc A. Hesse[b,c], and Steven L. Bryant[a,c]

[a]*Department of Petroleum and Geosystems Engineering*, [b]*Department of Geological Sciences*, [c]*Institute for Computational Engineering and Sciences, The University of Texas at Austin, Austin, Texas, U.S.A.*

Abstract

The transport of solutes through reactive porous media can be enhanced by the effect of pH-dependent adsorption/desorption processes. This effect can be particularly critical when the solutes, such as toxic metals and radionuclides, have hazardous impact on the environment and the health.

To study the effect of pH-dependent adsorption/desorption on the toxic metal and radionuclide transport, we have developed an analytical and a numerical model coupling transport and surface complexation for a single-phase incompressible fluid. We have analyzed a one-dimensional (1D) system of conservation laws to account of the pH-dependence and longitudinal dispersion and to gain a physical intuition of the phenomenon as well as a 2D system to include the effect of transverse dispersion and to investigate more realistic scenarios.

An important feature of these reactive transport problems is that one of the conservation laws is for the total hydrogen ($c_{ht} = c_h - c_{oh}$), also called acidity, which gives rise to an additional nonlinearity to the systems distinguishing them from other competitive adsorption/desorption problems where hydrogen is not involved.

Here, we present the results for two cases. A first one of pH-dependent transport of naturally-occurring toxic metals in shallow aquifers mobilized through a brine acidied by carbon dioxide (CO_2) leaking from deep geological storage [1,2]. A second case of pH-dependent transport of adsorbing radionuclides accidentally discharged into the subsurface from temporally storage pools [3,4]. In both cases, we have analyzed the transport through a reactive porous medium composed of hydrophilic iron-oxide.

In the first case, the 1D solution of the intrusion of a CO_2-acidified brine into a shallow aquifer has two retarded fronts separated by a spike of concentration of desorbed metal with a maximum much larger than the initial value. This is a classical behavior of pH-dependent adsorption, which occurs when the adsorption of the proton is more favorable than of the metal. In a 2D system, the structure of the plume resembles the 1D solution which, migrating into the aquifer with a maximum concentration much larger than the initial one, may present a health hazard.

In the second case, the 1D solution of the injection of an alkaline brine containing radionuclides into an aquifer consists of a retarded classical front and an unretarded non-classical pulse traveling at the average fluid velocity. The suitable initial and boundary conditions associated with the interaction of pH-dependent adsorption and the longitudinal dispersion are responsible for this phenomenon. In a 2D system in addition to the retarded front and the fast pulse, a fast spreading plume forms due to mixing induced by transverse dispersion. The unretarded pulse is a new transport phenomenon due to the interaction of pH-dependent adsorption and hydrodynamic dispersion, it leads to the migration of a contaminant with a velocity much faster than expected and it should be taken into account in the design of contaminated site remediation.

References

[1] Apps, J.A., Zheng, L., Zhang, Y., Xu, T., Birkholzer, J.T. (2010), Evaluation of potential changes in groundwater quality in response to CO_2 leakage from deep geologic storage. Transp Porous Med 82, 215-246.
[2] Kharaka, Y.K., Cole, D.R., Hovorka, S.D., Gunter, W.D., Knauss, K.G., and Freifeld, B.M. (2010), Gas-water-rock interactions in Frio formation following CO_2 injection: Implications for the storage of greenhouse gases in sedimentary basins. Geology 34, 577-580.
[3] Prigiobbe, V., Hesse, M. A., and Bryant, S. L. (2012a), Fast strontium transport induced by hydrodynamic dispersion and pH-dependent sorption. Geophys Res Lett 39 L18401.
[4] Prigiobbe, V., Hesse, M. A., and Bryant, S. L. (2012b), Anomalous reactive transport in the framework of the theory of chromatography. Transp Porous Med 91, 127-145.

(477d) Experimental Investigation Using a Contaminant-Monitoring Sytem in Storm Drains: A Dichotomous Preliminary Risk Assessment

Lackey, D., University of Alabama at Birmingham
Peters, R. W., University of Alabama at Birmingham
Originally presented on: 11/6/2013 13:12:00 - 13:26:00

As humans have built cities using concrete, asphalt, and large areas of impervious materials, they have potentially disturbed the normal hydrologic circle, and have learned that they must build storm drains to convey excess rainwater and prevent flooding. But these storm drains have a drawback in that they can also lead to pollution of natural environments if they carry contaminants such as antifreeze, fertilizers, paints, used motor oil, tossed pet waste, cigarette butts, and litter. Everything other than pure water is a potential contaminant that degrades water quality. Pollution sources entering storm drains include: automobile emissions, residential wastes, agricultural runoff, urban activities, construction runoff, and businesses [Hyperspective Studios, 2002].

In the summer of 2010, in conjunction with the UAB Occupational Health and Safety (OHS) Department, Dr. Robert W. Peters, and two of his graduate students conducted a project involving a risk assessment of contamination of runoff water flowing into storm drains on the UAB campus [Lackey *et al.*, 2010]. The project was successful enough to spin off into further studies but it had one key shortcoming: it considered only the effects of point sources on storm drains and largely ignored the effects of non-point sources. The details of that study, as well as the steps that will be taken to address that shortcoming, and what further studies the results will lead to, are listed below.

To begin the study, the UAB OH&S Department gave the civil engineering graduate research assistants a series of overhead maps of the UAB campus that have marked storm drain inlet locations from a previous study. The study took place over two main strategic areas on the UAB campus, one that contained the majority of storm drains on the western academic side of campus, and the other that contained the majority of the storm drains on the eastern side of campus which is more focused on medical research. The students were helped by the inlet locations provided on the maps by O&HS, but there were several differences between the locations given on each map and the actual locations of the inlets. Therefore, the project involved an update on the actual locations of each inlet as well as a risk assessment of the likelihood of dumping unwanted chemicals.

In the study, the students identified four main types of inlets:

1. Standard Covered Curbside: The type that contain a concrete or iron cover and engulf large volumes of water from adjacent roadways and parking lots during a rainstorm.
2. Uncovered Grate: These are often found in the middle of parking lots or on pedestrian

walkways. Occasionally, these are even found in grassy areas where irrigation is common.

3. Combination Curbside Grate: Essentially a combination of Types A and B, these are found in the same areas as Type A.
4. Elongated Grate: Found in the same types of areas as Type B, and extremely common at entrances of parking decks.

In addition to the types, each inlet was also assigned a risk assessment value of 0 to 3 based on the priority of its labeling. A color was assigned to each as well for easier readability:

0 (blue) – Marker already present on drain inlet

1 (green) – Low dumping risk

2 (yellow) – Moderate dumping risk

3 (red) – High dumping risk

Some of the criteria for determination of the risk as 1, 2, or 3 are as follows:

1. Default condition; generally assigned unless the inlet was near a probable pollution source such as a trash receptacle, dumpster, laboratory entrance, or parking space.
2. Assigned if the inlet was near a potential moderate pollution source, such as a laboratory entrance or trash can, or if the inlet was on or adjacent to a parking space.
3. Assigned if the inlet was next to a dumpster or other probable major pollution source, or two conditions or more that would result in "2" applied.

The major results of the study, in terms of number of inlets of each type and risk factor, are displayed below in Table 1 and Table 2.

Table 1. Summary of type and risk factor of inlets in Strategic Area 1 (Main Campus).

Area 1 – Main Campus

Risk Factor	Type A	Type B	Type C	Type D	TOTAL
Level 1	68	64	3	7	**142**
Level 2	11	20	1	0	**32**
Level 3	0	2	1	0	**3**
ALL	**79**	**86**	**5**	**7**	**177**

Table 2. Summary of type and risk factor of inlets in Strategic Area 2 (Medical Campus).

Area 2 – Medical Campus

Risk Factor	Type A	Type B	Type C	Type D	TOTAL
Level 1	112	49	0	20	**181**
Level 2	7	6	0	2	**15**

Level 3	1	4	0	3	**8**
ALL	**120**	**59**	**0**	**25**	**204**

In late 2012, the O&HS department decided that they not only want to install warning stencils on the UAB campus, but also sensors that would detect whenever harmful chemicals are present. The idea was expanded on to not only call for "artificial nose" chemical sensors but also a contaminant monitoring system that would map out the data from each sensor in real-time and send alerts to concerned parties whenever harmful chemicals are detected. Despite an initiative to cut costs as much as possible, the budget only calls for 50 of these "artifical noses". Therefore, priority needs to be assigned to the drains most in need of monitoring. When determining this, it turns out that anthropogenic point sources that result from dumping, while being the best criteria for whether or not to use a warning stencil, are not neccessarily the best criteria to decide whether a storm drain is a good candidate for a contaminant monitoring system, or at least not the only criteria. In order to complete the picture, non-point sources such as emissions from traffic must also be considered. O&HS has determined that 11th Avenue and University Boulevard are the roads on the UAB campus that have the highest traffic, and therefore the most at risk for contamination by traffic emissions. Some other adjacent roads have also been determined to have significant traffic.

Therefore, it would be best to expand the study to a two-phase, or dichotomous, risk assessment. The scale will be expanded from 1-5, and storm drains which are adjacent to high-traffic roads will have their risk factor raised by two, while storm drains which are adjacent to medium-traffic roads will have their risk factor raised by one. The new scale will become:

0: marked storm drain not on a high-or-medium-traffic road.

1: unmarked low point-source risk storm drain not on a high-or-medium-traffic road or marked storm drain on a medium-traffic road.

2: unmarked medium point-source risk storm drain not on a high-or-medium-traffic road, unmarked low point-source risk storm drain on a medium-traffic road, or marked storm drain on a high-traffic road.

3: unmarked high point-source risk storm drain not on a high-or-medium-traffic road, unmarked medium point-source risk storm drain on a medium-traffic road, or unmarked low point-source risk storm drain on a high-traffic road.

4: unmarked high point-source risk storm drain on a medium-traffic road, or unmarked medium point-source risk storm drain on a high-traffic road.

5: unmarked high point-source risk storm drain on a high-traffic road.

Storm drains which receive overall risk values of 5 will be the first to be monitored, followe.d by those with values of 4, and then those with values of 3, either until all 50 sensors have been placed or those of the 50 not meant to monitor "blank" base cases will be used. A "blank" would be one assigned a risk value of 0 or 1 that is used to determine if labeling or proximity to either point or non-point sources really does have an effect on contamination. And once the target drains are determined, the next phase of the project, which involves installing the sensors into the storm drains and beginning to analyze data generated by the contaminant-monitoring system, may proceed.

References Cited:

Hyperspective Studios, 2002. “Storm Drains”, website: http://protectingwater,com/storm-darins.html;

website accessed: October 16, 2012.

Lackey, D.A., A.N. Kajale, R.W. Peters, and J.D. Hagan, 2010. "The Effects of Labeling Storm Drains on the University of Alabama at Birmingham Campus", Poster paper presented at the Alabama Water Resources Conference, Orange Beach, Alabama (September 8-10).

(477e) Innovative Polymer Quantification and Management At a Large Scale Remediation Project

Grieco, S., O'Brien & Gere Engineers, Inc.
Redding, A.,
Lim, N.,
Originally presented on: 11/6/2013 13:26:00 - 13:40:00

Contaminated sediment remediation projects typically require a physical/chemical sediment pore-water treatment plant for metals and organics. In many cases, the sediments are thickened and dewatered for cost effective management and disposal. The polymer used for thickening can be problematic for downstream pore-water treatment. Specifically, this paper reports on a treatment system that experienced excessive "sticky" solids build-up in the clarifier and difficultly removing these solids through normal operations. The polymer also increased backwashing of Granular Activated Carbon (GAC), and decreased adsorption performance of Granular Activated Carbon (GAC).

In order to optimize thickening polymer management, this program identified the threshold of residual polymer that adversely impacted the pore-water treatment plant chemical precipitation & gravity settling unit operation. A mathematical relationship was established between a field-utilized China Clay Procedure and laboratory particle charge density (PSD) to establish an acceptable real-time simple procedure to validate the operating range and upper threshold for polymer dosing.

This program also quantified the impacts of residual polymer on the GAC. Equilibrium isotherms and rapid small scale column tests (RSSCTs) were conducted to confirm that the excess polymer was causing the observations in the full-scale GAC operations. Laboratory testing was used to quantify reduction of the bed volumes processed before both excess pressure drop and benzene breakthrough occurred. The program evaluated chemical types and dosages to dislodge and/or oxidize the excessive polymer that adhered to the carbon in the GAC beds.

The results of this testing were used in the field to optimize polymer dosing, which lowered operating costs and reduced system maintenance. The results also identified the impacts of GAC cleaning (both forward maintenance dosing as well as during backwash) on both pressure drop and adsorption capacity recovery.

(477f) Agile Multiple Water Network Designs for Soluble Contaminant Transport Identification and Site Remediation

Cai, T., Lamar University
Xu, Q., Lamar University
Originally presented on: 11/6/2013 13:40:00 - 13:54:00

Water network design (WND) is widely studied and employed in many industrial processes to minimize freshwater consumption and wastewater generation. Previous studies have been focusing almost exclusively on identifying one optimal WND for a studied process system. Actually, the quest for agile manufacturing by some multi-purpose industrial processes needs a comprehensive WND having the maximum operational flexibility to deal with various production requirements. Thus, the most desirable way for such agile operation is to couple various WNDs into one super design, and then through an effective control strategy to accomplish different production requests by switching among these WNDs. However, this has added the difficulty in contaminant transport identification once the network has been exposure to soluble hazards. In this paper, a novel methodology for the design of such a flexible super WND has been developed. It simultaneously considers the integration of multiple WNDs, network control switch based on different purposes and soluble contaminant transport identification in order to target timely site remediation. The objective of this optimization problem is a super WND coupled with all the recipe-based WNDs; the number of regular and three-way valves needed for the super WND; the optimal design of all of the valve placements in terms of the minimum total valve-switching times (TVST) to swap one pair of WNDs at a time; the control strategy (valve opening/closing strategy) to switch among WNDs in order to identify contaminant transport and reduce/terminate its further impact. The developed systematic methodology is general and thus applicable for many potential network integration problems. Case studies have demonstrated the efficacy of the development and its significance in the context of agile operation.

(477g) Modeling Electrokinetic-Soil Cleaning By Spatial Averaging

Tijaro-Rojas, R.,
Arce, P. E., Tennessee Technological University
Originally presented on: 11/6/2013 13:54:00 - 14:08:00

To predict the transport of ionic species into soils, such as in the acid mine drainages, a model will be implemented. An anisotropic porous medium and application of external forces as electric potential will be used in a multi-scale modeling approach. A minimum of key simplifications will be assumed, in order to capture the fundamental system behavior in model soils that mimic actual ones. Geometry of the control volume, medium properties, and intensity of electric potential constitute important variables to be considered. Since, the mass transport of the ionic species in soils takes place in a domain that displays different scales (molecular, microscopic, macroscopic and mesoscale), a multi-dimensional (spatial) averaging approach will be applied. Once, the spatial-averaged model become available, computational approaches will be used to illustrate the behavior of fundamental aspects. As a complementary part of this research, an experimental approach is being developed to guide and validate the predicted results.

(477h) Dynamics, Source Diagnosis, and Risk Assessment of Persistent Organic Pollutants in Water and Sediments of the Urban Aquatic Bodies

Kumar, R.,

Originally presented on: 11/6/2013 14:08:00 - 14:22:00

Abstract:

Organic compounds which are persistent, posses toxic characteristics, liable to bioaccumulate, prone to long range atmospheric transport and can cause adverse environmental and human health effects at locations near and far from their sources can be defined as persistent organic pollutants (POPs). Polycyclic aromatic hydrocarbons (PAHs) and organochlorine pesticides (OCPs) are the two important categories of POPs which are highly toxic to human beings, animals and plants. Polycyclic aromatic hydrocarbons are the product of any combustion process involving materials containing carbon and hydrogen namely coal, oil, petrol, wood etc. Emissions from the anthropogenic activities predominate, but some PAHs in the environment arise from natural combustion such as forest fires and volcanoes. PAHs occur naturally in crude oil and form a significant component of the petroleum products such as paints, creosote (used in wood preservation) and asphalt (used for road paving). There are some minor biogenic sources like plants, algae/phytoplankton and microorganisms and small amounts are formed by diagenesis in the lake sediments.

Organochlorine pesticides comprise a group of structurally diverse compounds with a large number of chlorine atoms. These pesticides have a tendency to bioaccumulate in the food chain and pose a great threat to the human health and the global environment. In spite of the ban on their use they are still being used in some of the developing countries, especially in the tropical regions, for health, agriculture and preservation purposes. Pesticides that are not bound in soils or taken up by the plants and animals can drain into rivers and lakes and move into the aquatic food chain.

The lakes and the rivers are two important types of surface water systems containing only 0.3% of the total global fresh water which is generally used for drinking purpose. These surface water resources, therefore, are precious and extremely sensitive and once degraded it would take immense efforts to revive them. In view of the above water and sediment samples were collected from two different types of urban surface water bodies of India. One of the water bodies is a lake system of Bhopal (Madhya Pradesh-a central state of India) and the other body is the stretch of river Yamuna at Delhi-the capital city of India. The aquatic systems under investigation are the typical urban aquatic systems and exist in the densely populated areas. Both of the systems are in different geographical locations. These surface water bodies being urban are expected to suffer with some similar types of anthropogenic activities. The various types of industrial, agricultural and domestic activities in the catchment area create different level of anthropogenic stress.

For the extraction of PAHs from water and sediment samples the method followed by Chen et al. [1] was adopted. The mobile phase was a gradient of acetonitrile and degassed water (50% acetonitrile held for 7 min; linear gradient to 80% acetonitrile in 7 to 20 min; 80% acetonitrile held till 25 min and linear gradient to 95% acetonitrile from 25 to 30 min) at a flow rate of 1.5 mL min^{-1}. The analysis was carried out at 254 nm and the individual PAH was identified and quantified by comparing the retention time and peak area with those of the standards. The procedure for extraction and cleanup of OCPs from both water and sediments is the same as outlined by Wan et al. [2]. The mobile phase was acetonitrile and water (70:30) with isocratic mode and a flow rate of 1.3 mL min^{-1}. The analysis of aldrin, endrin, pp'-DDT and α-endosulfan was performed at the wavelength 220, 228, 238 and 214 nm, respectively. The individual pesticide was identified and quantified on the basis of retention time and peak area. PAHs and OCPs were analyzed with HPLC equipped with a UV-detector (Agilent 1100, USA) and recorder. An octadecyl endcapped RP-C_{18} column (4.6 mm×250 mm, 5 μm particle size) was used.

The total PAHs (∑PAH) concentration in the water of the lakes and the river ranged from 4.85 to 18.2 ng/mL and 0.29 to 46.2 ng/mL, respectively. The profile of individual PAH in the water of both the aquatic systems is similar but different in sediments. The total PAHs concentration in sediments ranged from 952 to 5353 ng/g and 649 to 7910 ng/g for the lakes and river, respectively. The maximum pesticide concentrations in water and sediments of the lakes were 4.12 (aldrin), 2.26 (endrin), 6.17 (pp'-DDT), 5.98 ng/mL (α-endosulfan) and 167 (aldrin), 124 (endrin), 513 (pp'-DDT), 349 ng/g (α-endosulfan), respectively. The aldrin and endrin were absent at all the sites of the Yamuna. The maximum concentration of pp'-DDT and α-endosulfan in river water was 3.37 ng/mL and 3.55 ng/mL and in sediments 525 ng/g and 194 ng/g, respectively.

A perusal of the data points out that certain areas of the two aquatic systems are severely contaminated with PAHs and pesticides. The ∑PAH concentrations in water at some points of the lake system are above the lethal concentration and may cause acute toxicity to the exposed organisms. The data in the case of river indicate that both water and sediments are highly contaminated with PAHs. In both the cases the source of PAHs is mainly pyrogenic in nature. The potential toxicity due to total PAH concentrations will be rare in the lake system. The scenario is totally different in the case of river sediments where at most of the sites the toxic effects will be occasional. The apparent reasons for the deterioration of the said water bodies due to PAHs are dumping of untreated waste and a common practice of open burning of a variety of solid waste on the banks of the aquatic bodies. Apart from this, localized activity like plying of diesel operated motor boats is also contributing to the PAH pollution in the lake system. The major contributing activities responsible for PAHs contamination to the river through surface runoff and atmospheric transport are thermal power plants, heavy vehicular traffic load and domestic and industrial waste. A comparison of the two aquatic systems under study indicates that the ∑PAH concentrations in the river are about two times higher than those detected in the lakes. The PAH profiles in the water of the two systems are similar but the profiles in the sediments are different.

As far as the contamination due to pesticides is concerned aldrin is detected only at a few sites of the lake system and the other pesticides are observed only at sites surrounded by agricultural land. Endrin, pp'-DDT and α-endosulfan, wherever detected in water, are generally above the prescribed limit of acute toxicity. Aldrin and endrin are not detected in water and sediments of the river. Both the water and the sediments of the river are heavily contaminated with α-endosulfan and pp'-DDT. The biological impairments due endrin in the lakes and pp'-DDT in the river will be frequent as their levels in the sediments are above PEL. The accumulation of pesticides in the sediments may affect the natural bacterial composition and can lead to anaerobic conditions which may result in more toxic byproducts such as sulfide and ammonia.

References:

[1]. Chen, B., Xiaodong, X., Lizhong, Z., Jing, W., Yanzheng, G., Kun, Y., Xueyou, S. and

Baofeng, L., "Distributions of polycyclic aromatic hydrocarbons in surface waters, sediments and soils of Hangzhou City, China", Water Res. 38, 3558 (2004).

[2]. Wan, M.T., Kuo, J. and Pasternak, J., "Residues of endosulfan and other selected organochlorine pesticides in farm areas of the lower Fraser valley, British Columbia, Canada", J. Environ. Qual. 34, 1186 (2005).

(477i) Mass Transfer and Reaction Kinetics Limitations in Hazardous Waste (Organic and Metal) Treatment

Chawla, R., Howard University
Ighere, J., Howard University
Mlusu, J., Howard University
Farrell, A.,
Originally presented on: 11/6/2013 14:22:00 - 14:36:00

Improper storage and disposal practices have rendered hundreds of sites contaminated with hazardous substances. Some of the commonly found compounds include chlorinated organics such as trichloroethylene (TCE) and metals such as hexavalent chromium, Cr(VI), which occur simultaneously. This paper presents the feasibility of simultaneously treating TCE and Cr (VI), while addressing the problems related to (i) mass transfer limitations in chemical oxidation of TCE in soils, (ii) Effect of TCE oxidation on desired reduction of Cr (VI) to Cr (III), and (iii) removal of Cr(VI) using selected chelating agents.

Results of using alcohol cosolvents and household detergents for enhancing TCE solubility in water, thereby reducing mass transfer resistance in aqueous and soil systems will be presented. Redox reactions were studied using oxidizing agents, e.g. permanganate (MnO_4^-) for the remediation of TCE in the presence of Cr (VI) and using ferrous ions, Fe (II) for reduction of Cr (VI) to Cr(III) in the presence of TCE. In the redox reactions, TCE is oxidized to chloride, hydrogen, and carbon dioxide while permanganate is reduced to manganese dioxide (MnO_2). Similarly, Cr(VI) is reduced to Cr(III) and Fe(II) is oxidized to Fe(III). Batch experiments were performed to determine the reaction kinetics and extent of redox reactions of TCE and permanganate in the presence and absence of hexavalent chromium in aqueous systems. The reaction was determined to be second order overall, first order with respect to both TCE and MnO_4^- in the aqueous system. The TCE-Cr(VI) reaction was determined to be second-order overall, first order with respect to both Cr(VI) and TCE. The TCE-$KMnO_4$-Cr(VI) reaction at stoichiometric molar ratios (1:2:1), based on TCE reactions yielded 100% TCE oxidation. Cr(VI) was found to have a significant positive effect on the oxidation of TCE as it acted as an additional oxidant.

In the chelation study, batch experiments were performed to determine the degree and rate of chelation of Cr(VI) for three chelating agents- disodium dihydrogen ethylenediamine tetraacetate (Na_2H_2EDTA), ethylenediamine tetraacetic acid (H_4EDTA) and Nitrilotriacetic acid (NTA) Kinetic analysis indicated that the order of reaction was 1.32 with respect to Cr(VI) for Na_2H_2EDTA, 1.17 in the case of H_4EDTA and 2.38 for NTA. It was also found that the rate constants for the individual complexation reactions were 6.89×10^{-2} [mg Cr(VI)/g]$^{1-n}$hr^{-1}, 4.98×10^{-2} [mg Cr(VI)/g]$^{1-n}$hr^{-1} and 7.25×10^{-2} [mg Cr(VI)/g]$^{1-n}$hr^{-1} respectively.

(477j) A Simulation Of Weathering Of Colombian Crude Oils In The Colombian Caribbean Sea

Ramírez, J., Universidad Nacional de Colombia – Sede Medellín, Facultad de Minas, Bioprocesos y Flujos reactivos

Merlano, A., Universidad Nacional de Colombia – Sede Medellín, Facultad de Minas, Bioprocesos y Flujos reactivos

Guerrero-Gallego, J., Universidad Nacional de Colombia – Sede Medellín, Facultad de Minas, Grupo de investigación en Oceanografía e Ingeniería Costera - OCEANICOS

Peláez, D.,

Hernandez, J., Universidad Nacional de Colombia – Sede Medellín, Facultad de Minas, Grupo de investigación en Oceanografía e Ingeniería Costera - OCEANICOS

Osorio, A., Universidad Nacional de Colombia – Sede Medellín, Facultad de Minas, Grupo de investigación en Oceanografía e Ingeniería Costera - OCEANICOS

Molina, A., Universidad Nacional de Colombia – Sede Medellín, Facultad de Minas, Bioprocesos y Flujos reactivos

Originally presented on: 11/6/2013 14:36:00 - 14:50:00

A SIMULATION OF WEATHERING OF COLOMBIAN CRUDE OILS IN THE COLOMBIAN CARIBBEAN SEA

Juan Ramírez[1], Aura Merlano[1], Juan Guerrero-Gallego[2], Daniel Peláez[2], Juan Hernández[2], Andrés Osorio[2], Alejandro Molina[1,*]

[1]*Universidad Nacional de Colombia ? Sede Medellín, Facultad de Minas, Bioprocesos y Flujos reactivos, Medellín, Colombia*

[2]*Universidad Nacional de Colombia ? Sede Medellín, Facultad de Minas, Grupo de investigación en Oceanografía e Ingeniería Costera - OCEANICOS, Medellín, Colombia*

[*]*Corresponding author: amolinao@unal.edu.co*

A module, MEUN (*Módulo de Envejecimiento Universidad Nacional*), that describes the processes that occur due to the interaction, also known as weathering, of the crude in an oil spill with the atmosphere and the ocean, was developed. This module, programmed in FORTRAN, couples individual sub-models available in the literature to describe the evaporation, emulsification, dispersion and spreading processes that characterize the first week after an oil spill in the ocean. MEUN predicts as well the variation in density and viscosity over time as a result of the weathering processes.

In MEUN the evaporation submodel considered the pseudocomponent approach proposed by Jones, (1997). Knowledge of the True Boiling Point (TBP) curve and API gravity of the crude oil of interest, allowed the estimation, with Aspen Plus V7.2, of the thermodynamic properties (vapor pressure, mole fraction and mole volume) for each pseudocomponent. The weathering module includes the first-order kinetic expression proposed by Mackay et al., (1980) to model the rate of the emulsification process. The kind of emulsions formed (stable, mesostable, unstable or entrained water), the maximum water content of the emulsion and the evaporative thresholds to have a change in the type of the emulsion formed were interpolated from the experimental work developed by Fingas and Fieldhouse, (2004). MEUN couples the dispersion expression, developed by Delvigne and Sweeney, (1988), that takes into account the significant wave height and wave period to predict the amount of crude oil dispersed in the water column by the wave action. To predict the spreading rate of the oil slick, MEUN uses the equation developed by Lehr et al., (1984) based on observations in controlled oil spills. This expression considers the imbalance of forces acting on the oil slick and the effect of wind velocity to predict the spreading rate. The models developed by Lehr et al., (2002) and by Mackay et al., (1980), included in MEUN, predict the increase in density and viscosity, respectively, that results from the evaporation and emulsification processes.

All sub-model's constants were adapted to the specific requirements of Colombian crudes, particularly to Cusiana (°API 44.1) and Vasconia (°API 24.5). While the first one represents light crudes, the second one is an example of heavy oil. These two crudes have high production and require marine transport. The simulations results of the atmospheric model WRF (Weather Research and Forecasting Model), the wave model SWAN (Simulating WAve Nearshore) and the ocean model ROMS (Regional Ocean Modeling System) gave a typical ocean and atmospheric data of *Golfo de Morrosquillo* (latitude 09°29' N and longitude 75°46' W). Two scenarios, for high (4 m/s - 12 m/s) and low (0 m/s - 4 m/s) wind velocity, and the effect of time-averaging wind velocity, significant wave height and sea surface temperature were studied.

As an example of MEUN's predictions, Figure 1 shows the variation of the evaporated volume fraction with time after the spill for Cusiana and Vasconia oils. As expected, the lighter oil yields a higher evaporation fraction. Figure 1 also suggests that, while an average wind (discontinuous lines in Figure 1) yields reasonable predictions for a high wind condition, it significantly overestimate evaporation when the wind velocity (continuous lines in Figure 1) is low.

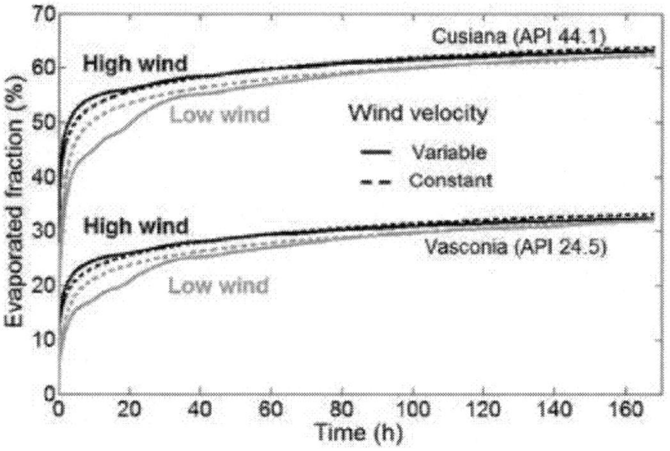

Figure 1. Predicted variation of the evaporated fraction for Cusiana and Vasconia crude oils after a spill in the Caribbean Sea under high and low wind velocity scenarios and variable and constant wind velocity.

Figure 2 summarizes MEUN's predictions as a plot of the volume fraction evaporated, dispersed and remaining on the sea surface. Because of differences in the physicochemical properties of the Colombian crude oils, Cusiana is more affected by evaporation and dispersion than Vasconia. According to MEUN, under high-wind conditions, a combination of evaporation and dispersion causes in Cusiana a decrease in the fraction of crude oil remaining on the ocean surface that is more than twice the decrease for Vasconia crude oil. While the results in Figure 1 and 2, as well as the fact that MEUN predicts an increase in viscosity of two to three orders of magnitude as the slick evaporates and the water makes an emulsion with the oil, agree with previous observations of weathering of oils in the ocean, a more complete validation of MEUN with experiments is currently under way using a wind tunnel and a rotating cylinder for evaporation and emulsion, respectively, at conditions typical of the Colombian Caribbean Sea.

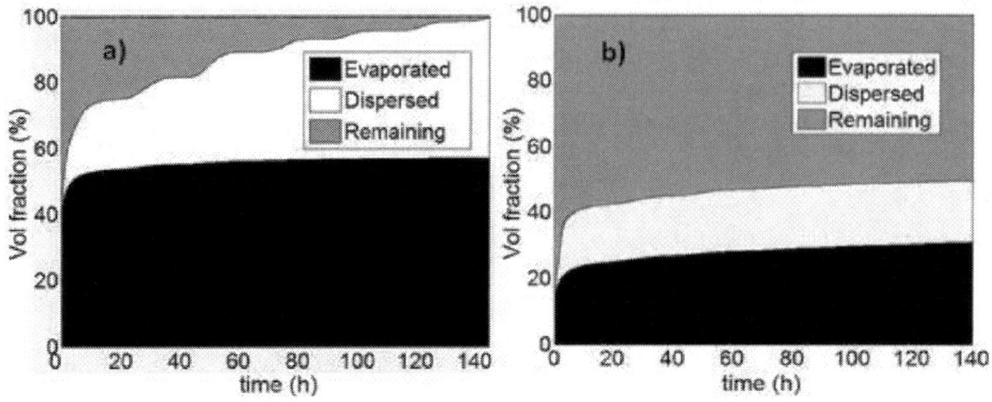

Figure 2. Distribution of the spilled crude oil between evaporated, dispersed and remaining on the ocean fractions for a) Cusiana and b) Vasconia oils. Predictions according to *MEUN*.

REFERENCES

Delvigne, G. A. L., & Sweeney, C. E. (1988). Natural dispersion of oil. *Oil and Chemical Pollution*, *4*(4), 281?310.

Fingas, M., & Fieldhouse, B. (2004). Formation of water-in-oil emulsions and application to oil spill modelling. *Journal of hazardous materials*, *107*(1-2), 37?50.

Jones, R. K. (1997). A simplified pseudocomponent oil evaporation model. *20th Arctic Marine Oilspill Program Technical Seminar.* (pp. 43?61). Ottawa, Ontario: Environment Canada.

Lehr, W. J., Fraga, R. J., Belen, M. S., & Cekirge, H. M. (1984). A new technique to estimate initial spill size using a modified fay-type spreading formula. *Marine Pollution Bulletin*, *15*(9), 326?329.

Lehr, W., Jones, R., Evans, M., Simecek-Beatty, D., & Overstreet, R. (2002). Revisions of the ADIOS oil spill model. *Environmental Modelling & Software*, *17*(2), 189?197.

Mackay, D., Buist, I., Mascarenhas, R., & Patterson, S. (1980). Oil Spill Processes and Models. Report No. EE/8. *Environment Canada* (p. 192). Ottawa, Ontario.

(477k) Increasing Strength of Thickened Oil Sand Tailings with Geopolymers

Xu, Y., Natural Resources Canada
Karimi, S.,
Originally presented on: 11/6/2013 14:50:00 - 15:04:00

The government of Alberta recently imposed new regulations requiring that oil sand tailings deposits must reach strength of 5 kPa one year after deposition. However, due to the limitations of gravitational settling and thickener operation, fine oil sand tailings can be thickened only to certain degree, and the strength may not meet the regulatory requirement. CanmetENERGY and Total E&P Canada (TEPCA) have conducted an investigation on the feasibility of using geopolymer additives to improve the strength of the deposit.

In this project, fly ash (type C) and cement were used as additives. Mature fine tailings (MFT) containing 35 wt% solids and thickened tailings (TT) with solids contents of 50 wt% were used as tailings material. Geopolymer was mixed with tailings and the treated tailings were distributed into 250-mL jars, some of which were closed, while others were left open. The solids content in the closed jars remained constant because no water evaporated, but increased in the open jars as the tailings dried in air at ambient conditions. The tailings strength was measured in each jar after a specified time.

These experiment results indicate that the addition of small amount of geopolymer increased the tailings strength significantly in both open and closed jars under the test conditions. Addition of about 5 wt% fly ash to TT increased the yield stress of the tailings to over 5 kPa after the solids content increased to around 60 wt% under natural drying. Similarly, adding around 5 wt% fly ash to the MFT increased the yield stress to over 5 kPa when the tailings were dried to about 55 wt% solids. Adding 3 wt% Portland cement also increased the yield stress of MFT or TT to more than 5 kPa when the solids concentration increased to 55 wt % for MFT and 65 wt% for TT. By comparison, untreated MFT or TT had to be dried to over 70 wt% and 85 wt%, respectively, in order for the yield stress to reach 5 kPa.

(484a) Selective Adsorption of CO2 On Flexible Ni/Dbm-Bpy MOF

Matranga, C., National Energy Technology Laboratory
Culp, J., National Energy Technology Laboratory
Shi, F., NETL
Originally presented on: 11/6/2013 12:30:00 - 12:50:00

Selective Adsorption of CO_2 on Flexible NiDBM-BPY MOF

Fan Shi[1,2,*],Â Jeffrey Culp[1,2] and Christopher Matranga[1],

(1)U.S. Department of Energy, National Energy Technology Laboratory, Pittsburgh, PA, USA (2)URS Corporation, South Park, PA, USA

* fan.shi@netl.doe.gov

To maintain fossil fuels as viable energy source, carbon capture and storage technology is essential. Recent novel sorbents for CO_2 capture [1-3], including Metal Organic Frameworks (MOFs), have attracted a lot of interests. Flexible MOFs have one attractive feature that their structure become dynamic only in the presence of suitable guests such as CO_2, CH_4, N_2, and H_2. These flexible materials can show interesting host-guest phenomena including crystal-crystal transformations and gated adsorption behaviors due to the structural transitions which occur between the guest free and guest-loaded states.[4-6] Such dynamic host-guest behavior is particularly attractive for applications involving sorbent-based gas separations and selective adsorption such as CO_2 and CH_4 separations.[7-9] In order to develop materials for these applications, a detailed understanding of the adsorption mechanisms of these flexible hosts is needed.

The material in this study is the linear chain coordination polymer catena-bis (dibenzoylmethanato)-(4,4′-bipyridyl)nickel(II), or referred as ?NiDBM-BPY?. This neutral one-dimensional polymer was propagated by a bridging 4,4′-bipyridine ligand bound to the z-axes of the nickel complexes which were bischelated in a planar fashion by two anionic DBM ligands. CO_2 breakthrough from binary mixtures, including CO_2/N_2, CO_2/H_2, and CO_2/CH_4, was performed in a fix-bed reactor with ID of 0.375? and a length of 12?. The average particle size of NiDBM-Bpy was around 100 mm to minimize external mass transfer limitations. The gas composition of effluent was analyzed by a *Mass Spectrometry* (*Pfeiffer Vacuum*). The breakthrough curve of CO_2 from a fixed-bed reactor packed with NiDBM-Bpy showed a step-shape curve, which also indicated the structure expansion above the structure transition threshold pressure. The selectivity for CO_2

over N_2, H_2, and CH_4 was calculated and results confirmed the sorbent's affiliation to CO_2 than other gases.

References

1. Kauffman et. al. *Angew. Chem. Int. Ed.* **2011**, *50, 10888 ?10892*

2. Millward, A. R.; Yaghi, O. M. *J. Am. Chem. Soc.* **2005**, *127*, 17998.

3. Wang, B.; Cote, A. P.; Furukawa, H.; O'Keeffe, M.; Yaghi, O. M. *Nature* **2008**, *453*, 207.

4. Culp, J. T.; Smith, M. R.; Bittner, E.; Bockrath, B. J. Am. Chem. Soc. **2008**, 130, 12427.

5. Takamizawa, S.; Nakata, E.; Yokoyama, H. Inorg. Chem. Commun. **2003**, 6, 763.

6. Uemura, K.; Kitagawa, S.; Fukui, K.; Saito, K. J. Am. Chem. Soc. **2004**, 126, 3817.

7. Maji, T. K.; Matsuda, R.; Kitagawa, S. Nat. Mater. **2007**, 6, 142.

8. Shimomura, S.; Horike, S.; Matsuda, R.; Kitagawa, S. J. Am. Chem.Soc. **2007**, 129, 10990.

9. Snurr, R. Q.; Hupp, J. T.; Nguyen, S. T. AIChE J. **2004**, 50, 1090.

(484b) Amine Impregnated N-Doped Mesoporous Carbon for Selective CO2 Capture

To, J., Stanford University
He, J., Stanford University
Gary, B., Stanford University
Stack, D., Stanford University
Wilcox, J., Stanford University
Bao, Z., Stanford University
Originally presented on: 11/6/2013 12:50:00 - 13:10:00

Carbon dioxide is a major component of the greenhouse gases. It is expected that the atmospheric CO_2 level will continue to increase over the current concentration of approximately 390 ppm in the near future, as fossil fuels remain the major source utilized to meet global energy demand. Carbon capture and storage is a technology developed to prevent release of large quantities of CO_2 into the atmosphere from power generation by capturing CO_2 and securely storing it away from the atmosphere. However, current state-of-the-art technology for CO_2capture is to use liquid phase amine scrubbing, which is corrosive and energy intensive due to high heating capacity of water during solvent regeneration. On the other hand, solid sorbent material adsorbs CO2 via physisorption, which requires less energy for regeneration. Therefore, porous carbon with high thermal conductivity is desirable for such application. With improved kinetics in adsorption and desorption processes, separation column required is smaller and the overall cost of the power plant is lower.

Mesoporous carbons are promising for CO_2 capture due to chemical inertness, low cost, high surface area and tunable pore structures. Its porous structure allows addition of chemical functionality by grafting or impregnation. Amine chemistry tells us that nitrogen functionalization plays an important role in surface chemistry to achieve high CO_2 adsorption capacity. We report here an ordered mesoporous carbon made using co-assembly of modified-pyrrole and triblock copolymer. To increase the nitrogen content of the adsorbent, nitrogen is incorporated into porous carbon framework and further functionalized with amines. Another potential benefit is that the thermal conductivity of mesoporous carbon is much higher than its silica counterpart, resulting in a faster regeneration step with enhanced stability using temperature swing process. These overall properties of mesoporous carbon made from conducting polymer made it a desirable material for CO2 capture.

(484c) Effect of Alumina Doping On Structure and Stability of CaO Based Sorbents for High Temperature CO2 Capture

Gunugunuri, K. R., University of Cincinnati
Qullin, S., University of Cincinnati
Smirniotis, P., University of Cincinnati
Originally presented on: 11/6/2013 13:10:00 - 13:30:00

In an age where environmental concerns have become increasingly prevalent, efficient and carbon neutral methods for satisfying the global energy demand have become necessities. Increasing levels of greenhouse gases are widely regarded as a key factor in the shifts observed in global climate patterns. In this study, we reported that Flame synthesized doped Calcium-based novel refractory sorbents have shown excellent performance in CO_2 adsorption. In our previous studies we incorporated several metals (Hf, W, Y, La, and Al) into calcium oxide lattice to improve stability of the sorbent during the CO_2 adsorption experiments. Among these dopants, Alumina doped CaO with molar ratio of 10:3 showed very high adsorption stability during 100 cycles of operation and better adsorption capacity than Ca/Zr sorbents (our previous best system). We reported that the formation of perovskite type of structure in CaO-Al sorbent is responsible for the observed stability. The present study mainly focuses on the Ca-Al system and obtains the optimum combination during extended carbonation-decarbonation operation cycles in the presence of highly concentrated CO_2. Various molar ratios of Ca: Al were synthesized using flame spray pyrolysis technique. The sorbents were characterized by BET, XRD, SEM and TEM techniques and analyzed using TGA. The optimal sorbent were tested under severe condition of high temperature. These interesting findings will be discussed in the presentation.

(484d) Enhanced Performance of Polymeric Amines Supported On Heteroatom-Incorporated Silica in CO2 Adsorption/Desorption

Kuwahara, Y., Georgia Institute of Technology
Copeland, J. R., Georgia Institute of Technology
Sievers, C., Georgia Institute of Technology
Kamegawa, T., Osaka University
Yamashita, H.,
Jones, C. W., Georgia Institute of Technology
Originally presented on: 11/6/2013 13:30:00 - 13:50:00

Silica-aminopolymer composites are promising materials that can be used to adsorb CO_2 from dilute gases, with the most commonly studied compositions being simulated flue gas from power plants (ca. 5-15% CO_2) and ambient air (ca. 400 ppm CO_2). The CO_2 adsorption characteristics of prototypical, class 1 [1] poly(ethyleneimine)-silica composite adsorbents are shown here to be significantly enhanced by altering the acid/base properties of the silica support via heteroatom incorporation into the silica matrix. [2] To demonstrate this, a series of poly(ethyleneimine)-impregnated mesoporous silica SBA-15 materials containing heteroatoms (Al, Ti, Zr, and Ce) in the silica matrix are prepared and characterized in adsorption experiments under conditions simulating flue gas (10% CO_2 in Ar) and ambient air (400 ppm CO_2 in Ar). The structure of the composite adsorbents, including local information concerning the state of the incorporated heteroatoms and the overall surface properties of the silicate supports, are investigated to develop a relationship between the adsorbent structure and CO_2 adsorption/desorption performance. The CO_2 adsorption/desorption kinetics are assessed by thermogravimetric analysis and *in-situ* FT-IR measurements. These combined results, coupled with data on adsorbent regenerability, demonstrate a stabilizing effect of the heteroatoms on the poly(ethyleneimine) contained within the support mesopores, enhancing adsorbent capacity, adsorption kinetics, regenerability, and stability of the supported aminopolymers over continued cycling. [3] It is suggested that the CO_2 adsorption performance of silica-aminopolymer composites may be further enhanced in future work by more precisely tuning the acid/base properties of the support, an approach that has not yet been widely exploited in the literature.

References

[1] P. Bollini, S. A. Didas, C. W. Jones, J. Mater. Chem. 2011, 21, 15100.

[2] Y. Kuwahara, D.-Y. Kang, J. R. Copeland, N. A. Brunelli, S. A. Didas, P. Bollini, C. Sievers, T. Kamegawa, H. Yamashita, C. W. Jones, J. Am. Chem. Soc. 2012, 134, 10757.

[3] Y. Kuwahara, D.-Y. Kang, J. R. Copeland, N. A. Brunelli, S. A. Didas, P. Bollini, C. Sievers, T. Kamegawa, H. Yamashita, C. W. Jones, Chem. Eur. J. 2012, 18, 16649.

(484e) Hydrophobic Polymeric Solvents for the Selective Absorption of CO2 From Warm Gas Streams That Also Contain H2 and H2O

Enick, R. M., University of Pittsburgh
Luebke, D., National Energy Technology Laboratory
Nulwala, H., National Energy Technology Laboratory
Warman, S., University of Pittsburgh
Stevenson, C., University of Pittsburgh
Koronaios, P.,
Originally presented on: 11/6/2013 13:50:00 - 14:10:00

Hydrophobic Polymeric Solvents for the Selective Absorption of CO_2 from Warm Gas Streams that also Contain H_2 and H_2O

Robert Enick, David Luebke, Hunaid Nulwala, Samantha Warman, Craig Stevenson, Peter Koronaios

Hydrophobic polymers may provide an alternative to physical solvents based on the hydrophilic solvent polyethyleneglycol dimethylether (PEGDME) for the pre-combustion capture of CO_2 from a warm, high pressure, post-WGS reactor gas stream that also contains H_2O and H_2. The candidate hydrophobic polymers are designed to absorb as much CO_2 as possible, and as little H_2 and water as possible. There are three promising candidates: polydimethyl siloxane (PDMS or silicone oil), PDMS-PEGDME hydrid solvent developed by our team, and a commercial PDMS/Fe compound with the addition of iron as an anti-oxidant to enhance its high-temperature performance. PDMS is commercially available over an extremely wide MW range (up to 500,000+). The PDMS-PEGDME hybrid solvent is also easy to synthesize from commercially available starting materials. Two silicone oils containing both methyl and phenyl groups were also tested and found to be much less promising candidates.

In this work, the solubility of hydrogen, CO_2 and water in PEGDME, PDMS, PDMS-PEGDME and PDMS/Fe is presented for pressures up to ~2000 psi and temperatures to 120°C. The results indicate that PDMS is completely hydrophobic over broad ranges of temperature and pressure, and PEGDME is completely miscible with water in all proportions. PDMS-PEGDME and PDMS/Fe are hydrophobic at ambient temperature but the hydrophobicity of both have not yet been measured over a broad temperature range. These three hydrophobic solvents absorb comparable amounts of CO_2 (on a wt% basis). The solubility of H_2 in the solvents at ~25°C is roughly three orders of magnitude less than the solubility of CO_2, with PEGDME absorbing the least H_2 and PDMS absorbing the most H_2 (about 2.5 times more than PPGDME). Unfortunately, because the solubility of CO_2 decreases with increasing temperature as the solubility of H_2 increases, the CO_2-selectivity of the solvents (as estimated by the ratio of CO_2 solubility:H_2 solubility at the same conditions) decreases with increasing temperature.

(484f) Synthesis and Characteristics of Amine Functionalized Biopolymers for CO2 Capture

Fujiki, J., Research Institute of Innovative Technology for the Earth
Yogo, K., Research Institute of Innovative Technology for the Earth
Originally presented on: 11/6/2013 14:10:00 - 14:30:00

The increase in the atmospheric concentration of CO_2 is considered as one of the major causes of global warming. Separation and recovery of CO_2 from large CO_2 emission point source have been investigated. Recently, there are many reports on amine loaded mesoporous silicas having high CO_2 adsorption capacity. However, toxic and expensive chemicals are used to synthesize mesoporous silicas. Additionally, most reports on amine loaded mesoporous silicas discuss the CO_2 adsorption characteristics for "powder" form mesoporous silicas. Mesoporous silicas seem to be less reliable now because the adsorbent must be in the form of bead, granule or pellet to use in actual CO_2recovery processes. Therefore, alternative adsorbent "beads" made from low-cost and eco-friendly chemicals should be developed.

In this study, highly porous amine functionalized biopolymer beads were synthesized and CO_2 adsorption performance was evaluated. As a result, it was found that the polyethyleneimine functionalized beads had high CO_2 adsorption capacity with high adsorption rate. The CO_2 adsorption capacity reached 2.3 mmol/g in the absence of water vapor and 3.6 mmol/g in the presence of water vapor at 313 K and 15 kPa of CO_2.

In general, amine impregnated materials show high CO_2 adsorption capacities but lower adosorption rate. On the other hand, amine grafted materials have the advantage of physical and chemical stability of loaded amines, high amine efficiency and adsorption-desorption rate. The adsorption characteristics of our material show advantages of both amine impregnated and grafted materials, i.e., relatively high adsorption capacity and fast adsorption rate. The results suggest that this material has potential for effective CO_2 adsorbent.

(484g) Removal of Sulfur Compounds in Methane Or Nitrogen By Nano-Sized Magnesium Oxide and Its Composite

Lee, C. H., Yonsei University
Kim, Y. H., Yonsei University
Kim, J. J.,
Tuan, V. A.,
Originally presented on: 11/6/2013 14:30:00 - 14:50:00

Removal of Sulfur Compounds in Methane or Nitrogen by Nano-sized Magnesium Oxide and Its Composite

Yo Han Kim, Vu Anh Tuan, Jae Jeong Kim, Chang Ha Lee*

Department of Chemical and Biomolecular Engineering, Yonsei University, Seoul, Korea

*E-mail: leech@yonsei.ac.kr, Tel: +82-2-2123-2762

Metal oxide particles are considered as reactive adsorbents for the decontamination of toxic materials. The important feature is to enhance the surface chemical reactivity toward incoming adsorbates. It is expected that the adsorption efficiency of nano-sized MgO is greater than commercially common used MgO.

Nano-sized magnesium oxides (MgOs) and MgO-SiO_2 composite were synthesized by polyol-meditation thermolysis, hydrothermal, and aerogel methods to remove sulfur compounds from methyl mercaptan/methane and CEES/nitrogen mixtures. And the difference in sorption capacity among the synthesized MgOs and between dry and wet gas conditions was compared by using dynamic adsorption method at different temperatures. In addition, the effects of calcination condition on adsorption capacity by calcination condition were investigated. The synthesized MgOs were characterized by XRD, BET, TGA, and TEM.

Keywords: Magnesium oxide, Composite, Desulfurization, Sol-gel

Reference

[1] S. Utamapanya, K.J. Klabunde, and J.R. Schlup, “Nanoscale metal oxide particles/clusters as chemical reagents. Synthesis and properties of ultrahigh surface area magnesium hydroxide and magnesium oxide,” Chem. Mater., 3 (1991), 175-181,

[2] S.H. Lim, E.J. Woo, H.J. Lee, C.H. Lee, “Synthesis of magnetite–mesoporous silica composites as adsorbents for desulfurization from natural gas,” Applied Catalysis B: Environmental, 85 (2008) 71–76.

(485a) Multimedia Environmental Distribution of Nanomaterials

Liu, H. H., University of California, Los Angeles
Cohen, Y., University of California, Los Angeles
Originally presented on: 11/6/2013 12:30:00 - 12:49:00

Nanotechnology has brought about many benefits and improvements to commercial and industrial applications and products, but its potential environmental impact must also be systematically assessed. In this regard, it is essential to assess the potential exposure levels (i.e. concentrations, exposure time, etc.) of ecological receptors to engineered nanomaterials (eNMs) in order to manage risks due to possible releases (chronic and/or accidental) of eNMs to the environment. Although a number of models have been developed to predict the multimedia environmental distribution of organics chemical contaminants, such models are not directly applicable to eNMs. For example, the concentration driving forces for transport of chemicals across environmental phase boundaries is typically constrained by equilibrium partitioning; in contrast, the transport of particulate matter is primarily driven by physical processes of driving forces that are unconstrained by equilibrium partitioning. In an effort to quantify the distribution of ENMs in the environment, "substance flow analysis" estimations have been employed in recent years as part of life cycle analysis (LCA) to assess the potential environmental impact of ENMs. Such approaches, however, are not a priori predictive and generally do not consider mechanistic descriptions of fundamental transport processes. Accordingly, in the present study, a generalized model of the multimedia environmental distribution of nanomaterials (MendNano) was developed to estimate the environmental distribution of nanomaterials subject to various release scenarios. In this model, the multimedia environment is considered as a collection of environmental compartments and subcompartments, with each compartment treated as either well-mixed (e.g. atmosphere, water), or non-uniform compartments (e.g. soil, sediment). Fundamental intermedia transport processes (e.g. dry deposition, rain scavenging, resuspension, runoff) link various compartments while considering the particle size distribution (PSD) of eNMs in both the atmospheric and aquatic compartments. In order to enable rapid "what-if?" scenario analyses, a web-based graphical user interface (GUI) was developed for the model as part of the MendNano cloud-based modeling system. The user defines the modeling scenario by specifying relevant properties of the eNM under consideration, geographical and meteorological parameters (i.e., size of regions, temperature, wind speed, rain rates, etc.), transport process parameters and source emissions (i.e., release rates, etc.). The time-evolution of eNM concentrations and mass in each environmental compartments are then calculated by solving the coupled series of differential equations and the results are displayed within the GUI. Simulations with MendNano for different geographical areas in the U.S. have shown that, when ENMs are released primarily to air the majority of their mass in the environment is accumulated in the soil and in certain cases in the soil compartment. Simulations confirmed that increased attachment efficiency of eNM to suspended solids resulted in significant increased concentration in the sediment compartment, and a modest concentration decreases for the water compartment. Additionally, for eNM with measurable solubility (e.g. Ag, ZnO), dissolution of the eNM had a significant impact in decreasing their accumulations in the water and sediment compartments. Lastly, risk indices were computed based on eNM concentrations and information from high throughput screening (HTS) of cellular toxicity studies, suggesting that ZnO, Fe_3O_4, and TiO_2 ranked high among 6 ENMs considered in the study. It is envisioned that the present multimedia analysis tool can assist regulators, industrial, and researchers in rapidly assessing the potential environmental implications of ENMs releases to the environment.

(485b) Origin, Cure and Control of Nanosilver Toxicity

Sotiriou, G. A., Harvard University
Fujiwara, K.,
Pratsinis, S. E., ETH Zurich

Originally presented on: 11/6/2013 12:49:00 - 13:08:00

Nanosilver is considered the most commonly used engineered nanomaterial, for example, in antibacterial textiles, polymer films for food packaging, paints and pigments, filters for water or air treatment to name a few applications (1). Concerns however about released Ag^+ ions during dispersion of nanosilver in liquids have limited its broad use. Here nanosilver supported on nanostructured silica is made with closely controlled Ag size both by dry (flame aerosol) and wet chemistry (impregnation) processes without, however, any surface functionalization that could interfere with its ion release (2). The dispersion method of nanosilver in solutions affects its dissolution rate but not the final Ag^+ ion concentration. By systematically comparing nanosilver size distributions to their equilibrium Ag^+ ion concentrations, it is revealed, that the latter correspond precisely to dissolution of one to two surface silver oxide monolayers, depending on particle diameter. When, however, nanosilver is selectively conditioned by either washing or H_2 reduction, the oxide layers are removed, drastically minimizing Ag^+ ion leaching and its antibacterial activity against *E. coli*. That way the bactericidal activity of nanosilver is confined to contact with its surface rather than to rampant ions. This leads to silver nanoparticles with antibacterial properties that are essential for medical tools and hospital applications. Complete cure of nano-Ag toxicity is achieved by its hermetic coating by nanothin inert silica films.

1. G.A. Sotiriou, S.E. Pratsinis, "Engineering nanosilver as an antibacterial, biosensor and bioimaging material", *Curr. Opinion Chem. Eng.* **1**, 3-10 (2011)

2. G.A. Sotiriou, A. Meyer, J.T.N. Knijnenburg S. Panke, S.E. Pratsinis, "Quantifying the origin of nanosilver ions and their antibacterial activity", *Langmuir*, **28**, 15929–15936 (2012).

(485c) Bacterial Colonization of Surfaces Displaying Adhered Silver Nanoparticles

Wirth, S. M., Carnegie Mellon University
Lowry, G. V., Carnegie Mellon University
Tilton, R. D., Carnegie Mellon University
Originally presented on: 11/6/2013 13:08:00 - 13:27:00

Silver nanoparticles (AgNPs) are among the most commonly employed engineered nanoparticles in consumer products, where the broad spectrum biocidal properties of silver are exploited to endow products with a measure of antimicrobial performance. Silver ions have well established microbial toxicity. In some cases AgNPs are embedded in a matrix; in others they are deposited on a surface. In either case, AgNPs provide a source of dissolved silver ions. The possibility exists as well for toxic consequences of direct microbe contact with the AgNPs, although this appears to be rather context-dependent. Often, materials that display AgNPs are envisioned as a means to inhibit biofouling or to enhance sterility. This can be particularly attractive in medical or food processing situations. AgNP leaching from consumer products under normal usage provides a pathway to environmental exposure, mainly via wastewater treatment plants. Concern over the possibly detrimental effects of engineered AgNPs in soil and aquatic ecosystems has motivated active research into the potential toxicity of these nanomaterials to microbes, small organisms and plants, as well as the chemical and physical transformation processes and transport processes that ultimately dictate the potential for broad environmental impact of the introduced nanoparticles.

While AgNPs can be highly toxic to planktonic bacterial cultures, recent work indicates that established bacterial biofilms, which represent the vast majority of naturally occurring populations, are considerably more tolerant to the introduction of AgNPs and the dissolved silver they release. Instead of AgNP effects on established biofilms, this presentation focuses on the ability of planktonic bacterial populations to colonize a surface that has been coated by adhesion of AgNPs from aqueous suspension before the introduction of bacteria to the system. Silver nanoparticles in this study were stabilized by poly(vinylpyrrolidone). Biofilm growth and viability were monitored on silica glass substrates, with or without adherent AgNPs, that were immersed in agitated suspensions of *Pseudomonas fluorescens*. The critical parameter that dictates whether or not the AgNP-decorated surface inhibits biofilm formation is the ratio of silver to planktonic biomass, or equivalently, the total cell population for a fixed amount of surface-adhered AgNPs. Biofilm formation is only prevented when the initial planktonic bacteria population is sufficiently low to suffer 100% lethality. Higher bacterial populations may initially suffer substantial loss of viability, but if this is less than 100%, the population recovers and establishes a thriving biofilm on the AgNP-decorated surface. An important aspect of this phenomenon is the binding of silver ions to suspended bacteria, which makes it unavailable to other bacteria. In this way, bacteria populations can overwhelm the release of toxic silver ions from AgNPs by sequestering silver ion in a non-bioavailable form. This enables the recovery from initial viability losses when the population grows at a faster rate than silver ion can be replenished in solution. Having sequestered most of the dissolved silver ions, the bacteria then readily colonize the AgNP-decorated surface. A particularly noteworthy effect is the preferential colonization of bacteria at early times on surface regions that have abundant AgNPs compared to AgNP-free regions. After several days, AgNP-rich and AgNP-free surface regions are indistinguishable in terms of biofilm biomass and viability. The AgNPs themselves provide no direct deterrence to biofilm formation. The system is dominated by dissolved silver toxicity, which can be readily overcome at high biomass:silver ratios. This work implies that even AgNPs that have managed to avoid any of the expected chemical transformations that are known to decrease silver toxicity, a "worst case scenario", may have little detrimental effect on bacterial biofilms in the environment.

(485d) Carbon Nanotubes Induced Cellular Biomeahanic Changes Is Depended On Treatment Time

Dong, C., West Virginia University
Eldawud, R., West Virginia University
Kashon, M. L., National Institute for Occupation Safety and Health
Lowry, D., National Institute for Occupation Safety and Health
Sargent, L. M., National Institute for Occupation Safety and Health
Dinu, C. Z., West Virginia University
Originally presented on: 11/6/2013 13:27:00 - 13:46:00

Carbon nanotubes (CNTs) are applied for a variety of applications from nanocircuits, to hydrogen storage devices, and from opticals fibers to conductive plastics. Recently, their functionalization with biomolecules offer opportunities for exciting biological and biomedical applications in drug delivery or bioimaging. However, because of their interactions with biological systems and their ability to translocate and persist into the circulatory and lymphatic systems and tissues, concerns about CNTs intrinsic toxicity have risen. It is thus necessary to develop significant and sensitive advances analysis technologies for real time investigation of CNTs toxicity. Our research aims at elucidating the mechanisms involved in CNTs internalization and association with cellular systems by evaluating the overall effects on of CNTs on cellular biophysical properties-biomechanics. To unravel the biomolecular mechanisms of such interactions, we used Fluorescence Activated Cell Sorting (FACS) and Atomic Force Microscopy (AFM). Our results suggest that cellular exposure to CNTs leads to higher stiffness and distinctive elastic modulus distribution for human lung epithelial cells, especially in cell nucleus region. Further, those changes are depended on CNTs uptake and cellular translocation time. Such findings provide necessary data for understanding the complex mechanotransduction processes within the cells interacted to nanomaterials which may be functionally linked to CNTs induced cytotoxicity and/or genotoxicity, as well as their potential for cancer progress.

(485e) A Simple Kinetic Model Describing Nanoparticle Interactions With a Tethered Lipid Bilayer

Liu, Y., Michigan State University
Worden, R. M., Michigan State University
Originally presented on: 11/6/2013 13:46:00 - 14:05:00

In this paper, we present a simple mechanistic kinetic model to describe interactions of engineered nanomaterials (ENM) with a tethered bilayer lipid membrane (tBLM). The model was developed to help elucidate molecular processes responsible for time-dependent changes in tBLM resistance (R_m) following ENM exposure. The model uses two kinetic constants to describe the rates of ENM binding to, and lipid removal from, the tBLM. The model was fit to R_m vs. time data for interaction of a tBLM formed using 1,2-dioleoyl-sn-glycero-3-phosphocholine (DOPC) with two types of ENM: functionalized polypropargyl glycolide nanoparticles and functionalized polystyrene nanoparticles. The model was able to reproduce diverse trends in the R_m vs. time data, including a continuous decrease in R_m and an initial increase in R_m followed by a decrease. The model also predicted that interaction of only a few nanoparticles could generate measurable changes in the tBLM's R_m value. Hierarchical clustering was applied to pairs of optimized kinetic constants for various ENM. The resulting dendrograms indicated that ENM having different properties (composition, size, surface charge) could be statistically distinguished using this approach. Results of this study provide insight into fundamental mechanisms by which ENM interact with biomembranes and may lead to improved methods to rapidly screen ENM libraries for desired functional and biosafety profiles.

Keywords: tBLM; engineered nanomaterials; model; kinetic constant; hierarchical clustering; lipid bilayer; resistance

(485f) Effect of Protein Corona On Nanoparticle Cellular Uptake

Nangia, S., Syracuse University
Jiang, W.,
Kamani, H.,
Originally presented on: 11/6/2013 14:05:00 - 14:24:00

Understanding the interface of nanomaterials and the environment is becoming increasingly important with the advances in biomedical applications of the nanoparticles. In the biological environment nanoparticles get exposed to serum proteins resulting in the formation of protein corona. Using chemically faithful coarse grained molecular dynamics simulations, the adsorption of specific serum proteins on a variety of nanoparticles have been studied to provide chemical insight into the into the dynamical equilibrium process. The role of physiochemical properties of nanoparticle such as size, shape, and surface charge on the corona formation has been investigated systematically. Furthermore, the dynamics of the nanoparticle-corona complex with the cell membrane is simulated to determine the mechanism of their cellular uptake. For comparison, we will provide estimates of the mechanistic and kinetic data for the cellular uptake of a various nanoparticle shapes both with and without protein corona. The impact of the corona on cell-adhesion and loss of cell specificity is evaluated.

(485g) Lessons From Nature: Analogies Between Protein and Nanoparticle Interactions With Lipid Membranes

Bothun, G., University of Rhode Island
Bobba, C., University of Rhode Island
Xi, A., University of Rhode Island
Kitchens, C. L., Clemson University
Originally presented on: 11/6/2013 14:24:00 - 14:43:00

Nanoparticle-cell membrane interactions play an important role in nanotoxicology and nanomedicine. Changes in membrane structure, phase behavior, and permeability due to nanoparticle binding are dependent upon nanoparticle size and surface functionality. Knowledge of nanoparticle-membrane interaction mechanisms can be used for the *a priori* design nanoparticles that are biocompatible or exhibit a desired membrane activity. Taking a lesson from nature, insight can be gained from our understanding of protein-membrane interactions. For example, theories describing lateral capillarity and changes in lipid phase behavior caused by hydrophobic proteins may be extended to hydrophobic nanoparticles embedded within lipid bilayers. Similarly, techniques used to evaluate interactions between hydrophilic (charged) proteins and membranes may be extended to hydrophilic nanoparticles. In this presentation we will describe the use of lipid bilayers and monolayers to determine nanoparticle interactions and draw analogies to proteins as a function of nanoparticle size and surface charge, and lipid composition and phase state. Our studies have been focused on engineered nanoparticles common to environmental and biomedical applications.

(485h) Reaction-Diffusion Model Describing Antioxidant Depletion in Polyethylene-Clay Nanocomposites

Ahmad, I.,
Hsuan, G.,
Li, C.,
Cairncross, R., Drexel University
Originally presented on: 11/6/2013 14:43:00 - 15:02:00

Polyethylene (PE) pipes are gradually replacing metal pipes in water and drainage transportations, but are inferior to metal pipes in mechanical properties. Many properties of PE are improved by adding clay to form nanocomposites. Long-term degradation of PE properties is primarily due to environmental oxidative degradation which is stabilized by antioxidants (AO). Although clay improves mechanical properties, in its presence AO fails to stabilize. Therefore a better understanding of underlying mechanisms of AO depletion in the nanocomposites can guide formulation of proper AO to ensure their long term durability. This paper presents research on modeling of AO (hindered-phenol) depletion in samples whose thicknesses are about 3 mm. The mathematical model considers reaction kinetic scheme for degradation and stabilization of neat PE and its nanocomposite. The model not only describes the AO depletion with time but also evolution of its profile throughout the depth of sample. Both chemical and physical losses of AO have been considered. Experimental 'oxidation induction time' (OIT) data showing variation in depletion with depth through the sample has guided development of the model. In nanocomposite samples there are two zones of AO depletion: a 'core zone' and 'depleted surface zone'. In the 'core zone' AO depletion is slower and uniform, and in the 'depleted surface zone', AO depletes more rapidly to produce a surface layer which is void of active AO.

In the model, alkyl free radical oxidizes rapidly compared to O_2 diffusion. Due to limited amount of alkyl free radicals present in the samples, all of them are oxidized in time less than 30 days. Once all alkyl free radicals are oxidized, O_2 diffuses into the sample and uniformly distributes throughout it. The role of AO is to stabilize all those oxidized free radicals, and thus stop their propagation. But continuous depletion of AO is brought about by generation of more free radicals. These additional free radicals are generated by decomposition of hydroperoxides due to catalytic effect of clay. The hydroperoxide decomposition keeps on supplying free radicals in a cyclic fashion resulting in continuous depletion of AO.

The surface depletion of AO can be brought about by evaporation of AO to the surrounding air combined with slower rate of diffusion in the nanocomposites. This can also be brought about by even lower diffusivity of AO in highly clay-orientation layer adjacent to sample surfaces.

In addition, the model predicts AO depletion profiles for a number of different experimental conditions: (1) thermal aging in stagnant air; (2) thermal aging in inert atmosphere; (3) initial AO concentration beyond saturation limit; and (4) thermal aging at a different temperature.

(499a) SiC Photoanodes for Solar Water Splitting

Bohn, C., National Institute of Standards and Technology
Sharifi, F.,
Kang, M. G.,
Szalai, V., National Institute of Standards and Technology
Originally presented on: 11/6/2013 12:30:00 - 12:49:00

AIChE 2013 Session: Nanomaterials for Hydrogen Production and Fuel Cells

Title: SiC Photoanodes for Solar Water Splitting

Authors: C.D. Bohn, M.G. Kang, F. Sharifi, V.A. Szalai

SiC ceramics offer a rich field for exploration for solar water splitting.Â Polytypes include 4H-SiC with a bandgap of 3.23 eV, 6H-SiC with a bandgap of 3.05 eV and 3C-SiC with a bandgap of 2.36 eV.Â Single crystal substrates of some of these materials were obtained from commercial vendors.Â The effect of nanoscale morphology and the addition of Pt nanocrystals at the interface between the SiC and electrolyte were systematically investigated in a pH=7 potassium phosphate buffer under AM 1.5 illumination.Â Gas chromatography provided insight into the chemical products formed by the n-type SiC photoanodes.Â Additionally, linear sweep voltammetry, impedance spectroscopy and scanning electron microscopy were used to monitor the structural evolution of the semiconductor surface.Â Results are discussed in the broader context of photoelectrochemical hydrogen production and recommendations for improving the efficiency of water splitting with SiC are proposed.

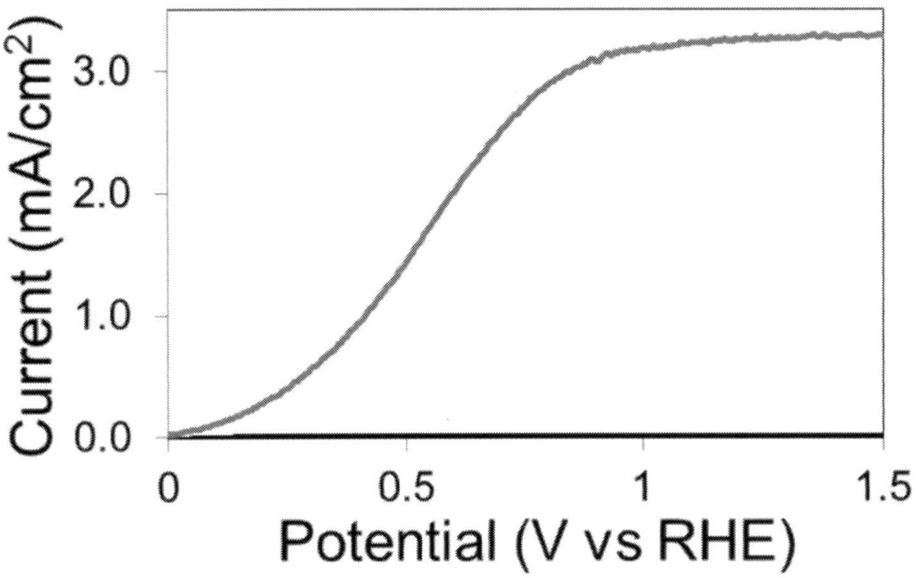

Keywords: solar water splitting, SiC, photoanode

(499b) Advanced Electrocatalysts From Organic Solution Synthesis

Wang, C., Johns Hopkins University

Originally presented on: 11/6/2013 12:49:00 - 13:08:00

Alloy nanoparticles (NPs) have attracted increasing interest due to their superior performance in various catalytic applications. For example, Pt alloys with transition metals (PtM with M = Fe, Co, Ni, etc.) have been found to be highly active for the oxygen reduction reaction in fuel cells. A lot of efforts have been dedicated to the synthesis of alloy catalysts, which are usually in the form of alloy NPs dispersed in high surface area carbon matrix. The approaches mostly include co-precipitation of metal salts in aqueous solution, impregnation, and electrodeposition. In spite of the successful preparation of various types of alloy catalysts, it yet remains challenging to synthesize monodisperse and homogeneous alloy catalysts with robust control over size and composition. Here we introduce our organic solvothermal approach towards high-quality alloy nanomaterials for electrocatalytic applications. Control over particle size, alloy composition and alloy homogeneity has been achieved by matching the different reduction and growth rates of the alloying elements, which has enabled systematic studies of the effects of these parameters on electrocatalysis. Our studies thus provide clues for the design and optimization of alloy nanomaterials for catalytic applications.

(499c) Nanostructured Ti –Fe2O3/Cu2O Bilayered Thin Films for Photoelectrochemical Water Splitting

Sharma, D., Dayalbagh Educational Institute,Dayalbagh

Upadhyay, S., Dayalbagh Educational Institute

Choudhary, S., Dayalbagh Educational Institute

Satsangi, V. R., Dayalbagh Educational Institute

Shrivastav, R., Dayalbagh Educational Institute

Dass, S., Dayalbagh Educational Institute

Originally presented on: 11/6/2013 13:08:00 - 13:27:00

Nanostructured Ti-Fe_2O_3/Cu_2O Bilayered Thin Films for Photoelectrochemical Splitting of Water

Dipika Sharma, Sumant Upadhyay, Surbhi Choudhary, Vibha R. Satsangi[?],

Rohit Shrivastav and Sahab Dass,

Department of Chemistry, [?]Department of Physics & Computer Sciences

Faculty of Science, Dayalbagh Educational Institute (Deemed University),

Agra-282110 (India)

Nanostructured thin films of pure Fe_2O_3, Ti ?Fe_2O_3, Cu_2O, and bilayered Fe_2O_3/Cu_2O, Ti-Fe_2O_3/Cu_2O were deposited on ITO glass substrate using spray pyrolysis method. Among various metal oxide semiconductors, cuprous oxide Cu_2O [Eg=2.4eV] and Iron oxide Fe_2O_3 [Eg=2.1eV] were
chosen for the present study on account of being abundant, non-toxic, environmentally safe, having low band gap and possess staggered type band edges which is suitable for efficient charge separation in water splitting process.

All samples were characterized using XRD, SEM and UV?Vis spectrometry. Photo electrochemical properties were also investigated in a three-electrode cell system. UV-Vis absorption spectrum for pure Fe_2O_3, Ti ?Fe_2O_3, Cu_2O, and Fe_2O_3/Cu_2O, Ti-Fe_2O_3/Cu_2O bilayered thin films shows absorption in visible light region. Nanostructured thin films for all the samples were used as photoelectrode in the Photoelectrochemical cell for water splitting reaction. Maximum photocurrent density of 2.22 mA/cm^2 at 0.95 V/SCE was exhibited by 454 nm thick Ti-Fe_2O_3/Cu_2O bilayered photoelectrode (Fig.1). Increased photocurrent density and enhanced photoconversion efficiency, exhibited by the bilayered thin films may be attributed to improved conductivity and efficient separation of the photogenerated charge carriers at the

$Ti-Fe_2O_3/Cu_2O$ interface.

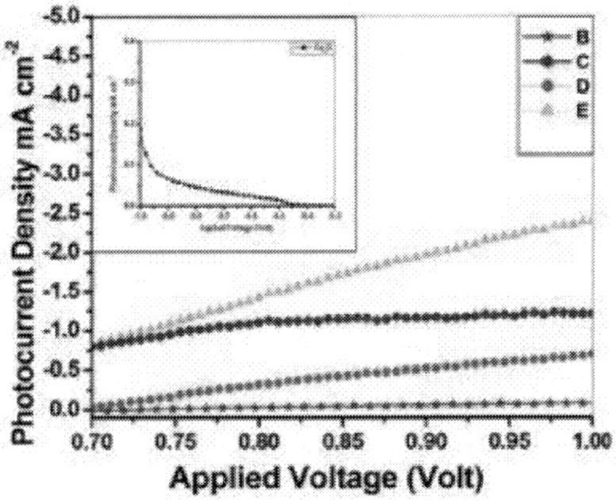

Fig. 1 Photocurrent density vs. Applied potential curve for pure Cu_2O(A), Fe_2O_3(B), Ti ?Fe_2O_3(C), and Fe_2O_3/Cu_2O(D), $Ti-Fe_2O_3/Cu_2O$(E) bilayered thin films respectively ,under visible light illumination in 0.1M NaOH electrolytic solution using 150W visible light source of irradiation 150 mW cm^{-2} at the position of sample.

Keywords: *Photoelectrochemical, Water splitting, Bilayered, Cu_2O, Fe_2O_3, Ti- Fe_2O_3, Spray-Pyrolysis*

(499d) Synthesis and Characterization of Mn2O3-ZnO Nanocomposites for Hydrogen Generation Via Photoelectrochemical and Photocatalytic Splitting of Water

Singh, N., Dayalbagh Educational Institute
Choudhary, S., Dayalbagh Educational Institute
Satsangi, V. R., Dayalbagh Educational Institute
Dass, S., Dayalbagh Educational Institute
Shrivastav, R., Dayalbagh Educational Institute
Originally presented on: 11/6/2013 13:27:00 - 13:46:00

Synthesis and Characterization of Mn_2O_3-ZnO Nanocomposites for Hydrogen Generation *via* Photoelectrochemical and Photocatalytic Splitting of Water

Nirupama Singh, Surbhi Choudhary, Vibha R. Satsangi[#], Sahab Dass and Rohit Shrivastav*

Department of Chemistry, [#]Department of Physics & Computer Science,

Dayalbagh Educational Institute, Dayalbagh, Agra-282 005, India

*E-mail: rohitshrivastav_dei@yahoo.co.in

The growing consumption of energy, which is now mainly produced from fossil fuels leads to a further depletion of the limited fossil reserves, increased dependency on fossil fuel imports from often politically unstable counties and furthermore gives rise to an increased concentration of green house gases mainly CO_2 in the atmosphere. The report deals with the synthesis of Mn_2O_3-ZnO nanocomposites, their characterization, and possible use in photoelectrochemical and photocatalytic splitting of water for hydrogen generation. Powdered samples of nanocomposites were prepared by co-precipitation at varying Mn/Zn molar ratios. XRD analysis revealed the evolution of wurtzite ZnO and Mn_2O_3 phases in samples, while optical absorption threshold was found shifted towards higher wavelength with increasing Mn amount. The Zn/Fe molar ratio in nanocomposites was optimized, yielding significantly enhanced photoelectrochemical current and photocatalytic hydrogen yield. Plausible explanations for observed change in photoresponse of samples have been offered.

(499e) High Performance Fluorine Doped Oxygen Evolution Reaction (OER) Electro-Catalysts for PEM Based Water Electrolysis

Kadakia, K., University of Pittsburgh
Datta, M. K., University of Pittsburgh
Velikokhatnyi, O.,
Kumta, P. N., University of Pittsburgh
Originally presented on: 11/6/2013 13:46:00 - 14:05:00

Identification of electro-catalysts containing non noble metal or significantly reduced amounts of expensive noble metals (e. g. RuO_2) is highly desirable. Development of such a catalyst with comparable electrochemical performance to the standard noble metal oxide counterpart for the proton exchange membrane (PEM) based water electrolysis would constitute a major advancement in generation of hydrogen by water electrolysis. We have accordingly developed a two-pronged theoretical first principles and experimental approach to generate a nano-structured solid solution of SnO_2:10wt.% F containing only 20at. % RuO_2 [*e.g.* $(Sn_{0.80}Ru_{0.20})O_2$:10F] that displays a remarkably similar electrochemical activity and moreover, comparable or even much improved electrochemical stability and durability compared to pure the noble metal counterpart, RuO_2. Results of these studies will be presented and discussed.

(499f) Multiwalled Carbon Nanotubes Drive the Activity of Metal@Oxide Core-Shell Catalysts in Modular Nanocomposites

Cargnello, M., University of Pennsylvania

Originally presented on: 11/6/2013 14:05:00 - 14:24:00

Multiwalled Carbon Nanotubes Drive the Activity of Metal@oxide Core-Shell Catalysts in Modular Nanocomposites

M. Cargnello,[1,2] M. Grzelczak,[1] B. Rodriguez-Gonzalez,[3] Z. Syrgiannis,[1] K. Bakhmutsky,[4]

V. La Parola,[5] L. M. Liz-Marzán,[3] R. J. Gorte,[4] M. Prato,[1] and P. Fornasiero[1]

[1] *Department of Chemical and Pharmaceutical Sciences, ICCOM-CNR, Consortium INSTM, University of Trieste, via L. Giorgieri 1, 34127 Trieste, Italy*

Â [2] Department of Chemistry, University of Pennsylvania, 231 S. 34th Street, Philadelphia, PA 19104, USA

[3] *Departamento de QuÃmica FÃsica, Universidade de Vigo, 36310 Vigo, Spain*

[4] *Department of Chemical and Biomolecular Engineering, University of Pennsylvania, 311A Towne Building, 220 S. 33rdÂ Street, Philadelphia, PA 19104, USA*

[5] *Istituto per lo Studio dei Materiali Nanostrutturati (ISMN-CNR), Via Ugo La Malfa 153, Palermo I-90146, Italy*

The ability to build hierarchical structures by arranging different building blocks with nanometer-scale precision is one of the most useful aspects of nanotechnology. This concept can produce novel materials with properties that are different from those expected from the simple sum of the individual blocks. Indeed, the interactions between the constituent parts in nanometer-scale ensembles can lead to novel electronic, optical, or catalytic properties not available in the initial building blocks. In the field of heterogeneous catalysis in particular, the interactions between the active phase, supports, and promoters is critical for obtaining high performance materials. These interactions can be both electronic and geometric. Furthermore, the demands in terms of geometry and binding energy of the active sites for different catalytic reactions can be very different.

In this contribution, rational nanostructure manipulation has been used to prepare nanocomposites in which multiwalled carbon nanotubes (MWCNTs) were embedded inside mesoporous layers of oxides (TiO2, ZrO2, or CeO2), which in turn contained dispersed metal nanoparticles (Pd or Pt). We show that the MWCNTs induce the crystallization of the oxide layer at room temperature and that the mesoporous oxide shell allows the particles to be accessible for catalytic reactions. In contrast to samples prepared in the absence of MWCNTs, both the activity and the stability of core-shell catalysts is largely enhanced, resulting in nanocomposites with remarkable performance for the water-gas-shift reaction, photocatalytic reforming of methanol, and Suzuki coupling. The modular approach shown here demonstrates that high-performance catalytic materials can be obtained through the precise organization of nanoscale building blocks.

(499g) Separation of Sulphur Dioxide and Oxygen in Thermochemical Hydrogen Production

Elder, R. H., University of Sheffield
He, G., University of Sheffield
Cumming, D. J.,
Allen, R. W. K., University of Sheffield

Originally presented on: 11/6/2013 14:24:00 - 14:43:00

Dense Yttria-stabilised-zirconia (YSZ) membranes have been shown to have great potential for separating SO_2 and O_2 in the sulphur family of thermochemical cycles for hydrogen production. Use of a catalytic YSZ membrane reactor in the sulphuric acid decomposition stage of these cycles could drive the reaction forward and significantly increase the yield of SO_2 and O_2. Whilst SO_2/O_2 separation has been shown to be feasible at temperatures between 750°C and 950°C, significant performance degradation is seen over time. This performance degradation is shown to be due to sulphur poisoning (from both sulphur adsorption and sulphide formation) of the triple phase boundaries. A similar effect is seen in solid oxide fuel cells, where ppm concentrations of sulphur severely diminish cell performance.

Here we investigate the effect of voltage pulsing on membrane performance recovery. Significant performance recovery is seen after the application of a short voltage pulse. The optimum pulse duration and magnitude were found to be 0.08 s and 10 V respectively. The lower the magnitude of voltage pulse, the longer the duration required to achieve maximum recovery. Impedance spectroscopy and SEM results are presented, showing the contribution of different resistances and the effect of the voltage pulsing on the membrane cell.

Consideration of oxygen reduction mechanisms at the triple phase boundary shows that as voltage across the membrane increases, the rate controlling step changes from oxygen adsorption and surface diffusion, to oxygen availability in the feed. The higher voltage pulses are beneficial to performance recovery as polarisation resistance and charge transfer resistance decrease due to the removal of sulphur from the surface. Pulses that have too great a magnitude, or are too long, dramatically promote oxygen reduction and lead to concentration polarisation due to lack of oxygen in the feed.

Longer term testing (20 hours) showed that repeated, regular voltage pulsing can significantly improve membrane performance compared to those that receive no voltage pulsing.

(499h) Chemisorption, Physisorption, and Hysteresis of Hydrogen On Carbon Nanotubes

Barghi, S., University of Southern California

Tsotsis, T. T., University of Southern California

Sahimi, M., University of Southern California

Originally presented on: 11/6/2013 14:43:00 - 15:02:00

We used a magnetic suspension balance, coupled with a residual gas analyzer, to investigate hydrogen adsorption/desorption on carbon nanotubes (CNTs) for pressures up to 100 bars at 25 °C. Such a combination of gravimetric and mass analysis systems, used for the first time, is capable of overcoming the shortcomings of the common hydrogen adsorption measurement techniques. The most crucial drawbacks of the conventional measurement methods are as follows. (i) The inability to distinguish between hydrogen adsorption and chemisorptions; (ii) the relatively large errors caused by the leakage of hydrogen out of the system, and (iii) the difficulty of measuring hydrogen uptake capacity under operating conditions that differ from those of hydrogen storage systems used in practice.

In the experiments multi-walled carbon nanotubes (MWCNTs) were exposed to hydrogen, and were subsequently kept under vacuum for 2 consecutive cycles, in order to study hydrogen adsorption/desorption in the nanostructured materials. The difference between the first and the second hydrogenation cycles was in the initial preparation of the MWCNTs. For the first cycle, the MWCNTs were first degasified at an elevated temperatures (120 °C). For the second cycle the MWCNTs were not degasified at the elevated temperatures and, therefore, hydrogen residue from the first adsorption/desorption cycle was still adsorbed on the MWCNTs at the beginning of the second cycle.

The first hydrogenation cycle was started after pretreatment of the CNT at the elevated temperature. Hydrogen adsorption kinetics for the cycle indicates that the adsorption equilibrium occurs within 4 - 5 hours. Careful examination of the recorded adsorption kinetics revealed the fascinating fact that a considerable part of hydrogen adsorption in the MWCNTs takes place *instantaneously*. The measured equilibrium data revealed some hysteresis for hydrogen desorption in this cycle, which does not vanish, even when the system's pressure approaches zero. This indicates that hydrogen adsorption in the first hydrogenation cycle is a combination of physisorption and chemisorption.

After the first cycle was finished, the CNTs were exposed to hydrogen for the second time without any pretreatments at the elevated temperature. Measurements indicated no hysteresis for during the second cycle. Therefore, only physisorption took place during the second cycle. Comparison between the first and the second hydrogen isotherms, together with the measured hydrogen adsorption kinetics, made it possible to quantitatively distinguish between the physisorption and chemisorption portions of hydrogen adsorption in the MWCNTs. According to our data, for pressures less than 20 bars physisorption and chemisorption are both important, with the former becoming the dominant mechanism as the pressure increases. To the best of our knowledge, this is the first time that hydrogen physisorption and chemisorption on carbon nanotubes have been measured separately.

(529a) Lignocellulosic Biomass Pyrolysis Product Comparison From a Micropyrolyzer, Batch and Continuous Reactor Systems

Garrido, R., Villanova University
Hammer, N. L., USDA
Zmiewski, A., Villanova University
Satrio, J. A., Villanova University
Originally presented on: 11/6/2013 15:15:00 - 15:31:00

Lignocellulosic Biomass Pyrolysis product comparison from a micropyrolyzer, batch and continuous reactor systems

Rene Garrido, Nicole L. Hammer, Alexander Zmiewski, and Justinus A. Satrio

Department of Chemical Engineering, Villanova University

ren.garrido@villanova.edu, nhammer@villanova.edu: justinus.satrio@villanova.edu

This project aims at advancing the current state of knowledge of biomass pyrolysis by developing an understanding of the pyrolytic reaction mechanisms of different feedstocks, and to find common grounds between different lignocellulosic biomass and system configurations.

To understand these differences three successive scales of pyrolysis equipment have been implemented; Stage I, a micro-pyrolyzer, Stage II, a small-scale tubular batch reactor, and Stage III, a fluidized-bed continuous reactor. Understanding the differences on reaction mechanisms for each of the stages is a key element to properly identify and determine the products obtained from pyrolysis. To separate primary from secondary reactions, a µg scale pyroprobe system combined with gas chromatography and mass spectrometry (Py-GCMS) is used to study the chemical composition of volatile products from previously hydrolyzed and torrefied biomass. Effects of the severity of hydrolysis pretreatment and torrefaction on the chemical composition and yield distribution of the fast pyrolysis products will be evaluated. The chemical composition study of pyrolyzed biomass would then aid in the understanding of the chemistry behind the different compounds to further enhance the process of upgrading bio-crude oil.

For a batch reactor, a tubular furnace is used to pyrolyze the samples, and the products will be compared with a continuous fluidizing bed reactor. Although the reaction mechanisms are different, and the biomass particles in the fluidization medium have greatly enhanced heat and mass transfer in fluidized beds reactors. This is the first step to understand and find common ground between the different feedstocks, and to move from lab scale to a pilot scale reactor.

This study investigates common Reed Grass *(Phragmites Australis)*, and Pinewood as examples of different sources of lignocellulosic biomass as potential feedstock for bio-oil production.

Phragmites represents grassy biomass; it is a prevalent waste on Villanova's campus and around the world. It is potentially a valuable energy and chemical feedstock due to its high yield (18-28 tons of biomass per acre per year). Pinewood represents hardwood biomass with high cellulose content and for our study it is the control system in which all other biomasses will be related to.

Based on the results obtained by analytical pyrolysis this study will focus on obtaining bio-crude oil from *Phragmites australis* and Pinewood using a batch reactor and compare the results from a fluidized bed continuous reactor. Through this study, groundwork is established as a preliminary attempt to understand how various feedstocks can be used for biomass conversion to more useful energy sources.

(529b) Spatio-Temporal Carbohydrate Characterization of Wood Particles During Fast Pyrolysis

Paulsen, A. D., University of Massachusetts Amherst
Hough, B.,
Schartz, D.,
Pfaendtner, J., University of Washington
Dauenhauer, P. J., University of Massachusetts Amherst
Originally presented on: 11/6/2013 15:31:00 - 15:47:00

Fast pyrolysis provides an inexpensive chemical process for converting lignocellulosic biomass including wood fibers and grasses to a liquid intermediate called bio-oil which can be transported and upgraded to chemicals and fuels. The complexity of fast pyrolysis arises from the feedstock which is comprised of several biopolymers (cellulose, hemicellulose, and lignin) integrated within a cellular microstructure. During rapid heating, cellulose and biopolymers rapidly degrade to a mixture of short-chain oligomers and eventually volatile components and aerosols which are collected as bio-oil [1,2,3]. When wood fibers are heated in pyrolysis, heat and mass transfer limitations within the cellular lignocellulosic structure produce dramatic temperature gradients, which ultimately dictate the local reaction conditions and the distribution of pyrolysis products. The complexity of characterizing the reacting particle chemistry and transport phenomena requires a large number of kinetic parameters and has been identified as one of the major fundamental challenges in research of fast pyrolysis of biomass [4].

In this work, we aim to improve reaction-transport pyrolysis models of reacting woody biomass by demonstrating a new experimental technique that characterizes the composition of woody biomass as it pyrolyzes to bio-oil. Using diffuse reflectance spectroscopy, the carbohydrate composition of Yellow Poplar (*Liriodendron tulipifera*) is characterized in position and time through a particle exhibiting heat transfer dominant in only one spatial dimension. Compositional profiles exhibit a propagating reaction front consistent with heat-transfer limited (high Bi) pyrolysis conditions. A range of reaction conditions are examined, and the resulting data are described using a one-dimensional reaction-transport model using existing lumped-kinetic reaction chemistry. Sensitivity analysis of the resulting model and comparison of the prediction with experimental spatio-temporal compositional data allows for robust characterization of wood fiber pyrolysis.

References

[1] M.S. Mettler, D.G. Vlachos, P.J. Dauenhauer, *Energy & Environmental Science* **2012**, 5, 7797-7809.

[2] V. Agarwal, P.J. Dauenhauer, G.W. Huber, S.M. Auerbach, *Journal of the American Chemical Society* **2012**, 134(36), 14958-14972.

[3] A.R. Teixeira, K.G. Mooney, J.S. Kruger, C.L. Williams, W.J. Suszynski, L.D. Schmidt, D.P. Schmidt, P.J. Dauenhauer, *Energy & Environmental Science* **2011**, 4, 4306.

[4] M.S. Mettler, D.G. Vlachos, P.J. Dauenhauer, *Energy & Environmental Science* **2012**, 5, 7797-7809.

(529c) Numerical Investigation of Multiphase Dynamic Effects in Catalytic Upgrading of Biomass Pyrolysis Vapor

Capecelatro, J. S., Cornell University

Desjardins, O., Cornell University

Pepiot, P., Cornell University

Jarvis, M., National Renewable Energy Laboratory

Foust, T. D., National Renewable Energy Laboratory

Originally presented on: 11/6/2013 15:47:00 - 16:03:00

A recurring challenge among the variety of existing biomass-to-biofuel conversion technologies is the need to ensure optimal and homogeneous contact between the various phases involved. One example can be found in the catalytic upgrading of bio-oil vapors, a bi-phasic process critical in pyrolysis-based conversion, typically performed in circulating fluidized bed (CFB) reactors. While the upgrading process is expected to strongly depend on the contact time between raw gas and catalysts, and on catalysts deactivation over time, the formulation of robust design rules from an empirical standpoint alone remains difficult, due to the wide range of granular flow regimes coexisting within the reactor, from near close-packing at the bottom of the reactor to more dilute conditions in the riser. The issue is amplified by the highly unsteady and turbulent nature of gas-solid flows. In particular, clusters of particles are known to form in risers, which can significantly decrease the overall catalytic activity in the reactor. In this work, a Lagrangian particle tracking tool recently developed in the large eddy simulation (LES) framework of NGA is used to quantify the role of cluster dynamics on biomass pyrolysis vapor catalytic upgrading in risers. This numerical framework has been extensively validated for a range of particle-laden flows, in particular, cluster fall velocities in risers of CFB reactors have been shown to match experimental correlations (Fig.1). Two configurations are investigated and compared: an homogeneous system for which the catalyst-pyrolysis vapor contact time is statistically uniform, and a pseudo-two-dimensional riser, such as shown in Fig.2, combining all granular flow regimes of interest.

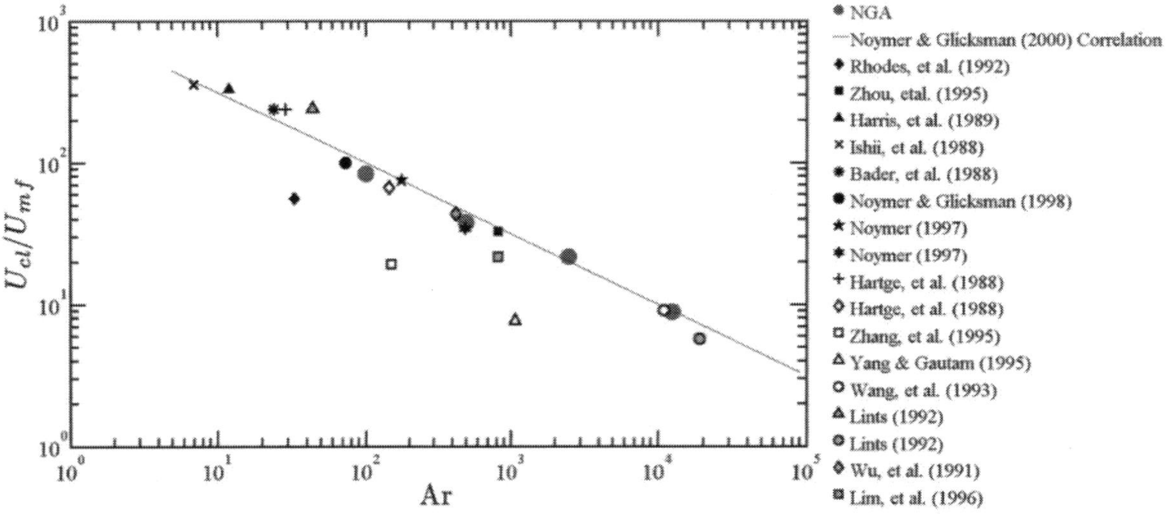

Figure 1: Mean cluster fall velocity normalized by the minimum fluidization velocity as a function of Archimedes number, a non-dimensional number that characterizes the flow inertia due to density differences between the gas and solid phases. The solid line indicates a correlation developed from a range of experimental studies by Noymer & Glicksman (2000).

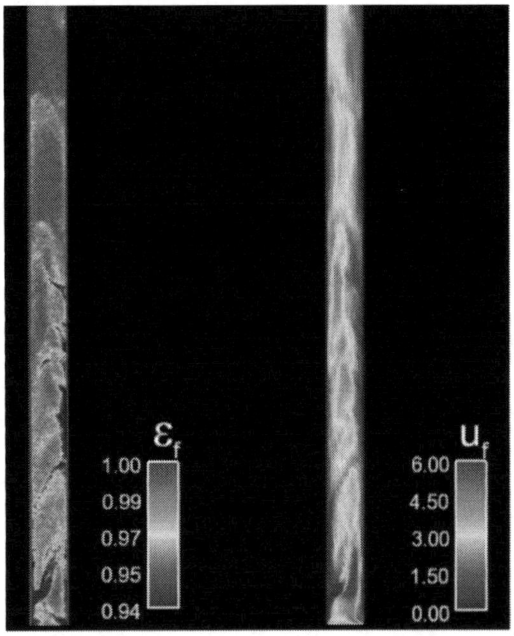

Figure 2: Pseudo two-dimensional simulation of a riser showing gas volume fraction (left) and gas velocity magnitude (right). Particles are injected at the bottom where the gas inlet is located.

(529d) Catalytic Effects of MgSO4 and ZnCl2 On Fast Pyrolysis of Pine for the Production of Bio-Oils

Yuan, L.,
Liu, Y. Q., Chevron Technology Ventures LLC
Lv, D. C., Xiamen University
Wang, D., Xiamen University
Originally presented on: 11/6/2013 16:03:00 - 16:19:00

The effects of $MgSO_4$ and $ZnCl_2$ as catalysts on fast pyrolysis of pine were investigated in a bench-scale bubbling fluidized bed reactor. The study showed that both $MgSO_4$ and $ZnCl_2$ could shift the pyrolysis of pine to a lower temperature zone, that is, most of pyrolytic reactions for cellulose and hemicellulose would happen at lower temperature while lignin at higher temperature, which results in different composition of bio-oils from that without adding any of these two salts. Compared to the bio-oils obtained without catalysts (base case), glucopyranose in bio-oils was increased to 20.69 % from less than 1% (base case) when 20 wt% of $MgSO_4$ was added; while ketones decreased from 26.71 % (base case) to 8.41 %, 2-Propanone, 1-hydroxy from 9.94 % (base case) to 0.62 %. When 30 wt% of $ZnCl_2$ is added, glucopyranose in bio-oils hiked to 20.21 % from less than 1% (base case); while phenols reduced to 6.25% from 27.09 % (base case). The pH of bio-oils rises whether $MgSO_4$ or $ZnCl_2$ was added. This study suggests that by adding specific salts into biomass, the composition of bio-oils obtained from fast pyrolysis can be of greatly different. Therefore, we can increase the content of certain chemicals in bio-oils on purpose, which could make the extraction of value-added chemicals from bio-oils more feasible and economical.

(529e) Pyrolysis of Alkali Pre-Treated Biomass and Biomass Constituents

Case, P., University of Maine
Stemmler, E., Bowdoin College
DeSisto, W. J., University of Maine
Wheeler, M. C., University of maine
Originally presented on: 11/6/2013 16:19:00 - 16:35:00

Although upgrading of pyrolysis oil has been extensively studied, it is still plagued with low product yields and significant catalyst deactivation. These issues are largely because of the instability of oxygenated species in pyrolysis oil, which leads to considerable char formation by polymerization at the elevated temperatures that are required for hydrodeoxygenation. In the current study, we eliminate compounds that cause polymerization by pre-treating sawdust with various alkali compounds before pyrolyzing it in a fluidized bed reactor. The resulting stabilized oils have significantly lower oxygen content than conventional pyrolysis oil. The higher heating values of the oils are in the range of 40 MJ/kg, and the energy yields exceed those commonly achieved in catalytic fast pyrolysis. Detailed analysis of conventional and pre-treated pyrolysis oils is presented and potential reaction mechanisms are discussed.

(529f) Study On the Aging Mechanism of Pyrolysis Bio-Oil

Meng, J., North Carolina State University
Tilotta, D., North Carolina State University
Adhikari, S., Auburn University
Kelley, S., North Carolina State University
Park, S., North Carolina state university

Originally presented on: 11/6/2013 16:35:00 - 16:51:00

Session 16004 Biomass Pyrolysis I

AIChE 2013

November 3 ~ 8, 2013

San Francisco, CA

Study on the Aging Mechanism of Pyrolysis Bio-oil

Jiajia Meng*,[1] David Tilotta,[1] Sushil Adhikari,[2] Stephen Kelley,[1] and Sunkyu Park[1]

[1] Department of Forest Biomaterials, North Carolina State University, Raleigh, NC 27695

[2] Biosystems Engineering Department, Auburn University, Auburn, AL 36849

*Presenting author: jmeng@ncsu.edu

The unstable nature of pyrolysis oil produced from biomass fast pyrolysis is one of the key issues that limit its application and upgrading, as aged bio-oil typically shows increased water content, viscosity, and phase separation. The aging phenomena are considered due to the polymerization of bio-oil components under acidic and thermal conditions catalyzed by the mineral compounds contained in char particles in the bio-oil. Current research on bio-oil aging is focused on characterizing the behavior together with seeking methods to slow down the aging speed. However, little research was reported on investigating the molecular-level polymerization mechanism of pyrolysis bio-oil during the aging. In this study, accelerated aging of the bio-oil produced from loblolly pine was performed with different aging time and temperature and with or without additives to access the bio-oil thermal stability by measuring water content, viscosity, acidity, molecular weight distribution, and chemical composition. With the chemical analysis of aged bio-oils, it was found that ether soluble fraction of the bio-oil was significantly reduced during the aging process and glycolaldehyde is the most reactive compound in that fraction. The pyrolytic lignin fraction, especially high molecular weight lignin, was found to gradually increase, which suggests the reaction between the compounds of ether soluble fraction and the lignin fraction. Because of the great reduction of glycolaldehyde concentration during the aging, synthetic bio-oil with different concentration of glycolaldehyde was mixed with pyrolytic lignin and aged at 80 °C to further study the role of glycolaldehyde on bio-oil aging chemistry. To elucidate the polymerization chemistry between ether soluble fraction and pyrolytic lignin fraction, model compounds, such as ^{13}C labeled glycolaldehyde and ^{13}C labeled phenolic compounds (phenol and guaiacol type), were reacted in a similar condition to that of accelerated aging. In addition, other suggested mechanisms of bio-oil aging will be reviewed and discussed in detail.

(529g) Effects of Post-Condensation Centrifugal Filtration On Pyrolysis Oil

Varadarajan, A., Mississippi State University
Walters, K. B., Mississippi State University
Originally presented on: 11/6/2013 16:51:00 - 17:07:00

The effects of post-condensation filtration of the crude pyrolysis oil with relatively high water content and solids content was studied. Tests were performed using a pilot plant-scale centrifuge filtration system. The centrifuge can be used as a three-phase separator [light liquid-heavy liquid-solids] or a two-phase clarifier [liquid-solid]. The testing was performed by varying the parameters such as separation time and displacement time which affects the separation efficiency of the centrifuge. In addition, the impact of increased temperature prior to and during the filtration process was examined. Effect of feed temperature on the separation efficiency of the centrifuge and the consequent physical and chemical property change on the filtered bio oil was performed. The filtered samples were collected and characterized. To determine the water content, the Karl Fischer titration method was utilized. Fourier transform infrared (FTIR) spectroscopy was used to determine chemical composition. Gel permeation chromatography was used to determine molecular weight distribution. Particle size distribution in the filtered samples was found using optical microscopy.Parameters were examined for the centrifugal filtration of post-condensation pyrolysis oil to optimize the particulate loading, processing time, and physicochemical stability for downstream processing, including upgrading

(529h) Transportable Fast-Pyrolysis Process for Distributed Conversion of Waste Biomass to Renewable Liquid Fuels

Yelvington, P. E., Mainstream Engineering Corporation
Schwartz, N., Mainstream Engineering Corporation
Zastrow, D., Mainstream Engineering Corporation
Amundsen, T., Mainstream Engineering Corporation
Originally presented on: 11/6/2013 17:07:00 - 17:23:00

In Florida alone, sugarcane bagasse, orange residue from juicing, and logging/mill residues are produced at a rate exceeding 16,000 dry tons per day. However, these agricultural residues are often spread over large areas and are difficult to access, particularly the logging residues. Rather than trucking large quantities of low-density biomass to a massive centralized biofuel production plant, the distributed conversion of agricultural residues into bio-oil is an attractive alternative for sustainable biofuel production. Opportunities also exist for the conversion of mixed waste to biofuel at forward-deployed military outposts, where the true delivered cost of fuel often exceeds $15 per gallon. Mainstream Engineering (MEC) recently commissioned a highly instrumented fast-pyrolysis pilot plant (1 ton/day) to demonstrate small-scale conversion of waste organic material to bio-oil. The pilot reactor was designed for flexibility and tight process control, as we explore the product yield and quality for a variety of feedstocks. Testing to date has used pine sawdust or a granulated mixed waste surrogate consisting of corrugated cardboard, office paper, and several plastics.

MEC is working to commercialize the portable process at a 10 ton/day scale for commercial applications and at the <1 ton/day scale for military applications. Our efforts have focused on improving the reactor yield by minimizing secondary cracking and improving the energetics with tight heat integration. Both improvements are geared towards achieving viable process economics at a small thoughput. We have also experimentally investigated the pre-processing (size reduction and drying) and end-use utilization of raw bio-oil in a burner/boiler. Results will be presented on our pilot reactor testing, techno-economic analysis of the portable reactor approach, and initial results from our efforts to mildly upgrade raw bio-oil to a renewable residual-grade fuel oil.

(529i) Biomasspyrolysisrefinery: Product Stability of Intermediates and Products

Schwaiger, N., Graz University of Technology
Feiner, R., Graz University of Technology
Pucher, H., Graz University of Technology
Steiner, L., Graz University of Technology
Ellmaier, L., BDI-Bioenergy International AG
Pucher, P., BDI BioEnergy International AG
Siebenhofer, M., Graz University of Technology
Originally presented on: 11/6/2013 17:23:00 - 17:39:00

The BiomassPyrolysisRefinery is a dual step concept for biomass liquefaction and hydrogenation, intending production of liquid energy carriers. First step of this concept is liquid phase pyrolysis (LPP). During pyrolysis lignocellulose is dehydrated to form biochar and pyrolysis oil (Schwaiger et al., 2011, 2012) The LPP process has been scaled up to a capacity of 100 kg/h biomass (BDI bioCRACK facility) at OMV refinery Schwechat. The second step of the BiomassPyrolysisRefinery is product upgrading. Pyrolysis oil can pass a dewatering step to raise the calorific value. It may also be deoxygenated and hydrogenated. Biochar is separated from the heat carrier after liquid phase pyrolysis and then hydrogenated.

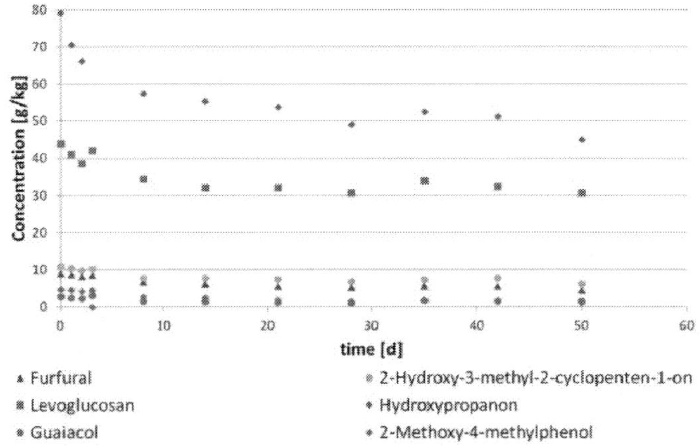

Figure 1: Stability of selected substances from liquid phase pyrolysis during storage at ambient temperature

During liquefaction biogenous material passes several heating and storing steps. Therefore sufficient chemical stability of intermediates and products is needed. Stability of pyrolysis oil for storing and upgrading purposes may be very limited (Fahmi et al., 2008; Kim et al., 2012), because of polymerization. The molar mass, viscosity and water content may rise significantly. The storing properties of liquid phase pyrolysis oil, upgraded liquid phase pyrolysis oil and liquified biochar at ambient temperature was monitored by GC-MS, GC-TCD, SEC and pH, and TAN analysis. Density and viscosity were recorded.

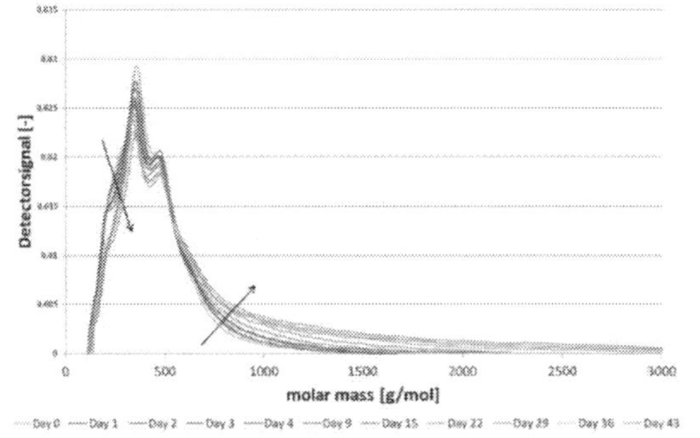

Figure 2: Liquid phase pyrolysis oil; molar mass versus storage time

Figure 1 shows the concentration of liquid key substances from liquid phase pyrolysis. During the first week of storage nearly one third of highly oxygenated substances vanish, afterwards degradation may cease. The same trend of property changes in the first days after pyrolysis is recorded with SEC-analysis. Figure 2 shows stabilization of molecular mass within day 15.

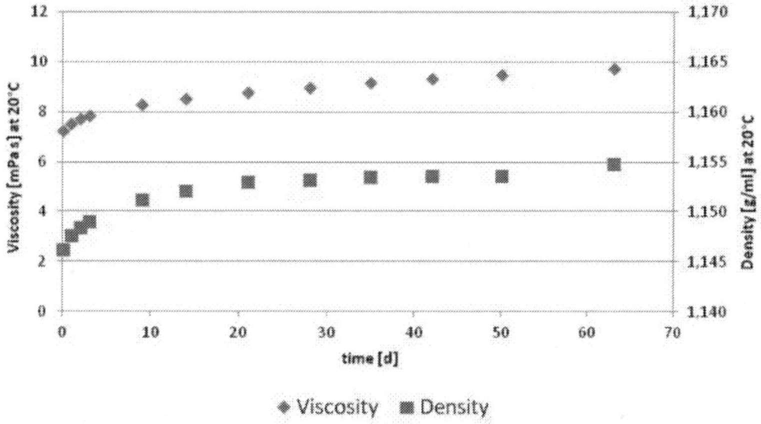

Figure 3: Viscosity and density at T = 20,°C of liquid phase pyrolysis oil

Viscosity and density trends, shown in figure 3, confirm the findings. Liquid phase pyrolysis oil undergoes ageing through repolymerization within the first 15 days of storage. After upgrading stability of pyrolysis oil and liquified biochar is significantly better. Compared with flash pyrolysis oil the stability of liquid phase pyrolysis oil is seemingly achieved through separation from char based particulate matter and inorganics.

Fahmi, R., Bridgwater, A., Donnison, I., 2008. The effect of lignin and inorganic species in biomass on pyrolysis oil yields, quality and stability. Fuel 87, 1230--1240.

Kim, T.-S., Kim, J.-Y., Kim, K.-H., Lee, S., Choi, D., Choi, I.-G., Choi, J.W., 2012. The effect of storage duration on bio-oil properties. Journal of Analytical and Applied Pyrolysis 95, 118--125.

Schwaiger, N., Feiner, R., Zahel, K., Pieber, A., Witek, V., Pucher, P., Ahn, E., Wilhelm, P., Chernev, B., SchrÃ¶ttner, H., Siebenhofer, M., 2011. Liquid and Solid Products from Liquid-Phase Pyrolysis of Softwood. BioEnergy Research 4, 294--302.

Schwaiger, N., Witek, V., Feiner, R., Pucher, H., Zahel, K., Pieber, a, Pucher, P., Ahn, E., Chernev, B., Schroettner, H., Wilhelm, P., Siebenhofer, M., 2012. Formation of liquid and solid products from liquid phase pyrolysis. Bioresource technology 124, 90--4.

(529j) Biomass Gasification: Effect of Sulfur Compounds On Catalytic Tar Removal Activity Using Nickel-Clay Catalysts

Vutukuri, J., University of Louisiana at Lafayette
Buchireddy, P., University of Louisiana at Lafayette
Bricka, R. M., Mississippi State University
Guillory, J., University of Louisiana at Lafayette
Bajpai, R., University of Louisiana at Lafayette
Zappi, M. E., University of Louisiana at Lafayette
Originally presented on: 11/6/2013 17:39:00 - 17:55:00

Biomass gasification is one of the thermochemical routes of converting biomass to fuel gas. This fuel gas can be used in various possible applications such as generation of heat or electricity, synthesis of liquid transportation fuels, production of hydrogen, synthesis of chemicals and generation of electricity in fuel cells. However, this fuel gas needs to be cleaned from the impurities like particulates, ammonia, hydrogen sulfide, hydrogen chloride, alkali metals, metals, and tars. The presence of tars in the fuel gas is one of the main technical barriers in the biomass gasification development. Tar buildup cause cracking in the pores of filters, forming coke and plugging the filters, and condensing in the cold spots and plugging the lines, resulting in serious operational interruptions.

Nickel (Ni) supported clays have been developed and evaluated for its activity towards tar removal at Mississippi State University. The results from these studies suggested that Ni-Clay catalysts have very good activity for tar removal. However, the influence of sulfur compounds, although, present in very low concentrations has not been evaluated. Hence, this study focuses on the effect of sulfur compounds on the catalyst performance for removal of tars from syngas. Literature suggests that the presence of sulfur in the fuel gas decreases the catalytic activity of the Ni catalysts due to strong chemisorption of sulfur on the Ni surface forming Nickel Sulfide, which is inactive. Although, the presence of sulfur has shown to have an inhibiting effect on the catalytic activity, the effect of sulfur on tar removal using Ni-Clays has not been investigated till date. This study investigates the influence of sulfur compounds on tar reforming activity of Ni Clay catalysts. A plug flow reactor has been designed and fabricated to perform these studies. Experiments are being conducted using naphthalene as a tar model compound and hydrogen sulfide as a sulfur compound in syngas. The effect of hydrogen sulfide partial pressure, temperature(650-850° C), steam to carbon ratio, and residence time on naphthalene removal activity is being evaluated, the results of which will be presented. Also, the influence of catalyst support, clay in this study is being investigated and will be compared with commercial silica/alumina supports.

(539a) Study of the Tcep-Zeolite System: Kinetics and Surface Chemistry Analysis Investigations

Grieco, S., O'Brien & Gere Engineers, Inc.
Ramarao, B. V., ESPRI, SUNY College of Environmental Science and Forestry
Originally presented on: 11/6/2013 15:15:00 - 15:35:00

Emerging contaminants (ECs) and micropollutants are becoming a global environmental problem as there is increasing information linking them to environmental and human health hazards. One micropollutant of interest is tris-2-chloroethyl phosphate (TCEP), an organo-phosphate ester used as a plasticizer and flame retardant. Recent studies by others have focused on the presence of TCEP in environmental and drinking waters, and several of these previous studies indicated that TCEP was the most frequently detected compound of the ECs investigated. Transport of TCEP from wastewater effluents to raw water sources is supported by research that identifies the presence of TCEP in surface waters similar to those detected in wastewater effluents. This supports the premise that the TCEP is not being removed from the water cycle.

In our previous work presented last year, adsorption of TCEP to zeolite was studied. Equilibrium adsorption isotherms and surface hydrophobicity/energy measurements were conducted. These results demonstrated excellent removal of TCEP using laboratory water and zeolite (>99% removal; Langmuir sorption capacities of 23 mg/g to 103mg/g). Adsorption was determined to be a function of both solution pH and adsorbent type. The observations of pH-dependant adsorption (given a non-ionizing adsorbate) and the additional lack of adsorption related to low zeolite concentrations and higher pH values suggested a complex function of pH and hydrophobic/hydrophilic interaction which required further investigation.

This paper presents the continued research efforts to better understand the adsorption mechanisms of the TCEP-zeolite system. Kinetic modeling and measurements at various TCEP-zeolite ratios were conducted. These data suggest a system with fast kinetics. The modeling was used to establish characteristics of possible reaction mechanisms. Also zeta potential and surface area measurements were conducted to support the kinetic data and better understand the effects of pH on adsorption. These results will be presented.

Thermo Gravimetric Analyses (TGA), Nuclear Magnetic Resonance (NMR), High Perforamnce Liquid Chromatography (HPLC), and Fourier Transform Infrared Technology (FTIR) were used to characterize the TCEP-zeolite adsorption system and postulate the potential adsorption mechanisms occurring. The results of these data will be presented and discussed.

(539b) Adsorption and Degradation of Perfluorooctanoic Acid

Almquist, C. B., Miami University
Berberich, J. A., Miami University
Originally presented on: 11/6/2013 15:35:00 - 15:55:00

PFCs are anthropogenic industrial compounds used in many applications, including surface coatings, surfactants, and flame retardants. They are, however, emerging contaminants due to their recalcitrant nature; they are persistent and very stable in air, water, and soil environments. As a result of their environmental stability, PFCs have been detected in water and soil samples all over the globe, and they have been reported as bioaccumulative, with detectable concentrations in humans and in animals. For reasons of their stability and bioaccumulative nature, and for possible health effects on humans, PFOA has been added to Annex B of the Stockholm Convention on Persistent Organic Contaminants.

In this research, the adsorption of PFOA has been studied using both commercially-available and lab-synthesized adsorbents in combination with both biocatalysts and inorganic catalysts. Immobilizing PFOA using adsorption combined with chemical or enzymatic degradation of PFOA was investigated and found to be superior to each technology alone.

(539c) Removal of Arsenic in Water With An Adsorbent Derived From Waste Concrete Sludge

Sasaki, T., Seikei University
Watanabe, M.,
Hongo, T., Seikei University
Iizuka, A., Tohoku University
Yamasaki, A., Seikei University
Originally presented on: 11/6/2013 15:55:00 - 16:15:00

Introduction:

Novel adsorbents for removal of arsenic in waster were prepared from concrete wastes including waste cement powder and concrete sludge. The waste cement powder is fine particle of which the main component is hydrated cement, which is generated as waste in the recycling process of aggregates from waste concrete. The concrete sludge is waste of raw concrete generated and disposed of from construction sites.

Experimental:

(1) Adsorbent from waste cement. The waste cement particle supplied from a waste concrete recycling plant was used as an adsorbent for the arsenic removal experiment after sieving. The diamter range was 53 to 106 μm.

(2) Adsorbent from concrete sludge. The concrete sludge, which is generated in a concrete product plant, is first diluted with water; this pretreatment could avoid the hardening reaction of concrete. Then solid was separated from aqueous phase, and dried under various temperatures. The dried solid was then pulverized and thieved. The adsorbent so prepared with the diameter range of 53 to 106 μm was used for the arsenic removal experiment.

The arsenic removal experiment was conducted in a batch reactor with stirring under room temperature. The concentration (As (V)) of the solution was changed in the range of 10 to 700 mg-As/L. The ratio of solution to the adsorbent was changed in the range of 67 to 400. The stirring rate was fixed at 350 rpm, which is considered to be high enough to neglect the mass transfer resistance in the aqueous phase. The solution was sampled with filtration, and the concentration change was measured with ICP-AES.

Results and discussion:

It was confirmed that the arsenic concentration decreased with time for all the adsorbents studied. When adsorbent derived from concrete sludge was used, the final concentration of arsenic after 24 h for the initial concentration at 100 mg/L was lower than 0.1 mg/L, which is below the effluent standard for the waste water in Japan. In addition, the removal rate was much higher with the adsorbents from concrete sludge than those from waste cement. The capacity of the arsenic removal with the adsorbent prepared from concrete sludge was about 170 mg-As/g-adsorbent based on the removal experiments of various initial concentrations. This capacity is comparable with the ones of the conventional adsorbents. The arsenic was removed mainly in the form of calcium salts according to XRD analysis of the solid phase after removal. By considering the preparation cost, which is derived from waste, the present adsorbents could be widely applicable to the water treatment, especially in developing countries.

(539d) Effect of Organic Anion On Selective Adsorption

Koong, L. F., The Hong Kong University of Science and Technology
Lam, K. F., The Hong Kong University of Science and Technology
Barford, J., The Hong Kong University of Science and Technology
Mckay, G., The Hong Kong University of Science and Technology
Originally presented on: 11/6/2013 16:15:00 - 16:35:00

Metal containing waste such as electronic waste is the fastest growing waste in the world and they pose a great challenge to waste management and disposal. Today we are not only interested in metal removal, but also recovery especially from waste, due to the dwindling metal reserves. Many researches have looked into leaching metals from batteries, printed circuit boards, magnet etc. Adsorption could make a good application in metal recovery from leachate due to its simple application and ability to operate at low concentration levels. Selective adsorption will be especially useful in recovering individual metal, especially in wastewater or leaching medium where multiple metal species can be present. The selectivity is strongly dependent on the interactions between the functional groups on the adsorbent and the metals. Another factor that could affect selectivity is the metal speciation. Organic anions are able to form complexes with metals, which can then affect the adsorption capacity, and most importantly, selectivity. Organic anions can be naturally found in water environment, the organic acids and derivatives also find wide applications in industry, such as textile, drug, paper, electroplating and so on, and therefore it is important to investigate the influence of anion on adsorption mechanism. This paper aims to demonstrate the effect of organic anions on selectivity of metals using the example of Copper and Cadmium in different anion system: nitrate, sulfate, acetate and formate salt. Iminodiacetic acid resin and chitosan are used in this study. It was found that when organic anions of COOH-based are present, the adsorption capacity is affected but not the selectivity, which Cu is still preferentially adsorbed. However, when amine based organic anions are present, the selectivity can be affected due to stronger complexation between the metal and anions. To demonstrate this EDTA is added to into the nitrate system of Cu-Cd and it is observed that Cd can be selectivity adsorbed at a right concentration of EDTA. EDTA is also removed by adsorption to return the selectivity back to Cu. Total Organic Carbon Analyzer is used to measure the concentration of EDTA in this case.

(539e) Optimization of Commercial Para-Xylene (PX) Simulated-Moving-Bed (SMB) Simulation Using Extended Langmuir Isotherm

Dang, T. T., Hankyong National University
Lim, Y. I., Hankyong National University
Originally presented on: 11/6/2013 16:35:00 - 16:55:00

In the petrochemical industry, Para-xylene is the high valuable compound, which exits in the mixture of ethylene benzene (EB), toluen and xylene isomers (para-xylene (PX), meta-xylene (MX) and ortho-xylene (OX)) after the catalytic reforming reaction. It is very difficult to separate PX from this mixture by the conventional distillation method because of components's closed boiling temperatures. SMB is the continuous counter-current multicolumn chromatographic process, which has advantages to achieve the high PX's purity. The objective of this study is to optimize the operating conditions of SMB unit using the design software, FAST-CHROM/SMB with the Extended Langmuir Isotherm (ELM). To get this objective, the dead volume and flushing scheme is developed and a novel Fowler isotherm adsorption is applied. The ELM parameters are obtained from single adsorption and binary adsorption experimental data. The optimization procedure comprises two levels to optimize bed-configuration parameters and flow rates. Optimal operating conditions of 7 and 8-zone are found by parametric study and maximizing recovery with a desired purity more than 99.7%.

(539f) Synthesis of Mesoporous Carbons: Oxidative Modification and Hydrothermal Carbonization for Efficient Chemical Activation

Jain, A., National University of Singapore
Srinivasan, M. P., National University of Singapore
Balasubramanian, R., National University of Singapore
Originally presented on: 11/6/2013 16:55:00 - 17:15:00

Preparation of mesoporous carbons and activated carbons in general has always been a resource intensive process. Generally, chemical activation is carried out by simple mixing and drying of the activating agent with the raw material, which might lead to ineffective utilization of the former. Improvised chemical activation using hydrothermal treatment and incorporation of oxygen functional groups can lead to more effective chemical utilization and therefore render the process less resource-intensive. In this work, the effect of oxygen functional groups present in the hydrothermally treated precursor on porosity and mesoporosity was investigated. The amount of oxygen containing functional group was varied by appropriate changes to the processing and activating conditions. Coconut shell precursors were subjected to hydrothermal treatment in the presence of hydrogen peroxide as the oxidizing agent at various temperatures. The hydrothermal pre-treatment was followed by chemical activation with $ZnCl_2$ accompanied by pyrolysis. The samples were characterized by surface area methods, titrations and microscopy. Hydrothermal treatment in the presence of H_2O_2 results in significant improvement in the mesoporosity of the final product compared to conventionally processed samples. Samples pretreated with H_2O_2 at 200^0C resulted in highest mesoporosity and this was consistent with the observation of the highest amount of oxygen functional groups in precursor. By the use of H_2O_2 and hydrothermal treatment there is an increase in mesoporous area by 144% and BET surface area by 35%. The hydrothermal pretreatment and surface oxygen functional groups in the precursor, coupled with the use of H_2O_2 as the oxidizing agent, together contribute to improved chemical activation, thereby increasing mesoporosity.

(539g) Biosorption of Heavy Metals Using An Acid Activated Walnut Shell

Andrade-Espinosa, G., Universidad Autonoma de San Luis PotosÃ
Berber-Mendoza, M. S.,
Bernal-Jacome, L. A., Universidad Autonoma de San Luis Potosi
Medellin-Castillo, N. A., Universidad Autonoma de San Luis Potosi
Originally presented on: 11/6/2013 17:15:00 - 17:35:00

In this research the removal of heavy metals [Pb (II), Cd (II), Zn (II)] ions from aqueous solutions by chemically modified walnut shell was studied. A response surface design analysis was performed to screen the variables affecting heavy metals removal efficiency. The effects of citric acid concentration, temperature and time of modification on the adsorption capacity of metals were examined in batch reactors. The walnut shell was characterized before and after modification by Scanning Electron Microscopy (SEM), Fourier Transform Infrared Spectroscopy (FTIR), Thermogravimetric Analysis (TGA) and acid/base titration. Also, charge distribution of the walnut shell was determined by equilibrium. Plus pH and temperature effects on the adsorption capacity of Pb, Cd and Zn were investigated by adsorption isotherms that were conducted in batch adsorbers. The results of the surface design showed that metal ions adsorption is highly dependent on the modification time and less dependent on the modification temperature and citric acid concentration, accordingly. On the other hand, SEM micrographs revealed no significant changes in the surface of the walnut shell after modification. This indicates that the degree of modification does not impair the physical properties of the shell. In addition, total acid sites content of modified walnut shell decreased from 2.6 to 2.4 $mmolg^{-1}$ which was reflected in a less amount of carboxylic sites (from 0.78 to 0.56 $mmolg^{-1}$). These last findings were supported by FTIR and charge distribution; treated material showed less accentuated energy absorption peaks at wavelengths between 1750 and 1700 cm^{-1} and the point of cero charge (PCC) changed 1 unit to more basic values of pH, which is attributed to lower concentration of carboxylic acids (COOH). TGA analysis showed a small shoulder at around 380 °C and an intensive peak at around 350 °C. These minor and major reactions can be attributed to the hemicelluloses and cellulose decomposition, respectively (Acikalin, 2011). In the other side, the sorption experiments reported, that, the heavy metals removal increased as pH rise from 3 to 7; resulting a maximum sorption capacity of 18, 12, 4 mgg^{-1} of Pb (II), Cd (II) and Zn (II), respectively, for untreated walnut shell. The sorption capacity of modified walnut shell was 2% higher than the untreated material. Finally, it was found a slight increase onto the sorption capacity as the temperature rise from 15 to 35 °C. These findings indicate that the untreated and modified walnut shell studied herein have a potential application in the removal of contaminants presents in aqueous phase.

(540a) Biomimetic Properties of Carbon Nanotubes In Vivo

Villegas, J. C.,
Valiente, R.,
Rodriguez-Fernandez, L.,
Gonzalez, J.,
Fanarraga, M. L.,
Originally presented on: 11/6/2013 15:15:00 - 15:37:00

Multiwalled carbon nanotubes (MWCNTs) are well documented to invade the cytoplasm of eukaryotic cells and to interact with actin microfilaments and microtubules due to the many attributes they share with these cytoskeletal filaments. In particular, MWCNTs and microtubules have a similar length-to-with ratio and resiliency; they are both hollow cylinders; they both self-assemble, and form bundles to enhance their resistance. These features that provide MWCNTs with the intrinsic ability to interact with microtubules have also been suggested to lead, among others, to severe proliferative defects resulting in aneuploidy and cell death.

Our work demonstrates how MWCNTs display bio-mimetic interactions with microtubules at the molecular level assembling biosynthetic filaments *in vivo*. As a result of the different dynamical behavior of MWCNTs and microtubules, their interaction gives rise to the stabilization of the tubulin polymers. In fact, carbon nanotubes are exceptionally stable, while microtubules are highly dynamic polymers that are continuously undergoing assembly/disassembly cycles *in vivo*. Therefore, these interactions severely perturb microtubule dynamics, disorganizing the centrosomal microtubule pattern in the cell leading to aberrant mitotic spindle assemblies, chromosomal malsegregation between daughter cells, and often to mitotic catastrophe and cell death. Thus, our work gives an consistent explanation aneuploidy and mitotic failure in cells exposed to MWCNTs.

(540b) Differential Analysis Of Single Wall Carbon Nanotubes Cellular Uptake Mechanism

Eldawud, R., West Virginia University
Dong, C., West Virginia University
Sargent, L. M., National Institute for Occupation Safety and Health
Rojanasakul, Y., West Virginia University
Dinu, C. Z., West Virginia University
Originally presented on: 11/6/2013 15:37:00 - 15:59:00

Biomedical applications of Carbon nanotubes (CNTs) are currently being constrained by many uncertainties regarding the cellular mechanisms associated with their uptake and delivery as well as their fate inside the biological systems. In this research, a new approach is used to quantify the cellular mechanisms of a wide library of surface modified CNTs incubated with human lung epithelial cells (BEAS-2B) in real time. This non-invasive technique relies on an electrical cell impedance sensing system (ECIS) used as a proxy to measure morphological changes and cellular interactions upon exposure to different concentrations and functionalities of CNTs. Such measurements are correlated to the cellular behavior under the influence of different parameters like endocytotic inhibitors or electroporation and further complemented by classical microscopic, apoptotic and flow cytometry assays to derive cellular structure-function relationships. Our real time results provide an insight to the underlying cellular mechanisms responsible for the uptake and delivery of CNTs to the biological systems and promises to extend our understanding of such cellular-based interfaces to advance the biomedical applications of CNTs.

(540c) Development of in-Vivo Toxicity Screening Benchmarks for Complex Engineered Nanomaterials

Mahoney, S., Univeristy of Wyoming
Najera, M., Mascaro Center for Sustainable Innovation, University of Pittsburgh
Bai, Q., University of Pittsburgh
Burton, E., University of Pittsburgh
Veser, G., University of Pittsburgh
Originally presented on: 11/6/2013 15:59:00 - 16:21:00

Functional nanomaterials are developing at an ever accelerating pace and are already finding widespread application in consumer products including cosmetics, sporting goods, clothing, and consumer electronics. However, increasing evidence indicates that nanomaterials are showing elevated toxicity, motivating the need to establish appropriate toxicity tests for these emerging technology. Silica nanomaterials have been identified for a wide range of applications including biomedical imaging, drug delivery, catalysis, separations, and sensors, as they constitute a versatile, easily modifiable template that has been established as being non-toxic. However, the rapidly increasing volume of nanomaterials research and production requires development of modern high-throughput *in vivo* nanotoxicity screenings for these nanomaterials. Zebrafish (*Danio rerio*) is rapidly emerging as a well-suited model for such studies due to prolific breeding, embryo tissue transparency, rapid development time, and the relative simplicity of controlled toxin exposure.

Porous silica nano- and microparticles are accepted to be non-toxic, thus leading to the hypothesis that embedding metal nanoparticles in silica could reduce or entirely mitigate nanoparticle toxicity while still providing accessibility of the metal nanoparticle surface and hence maintaining the functionality of the embedded nanoparticle. Three different complex nanostructures are investigated in this work including surface-deposited metal nanoparticles on silica nanoparticle supports, embedded metal nanoparticles in silica and encapsulated metal nanoparticles in silica. In the present work, the toxicity of carefully synthesized Ni-silica nanomaterials with these complex, but well-controlled nanostructures is investigated using zebrafish as the *in vivo* toxicity model.

Comparing the different structures, our results from the *in vivo* zebrafish toxicity study demonstrate that all three silica configurations significantly reduce the toxicity of metal nanoparticles compared to the equivalent dosing of an analogous metal salt. While the engineered nanomaterials showed limited dissolution of metal ions from the structured materials into the media solution, nickel uptake was still detectable at ng levels within the zebrafish larvae tissue. Interestingly, relative to free metal ion concentrations, metal uptake was significantly enhanced for the engineered nanomaterials, suggesting the presence of a "trojan horse" transport mechanism into the fish. All three engineered nanomaterials showed high zebrafish survival rates as compared to the Ni salts, hence necessitating more sensitive means for detection of possible toxic effects. Therefore, the studies were extended to include motility analysis which allowed for a much enhanced sensitivity compared to endpoints of survival and developmental deformations, yielding statistically significant dose-dependent results for the different nanomaterials. Our current work includes extension of these studies to a range of nanoparticle sizes and metals to further investigate the effect of nanostructuring on nanomaterial toxicity.

(540d) NLRP3 Inflammasome Activation Induced By Long Aspect Ratio Engineered Nanomaterials: Role of Oxidative Stress

Sun, B., University of California, Los Angeles
Xia, T., University of California, Los Angeles
Originally presented on: 11/6/2013 16:21:00 - 16:43:00

NLRP3 inflammasome is an intracellular multi-protein complex assembled and activated upon exogenous stimuli including engineering nanomaterials (ENMs). Its activation has been increasingly implicated in nanomaterial-induced toxicity. However, the detailed mechanisms of inflammasome activation induced by ENMs are still not well understood. In this study, we examined the interaction mechanisms of ENMs with NLRP3 inflammasome using a library of long aspect ratio ENMs. We identified that besides the lysosomal damage and cathepsin B release, ROS generated through NADPH oxidase was found to play an important role in long aspect ratio ENM-induced NLRP3 inflammasome activation. The ROS generation could further lead to secondary cytosolic as well as mitochondrial ROS. In the meanwhile, we demonstrated that suppression of oxidative stress could attenuate inflammasome activation *in vitro* and the development of lung fibrosis upon ENM exposure *in vivo*. This study delineates mechanisms of NLRP3 inflammasome activation induced by long aspect ratio ENMs and identifies the role of oxidative stress in ENM-induced chronic lung inflammation and fibrosis.

(540e) Effects of Aerosolized Sub-Micron Particles On The Interfacial Properties of Lung Surfactant Models

Farnoud, A. M., University of Iowa
Fiegel, J., The University of Iowa
Originally presented on: 11/6/2013 16:43:00 - 17:05:00

Pulmonary surfactant plays an important role in lung stability by reducing the surface tension of the alveolar fluid and avoiding alveolar collapse. Changes to the interfacial properties of this surfactant can lead to potential health implications such as increased work of breathing and impaired gas exchange. Inhaled particles that deposit in the alveolar region of the lung come into contact with the pulmonary surfactant and might alter its interfacial properties. A few recent *in vitro* studies have reported surfactant dysfunction following exposure to small particles. However, these studies have been focused on colloidal particle suspensions and little is known about particle-surfactant interactions when dry particles are aerosolized onto surfactant films. The present study seeks to improve the current understanding of the effects of inhaled particles on surfactant function by examining particle-surfactant interactions after exposing surfactant films to aerosolized particles.

Langmuir films of dipalmitoyl phosphatidylcholine (DPPC) and Infasurf (calf lung surfactant extract) were used as lung surfactant models and their interactions with carboxyl-modified polystyrene particles (200 nm) were studied. Surfactant films were spread in a Langmuir trough and their dynamic surface tension was probed using a Wilhelmy plate balance. A Dry Powder Insufflator^{TM} was used to aerosolize dry particles onto surfactant films. Following exposure to particles, surfactant films were subjected to two modes of surface area changes: 1. Surface compression from the surface tension of pure subphase until film collapse and 2. Compression and expansion within the range of physiologically relevant surface tension values to mimic breathing cycles.

Aerosolization of particles on both DPPC and Infasurf films resulted in a shift in the surface area of surfactant collapse and a slight reduction in the minimum surface tension. However, particles induced significant changes (~20 mN/m) in the minimum surface tension of DPPC when aerosolized in the lung relevant surface tension range. In contrast, only minor changes were observed in the maximum surface pressure of Infasurf following aerosolization in the physiologically relevant surface tension range suggesting that Infasurf films are more stable than DPPC in response to particle exposure. Further experiments with different particle-surfactant interaction times showed that particles effects on Infasurf interfacial properties changed as a function of time suggesting that particles affect surfactant function by adsorption of surfactant components. These results provide information on the effects of particles on surfactant function following inhalation, which is of great importance for the fields of pulmonary drug delivery and environmental and occupational health.

(540f) Carbon Nanotubes Induce Invasion of Human Mesothelial Cells Through Matrix Metalloproteinase-2

Lohcharoenkal, W., West Virginia University
Dinu, C. Z., West Virginia University
Stueckle, T., National Institute for Occupational Safety and Health
Wang, L., National Institute for Occupational Safety and Health
Rojanasakul, Y., West Virginia University
Originally presented on: 11/6/2013 17:05:00 - 17:27:00

Carbon nanotubes (CNT) have increasingly been used in a wide variety of applications. However, because of their needle-shape morphology, biopersistence and mode of exposure similar to those of asbestos which is a known carcinogen causing lung mesothelioma, there has been a great concern about their potential carcinogenicity. In this study, we investigated the effect of long-term exposure of CNT on proliferative and invasive properties of human pleural mesothelial cells. We found that sub-chronic exposure of human mesothelial MeT5A cells to low-dose non-cytotoxic concentration of single- and multi-walled CNT (0.02 $\mu g/cm^2$ for 6 months) in culture induced malignant transformation of the cells as indicated by their increased cell invasion and migration. An up-regulation of several key genes known to be important in cell invasion, notably matrix metalloproteinase-2 (MMP-2), was observed in the exposed mesothelial cells as determined by real-time PCR. Western blot analysis of protein expression confirmed the up-regulation of MMP-2 in the exposed cells. Whole genome microarray analysis further indicated the importance of MMP-2 in the invasion gene signaling network of the cells. Knockdown of MMP-2 in the CNT-transformed cells by short-hairpin (sh)RNA stable transfection effectively inhibited MMP-2 activity and expression in these cells as well as their invasive and migratory activities. Preliminary animal studies demonstrated colonization of CNT-transformed cells in the lungs of mice which was absent in mice receiving shMMP-2 knockdown cells. These results indicate MMP-2 as a key regulator of the aggressive phenotype of CNT-transformed cells. This study fortifies the carcinogenic potential of CNT and reveals a novel mechanism that may be important in CNT-induced mesothelioma.

(540g) Neoplastic-Like Transformation Ability of Carbon Nanotubes On Small Airway Epithelial Cells: Linking Toxicogenomic Signatures and Lung Cancer Hallmarks

Stueckle, T., National Institute for Occupational Safety and Health
Mishra, A., National Institute for Occupational Safety and Health
Derk, R., National Institute for Occupational Safety and Health
Meighan, T., National Institute for Occupational Safety and Health
Castranova, V., National Institute for Occupational Safety and Health
Wang, L., National Institute for Occupational Safety and Health
Rojanasakul, Y., West Virginia University

Originally presented on: 11/6/2013 17:27:00 - 17:49:00

Concern for increased risk of CNT-associated lung carcinogenesis has been raised due to asbestos-like high aspect ratio, deep pulmonary deposition, and biopersistance. Studies evaluating long-term human health risks associated with chronic pulmonary CNT exposures compared to other well-characterized particulates (e.g. asbestos) are lacking. To address this knowledge gap, we conducted subchronic *in vitro* exposures of dispersed single walled CNT (D-SWCNT), multi-walled CNT (D-MWCNT) and crocidolite asbestos (ASB) to human small airway epithelial cells (SAEC) to assess their neoplastic transformation potential, an early marker of carcinogenesis. Ultrafine carbon black (D-UFCB) and dispersant-only exposed cells (DISP) served as negative controls. SAEC were exposed for 25 weeks to 0.02 $\mu g/cm^2$ and evaluated for morphological transformation, cancer cell phenotype and subjected to whole genome expression signature profiling. Subchronic *in vitro* exposure to D-SWCNT and D-MWCNT resulted in a more aggressive neoplastic-like transformation phenotype of exposed SAECs compared to asbestos. Furthermore, genome expression signature analysis identified oncogene signaling mechanisms in CNT-exposed SAEC that was substantially different from pro-inflammatory signaling in asbestos-exposed SAEC. Correlation feature selection strategies identified known lung cancer markers and robust subsets of genes that could be used to develop specific gene bio-signatures for CNT-exposed cells. In conclusion, toxicogenomic signature profiling in a CNT neoplastic-like transformed lung cell model identified particle-specific gene markers and known lung cancer markers which can potentially aid in assessing CNT exposure and detection of early disease bio-signatures.

Disclaimer: The findings and conclusions in this abstract are those of the authors and do not necessarily represent the views of the National Institute for Occupational Safety and Health.

(546a) Controllability of Complex Networks for Sustainable System Dynamics

Benavides, P. T., University of Illinois at Chicago
Diwekar, U. D., Vishwamitra Research Institute
Cabezas, H., U.S. Environmental Protection Agency
Originally presented on: 11/6/2013 15:15:00 - 15:37:00

Successful implementation of sustainability ideas in ecosystem management requires a basic understanding of the often non-linear and non-intuitive relationships among different dimensions of sustainability, particularly the system-wide implications of human actions. This basic understanding further includes a sense of the time scale of possible future events and the limits of what is and is not likely to be possible. With this understanding, systematic approaches based on control theory can then be used to develop policy guidelines for the system. Therefore, controllability of the system is very important to determining long term sustainability of the system. A recent article in the journal Nature (1) presents a new analytical approach to study the controllability of complex systems. We apply this approach to three dynamic systems developed to study the sustainability of our planet. These three systems consist of an ecosystem based on wild and domesticated compartments, the ecosystem model with an industrial system and a simple economic model, and an integrated model involving the ecosystem, industrial systems, and energy producers. We argue that controllability of this system is linked with the long term sustainability, and we present our arguments in the light of previous studies of these systems.

(1) Liu, Y. Y., Slotine, J. J., Barabási, A. Controllability of complex networks. *Nature.* **2011**, *473*, 167

(546b) Sustainable Process Decision Making With Greenscope

Smith, R. L., U.S. Environmental Protection Agency
Ruiz-Mercado, G. J., U.S. Environmental Protection Agency
Gonzalez, M. A., U.S. Environmental Protection Agency
Originally presented on: 11/6/2013 15:37:00 - 15:59:00

The U.S. Environmental Protection Agency is developing a method and tool called GREENSCOPE (Gauging Reaction Effectiveness for the Environmental Sustainability of Chemistries with a multi-Objective Process Evaluator) to support sustainable processes. GREENSCOPE can be used to evaluate existing processes or in the design of new ones. With 140 indicators for objectives in four basis areas: environment, economics, energy, and efficiency, GREENSCOPE provides information on process sustainability. Results for the various objectives show where processes are performing well and where improvements could be made.

One of the most difficult issues of dealing with multi-objective results is determining a scientific way to incorporate the process information and user preferences. These preferences represent the values of the decision maker, and often scientists and engineers are uncomfortable with the methods available for integrating values into objective and quantifiable science. While the existence of values in decision making cannot be eliminated, one can navigate a path that provides scientific clarity.

Much of the effort needed to provide clarity in decision making has already been incorporated into the GREENSCOPE methodology. Previous work has developed absolute sustainability limits for the indicators in GREENSCOPE [1], which provides ranges that delineate worst case and target values. Decision makers can define marginal rates of substitution (i.e., points of equal tradeoffs) between the range-bound indicators. Inputs such as the marginal rates of substitution and GREENSCOPE indicator results provide equations that describe decision making for multi-objective process evaluations, with a strictly mathematical method solving the equations for various inputs [2]. The method will be exemplified through various results for a terephthalic acid process.

[1] Ruiz-Mercado, G. J.; Smith, R. L.; Gonzalez, M. A., Sustainability Indicators for Chemical Processes: I. Taxonomy. *Ind. & Eng. Chem. Res.,* 51, 2309-2328 (2012).

[2] Smith, R.L.; Ruiz-Mercado, G.J., A Method for Decision Making using Sustainability Indicators. Submitted, *Clean Tech. & Env. Policy* (2012).

(546c) Innovative Chemical Engineering Applications to Control Air Emissions From Oil Refineries

Nair, D. C., Belco Technologies Corporation

Originally presented on: 11/6/2013 15:59:00 - 16:21:00

Wet gas scrubbing technology is a well proven and reliable method for controlling flue gas emissions of particulate, SOx and NOx to the atmosphere from fluid catalytic cracking units (FCCUs) used in oil refineries.

This paper will show that there has been a viable technology for the past several years that can minimize environmental impacts using innovative chemical engineering practices along with design of unique proprietary equipment that provides non-stop operation for at least five years. The paper will provide details of the unique design that have been used at well over 50 installations at oil refineries in the U.S and overseas. A description of the wet scrubbing system utilized for this project is described. Specifically, this description will address the methodology for removing particulate, SO_2 and NOx and the process for treating the liquid effluent from the wet scrubbing system. Also addressed are the aspects of three designs that allow for achieving different levels of particulate emissions.

This paper is also an attempt to compare a presentation made at the AIChE Spring meeting that a Genetic Algorithm Approach for Optimal Control of Air Emissions from an Oil Refinery was possible.

The reduction of environmental impacts is an essential step towards the improvement of sustainability for any given producer.

(546d) Direct and Indirect Emissions From Mobil Air Conditioner Units in Transportation By Truck and Rail

Blowers, P., The University of Arizona
Dougless, A., U of Arizona
Gerwe, B., U of Arizona
Mesta, A., U of Arizona
Rugel, C., U of Arizona
Originally presented on: 11/6/2013 16:21:00 - 16:43:00

The U.S. EPA has announced repeatedly that hydrofluorocarbons (HFCs) will become regulated, not for their direct impacts on the environment, but because of their contributions to global climate change. HFCs are used in many industrial processes as solvents and refrigerants because they are relatively inert, inexpensive and have physical properties that allow them to function well. The use of HFCs is largest in the refrigeration sector. R-134a (1,1,1,2-tetrafluoroethane is the dominant refrigerant used currently in mobile air conditioner systems and has emissions of about 38.6 x 10^6 kg/year per year from all sources, with approximately 47% of that being due to vehicle emissions. The European Union took steps in January 2011 to eliminate fluorinated greenhouse gases in certain classes of vehicles with 100 year GWP's greater than 150 as defined by the third assessment report adopted by the IPCC, including R-134a. Currently passed legislation in Europe also calls for a complete phase-out by January 2017.

Mobile air conditioning units (MAC) are in use in several applications, with their use in passenger cars having been evaluated for greenhouse gas emissions in the prior literature. If HFCs are banned as a class of compounds, alternatives will need to be developed or exploited. While MAC and environmental impacts have been investigated for passenger cars, there are other MAC units where the temperature constraints and demands may lead to a more restricted range of possible refrigerants. Mobile air conditioning units used in the shipping of temperature sensitive products, like foods and medication, is common on both tractor trailers and on refrigeration, reefer, units installed on rail cars. With the large number of US and worldwide refrigerated shipping units, evaluations of the emissions for this application of refrigeration need to be obtained before deciding if banning hydrofluorocarbons is indeed better for the environment.

This work uses chemical engineering heuristics and design parameters, combined with actual operating conditions and design specifications of manufactures to estimate the direct and indirect greenhouse gas emissions for MAC using diesel powered reefer units in transport. We find that many refrigerants that can be used in dedicated non-mobile units where water cooling is possible fail for this application. The remaining refrigerants are small in number and it appears that banning HFCs would be unattractive due to the indirect emissions.

(546e) Multi-Objective Optimization of International Economies Via Multi-Regional Input-Output Analysis: Application to the US Economy

Pascual, J.,

Guillén, G., Universitat Politécnica de Catalunya

Jimenez, L., University Rovira i Virgili

Cortés, D., University Rovira i Virgili

Originally presented on: 11/6/2013 16:43:00 - 17:05:00

Multi-objective optimization of international economies via multi-regional input-output analysis: Application to the US economy

J. Pascual-González[a], G. Guillén-Gosálbez[a][*] and L. Jiménez[a]

[a]Departament d'Enginyeria Química, Universitat Rovira i Virgili, Av. Països Catalans 26, Tarragona 43007, Spain

Abstract

In recent years, new environmental policies have been developed to reduce the environmental impact of human activities. Designing these policies in an effective manner is challenging because it requires a detailed knowledge of the whole economic and environmental system and the way in which they interact with each other.

In terms of economics, input-output analysis (IO) is widely used to predict changes in an entire economy according to changes in the demand of a single (or several) sectors. This method studies the effects that different sectors have on the economy as a whole. Purely economic IO tables can be modified so as to include environmental aspects, giving rise to environmentally extended input-output tables (EEIO). These tables contain information on the environmental assessment of the economic transactions taking place in a region, but provide no guidelines on how to reduce such an impact.

We present in this work a systematic approach to reduce the environmental impact of a whole economy. Our strategy combines multi-objective optimization and multi-regional input-output models within a single unified framework that allows identifying the activities whose control and regulation can lead to larger reductions of impact at a minimum decrease in gross domestic product (GDP).

We have applied our tool to the economy of the United States in order to identify the best policies to be implemented in practice for optimizing simultaneously the economic and environmental performance. We have employed in the calculations the World Input-Output Database (WIOD), which was originally developed to analyze the effects of globalization on trade patterns, environmental pressures, and socio-economic development across a wide set of countries. WIOD describes the economic inputs and outputs in monetary terms of 35 manufacturing sectors covering 27 EU countries and 13 other major countries in the world for the period 1995 to 2009.

Our approach provides valuable insight into how to reduce the impact globally. By adopting this birds' eye view of a whole economy, it is possible to determine how the final demand (and the total production) of every sector in an economy should be modified so as to improve simultaneously the economic and environmental performance of an economy.

---

[*] Corresponding author: tel. +34-977 55 86 18 email: gonzalo.guillen@urv.cat

(546f) Urban Systems and Sustainability: Comparative Analysis of Multiple Metropolitan Statistical Areas in Ohio, USA

Gonzalez-Mejia, A., US EPA - ORISE
Eason, T., U.S. Environmental Protection Agency
Cabezas, H., U.S. Environmental Protection Agency
Suidan, M., University of Cincinnati
Originally presented on: 11/6/2013 17:05:00 - 17:27:00

Urban systems have a number of factors (e.g., economic, social and environmental) that can potentially impact their growth, change and transition. As such, assessing and managing these systems is a complex challenge. While, tracking trends of key variables may provide some insight, identifying the critical characteristics that truly impact the dynamic behavior of these systems is difficult. As an integrated approach to evaluate real urban systems, this work contributes to the research on scientific techniques for assessing sustainability. Specifically, it proposes a practical methodology based on the estimation of dynamic order, for identifying stable and unstable periods of sustainable or unsustainable trends using Fisher Information (FI). In this work, the dynamic behavior of main six Metropolitan Statistical Areas (MSAs) in Ohio was evaluated by using 29 social and 11 economic variables to characterize each system from 1970 to 2009. For a complete regional sustainability analysis, these Ohio MSAs were compared to each other according their stability and population size over the period of study, to assess the impact of social and economic characteristics on system stability at geographic scales, which are critical to policy and management. Results from this multi-year assessment indicated that the Columbus MSA is the most stable of the MSAs understudy, followed by the Cincinnati MSA. Further, while there were unique trends in each MSA, changes in family structure and income appeared to correspond with the patterns experienced, particularly near the end of the study period (2000 – 2009). While the approach presented is not a cause and effect analysis, it points out characteristic features of complex systems that are otherwise very difficult to see by inspecting the individual variables, which characterize the system over time. Further, this approach intends to aid in locating significant deviations in the system's trajectory and determine whether these shifts corresponds to a new path characterized by sustainable (e.g., desirable growth rate) or unsustainable trends (e.g., decrease in the income growth rate). This work contributes holistically to the research on evaluating complex urban systems and affords the ability to identify and distinguish local and regional drivers of change. The methodology described can be used by scientists, planners and policy makers interested in evaluating dynamic changes in complex systems and identifying potential drivers at multiple spatial and temporal scales.

(546g) Environmental Impact Minimization of Vaccine Supply Chain for Developing World

Mousavi, B., University of Colorado Denver
Karunanithi, A. T., University of Colorado Denver
Originally presented on: 11/6/2013 17:27:00 - 17:49:00

People in low and middle income countries are most vulnerable to environmental, climate change and human health impacts. Vaccines offer enormous public health benefits (avoid disease and reduce treatment costs) while emissions from vaccine supply system result in environmental/health impacts (Green house gases and criteria pollutants) in particular, immunization supply systems of low and middle-income countries are not optimized to reduce environmental impacts. Current immunization supply systems are optimized to minimize direct costs. Such an approach usually results in higher levels of pollutants leading to lower public health outcomes which in terms results in higher treatment costs to the society. This nexus is not fully understood or recognized by public health programs. A truly optimized immunization supply system will not only consider direct costs but also indirect treatment costs averted by minimizing public health burdens to the society.We will present an approach that utilizes LCA methodologies to optimize the life cycle of vaccines that include production, packaging, transport, cold storage, use and disposal in the context of developing world.

(564a) Strategies in the Sustainable Design of Active Pharmaceutical Ingredients for Solvent Recovery Systems

Savelski, M. J., Rowan University
Slater, C. S., Rowan University
Cavanagh, E. J., Rowan University
Originally presented on: 11/6/2013 15:15:00 - 15:40:00

The Chemical Engineering Department at Rowan University has partnered with several pharmaceutical companies and the US EPA to explore more sustainable pathways in the manufacturing of active ingredients and to educate future generations of engineers and decision makers in the importance of green engineering in pharmaceutical syntheses. The conducted research has evaluated greener designs for solvent recovery targeting not only the reduction of the overall solvent consumption, but the associated waste, and life cycle emissions from cradle to gate. These studies have explored greener separation processes and greener solvents using both LCA and economic metrics. Software has been developed to guide decision making and optimize processes based on these desired outcomes. When integrated correctly, solvent recovery is a viable option to reduce significantly the carbon footprint of drug synthesis as well as operating costs. Life cycle analysis of pharmaceutical manufacturing routes involving solvents such as acetonitrile, isopropanol, methanol, tetrahydrofuran, and methyltetrahydrofuran have been conducted. This paper will present the approaches taken to determine the optimal application of greener technologies to minimize emissions and make pharmaceutical manufacturing more sustainable.

(564b) Towards a Complete Solution of Allocation in Life Cycle Inventories

Hanes, R., The Ohio State University
Cruze, N.,
Goel, P. K., The Ohio State University
Bakshi, B. R., The Ohio State University
Originally presented on: 11/6/2013 15:40:00 - 16:05:00

The need for allocation arises in life cycle inventories when a process results in multiple products. In such a situation, resources and emissions embodied in the inputs need to be partitioned between the products. Popular approaches for accomplishing this task involve partitioning the embodied resources in proportion to characteristics of the outputs such as mass, energy, exergy or cost. Given the subjectiveness of this decision, many practitioners apply multiple allocation methods. Recent efforts have proposed the use of regression methods such as ordinary least squares and total least squares to solve the allocation problem. Allocation may be avoided by expanding the system boundary, but this approach can be tedious due to the need for more data.

We show that allocation can be understood as an ill-posed problem, analogous to the situation where the number of variables in simultaneous equations is larger than the number of independent equations. In such a situation, the number of solutions is infinite, and without additional information or assumptions, a single "correct" solution cannot be determined. Using a matrix framework for life cycle inventories, we devise a method of calculation whereby results from all possible allocation methods can be examined.

We use this insight to evaluate the special case where LCA is used to compare products that have the same use but are produced by different processes. Examples of such a case include comparing ethanol from various sources, gasoline, and other transportation fuels. We develop an approach to determine the sensitivity of the comparative LCA to allocation weights and show that sometimes the better product may be independent of the allocation approach. For cases when this does not happen, we propose an optimization approach to determine the "tipping curve" or the boundary across which the product preference changes.

The practical usefulness of this approach is demonstrated with two case studies. The first study compares the LCIs of 1,3-propanediol (PDO) produced from fossil fuels and from biomass. This comparison was found to be robust to allocation for nine resources and emissions out of the eleven considered: biomass PDO was preferred over fossil PDO for all choices of allocation weights. The second case study compared ethanol produced from corn stover hydrolysis, corn stover gasification, and corn grain fermentation. This comparison was not robust to allocation, although the two corn stover processes were preferred over the corn grain process for most choices of allocation weights. More information is needed to decide conclusively which process is preferred. Extension of the framework to incorporate this additional information is proposed.

(564c) Hybrid Techno-Economic Modeling Tool for Greener Chemicals Supply Chains

Yao, Y., Northwestern University
Chang, Y., Northwestern University
Thwaites, F., Northwestern University
Masanet, E., Northwestern University
Originally presented on: 11/6/2013 16:05:00 - 16:30:00

The demand of chemical products is rapidly increasing due to the growth of global economy. At the same time, the chemicals industry is a major source of global greenhouse gas emissions (GHG) because of its high energy consumption(1). As such, reducing the energy and emissions footprints of chemicals is a critical priority for environmentalists, manufacturers and policy makers. Promoted by this question, green supply chain management is receiving increasing attention because it can help chemical industry to reduce the overall environmental intensity and thereby offer "greener" products. Most supply chain initiatives and related studies apply process-based life-cycle assessment (LCA) techniques to model and quantify the environmental impacts and sustainability performance of chemicals(2). These LCA studies help people to understand the current environmental footprints of many chemicals, but they are not able to answer the question: how and where can we reduce the life-cycle energy and pollution footprints of these chemicals? This question requires the support of techno-economic technology potentials analysis, which is a vibrant facilities engineering and analysis field that has operated largely independent of LCA to date(3, 4).

A comprehensive decision support tool incorporating techno-economic analysis methods and data into the life cycle assessment (LCA) framework is presented in this paper. The techno-economic analysis includes detailed characterizations of discrete technologies for reducing energy use and environmental impacts, their remaining installation potential in various plants and economic sectors, and their relative cost and environmental performance when deployed in combination. With this analysis, the model can identify and incentivize cost effective improvements for green supply chains within the chemicals industry to produce maximal energy reduction at minimal environmental burden. It can provide chemical manufacturers with comprehensive understanding of the greatest opportunities and potential cost requirements for supply-chain energy savings and environmental improvements. Also, the adoption of new process and material can be easily incorporated into this model, which can help designers to prescreen the sustainability performance of their design on a higher level. This tool also provides policy makers with cost-effective and efficient policy options to promote the development of more sustainable chemical product supply chains.

References

1. ICCA. 2009. *Innovations for Greenhouse Gas Reductions* ICCA

2. Goldberg A, Holdaway E, Reinaud J, O'Keeffe S. 2012. *Promoting Energy Savings and GHG Mitigation Through Industrial Supply Chain Initiatives* Institute for Industrial Productivity

3. Masanet E, Matthews HS, Carlson D, Horvath A. 2012. *Retail Climate Change Mitigation: Life-cycle Emission and Energy Efficiency Labels and Standards : Final Report*: University of California, Berkeley

4. McKinsey & Company. 2009. *Unlocking Energy Efficiency in the U.S. Economy.*, McKinsey Global Energy and Materials New York, New York

(564d) Design Decision-Making Under Uncertainty to Identify Sustainable Integrated Forest Biorefinery Strategies

Sanaei, S., Ecole Polytechnique de Montreal
Stuart, P. R., Ecole Polytechnique de Montreal
Originally presented on: 11/6/2013 16:30:00 - 16:55:00

Integrated biorefinery into existing pulp and paper facilities is a potential game-changing solution for the transformation of North American forestry industry, and the manufacture of an expanded product portfolio. However not all biorefinery strategies are sustainable and not have the same level of risk. Thus it is crucial that unsustainable strategies be systematically screened-out from the list of possibilities at the early design stage. The methodology presented in this study assesses candidate biorefinery alternatives from economic, competitiveness and environmental perspectives using different process systems analysis tools (e.g, techno-economics, Life Cycle Analysis, etc.) for both deterministic and uncertain conditions. The results of these assessments are used to evaluate a set of sustainability criteria, despite of the conventional analyses which generally use only short-term profitability metrics for decision-making. These conflicting criteria were aggregated into a unique sustainability score for each candidate biorefinery strategy by conducting a multi-criteria decision-making (MCDM) panel by which candidate strategies are ranked using a sustainability perspective.

However the candidate biorefinery strategies imply different levels of technology and market risk due to different sources of uncertainties which should be addressed in decision making. Thus to make a realistic strategic decision, the sustainability of biorefinery strategies should be assessed under uncertainty. In this study, different sources of uncertainty including market, technology and panel uncertainties are considered in biorefinery decision-making. In order to address this objective, a stochastic risk analysis method (Monte-Carlo Analysis) was used to develop probability distribution functions for sustainability criteria, which were then employed in a second MCDM panel. The proposed methodology in this study not only takes into account a complete set of sources of uncertainty, but also reflects the risk attitude of decision-makers in utility functions applying risk aversion theory. Comparing the results of deterministic and uncertain conditions shows that by evaluating uncertainty and quantifying risk attitude of panelists, the basis for the panel decision changes. This confirms the necessity of systematically addressing risk in biorefinery strategic decision making.

(564e) MILP-Based Clustering Method for Multi-Objective Optimization: Application to Environmental Problems

Oliva, D. G., Universitat Rovira i Virgili
GuillÃ©n-GosÃ¡lbez, G., Universitat Politecnica de Catalunya
JimÃ©nez-Esteller, L., Universitat Rovira i Virgili
Originally presented on: 11/6/2013 16:55:00 - 17:20:00

Multi-objective optimization (MOO) has recently emerged as a useful technique in environmental engineering. One major limitation of this approach is that its computational burden grows rapidly with the number of environmental objectives, which causes difficulties regarding the computation and visualization of the Pareto solutions. In this work, we present several theoretical and algorithmic developments for grouping environmental objectives into clusters on the basis of which the multi-objective optimization can be performed, thereby facilitating the computation and analysis of the Pareto solutions. Particularly, we study how to group objectives of a MOO problem into meaningful clusters with the property that the minimization of any objective within a cluster will result in the minimization of the rest of objectives within the same cluster.

Our method is based on a novel mixed-integer linear program (MILP) that identifies in a systematic manner groups of objectives that behave similarly. The capabilities of the approach presented are tested through its application to several case studies, in which we compare the results of the MILP with those produced by a statistical method. The statistical method consists of a standard clustering algorithm.

The first case study deals with the design of petrochemical supply chains considering several environmental indicators simultaneously (Pozo et al., 2012). The second example deals with the design of bio-ethanol supply chains in Argentina considering 6 objectives (Kostin et al., 2012). The third example addresses the multi-objective optimization of hydrogen supply chains for vehicle use in Spain considering several Life Cycle Assessment impacts (Sabio et al., 2012). In the last example, we study an artificial dataset with 10 objectives and 41 points generated randomly. The goal is to test the performance of the statistical method when the objectives are poorly correlated.

Numerical results show that when the objectives are highly correlated, both methods (i.e., the MILP and the statistical method) lead to similar results, being the statistical approach the one with lower CPU times. In contrast, the statistical method leads to clusters with larger errors that differ from those produced by the rigorous MILP approach when the correlation between objectives is low. Our rigorous MILP-based approach is the only one that guarantees the minimum approximation error.

(564f) Emergy Method Analysis On Sludge Residues For Use From Waste-WATER Treatment Plant

Cano, N. A., Universidad Nacional de colombia
Gallego, D.,
Velasquez, H. I.,
Originally presented on: 11/6/2013 17:20:00 - 17:45:00

EMERGY METHOD ANALYSIS ON SLUDGE RESIDUES FOR USE FROM WASTE-WATER TREATMENT PLANT

ABSTRACT:

It was confirmed that the emergy analysis method allows comparison between the two alternatives for biosolids disposal while integrating disposition the energy, economic and environmental aspects. Therefore, by using the method of emergy analysis as a tool for sustainability assessment for environmental decision making, and understanding the sustainability of a disposal process that would reach a greater economic benefit with minimal environmental load, the actual benefit of biosolids "using biosolids as Soil Fertilizer" was evaluated and compared against a suitable alternative referenced in the literature as "energy recovery from Biosolids using fluidized bed gasification" under two scenarios: biosolids considered as a renewable resource and as an imported resource.

Being the best alternative according to the sustainability indices "Energy Recovery from Biosolids Using Fluidized Bed Gasification", ie, this alternative is energetically profitable, economically viable and environmentally acceptable; It does not mean that the actual use of the biosolids is economically or environmentally inappropriate. On the contrary, it does not produce environmental stress because its economic performance is not being taken advantage of at its full potential.

The analysis of emergetic levels not only allow us to evaluate the sustainability of a project, It also offers the possibility to improve the practice, so for example for the current use of biosolids the sustainability index allows us to say that a pretreatment of sewage sludge before its disposal improve to the alternative and become the most acceptable [1]

KEYWORDS: Emergy, sustainability, indicators, waste water, disposal, sludge

In Colombia, it is not uncommon to rely on plants for waste-water treatment purposes followed by the anaerobic stabilizing of biosolids as an alternative source of treatment. This practice is due to the many topographical limitations and conditions that do not allow mainstream treatment such as the use of oxidation ponds.

However, during waste-water treatment processing, large significative quantities of highly putrescible bio-solids with a higher energy concentration as well as organic charge/volume are produced; such is the case-example of the waste water treatment plant in San Fernando (an EEPPM-owned plant) where more than 70 tons of bio-solid material are generated, to be used to treat the 1.100m l/s of water needed to be released on a permanent basis [2].

Although bio-solid materials processed in this particular example is known for carrying similar traits to those found in urban areas, as well as being defined as neither toxic nor hazardous, they do in fact carry contaminants that required treatment and profitable use of such. Based on these findings, many alternative traditional methods for treatment and use of bio-solids have been implemented, those being for energy production, materials for construction purposes, restoration of natural habitat as well as agricultural uses [3]. However they have proven to be ineffective in their result.

In retrospect; it can be said that management of waste-water treatment is ultimately lacking and inconclusive, lest treatment of its additional waste product is carried-on in a sustainable fashion. defining sustainability of course, as the process of obtaining the highest financial return with a minimum effect in the environment; for such reason, the use of the emergy analysis method is a reliable gauging tool to measure level of sustainability to take full advantage of such, as proposed by [4] [5] [6] [7], as the most prominent and noteworthy method used in other countries [8] [9] [10] [11] [12] for sustainability based practices, in a manageable and equitative way in the economic, energy and environmental areas.

The emergy analysis approach is known as an analytical method, that through its integration of economy, ecology and thermodynamics concepts allows the assessment of economic, social and environmental impacts in a theoretical form, on development-driven projects, thus becoming a stronghold system in the fields ecological economy and environmental engineering.

As an emergy analysis method-based research, its ultimate intent is to weigh-in the efficient level of sustainability of any processing system used to make decisions concerning the environment. this method will be used in a way whereby a correct approach to evaluate all sludge residues found in any waste-water treatment plant can be identified in order to choose an assessment alternative that is ´applicable´. that is to say, an alternative that is ´´energy-suitable´´, ´´financially viable´´, and ´´environmentally friendly´´, in accordance to sustainability indicators; this research means to assess the use of the emergy analysis method as a sustainability appraisement tool of bio-solids production at a waste-water treatment plant in Colombia, while being able to satisfy all economic, environmental and energy levels of sustainability. Having as contrast, an already existing method of assessment found in the table of reference, allowing evidential proof of its utility on projects of a similar nature, as well as measuring if current alternative methods of waste-water treatment used in Colombia fit the standards of sustainability, by using the following parameters.

In order to meet such general objective, the application and results of the emergy analysis method used in a waste-water treatment plant in Colombia is analyzed, followed by an analysis of the system of sustainability. Then, by comparing the assessment of the two chosen alternatives of bio-solids used, through the estimate of emergy of both alternatives, the better method for bio-solid analysis is defined and accepted as the best one.

Thus, based on the sustainability indicators, the recovery of bio-solids energy through fluidized bed gasification becomes the best alternative for sustainability. That is to say, this alternative, meets all three standards of sustainability when compare to the alternative of using bio-solids as ground fertilizer.

REFERENCES

[1] Cano Londoño N. A. (2012). Análisis mediante el mÃ©todo emergÃ©tico de la disposición de los

lodos producidos en una planta de tratamiento de aguas residuales. (Aplicación a una PTAR en el Área Metropolitana del Valle de Aburrá). [Tesis MSc.]. Universidad Nacional de Colombia Sede Medellín.

[2] Velez, J. (2007). Biosolids: A problem or a solution? *Producción mas Limpia*, *2*, 57-71.

[3] Fytili, D., & Zabaniotou, A. (2008). Utilization of sewage sludge in EU: application of old and new methods—A review. *Renewable and Sustainable Energy Reviews*, *12*(1), 116-140.

[4] Agostinho, F., Diniz, G., Siche, R., & Ortega, E. (2008). The use of emergy assessment and the Geographical Information System in the diagnosis of small family farms in Brazil. *Ecological Modelling*, *210*(1-2), 37-57.

[5] Cherubini, F., Bargigli, S., & Ulgiati, S. (2009). Life Cycle Assessment (LCA) of waste management strategies: Landfilling, sorting plant and incineration. *Energy*, *34*(12), 2116-2123.

[6] Cohen, M. J., Brown, M. T., & Shepherd, K. D. (2006). Estimating the environmental costs of soil erosion at multiple scales in Kenya using emergy synthesis. *Agriculture, Ecosystems & Environment*, *114*(2-4), 249-269.

[7] Odum, H., & Odum, B. (2003). Concepts and methods of ecological engineering. *Ecological Engineering*, *20*(5), 339-361.

[8] Geber, U., & Bjo, J. (2001). The relationship between ecosystem services and purchased input in Swedish wastewater treatment systems — A case study. *Ecological Engineering, 18, 39-59.*

[9] Liu, Y., Kong, S., Li, Y., & Zeng, H. (2009). Novel technology for sewage sludge utilization: preparation of amino acids chelated trace elements (AACTE) fertilizer. *Journal of hazardous materials*, *171*(1-3), 1159-67.

[10] Siracusa, G., & La Rosa, A. D. (2006). Design of a constructed wetland for wastewater treatment in a Sicilian town and environmental evaluation using the emergy analysis. *Ecological Modelling*, *197*(3-4), 490-497.

[11] Zhou, J. B., Jiang, M. M., Chen, B., & Chen, G. Q. (2009). Emergy evaluations for constructed wetland and conventional wastewater treatments. *Communications in Nonlinear Science and Numerical Simulation*, *14*(4), 1781-1789.

[12] Zhou, S. Y., Zhang, B., & Cai, Z. F. (2010). Emergy analysis of a farm biogas project in China: A biophysical perspective of agricultural ecological engineering. *Communications in Nonlinear Science and Numerical Simulation*, *15*(5), 1408-1418.

(592a) Catalytic Fast Pyrolysis of Forest Thinnings in a Bubbling Fluidized Bed Reactor With ZSM-5 Catalysts

Paasikallio, V., VTT Technical Research Centre of Finland
Agblevor, F., Utah State University
Originally presented on: 11/7/2013 8:30:00 - 8:48:00

Catalytic pyrolysis is a technique which has been receiving increasing attention in the recent years due to its ability to produce partially deoxygenated pyrolysis oil with improved fuel properties. This change in the chemical composition of pyrolysis oil can be seen as a shift from its typical polar nature more towards conventional petroleum-derived fuels. This transformation in turn opens up possibilities for the utilization of pyrolysis oil in new applications such as co-processing in conventional oil refineries. While a multitude of biomass types have been subjected to catalytic pyrolysis, a low-cost and abundantly available feedstock is one of the main prerequisites for a full-scale industrial process. In countries with strong forest industry backgrounds, forest residues and forest thinnings are an attractive source of biomass with otherwise low economic value.

In this work, conventional and catalytic fast pyrolysis of Finnish forest thinnings was carried out in a 200 g/h bench-scale bubbling fluidized bed reactor using silica sand and two ZSM-5 catalysts as the fluidized bed material. The first catalyst was a ZSM-5 based fluid catalytic cracking additive, whereas the second catalyst was a typical ZSM-5 zeolite supplied as extrudates. The resulting pyrolysis oils, which were collected both at the middle and at the end of the catalytic two hour experiments, were characterized for their moisture content, elemental composition, heating value, viscosity and pH. Chemical composition of the pyrolysis oils was analysed using ^{13}C nuclear magnetic resonance spectroscopy.

When compared to the silica sand at a reaction temperature of 475°C, the ZSM-5 based fluid catalytic cracking additive did not induce major changes in the physical properties or the chemical composition of the pyrolysis oil. The second ZSM-5 zeolite proved to be more effective in converting the pyrolysis vapours, and produced pyrolysis oil with oxygen content as low as 20 wt-%. The change in elemental composition was accompanied by an increased heating value and lower viscosity. Elimination of carbohydrate degradation products resulted in an increased proportion of aromatic compounds in the pyrolysis oil. The second catalyst was selected for further experiments where the effect of varying the reaction temperature was investigated. Decreasing the reaction temperature to 400°C caused a clear increase in the yield of organic liquids, which was accompanied by an increase in the oxygen content of the pyrolysis oil. Compared to 475°C, the pyrolysis oil produced at 400°C exhibited lower viscosity, which cannot therefore be explained by elimination of oxygenated molecules. On the other hand, increasing the reaction temperature to 550°C caused an increase in the viscosity of the pyrolysis oil. Other physical properties were not affected to such a large extent by the temperature increase.

(592b) Effects of Reactor Configurations and Catalyst Properties for Vapor Upgrading in HZSM5 Catalyzed Pyrolysis Vapor Upgrading

Wan, S., University of Oklahoma
Waters, C., University of Oklahoma
Gumidyala, A., University of Oklahoma
Jentoft, R., University of Oklahoma
Lobban, L., University of Oklahoma
Crossley, S., University of Oklahoma
Resasco, D. E., University of Oklahoma
Mallinson, R. G., University of Oklahoma
Originally presented on: 11/7/2013 8:48:00 - 9:06:00

Effects of Reactor Configurations and Catalyst Properties on HZSM5 Catalyzed Upgrading of Biomass Pyrolysis Vapors

Shaolong Wan, Christopher Waters, Abhishek Gumidyala, Rolf Jentoft, Lance Lobban, Steven Crossley, Daniel Resasco, and Richard Mallinson

Since the mid 1980s, HZSM-5 has been shown to be a superior catalyst for upgrading the vapors from biomass pyrolysis to produce aromatic hydrocarbons.Â However, this comes with relatively low liquid yields, significant coke and gas formation and rapid deactivation.Â We have studied HZSM-5 catalysis of biomass pyrolysis vapors in a custom CDS pyroprobe with a separate reactor in order to understand how to increase liquid yield and improve catalyst stability.Â Compared with pyrolyzing the catalyst mixed with the biomass, different product selectivities are observed with the separate reactor due primarily to differences in residence time.Â The pyroprobe pyrolysis of a biomass-catalyst mixture lasts for seconds and results in high conversion of a primary cellulose decomposition product, levoglucosan, as well as light oxygenates including acetic acid.Â When configured in a separate reactor with much shorter residence time, high conversion of the light oxygenates is obtained, but low conversion of levoglucosan is observed.Â In this separate configuration, multiple pulses of pyrolysis vapors may be passed over the catalyst, thus allowing the deactivation characteristics to be studied.Â For example, at a constant number of acid sites in the catalyst bed, the deactivation rate was found to be much faster with a catalyst containing a higher acid site density.Â Different HZSM-5 materials have been tested, with different separate reactor temperatures.Â These have included variation in the Si/Al ratio from 40 to 11.5 and the use of core-shell materials with a core of silicalite, an outer layer of HZSM-5, and also a material with a further outer layer of silicalite.Â Additionally, the effect of the passivation of surface acidity has been examined.

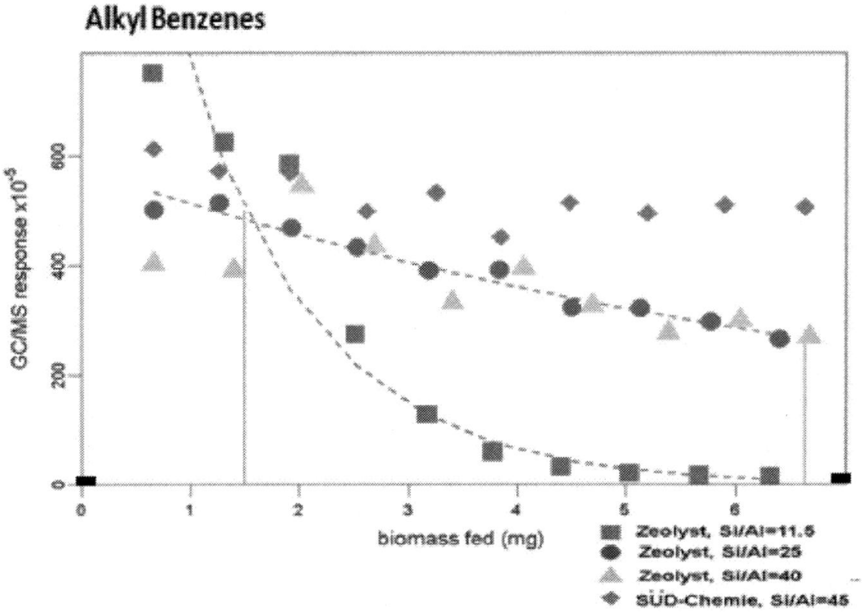

(592c) CFD Modeling of the Catalytic Pyrolysis of Biomass With Zeolite in Fluidized Bed

Sun, D.,
Papadikis, K.,
Gu, S.,
Originally presented on: 11/7/2013 9:06:00 - 9:24:00

Fast pyrolysis of biomass using zeolites as the catalyst material is a direct method to improve the quality of bio-oil in the pyrolysis reactor. In this paper, catalytic pyrolysis of sawdust is simulated in a fluidized bed. Two different cases, one with zeolite catalyst and one without catalyst are modeled and compared.

The flow behavior in the fluidized bed is modeled by means of gas-solid two-fluid model with kinetic theory of granular flow for the granular phase. Two kinds of granular are used in the reactor, i.e. the catalyst and biomass particles. The momentum transfer between the solid phases can be modeled by the drag coefficient simplified for the kinetic theory by Syamlal(1987). Kinetic scheme for catalytic pyrolysis is based on the kinetic research by Atutxa et al.(2005). The kinetic experiment has been carried out to study the catalytic pyrolysis of sawdust with zerlite HZSM-5 in a conical spouted bed reactor at the temperature 400°C. The effect of space time is also incorporated in the kinetic mechanism.

The predicted product yield distribution from catalytic pyrolysis is in good agreement with the experiment data of Atutxa et al. (2005). The gas yields, due to the cracking of tar, increase with an increasing amount of catalyst. Gas, tar and char products vary along the height of the reactor, which give a good prediction of the residence time of biomass into the bed. Along the height, concentration of tar increases from bottom of the riser result from the pyrolysis reaction of biomass and then reduced by the catalytic reaction to gas. Three regions can be divided to indicate the biomass heating, pyrolyzing and catalyzing process. The simulation also shows the velocity of two solid phases in the riser. The velocity of catalyst is higher than the sawdust because of the larger diameter. This velocity difference between them can enhance the collision occasions and promote the turbulence intensity and solids distribution in the riser, hence benefiting chemical reaction and consequently the final pyrolysis yields.

(592d) Deoxygenation of Bio-Oil Intermediates: Examination of Methane As a Direct Reducing Agent

Oberg, A. A., Pacific Northwest National Laboratory
Miller, D. J., Michigan State University
Originally presented on: 11/7/2013 9:24:00 - 9:42:00

Thermochemical conversion of biomass into bio-oil is a promising way to create renewable liquid transportation fuels. However, untreated bio-oil is not suitable for use a fuel, because it is thermally unstable, corrosive, has low heating value, and is often immiscible with hydrocarbon fuels. Upgrading of bio-oil to remove oxygen, saturate double bonds, and increase energy density is therefore required; such upgrading currently is done via catalytic hydrodeoxygenation with molecular hydrogen. Another possible source of reducing equivalents for upgrading is methane, available either from anaerobic digestion of biomass or from abundant fossil reserves. Each mole of methane is capable of providing up to four reducing equivalents; however, methane is the most stable hydrocarbon having strong C-H bonds (104 kcal/mol), stable tetrahedral geometry, and no reactive functional groups to facilitate reaction.

The present work examines methane activation over supported metal catalysts as a first step in using methane directly for bio-oil upgrading. In addition to methane as a source of reducing equivalents, the possibility of methane incorporation into bio-oil species, which would further increase fuel value because of methane's high energy content (~50 MJ/kg), is explored. Initial studies of methane activation over Pt and Ru catalysts at 150°-250°C showed that methane readily undergoes H-D exchange between CH_4 and D_2, and between CH_4 and CD_4 to give mixed CH_xD_{4-x} species. The $CH_4 - CD_4$ exchange takes place without H_2 or D_2 present.

Direct incorporation of methane into benzaldehyde and other species as model bio-oil compounds over Pt and Pt/Cu catalysts was not observed. However, it was shown that over Pt/Cu/SiO_2 benzaldehdye was quantitatively converted to toluene in the presence of hydrogen, and into benzene in the presence of methane or nitrogen. Copper was confirmed as the active catalyst species for deoxygenation.

(592e) Fast-Hydropyrolysis and Catalytic Hydrodeoxygenation for Conversion of Biomass to Liquid Fuel

Venkatakrishnan, V. K., Purdue University
Degenstein, J., University of North Dakota
Delgass, W. N., Purdue University
Agrawal, R., Purdue University
Ribeiro, F. H., Purdue University
Originally presented on: 11/7/2013 9:42:00 - 10:00:00

The production of high volumetric energy density fuels from a sustainable carbon source, such as lignocellulosic biomass, to supply the liquid fuel needs of the transportation sector is a grand challenge to be addressed. As one of the plausible solutions, a novel process, *H_2Bioil*[1], based on biomass fast-hydropyrolysis and catalytic hydrodeoxygenation (HDO), has been proposed for conversion of lignocellulosic biomass to liquid fuel with high estimated energy and carbon efficiencies. The *H_2Bioil* process integrates fast-pyrolysis thermochemical conversion, involving rapid heating of biomass, in a high-pressure (up to 50 bar) hydrogen environment, to produce fast-hydropyrolysis vapors that are subsequently catalytically upgraded in the vapor phase and quenched to form a deoxygenated liquid product with high energy density.

This presentation summarizes our work on *H_2Bioil* process development, using laboratory-scale high-pressure fast-hydropyrolysis and catalytic hydrodeoxygenation with cellulose, as a model biomass feedstock, and lignocellulosic feedstocks. To identify and quantitatively compare major chemical species present in the liquid product, we have developed and utilized a novel liquid chromatography-mass spectrometry (LC-MS) analytical method which accounts for up to 95% by mass of the cellulose hydropyrolysis liquid products. We have also used gas chromatography-mass spectrometry (GC-MS) and other analytical techniques such as elemental (C, H, O) and Karl Fisher water content analysis to characterize the liquid product. The effects of process conditions such as pyrolysis temperature (480°C-580°C), hydrogen partial pressure (25 bar, 50 bar) were studied with a cellulose model feedstock. The major product of high-pressure fast-pyrolysis and fast-hydropyrolysis of cellulose is the monomer (levoglucosan), which accounts for ~35-40% of the starting mass of cellulose pyrolyzed. As the pyrolysis temperature is increased, the product distribution shifts towards the lower molecular weight compounds, like glycolaldehyde and formic acid, which is attributed to increased thermal cracking of pyrolysis vapors. The first stage cellulose fast-hydropyrolysis does not lead to significant deoxygenation, even up to 50 bar hydrogen partial pressure, but presence of HDO catalysts leads to improvements in deoxygenation in the liquid product. The results from testing of candidate Pt-based and Ru-based catalysts for on-stream vapor-phase catalytic hydrodeoxygenation suggests that a balance of metal and acid catalytic functions is necessary for improving the extent of deoxygenation along with high carbon recovery in the liquid product and avoiding coking on the catalyst.

[1]Agrawal R, Singh NR. Synergistic Routes to Liquid Fuel for a Petroleum-Deprived Future, *AIChE J.* 2009; 55: 1898-1905

(592f) Low-Oxygen Bio-Crude From Hydropyrolysis in a Pressurized Fluid-Bed Reactor

Carpenter, J. R., RTI International
Hlebak, J., RTI International
Farmer, J., RTI International
Dayton, D. C., RTI International
Originally presented on: 11/7/2013 10:00:00 - 10:18:00

Pyrolysis oil prepared by conventional (non-catalytic) fast pyrolysis is not suitable for integrating into the current fuel infrastructure or into a petroleum refinery because it is : a) thermally unstable with a high fouling tendency, b) corrosive due to high carboxylic acid content (pH 2.2 – 2.4 typically), and c) not miscible with refinery feedstocks. Fast pyrolysis bio-oil tends to have high oxygen content similar to the solid biomass feedstock used as the starting material. The thermal instability and high water content require specialized hydroprocessing to lower the oxygen content of the bio-oil before use. Hydropyrolysis is a biomass thermal liquefaction process option that combines biomass, hydrogen, and a catalyst to integrate pyrolysis and a degree of hydroprocessing into a single unit operation.

The National Advanced Biofuels Consortium (NABC) is working to develop sustainable, cost-effective processes to produce advanced "drop-in" biofuels that are compatible with today's transportation infrastructure. As part of the NABC, the hydropyrolysis of loblolly pine has been explored to produce low-oxygen content bio-crude. A pressurized fluid-bed reactor was designed and fabricated to test catalysts and reaction conditions (T, P, and pH_2). Testing was performed with a nominal loblolly pine feed rate of 200 g/hr at temperatures between 350°C to 500°C and pressure up to 300 psig with varying hydrogen concentrations. Full mass closures are obtained by measuring all liquid, solid, and gas products. Bio-crudes are characterized by ultimate analyses, Karl-Fischer titration, total acid number (TAN), pH, and GC/MS. In addition to parametric testing, several catalysts were screened and performance with a corn stover was evaluated.

A selected catalyst was tested over an extended trial of 32h on stream with over 8kg of biomass. Consistent activity was observed based on product yields and composition when catalyst regeneration occurred after conversion of 1 kg of biomass. The liquid product separated into an aqueous layer that was 97wt% water and a bio-oil that was less than 1wt% water and had an oxygen content of 2.5wt%. The bio-crude had high levels of aliphatic and aromatic compounds with reduced concentrations of acids, sugars, and ketones in comparison with conventional fast pyrolysis bio-oil. Hydropyrolysis has the potential to produce bio-oils with improved properties and suitability for integration into refinery processes.

(592g) Catalytic Fast Pyrolysis of Isotopically Labeled Biomass and Plastic Blends

Dorado, C., Eastern Regional Research Center, Agricultural Research Service, USDA
Mullen, C., USDA
Boateng, A. A., USDA Agricultural Research Service
Originally presented on: 11/7/2013 10:18:00 - 10:36:00

This study aims to address two problems important to agriculture, production of drop-in fuels from biomass and disposal of agricultural plastics via co-pyrolysis of both materials. Mixtures of biomass (cellulose and switchgrass) and polypropylene (PP) or polyethylene terephthalate (PET) were subjected to catalytic fast pyrolysis (CFP) at 650°C in the presence of H-ZSM5 using a micro pyrolyzer coupled with GC/MS (py-GC/MS). Yields of the aromatic hydrocarbons, toluene, naphthalene and p-xylene, were enhanced for the CFP of mixtures of cellulose and plastic versus the CFP of biomass or plastic alone, suggesting a synergistic effect that enhanced the carbon efficiency of the CFP process. To further probe this effect, CFP of uniformly labeled ^{13}C biomass with non-labeled (^{12}C) PP and PET in the presence of H-ZSM5 were co-pyrolyzed. The resultant aromatic compounds had varying $^{13}C_x{}^{12}C_y$ compositions, proving that each reactant provides carbon to a pool from which aromatics are formed. Further analysis allowed for a more detailed picture of the source of carbon for each individual product. For example, it was found that for CFP of ^{13}C biomass and PP pyrolysis, 69% of the toluene product was composed of at least four carbon atoms from PP. In this presentation the results of the catalytic co-pyrolysis of biomass and plastics and the isotopic labeling experiments will be discussed in detail in terms of yields of individual products, product selectivities and implications for the operative CFP mechanisms.

(592h) Mild Pyrolysis of P3HB/Switchgrass Blends for the Production of Bio-Oil Enriched With Crotonic Acid

Mullen, C. A., USDA Agricultural Research Service
Boateng, A. A., USDA Agricultural Research Service
Dorado, C., Eastern Regional Research Center, Agricultural Research Service, USDA
Schweitzer, D., Metabolix Inc.
Sparks, K., Metabolix
Snell, K., Metabolix, Inc.
Originally presented on: 11/7/2013 10:36:00 - 10:54:00

Poly-3-hydroxybutyrate (P3HB) is a naturally occurring polyester that is produced by some microorganisms as a method of carbon and energy storage. In addition to being utilized as a renewable, biodegradable substitute for plastics produced from petroleum, it can be thermally degraded *in situ* to crotonic acid in high yield. Crotonic acid can be used as a platform chemical enabling an alternative access to high-valued industrial and fine chemicals. For example, it can be reduced to *n*-butanol for chemical or fuel applications or oxidized to maleic anhydride. Other targets available through simple transformations include propylene and acrylic acid. In the work presented in this presentation, the mild pyrolysis of switchgrass/P3HB blends that mimic P3HB-producing switchgrass lines was studied in a pilot scale fluidized bed reactor with the goal of simultaneously producing crotonic acid and switchgrass-based pyrolysis oils to be further converted to biofuels. While pure P3HB decomposes under distillation conditions to provide a near quantitative yield of crotonic acid, pyrolysis of P3HB-containing switchgrass requires an additional crotonic acid purification step. Factors such as pyrolysis temperature, residence time, flow rates and particle size of the P3HB were studied for their effects on the recovery of crotonic acid as a component of the pyrolysis oil produced from the mixture. Effects of the presence of P3HB and crotonic acid on the switchgrass bio-oil in terms of chemical composition and properties will also be examined. Furthermore, efforts to isolate pure crotonic acid from the switchgrass bio-oil will be discussed. Finally, an example of the utility of crotonic acid as a chemical feedstock will be briefly discussed. Hydrogenation of the *n*-butyl ester of crotonic acid was found to produce *n*-butanol in excess of 90 % yield over commercial Cu-based catalysts under the optimized conditions.

(600a) Mixed-Conductor Supported Ferrites for Chemical Looping Conversion of Methane

Chen, Y.,
Li, F., North Carolina State University
Originally presented on: 11/7/2013 8:30:00 - 8:51:00

The chemical looping strategy represents a promising option for carbon dioxide capture from fossil fuels. The chemical looping processes utilize solid oxygen carriers to indirectly convert carbonaceous fuels to concentrated CO_2 and carbon free products via redox reactions. Due to its low cost and abundance, iron oxide is a desirable option as an oxygen carrier. The challenge for iron oxide based oxygen carriers resides in their low activity for methane and solid fuel conversions. Our recent studies have indicated that mixed ionic and electronic conducting (MIEC) oxides, when used as the "support", can significantly enhance the redox activity of ferrite based oxygen carriers. In the current study, ferrite based oxygen carriers supported with various MIEC perovskites are synthesized. The redox activities of these oxygen carriers are evaluated using a thermogravimetric analyzer (TGA) with H_2 and CH_4 as the reducing gases. The phase structures of these samples, before and after redox reactions, are determined using X-ray diffraction (XRD) and the morphologies of the samples are characterized by SEM. The results confirm the high redox activity of the MIEC supported ferrites. A general correlation between the mixed conductivity of the support and the redox activity of the oxygen carriers are observed. Long term redox reactions and microscopic/spectroscopic characterizations are also carried out to determine relationships between support composition/structure and the long term stability of the ferrite-based oxygen carriers.

(600b) A Novel Mixed Metallic Ni Based Oxygen Carrier for Chemical Looping Combustion

de Lasa, H., University of Western Ontario
Quddus, M. R., The University of Western Ontario
Hossain, M. M., King Fahd University Petroleum and Minerals
Originally presented on: 11/7/2013 8:51:00 - 9:13:00

This study reports the development of a high surface Ni based oxygen carrier (OC) and the investigation of its reactivity with gaseous fuel for Chemical-Looping Combustion (CLC). High surface area facilitates high dispersion of metal particle and increases the oxygen carrying capacity. On the other hand, high surface area also enhances the formation of irreducible species like $NiAl_2O_4$, which minimize the reactivity and oxygen transport capacity. Therefore, to achieve the high surface OC while preserving its reactivity, stability and fluidizability, a Ni based OC was developed in this study over Co promoted La modified γ-Al_2O_3 using incipient wetness impregnation technique. The γ-Al_2O_3 was first modified with addition of 5wt% La. The resulting La/γ-Al_2O_3 was then impregnated with 1wt% Co. Finally, a total of 20wt% Ni was loaded to the OC precursor via successive impregnations of 5wt% Ni. The resultant paste after each successive metal loading was reduced under three controlled gas flow rates containing 10% H_2 with the balance being He. The surface area was found to vary considerably based on the gas flow rates used during the preparation stages. 50-120 m^2/g surface area was found for OCs prepared under low reducing gas flow. These synthesized OCs display excellent reducibility and stability in successive TPR/TPOs and H_2 chemisorptions. TPR profiles for the two OCs consistently show 95% reducibility at/below 750 °C. These results are encouraging given that they demonstrate the oxygen in the OC is very close to the expected amount. These results also show that the OC reducible phase is primarily NiO with small fractions of $NiAl_2O_4$. The controlled gas flow rates help the formation of easily reducible NiO species minimizing Ni-support interactions and formation of $NiAl_2O_4$. It was also found that the reduction behavior of these OCs resemble the chemical properties of δ- and θ-Al_2O_3 for low gas flow rates. In contrast, the structural integrity and characteristic chemical properties of Ni over γ-Al_2O_3 was observed in the case of high gas flow rates. Considerable amount of Ni was observed to be present as $NiAl_2O_4$ in this case. However, the absence of θ- and/or δ-phase of Al_2O_3 in the XRD experiments suggest that the phase transformation might occur at the interfaces between the Ni crystallites and alumina support surfaces, whereas the bulk structure of the support remained in the γ-phase. The SEM/EDX and H_2 pulse chemisorption results confirmed that all the three prepared OCs were highly stable in consecutive redox cycles. A well dispersed state of Ni crystals and the absence of agglomerations were observed for both the fresh and used OCs.

The reactivity of these prepared OCs was studied under the expected conditions (turbulent fluidized bed and high temperature) of an industrial scale CLC unit in CREC Riser Simulator. The CREC Riser Simulator is a bench scale mini-fluidized bed reactor capable of simulating circulating fluidized bed reactions conditions. The reactivity study was performed with stoichiometric amount of CH_4. The CH_4 conversion was found to be 81% ($A_{BET} = 50\ m^2/g$) and 86% ($A_{BET} = 120\ m^2/g$) for the OCs prepared under low gas flow rates. Maximum 88% CH_4 conversion was observed for the OC with high surface area ($A_{BET} = 140\ m^2/g$). The increased conversion of CH_4 was resulted from the increased oxygen transport capacity with the surface area of the carrier particles. However, the selectivity of CO_2 was found to decrease from 97% to 82 % with the increasing surface area. This was mainly due to the increased catalytic effect resulting from highly dispersed Ni crystals over high surface area.

This systematic investigation of the reactivity of the Ni particles over wide range of crystal sizes reveals the fact that the low surface and low dispersed Ni based OC is favorable for CLC operation, whereas highly dispersed Ni based OC is suitable for CLR process. The oxygen transport capacity can be significantly increased with increasing surface area.

(600c) Low-Cost Oxygen Carrier Materials for Chemical-Looping Combustion of Biomass

Schwebel, G.,
Vilela Perez, G., Chalmers University of Technology
Knutsson, P., Chalmers University of Technology
Leion, H., Chalmers University of Technology
Originally presented on: 11/7/2013 9:13:00 - 9:34:00

Chemical-looping combustion (CLC) is one of the most promising methods to implement carbon capture technologies to mitigate CO_2emissions from thermochemical fuel conversion.

The solid oxygen carrier is a core component of every CLC system and the choice of the oxygen carrier depends on the fuel and operation conditions. When it comes to conversion of solid fuels, low-cost oxygen carriers seem to be more suitable. This is because the lifetime of an oxygen carrier in a CLC system may be limited by side reactions with fuel ash, or by carryover losses in the ash separation.

Hence the use of low cost oxygen carrier materials is preferred, lowering the economic and technologic effort connected to higher demands of make up.

Due to its carbon neutrality, the use of biomass in carbon capture strategies leads to net negative emissions. However, the ash of biomass is known to be more reactive and has generally lower softening and melting temperatures than most coals. This makes biomass combustion units more prone to defluidization, slagging and fouling and corrosion. If biomass should be used as fuel the temperature is usually kept at lower levels to overcome these problems.

In the present contribution different low cost oxygen carrier materials are investigated with respect to their reactivity at different temperatures. The data is obtained from experiments carried out in a quartz glas reactor under fluidized bed conditions. The bed temperature was set to 700, 800, 850 and 950°C, respectively. The perforance was tested with volatile components and in selected cases the influence of the char gasification has been investigated separately. Before the actual experiments, the oxygen carrier materials were activated at 850 and 950°C using a mixture of H_2 and CO as fuel.

(600d) Fe and Mn-Based Oxygen Carriers for Chemical Looping Combustion

Bhavsar, S., US DOE-National Energy Technology Laboratory, Pittsburgh
More, A.,
Veser, G., University of Pittsburgh
Originally presented on: 11/7/2013 9:34:00 - 9:56:00

Chemical looping combustion (CLC) is an emerging technology for clean energy-production from fossil and renewable fuels based on cyclic oxidation and reduction of an oxygen carrier. These carriers – typically metal oxides – have to show sufficient thermal and mechanical stability to withstand the demanding conditions of CLC while at the same time yielding high carrier and fuel conversion. Among the wide range of materials investigated, extensive work has recently been focused on iron and manganese oxides as oxygen carriers due to their abundance, low-cost, and environmentally benign nature. However, both metals show comparatively low reactivity and only moderate selectivity for total oxidation products.

In the present study, we investigated the use of Mn-Fe mixed-metal carriers for CLC. We had previously shown that doping Fe-based carriers with Ni can yield significant enhancements in reactivity. Ceria was chosen as support based on our previous findings that ceria can result in strong enhancements of carrier utilization and stability of the bimetallic metal phase.

Fe/Mn mixed oxides were synthesized with various Fe/Mn ratio. All carriers were synthesized with 40wt% metal by simple incipient wet impregnation technique on ceria supports. The carriers were characterized (as synthesized and post-reactive tests) using TEM, XRD, BET, and EDX. In a first step, the carriers were tested in a thermogravimetric analyzer (TGA) with H_2 and CH_4 as fuels. All FeMn carriers showed good reactivity and thermal stability with no signs of metal-support spinel formation. Increasing the Fe/Mn ratio resulted in improved oxygen carrying capacity beyond a simple linear addition of the capacity of the pure metal components. Next, selectivity and fuel conversion was further evaluated in fixed-bed reactor experiments with CH_4 as fuel. For both pure metals, a transition in selectivity was observed from total oxidation at high carrier oxidation states to partial oxidation over the reduced carriers. Interestingly, the bimetallic carriers showed a combination of the high syngas selectivity of Mn-based carriers with the widened operating window of Fe-based carriers, rendering this bimetallic carrier material of particular interest for partial oxidation of methane via chemical looping.

(600e) Experimental and Model Comparison of Mixed Metal Ferrites for H2 Production Using Chemical Looping

Aston, V. J., University of Colorado at Boulder
Evanko, B. W., University of Colorado at Boulder
Muhich, C. L., University of Colorado at Boulder
Scheffe, J., University of Colorado at Boulder
Weimer, A. W., University of Colorado at Boulder
Originally presented on: 11/7/2013 9:56:00 - 10:17:00

The chemical looping hydrogen (CLH) production process is an alternative method to water-gas shift to generate pure H_2, as well as a separate stream of pure and easily-sequestered CO_2, from any synthesis gas without complicated and expensive separations equipment. Synthesis gas is used to reduce a metal oxide material at moderate to high temperatures (500°C to 1000°C) as it leaves a gasifier, producing a stream of CO_2 and H_2O. The resulting reduced metal is then oxidized with H_2O to produce H_2 and regenerate the metal oxide. In this study, the mixed metal ferrites $CoFe_2O_4$ and $NiFe_2O_4$ were investigated experimentally and shown to meet the criteria for CLH metal oxides. A comparison was made with Fe_2O_3, which is the current state of the art CLH metal oxide. At low solid conversions the mixed metal ferrites allowed the recovery of nearly three times more H_2 during H_2O oxidation than was recovered using Fe_2O_3 under the same conditions. This H_2 recovery required additional steam generation due to the low H_2O to H_2 conversion during the oxidation of near-stoichiometric $CoFe_2O_4$ and $NiFe_2O_4$. A system model developed based on the mixed metal ferrite equilibrium with the H_2/H_2O and CO/CO_2 systems was used to investigate the effects of syngas composition, temperature and a supplementary O_2 oxidation on the steam production required for complete oxide regeneration and the overall efficiency of the system. The $CoFe_2O_4$ and $NiFe_2O_4$ materials offer significant benefit over conventional Fe_2O_3 at lower reduction temperatures and for lower quality feed syngas.

(600f) Sol-Gel Derived Fe2O3-Based, Al2O3 Supported Oxygen Carriers for Hydrogen Production Via a Chemical Looping Cycle

Yuzbasi, N. S., ETH Zurich
Imtiaz, Q., ETH Zurich
Muller, C. R., ETH Zurich
Originally presented on: 11/7/2013 10:17:00 - 10:39:00

Sol-gel derived Fe_2O_3-based, Al_2O_3 supported oxygen carriers for hydrogen production via a chemical looping cycle

Nur Sena Yüzbasi, Qasim Imtiaz, Christoph R. Müller*

Laboratory of Energy Science and Engineering, ETH Zurich, Leonhardstrasse 27, 8092

Zurich, Switzerland

***Corresponding author: muelchri@ethz.ch**

Keywords: Hydrogen production, sol-gel, iron oxide, iron precursors.

Current technologies for hydrogen production such as the steam methane reforming (SMR) process have high production capacities of~15 kte H_2/year or greater. However, due to issues associated with the storage and distribution of H_2, processes that allow the distributed production of H_2 on a small scale may be advantageous.[1] For example, the distributed production of high purity hydrogen from biomass (with simultaneous CO_2 capture) can be achieved by modifying the conventional thermo-chemical looping combustion process. Here, biomass is first gasified with steam to produce a synthesis gas i.e. a mixture of predominantly CO, H_2, CO_2 and H_2O. The synthesis gas subsequently reduces iron oxide to a lower oxidation state, thereby producing CO_2 and H_2O which can be separated via condensation. Subsequent re-oxidation of the reduced iron oxide with steam yields high purity H_2. One of the key challenges of the above described looping process is the synthesis of iron oxide-based oxygen carriers that possess (i) a high reactivity, (ii) stable redox characteristics over many cycles and (iii) tolerance to typical synthesis gas impurities e.g. sulphurous components or tarry substances. Recent studies have demonstrated that pure iron oxide, i.e. unsupported iron oxide, deactivates after only a few cycles when reduced completely to Fe. [1, 2]

A possibility to stabilize the redox characteristics of iron oxide based oxygen carriers is to stabilize Fe_2O_3

with a support material e.g. Al_2O_3,[3] ZrO_2,[4] $MgAl_2O_4$.[5] With regards to synthesis techniques, sol-gel methods have received some attention recently since they ensure a homogeneous mixing of the different components and allow the control of the pore structure and surface area of the material. However, there is only limited work on Fe-based oxygen carriers synthesized via sol-gel techniques.[3, 7] In this study, we prepared different Fe_2O_3-based, Al_2O_3 supported oxygen carriers (containing 60 wt. % Fe_2O_3) to assess critically the influence of the iron precursor, *viz.* iron nitrate ($Fe(NO_3)_3{\cdot}9H_2O$), iron chloride ($FeCl_3$) or iron acetylacetonate ($Fe(C_5H_7O_2)_3$), on the yield of H_2. In addition, the effect of the pH on the structure of the synthesized materials was studied. For comparison, pure iron oxide, i.e. unsupported iron oxide manufactured via mechanical mixing technique, was used. The oxygen carriers were characterized using (i) a thermogravimetric analyser (TGA), (ii) a packed bed reactor, (iii) X-ray diffraction (XRD), (iv) N_2 adsorption, (v) high resolution scanning electron microscopy (HR-SEM) and (vi) Fourier transform infrared spectroscopy (FTIR). It was found that the oxygen carriers synthesized using iron chloride or iron acetylacetonate as the iron precursor possessed the highest surface areas and pore volumes. In addition, SEM images showed that the use of iron nitrate or iron acetylacetonate as the iron precursor resulted in oxygen carriers that possessed a homogenous and nano-structured morphology. The cyclic H_2 production of the oxygen carriers was tested over 15 cycles in a fixed bed reactor. Pure Fe_2O_3 possessed a high initial H_2 yield of i.e.~ 13 mmol/g oxygen carrier. However, after the first cycle the quantity of H_2 produced decreased substantially. On the other hand, Al_2O_3-supported oxygen carriers showed stable redox characteristics over 15 cycles. Oxygen carriers synthesized using iron nitrate as the iron precursor showed the highest H_2 yield, i.e. ~ 6 mmols/g oxygen carrier which is nonetheless less than the theoretically expected H_2 yield. The lower H_2 yield is linked to formation of hercynite, $FeO{\cdot}Al_2O_3$ (as confirmed by the XRD) since its re-oxidation with steam is thermodynamically limited.[7]

References:

1. C. D. Bohn, C. R. Müller, J. P. Cleeton, A. N. Hayhurst, J. F. Davidson, S. A. Scott, J. S. Dennis, Ind. Eng. Chem. Res., 47 (2008) 7623-7630.
2. C. D Bohn, J. P. Cleeton, C.R. Müller, S. Y. Chuang, S. A. Scott, J. S. Dennis, Energy Fuels, 24 (2010) 4025-4033.
3. A. M. Kierzkowska et al., Ind. Eng. Chem. Res., 49 (2010) 5383-5391.
4. W. Liu, J. S. Dennis, S.A. Scott, Ind. Eng. Chem. Res., 51 (2012) 16597-16609.
5. J. Adanez et al., Progress in Energy and Combustion Science, 38 (2012) 215-282.
6. F. Li, H. R. Kim, D. Sridhar, F. Wang, L. Zeng, J. Chen, L. S. Fan, Energy and Fuels, 23 (2009) 4182-4189.
7. P. R. Kidambi, J. P. E. Cleeton, S. A. Scott, J. S. Dennis, C. D. Bohn, Energy and Fuels, 26 (2012) 603-617.

(600g) NiO/Ce-γAl2O3 Oxygen Carrier for Chemical-Looping Combustion: Reactivity and Stability

Mohamed, S.,
Razzak, S., King Fahd University of Petroleum and Minerals
Quddus, M. R., The University of Western Ontario
de Lasa, H. I., University of Western Ontario
Hossain, M. M., King Fahd University of Petroleum & Minerals
Originally presented on: 11/7/2013 10:39:00 - 11:00:00

This study deals with the development of a Ce modified NiO/Ce-γAl_2O_3 oxygen carrier suitable for a CLC process. In CLC, the most important characteristics of an oxygen carrier are the reactivity, oxygen carrying capacity and stability under cyclic operation [1, 2]. To investigate these matters successive TPR and TPO experiments were developed in the context of the present study. In cyclic TPR/TPO experiments, the reduction and oxidation profiles of NiO/Ce-γAl_2O_3sample remains unchanged and the amount of Ni reduction also shows a stable behavior. Approximately 90 % nickel oxide conversion was achieved using the Ce modified sample, while the value was in the 75 % range for the unmodified carrier. This observation suggests that the addition of Ce helps the formation of easily reducible nickel oxides on the Ce-modified alumina support and minimizes the formation of nickel aluminates (reduces at high temperature only) [3].

The hydrogen pulse chemisorption results (after each TPR cycle) further confirmed the stable behavior of the Ce modified sample. A higher metal dispersion was observed in the Ce modified sample although both the unmodified and modified samples showed a stable metal dispersion with successive redox cycles. This indicates that Ce helps to increase the dispersion of nickel. Thus, the addition of Ce alters the metal surface modifying the degree of interaction between Ni and alumina support, maintaining a consistent metal dispersion during the repeated recox process. It is also apparent that the metal crystal size of the NiO/Ce-γAl_2O_3sample remained unchanged over repeated cycles, an indication of the absence of agglomeration of nickel crystals.

References

1. M.M. Hossain, H.I. de Lasa, Chem. Eng. Sci., 63 (2008), 4433-4451
2. M.M. Hossain, K.E. Sedor, H.I. de Lasa, Chem. Eng. Sci., 62 (2007) 5464.
3. M. R. Quddus, M.M. Hossain, de Lasa, H.I., Catal. Today (In press).

(619a) Nanoscale Interfacial Design Concept Towards High Flux CO2 Separation Maintaining High Selectivity and Longevity

Kocherlakota, L. S., University of Washington
Pham, T. H., University of Washington
Overney, R. M., University of Washington
Originally presented on: 11/7/2013 8:30:00 - 8:55:00

Nanoscale Interfacial Design Concept Towards High Flux CO_2 Separation Maintaining High Selectivity and Longevity

Lakshmi S. Kocherlakota, Tiep Pham, and Rene M. Overney*

Contemporary environmental and energy production areas are driving current research for pursuing green and sustainable technologies for an economically viable and energy efficient solutions. Membrane based separation technology offers a less energy intensive and environmentally friendly alternative for gas purification processes. Particularly novel organic-inorganic nanocomposite membranes have been shown to push the gas permeability and selectivity capabilities of polymeric membrane materials into the attractive regions based on commercial separation requirements. One of the most notable permeation materials in regards of its high permeation values of up to several thousand Barrer, and its reverse-selectivity towards the lighter component in H_2/CO_2 and N_2/CO_2 is poly(l-trimethylsilyl-1-propyne) (PTMSP), a high free volume glassy polymer. Problematic in nanocomposites are particle dispersion and environmental health concerns due to the small particle size. These two shortcomings are resolved potentially by switching to "inverse composites", in which the polymer matrix is replaced by a porous membrane that provides the interfacial/dimensional constraints, while the polymer itself is injected into the pores. Porous membranes also provide mechanical and temporal stability. In this paper, we discuss how beneficial interfacial constraints, concerning both permeation and selectivity in PTMSP (found in ultrathin film membranes[1]), can be imposed controllably in macroscopic inverse composites, thereby maintaining the enhanced transport properties of thin films, and provide relative improved longevity. PTMSP embedded nanoporous anodic aluminum oxide (AAO) hybrid membranes reveal gas permeabilities for helium, nitrogen and CO_2 that exceeded 5 to7 times the values of pristine bulk PMTSP. Most notably, the gas selectivities of CO_2/He and CO_2/N_2 could also be enhanced by ~40%, and physical aging caused by free-volume collapse, could be reduced 20- fold compared to ultrathin film systems due to pore wall stabilization. Our current studies on polyethylene oxide (PEO) based polymeric materials which are being extensively studied for CO_2 separation are also revealing improved separation capabilities in AAO-confined hybrid polymeric structures. We found that AAO-confined PEO exhibits CO_2/He separation of ~10, which represents a 40% enhanced selectivity over the bulk membranes.

1. Kocherlakota LS, Knorr DB, Foster L, and Overney RM. Polymer 2012;53(12):2394-2401.

(619b) Investigation of Water Quantification in Ionic Liquids Utilizing Near Infrared Absorption of Water

Fagnant, D. Jr.,
Goff, G. S., Los Alamos National Laboratory
Runde, W., Los Alamos National Laboratory
Brennecke, J. F., University of Notre Dame
Originally presented on: 11/7/2013 8:55:00 - 9:20:00

Ionic Liquids (ILs) are potential candidates to replace traditional organic solvents for a variety of separation processes. Some ILs are capable of forming aqueous biphasic systems which allow for solvent extraction of various solutes from aqueous solutions. The ability to quantify water content in an IL phase is important to implementing an ionic liquid as a solvent in a real world process. The water content of an IL affects physical properties such as density, viscosity and conductivity and also significantly affects energy costs for heating and cooling.

Here we investigate and demonstrate the use of near infrared absorption spectroscopy to quantify water content in an imidazolium-based IL. Near infrared absorption offers an accurate, rapid and in situ method for water quantification in the ionic liquid phase, which provides a non-destructive alternative to Karl-Fischer titration/coulometry. The change in water absorption at different water concentrations in the IL is investigated.

(619c) Mixed Gas Sorption of CO2/CH4 Mixtures in PIM-1 and PTMSP Membranes: Experiments and Modeling

Vopicka, O.,
De Angelis, M. G., University of Bologna
Sarti, G. C., Alma Mater Studiorum - Universita di Bologna
Guiver, M. D., National Research Council Canada
Originally presented on: 11/7/2013 9:20:00 - 9:45:00

Sorption of pure methane, carbon dioxide and their binary mixtures in two glassy polymers, poly(1-trimethylsilyl-1-propyne) (PTMSP), and the first polymer of intrinsic microporosity (PIM-1), has been studied experimentally and theoretically, at 35.0 °C.

Measurements were obtained on a newly designed pressure decay sorption apparatus for mixtures of gases, having the basic construction according to Sanders et al. [1], but with a more versatile procedure than that used in [1], which allowed to measure sorption isotherms at constant partial pressure of one component of the gas mixture. Indeed this novel method allows one to measure sorption isotherms *i*) at constant composition of the gaseous phase, *ii*) at constant fugacity of one component or *iii*) at constant equilibrium pressure. The first protocol, in particular, allows to mimic better the real constraints faced when dealing with a membrane separation process, where one has a gas stream of fixed composition, in which only the total pressure can be varied, by compression. The pressure decay apparatus is coupled to a gas chromatograph Varian CP-4900 Micro-GC equipped with a capillary column and with a thermal conductivity detector for analysis of the gas phase composition.

In the case of PTMSP, the mixture n-C_4/CH_4 was initially considered, to provide a direct comparison with literature data [2] and validation of the method. Indeed, the sorption of n-C_4/CH_4mixtures showed a reasonable agreement with the existing mixed gas sorption data [2].

On the other hand, the CO_2/CH_4mixed sorption data in PTMSP are completely new, and were measured in the range from 0 to 33 atm of total equilibrium pressure, and from 5 to 90 mol.% of carbon dioxide in the gaseous phase.

Furthermore, the same characterization of CO_2/CH_4 mixed sorption was performed in PIM-1: the pressure range inspected was the same as in PTMSP, while the composition of CO_2 ranged from 10 to 50 mol.%.

PTMSP membrane was cast from a solution of toluene, immersed in methanol and then dried under vacuum before characterization; its density was 0.77±0.01 g/cm^3.

The PIM-1 membrane was prepared from a filtered ca. 2.0 wt.% chloroform solution of PIM-1 and heated in vacuum at 70 °C, then submerged in methanol and dried under vacuum at 70 °C. The density of pure PIM-1 was (1.143±0.008) g/cm^{3}at 25°C.

For both PTMSP and PIM-1, the mixed gas solubility differs significantly from the pure gas value, and, in particular, the solubility of both components is depressed by the presence of the second one, as it often happens in glassy polymers.[3] The solubility selectivity ranges between 2 and 6 for PTMSP and between 5 and 10 for PIM-1. The methane solubility, however, is more significantly depressed by CO_2 than that of CO_2 is decreased by CH_4, therefore the real solubility selectivity (CO_2/CH_4) for PTMSP and PIM-1 is higher than the ideal solubility selectivity. Such effect becomes more significant with increasing the mole fraction of CO_2 in the gaseous phase and with pressure, and is more significant for PIM-1 than for PTMSP. Indeed, the real solubility selectivity becomes 3 times higher than the ideal one in PTMSP for a fraction of 70 mol.% of CO_2 in the gas phase, while for PIM-1 such point is reached with a lower concentration of CO_2(50 mol.%).

Both results indicate the presence of a competition for available polymer matrix sites, which is not surprising due to the nature of physical sorption in glassy matrices, and possibly also of different interactions between polymer and penetrants.

To investigate that behavior, the Non-Equilibrium Lattice Fluid model (NELF) was used [3], while the widely used Dual Mode Sorption (DMS) model was also considered as a reference tool. The NELF model, as well as the DMS, does not require additional parameters for the prediction of the mixed gas behavior, and is fully predictive provided a few pure gas sorption data in the polymer matrix. Indeed, binary interaction parameters are the same as in the pure gas case, and the swelling induced by the mixture is estimated from pure gas swelling. Remarkably, in the DMS model, only competition (depression) effects are accounted for, because the mixed gas additional term (positive) appears only in the denominator of the expression for solubility.

The NELF model provided quantitative predictions of the mixed gas sorption of CO_2 and CH_4under pure- and mixed-gas conditions in PTMSP and in PIM-1. The solubility selectivity is also predicted, although with less accuracy, by the NELF model. The DMS model works fairly well in the case of PTMSP, but provides poorer predictions than the NELF model of the mixed gas solubility in PIM-1.

Sorption of mixtures of CO_2 and CH_4 in PTMSP and in PIM-1 was predictable with the NELF model with an accuracy that is comparable to the experimental one, which could reduce the need for the laborious measurements of mixed gas sorption in polymers. Better insights and interpretation of the mixed gas sorption mechanism can also be obtained by using the NELF model.

References:

[1] E.S. Sanders, W.J. Koros, H.B. Hopfenberg, V.T. Stannett J. Membr. Sci. 18 (1984) 53-74.

[2] R.D. Raharjo, B.D. Freeman, E.S. Sanders, Polymer 48 (2007) 6097-6114.

[3] M. Minelli, S. Campagnoli, M.G. De Angelis, F. Doghieri, G.C. Sarti, Macromolecules 44 (2011) 4852-4862.

(619d) Diffusive Separation of Propylene/Propane With ZIF-8 Membranes

Hara, N., The University of Tokyo
Yoshimune, M., National Institute of Advanced Industrial Science and Technology (AIST)
Negishi, H., National Institute of Advanced Industrial Science and Technology (AIST)
Haraya, K., National Institute of Advanced Industrial Science and Technology (AIST)
Hara, S., National Institute of Advanced Industrial Science and Technology (AIST)
Yamaguchi, T., Tokyo Institute of Technology
Originally presented on: 11/7/2013 9:45:00 - 10:10:00

Metal-organic frameworks (MOFs) have been recognized as a new candidate for gas and vapor separation. Among several varieties of MOFs, ZIF-8 is one of the most common materials for separation membranes. ZIF-8 is composed of zinc atoms and 2-methyl imidazole as ligands, forming the sodalite structure with large cavities, 1.16 nm in diameter, which are connected with a small pore aperture of 0.34 nm diameter [1]. Recently, several reports have focused on the potential of ZIF-8 for propylene/propane separation on the basis of their difference of diffusivity [2].

The most common preparation methods of ZIF-8 membranes are the in situ growth method and the secondary growth method. Pan et al. reported the separation of propylene/propane mixture, in which the ZIF-8 membrane is prepared by the secondary growth method with a separation factor of up to 50 for propylene/propane binary mixture [3]. Kwon et al. reported the preparation of ZIF-8 membranes using microwave-assisted seeding and secondary growth with the separation factor of up to 40 for propylene/propane binary mixture [4].

To improve the propylene/propane separation properties, diffusivity and solubility separation factors should be separately investigated from the single-component permeation measurements. In this study, we clarified the contribution of diffusive separation for propylene/propane with ZIF-8 membrane based on the analysis of diffusivity and solubility obtained from the membrane permeation properties. Permeation properties were studied using the single-component gas permeation properties, and superior separation properties for propylene/propane were achieved for ZIF-8 membranes with uniform selective layer. Permeability, diffusivity, solubility, and their dependencies on temperature were studied, and the contribution of diffusivity on propylene/propane separation is thoroughly discussed.

(295 words)

[1] K. S. Park, Z. Ni, A. P. Cote, J. Y. Choi, R. D. Huang, F. J. Uribe-Romo, H. K. Chae, M. O'Keeffe and O. M. Yaghi, Exceptional chemical and thermal stability of zeolitic imidazolate frameworks, Proc. Natl. Acad. Sci. USA 103 (2006) 10186-10191.

[2] K. H. Li, D. H. Olson, J. Seidel, T. J. Emge, H. W. Gong, H. P. Zeng and J. Li, Zeolitic Imidazolate Frameworks for Kinetic Separation of Propane and Propene, J. Am. Chem. Soc. 131 (2009) 10368-10369., C. Zhang, R. P. Lively, K. Zhang, J. R. Johnson, O. Karvan and W. J. Koros, Unexpected Molecular Sieving Properties of Zeolitic Imidazolate Framework-8, J. Phys. Chem. Lett. 3 (2012) 2130-2134.

[3] Y. C. Pan, T. Li, G. Lestari and Z. P. Lai, Effective separation of propylene/propane binary mixtures by ZIF-8 membranes, J. Membr. Sci. 390 (2012) 93-98.

[4] H. T. Kwon and H. K. Jeong, Highly propylene-selective supported zeolite-imidazolate framework (ZIF-8) membranes synthesized by rapid microwave-assisted seeding and secondary growth, Chem. Commun. 49 (2013) 3854-3856.

(619e) Simulation of Propylene/ Propane Plasticization in 6FDA-Based Polyimide Membranes

Halitoglu-Velioglu, S., Istanbul Technical University
Ahunbay, M. G., Istanbul Technical University
Tantekin-Ersolmaz, S. B., Istanbul Technical University
Originally presented on: 11/7/2013 10:10:00 - 10:35:00

The permeation properties of hydrocarbon gases through polymeric membranes are of interest for the development of membrane processes for separation of olefin/paraffin mixtures, such as propylene (C_3H_6)/propane (C_3H_8), in the petrochemical industry. Propylene is largely a byproduct of ethylene production, which is the largest volume organic produced in the petrochemical industry. Furthermore, propylene is produced in propane dehydrogenation, and natural gas to olefins and other olefin conversion processes. While rubbery polymers have very low selectivity for propylene/propane, glassy polyimides (PIs) and polyimide-co-polypyrrolones based on 6FDA (4,4'-(hexafluorisopropylidene)dipthalic anhydride) have been reported to exhibit the best performance [1]. However, a major issue for these polyimides is the plasticization of the membrane during the permeation of these condensable hydrocarbons.

Propylene and propane, both being condensable, tend to plasticize polymeric membranes, even at partial pressures as low as 2 bars. There are several experimental studies investigating plasticization effect of propane and propylene in polyimides reported in literature [1-4]. The observed plasticization pressure among these studies was between 2 and 5 atm. In order to suppress plasticization, cross-linking and thermal treatment have been used. On the other hand, there are no molecular simulation studies reported in the literature that focus on propane and propylene-induced plasticization in polyimides.

In this study, molecular simulation techniques are used to estimate the degree of plasticization of 4,4-hexafluoroisopropylidene-diphthalic anhydride (6FDA)-based PI membranes induced by sorption of propane (C_3H_8) and propylene (C_3H_6). The diamines used to build the polyimide structure are 2,4,6-trimethyl-m-phenylene diamine (DAM), 2,5-dimethyl-p-phenylenediamine (DPX), and 4,4-oxydianiline (ODA). To the best of our knowledge, this is the first simulation study examining propane/propylene plasticization in 6FDA based PIs. The sorption simulations are carried out in the Grand Canonical Monte Carlo (GCMC) ensemble. To reproduce the hydrocarbon-induced plasticization effect, sorption-relaxation cycles are applied until C_3H_6 or C_3H_8 concentration converges. The increase in the FFV of the resulting polymer structure is considered as the extent of plasticization. The change in the sorption selectivity is determined for both pure gases and binary C_3H_6/C_3H_8 mixtures for % 50/50 feed composition. The glass transition temperatures (Tg) of the polymers are estimated by observing the volume change of the simulation cells during cooling down from 760 K to 380 K. Rotational time autocorrelation functions of the unit vectors defining to the distance between the diamine and dianhydride moities are also calculated in order to analyze changes in the local dynamics of the polymers due to hydrocarbon sorption. Our recent work [5] on the CO_2-sorption induced plasticization of these polymers have shown that the CO_2 sorption capacity at 35°C and 10 bars depend on the FFV of the swollen PI, yielding similar CO_2 concentrations for similar FFV values. Thus, particular interactions between CO_2 and sorption sites have limited differential contribution on sorption capacity. However, no such relationship was obtained for propane and propylene at 25°C and 1.13 bar due to changing site preferences, indicating the strength of sorbate-site interactions is significant on sorption capacity, as well as the FFV.

[1] Das, M., Koros, W.J., *J. Memb. Sci.,* 365, 399-408, 2010.

[2] Staudt-Bickel, C., Koros, W. J., *J. Memb. Sci.,* 170, 205, 2000.

[3] Tanaka, K., Taguchi, A., Hao, J. Q., Kita, H., Okamoto, K., *J. Memb. Sci.,* 121, 197, 1996.

[4] Jiang, L. Y., Chung, T. S., Rajagopalan, R., *Chem. Eng. Sci.,*63, 204-216, 2008.

[5] Velioglu, S., Ahunbay, M. G., Tantekin-Ersolmaz, S.B., *J. Memb. Sci.,* 417, 217-227, 2012.

(641a) Algal Biorefinery: Production of Biocrude Oil & Byproducts

Reddy, H. K., New Mexico State University
Selvaratnam, T., New Mexico State University
Muppaneni, T., New Mexico State University
Nirmalakhandan, N., New Mexico State University
Schaub, T., New Mexico State University
Dungen, B., New Mexico State university
Appuhamilage, N. S.,
Lammers, P., New Mexico State University
Voorhies, W., New Mexico State university
Deng, S., New Mexico State University
Originally presented on: 11/7/2013 8:30:00 - 8:52:00

We have developed a strategy needed for the production of sustainable algal biofuels with hydrothermal liquefaction of three different algal species. Hydrothermal liquefaction experiments with 10% solid loading of *Chlorella sp., Nannochloropsis sp.*, or *Galdieria sp.* were carried out between 180°C-300°C and 30 min. of reaction time to investigate the effects of HTL conditions on the important liquefaction reactions and final products. The liquid biocrude products were analyzed with GC-MS and FT-ICR for their chemical compositions, and micro bomb calorimeter for their high heating values. The composition of residual algal biomass was analyzed after each experimental run to identify the biochemical composition of the biomass. The residual water collected after liquefaction was also analyzed to quantify the nutrients and dissolved organic compounds in the water phase. This water phase produced in the HTL process was recycled for cultivation of *galdieria sp.* at different levels of dilution. An integrated processing strategy was developed for individual algal species based on the values of three products (bio-crudes, solid residues and water phase) obtained in the HTL process.

(641b) Biodiesel Production Through Microwave Assisted Transesterification of Microbial Cells

Cui, Y., Southern Illinois University

Originally presented on: 11/7/2013 8:52:00 - 9:14:00

One strain of oleaginous yeasts, Cryptococcus curvatus (ATCC 20509) has been studied to grow on several substrates including biodiesel production byproduct crude glycerol and sweet sorghum juice. After cultivation, yeast cells were heated under microwave radiation to extract lipid and produce biodiesel through in-situ transesterification.

Firstly, the yeast growth with crude glycerol was studied.On the second step, an optimization of yeast fermentation with crude glycerol was conducted. The resulting parameters of the response surface method optimization were used in a fed-batch fermentation where crude glycerol was automatically pumped in responding to exhausted oxygen levels in the fermentor.At the end of 12 days, the biomass density and lipid content were 44.53 g/L and 49%. Compared with our previous fed-batch experiment which was conducted under un-optimized condition, the yield of biomass and lipid increased 35.26% and 25.29%.

Considering its capability to utilize a broad range of substrates, C. curvatus was tested on its lipid productivity on sweet sorghum juice. When cultured in a fed batch process for 3 days, the final biomass density was 23.6 g/L with a lipid content of 51%. To extract lipids from cells in an effective and fast fashion, a domestic microwave oven was used with different solvents. With only methanol, a lipid yield of 33.2% of yeast cells was obtained in 4 min.

In the end, to convert yeast lipids to biodiesel directly without the step of lipid extraction,the in-situ transesterification method used microwave irradiation on the simultaneous extractionand transesterification of wet yeast biomass to biodiesel. Based on the experimental results and RSM analysis, the optimal conditions for this process were determined as: methanol to yeast biomass (v:w) ratio of around 50:1, catalyst concentration about 5 wt.%, and reaction time of 2 min. The biodiesel samples were analyzed with GC and the FAME content in biodiesel is about 50%. Compare the with other researches, this process proved to be a fast and easy method to produce biodiesel from wet yeast biomass.

(641c) Design Optimization By Response Surface Methodology for Continuous Fermentative Production of 1,3 Propanediol From Waste Glycerol By Product of Biodiesel Processes

Kanjilal, B., University of Connecticut
Noshadi, I., University of Connecticut
Intoci, N., University of Connecticut
Dowding, M.,
Hale, W.,
Brendel, B., University of Connecticut
Srivastava, R., University of Connecticut
Parnas, R., University of Connecticut
Originally presented on: 11/7/2013 9:14:00 - 9:36:00

The production of ~10 gallons of crude glycerol for every 100 gallons of biodiesel has led to a current glycerol excess of ~2.52Bn pounds per annum by the US biodiesel industry and non biodiesel sources. The bio-fermentation of crude glycerol to 1, 3-propanediol, an emerging intermediate platform in biorenewable chemicals with a nascent market of $575 million worldwide, as a value added product gives a recourse to capitalizing on the current over capacity of glycerol. This work presents the design optimization for a green and efficient process for the batch and continuous production of 1,3 propanediol from industrial waste glycerol using soil based bacterial inoculums. The optimal process parameters for the best possible yield of 1,3 propanediol were determined using a design of experiment by varying paramaters such as pH, glycerol concentration and yeast concentration. Response surface methodology (RSM), an important branch of experimental design, was used to find the relationship between experimental parameters and the yield and biomass growth in order to find the optimal conditions for the process. The optimization of 1,3 propandeiol production, in this way, with several inputs, helps to enhance the conversion in an economic manner. pH was seen to have the most profound effect on the process both in terms of yeild as well as bacterial growth. The optimised conditions were used to run a continous fermentation and arrive at a steady state for a given dilution rate, while the inhibitory solvent effects were mitigated by cycling between acetate and butyrate laced batches. While the addition of acetate enhanced biomass growth, butyrate addition enhanced 1,3 propanediol production with distinct inhibitory effects on the growth of bacteria. The effects were utilized to systematically maximize the survival of the mixed inoculum in a continuous run.

(641d) Review of Bio-Jet Fuel Conversion Pathways

Wang, W. C., National Renewable Energy Laboratory
Tao, L., National Renewable Energy Laboratory
Originally presented on: 11/7/2013 9:36:00 - 9:58:00

Bio-derived jet fuel has become a key subject in the aviation industry strategy to address the challenges from reducing operating cost, environmental performance and impacts to the global ecosystem. For producing a long-term renewable aviation fuel, economic and environmental sustainability are necessary. Researchers from oil refining industry, commercial aviation industry, government, biofuel companies, agriculture organizations and academic systems are looking forward to developing a commercial viable process that has not only low production cost but low GHG emissions. Additionally, the fuel produced from the process has to meet the ASTM specifications and is intended to be 100% drop-in for the replacement of current petroleum jet fuel. This study reviewed the current major technologies for converting bio-based materials into renewable aviation fuels. All the conversion pathways are classified to four categories: (1) syngas-to-jet, includes Fischer-Tropsch(F-T) synthesis and gas fermentation; (2) renewable-oil-to-jet, includes hydroprocessed renewable jet (HRJ) via plant oil and algal oil, catalytic hydrothermolysis (CH) and hydrotreated depolymerized cellulosic jet (HDCJ, also named pyrolysis) via pyrolysis oil; (3) alcohol-to-jet, includes the conversion of ethanol, butanol and methanol to jet fuel from lignocellulosic biomass feedstocks; (4) sugar-to-jet, includes fermentation of sugars to hydrocarbon or hydrocarbon precursors as well as catalytic conversion of sugars or sugar intermediate to fuels. Discussions for each pathway are primarily focused on the available feedstocks, conceptual process design, process economics, life-cycle assessments and commercial readiness, based on the literature studies. With the help of this comprehensive review on bio-derived jet fuel conversion technologies in recent literatures, it is believed that, whereas there are significant issues related to feedstock price and availability and energy consumption of the process, bio-derived jet fuels could progressively replace a great portion of conventional jet fuel required to meet the commercial and military demand.

(641e) Pellets From Lignocellulosic Materials: Briquetting Process Evaluation

Gonzalez, Y., Universidad Libre
Rincon, S., Universidad Libre
Camargo, G., Universidad Libre
Sierra, F., Universidad Nacional
Forero N, C. A., Universidad Nacional
Originally presented on: 11/7/2013 9:58:00 - 10:20:00

Colombia is a palm oil producer country and the implementation of biodiesel industry has generated huge quantities of the residues. On another hand, there are several regions in Colombia without a sufficient energy supply for a better and faster development. The purpose of this research is investigate the use of agro industrial residues coming from palm oil for obtain briquettes that can be used as fuel in remotes region in Colombia. Different parts of the agricultural waste they be employed in the manufacture of briquettes, the fibers and the seed. The biomass was characterized by ultimate analysis and in the densification process two different levels of pressure was employed. Finally densities, mechanical strength, heat of combustion were evaluated for the briquettes. The initial results shown that the waste from the seed materials presents a better performance for the densification process than the fiber materials and the briquettes elaborated a higher pressure exhibit more elevated calorific power

(641f) Pyrolytic Liquid Yields From Bamboo Species (Bambusa Oldhamii and Bamboo Angustifolia)

Pliego-Bravo, Y. S., Instituto Tecnologico de Orizaba
Cervantes- Pesado, L., INSTITUTO TECNOLOGICO DE ORIZABA
Bolanos-Reynoso, E., Instituto Tecnologico de Orizaba
Gonzalez-Garcia, P.,
Originally presented on: 11/7/2013 10:20:00 - 10:42:00

PYROLYTIC LIQUID YIELDS FROM BAMBOO SPECIES (*Bambusa Oldhamii and Bamboo Angustifolia*)

Abstract

A characterization of the species of bamboo (*Bambusa Oldhamii and Guadua Angustifolia*) with respect to compositional analysis, elemental analysis and thermochemical calorific using as a carrier gas, nitrogen, to identify desirable traits for subsequent use in a thermal conversion process such as pyrolysis, this energetic process provides greater value to convert biomass into a liquid fuel. Moisture content lignin degradation temperatures and ash content are indicators to determine the ideal raw material for conversion to a renewable process.

Introduction

Among the new alternatives that have been implemented for reducing the amount of emissions of CO, SOx, NOx, there being a source of biomass energy that partially replaces fossil fuels and decreases the concentration of these gases emitted into the atmosphere. (Zhang *et al*, 2007). Immediate analysis, elemental and calorific thermochemical analysis are useful in the understanding of the properties of the species to consider a material as an energy source, allowing to know the effective thermochemical process for producing a biofuel (Bridgwater, 2001 and Brown 2011). In this paper we characterize the species of bamboo (GA and BO) to propose an alternative to using craft or promote the efficient use of waste generated by the industry based on the characterizations mentioned above in order to obtain a liquid fuel

Methodology

The proximate analysis, elemental, calorific power and thermogravimetric were performed these are described below.

Materials

In lignocellulosic analysis were used: Ethanol analytical reagent grade (99.5% purity), analytical reagent grade sulfuric acid (98.2% purity), analytical reagent grade toluene, analytical reagent grade nitric acid (69% purity) all of J. T. Baker and distilled water. In thermochemical analysis was used 99% pure nitrogen and air.

Immediate analysis
Establishes the relative amounts of moisture (ASTM D 2867-99), volatile matter (ASTM D 5832-98), fixed carbon, ash (ASTM D 2866-99). Get the amount of carbon fixed subtracting the amounts of ash and volatile matter of a total of 100%.

Elemental analysis and calorific power
The amounts of C, H, N and S of the constituents of bamboo species were estimated experimentally using elemental analyzer (Vario EL III) based on moisture and ash free (LHC), the oxygen obtained by difference. The calorific value was estimated by the equation Reed and Das (1998) expressed in MJ / Kg.

Thermochemical analysis
Thermogravimetric analysis were performed with a non-isothermal model held in the TG Mettler Toledo 851e with software version Star. Samples between 6-10 mg were analyzed, studies were carried out using nitrogen and air, and before each experiment was performed a nitrogen purge for 10 min to all samples with a flow of 100 mL / min.

Getting pyroligneous liquids
For testing proceeded with an experimental 3^2 where the factors were time (15, 30 and 60 min) and the temperature (350, 500 and 600 ° C) for each of the species studied. These were carried out on a laboratory scale reactor with stainless steel fixed bed capacitors and nitrogen supply 200 mL / min.

Results and discussion

Moistures were between 16-17%, whereby the raw material was suitable for a thermochemical process. The high ash contents cause slag deposits on equipment deteriorating and decreasing the life thereof. The species studied showed percentages of 4.92 of ash for BO and 3.33 for GA making it attractive for use in these applications. Low fixed carbon content was related to the age of the material, in young bamboo lignin was removed easily due to low contents of total polysaccharides and cellulose, whereas an older kind more lignin is harder removing (Garcia *et al*, 2009).

The calorific value was obtained by correlations of Reed and Das (1998). In thermochemical analysis TGA curves showed that the decomposition of the species begins about 300 ° C, the displacement of the curves to the right with increasing temperature. For Bambusa oldhamii degradation stabilized 550 ° C to 600 ° C heating speeds of 15 and 25 ° C / min, while at 5 ° C / min stabilized at 500 ° C. In the case of Guadua angustifolia behavior was observed for more stable speeds of 5 and 15 ° C / min stabilizing at approximately 480 ° C whereas at 25 ° C / min from around 540 ° C. For older BO liquid yields were achieved at 600 ° C with 46 ± 0.5% in the times of 30 and 60 min, for the species GA best performance was obtained at 600 ° C for a time of 60 min of 43 ± 2%.

Referencias

[1] Bridgwater, A., V. (2001). *Progress in Thermochemical Biomass Conversion*. Backwell Science. Volume 2 IEA bioenergy.

[2] Brown R. C., *Thermochemical Process of Biomass*. Wiley, U.S.A. (2011).

[3] García, L, Marquez, F, Aguiar, L, Arauso, J, Carballo, L, Orea, U Zanzi, R (2009). *Rendimiento de los productos de la descomposición tÃ©rmica de la madera Eucalyptus saligna Smith a diferentes alturas del fuste comercial.* Revista Chapingo. Serie ciencias forestales y del ambiente, Vol. 15, Núm. 2. Julio_diciembre. pp. 147-154

[4] Higman C. y van der Burgt M., *Gasification*. Segunda edición. Elsevier. U.S.A. (2008).

[5] Reed T. B., y Das A., *Handbook of biomass downdraft gasifier engine system*. Solar Energy Research Institute, U.S. department of energy. U.S.A. (1998).

[6] Zhang Qi, Chang Jie , Wang Tiejun , Xu Ying (2007). Review of biomass pyrolysis oil properties and upgrading research. Energy Conversion and Management 48 (2007) 87–92.

(641g) The Impact of Ethanol and Iso-Butanol Blends on Regulated Emissions, Air Toxics, and Particle Emissions from SIDI Vehicles

Karavalakis, G., UC Riverside
Short, D., UC Riverside
Vu, D., UC Riverside
Durbin, T., UC Riverside
Asa-Awuku, A., UC Riverside
Originally presented on: 11/7/2013 10:42:00 - 11:04:00

Spark-ignition gasoline direct injection (SIDI) vehicles have been the subject of several studies, with much of this work has been done outside the U.S., as this is where GDI vehicles have been more prominent. The penetration of SIDI vehicles in the U.S. market is rapidly growing with most domestic automotive manufacturers offering direct injection versions in their existing models. The SIDI technology has emerged from the need to improve the thermodynamic efficiency (and thus reduce fuel consumption) and power output of spark ignited engines. The improved fuel efficiency compared to the conventional port fuel injection (PFI) vehicles could potentially provide the manufacturers the means to meet the target set in the U.S. on the fleet-average carbon dioxide (CO_2) emissions of future vehicles. It is foreseen that this category of vehicles will dominate the gasoline market, eventually replacing the conventional and less efficient PFI vehicles. However, SIDI engines are known to emit more particulate matter (PM) emissions than homogeneous-charge PFI spark-ignited gasoline engines and diesel engines equipped with diesel particle filters (DPFs) in the exhaust system. In particular, it is likely that SIDI vehicles will be higher than 3 mg/mile CARB LEVIII PM Standard through 2021, whereas PFI vehicles generally emit below 1 mg/mile, and often below 0.5 mg/mile, providing the potential to more than double PM emissions from new light-duty vehicles. The PM characteristics of direct injected spark ignition engines in comparison to conventional PFI engines can be mainly attributed to the injection characteristics. There is a growing consensus amongst the health experts that particles in the ultrafine range (smaller than 100 nm) are potentially more toxic and have greatest adverse health effects on human health.

Limited studies have investigated the influence of alternative fuels on the exhaust emissions, especially particulate emissions, from SIDI vehicles. The fuels employed in these studies were mainly low and mid-level ethanol blends. Currently, ethanol is the most widely used renewable fuel for transportation in the U.S. and with the push to use increasingly higher levels of renewable fuels, there has been an accompanying push to further increase the ethanol level in gasoline. The EPA Renewable Fuel Standard version 2 (EPA-RFS2) and the California Low Carbon Fuel Standard (CA-LCFS) are driving the U.S. market for the development of biofuels. The EPA-RFS2 requires that 36 billion gallons of renewable fuel are available in the U.S. market by 2022, with ethanol expected to make up the majority of this requirement.

Another pathway that could be used to reach the congressionally mandated biofuel volumes is the use of bio-butanol. Analogous to ethanol, butanol can be produced both by petrochemical and fermentative processes. Biomass-derived butanol can be produced by alcoholic fermentation of biomass and agricultural feedstocks, such as corn, wheat, sugar beet, sugarcane, etc. Butanol offers a number of advantages over ethanol for transportation use. Butanol has considerably higher energy content than ethanol. Butanol is less hydroscopic and less corrosive than ethanol, therefore, it is more suitable for distribution through pipelines, whereas ethanol must be transported via rail or truck. Moreover, in blends with gasoline, butanol is less likely to separate from the base fuel than ethanol if the fuel is contaminated with water. Butanol has a lower vapor pressure and a higher flash point than ethanol. The heat of vaporization of butanol is also less than half of that of ethanol.

The literature is particularly poor with respect to studies on butanol/gasoline blends on SIDI engines. In addition the effect of oxygenated fuels especially on particulate emissions from SIDI engines is not well understood. The purpose of this investigation is to elucidate the effects of fuel type and blend concentration for ethanol and iso-butanol on the exhaust emissions from two different modern technology stoichiometric SIDI wall-guided light-duty vehicles fitted with three-way catalysts (TWC). The vehicles were tested over the Federal Test Procedure (FTP) and the Unified Cycle (UC) on light-duty chassis dynamometer. The fuel blends included E10, E15, E20, Bu16, and a E10/Bu8 blend. A total of seven fuels were employed in this study, including an E10 fuel, which served as the baseline fuel, and two more ethanol blends, namely E15 and E20. For this study, iso-butanol isomer was blended with gasoline at proportions of 16, 24, and 32% by volumes, which are the E10, E15, and E20, respectively, equivalent based on the oxygen content. The seventh fuel was an alcohol mixture consisting of 10% ethanol and 8% iso-butanol. This mixed alcohol formulation was equivalent to E15 based on the oxygen content. Each vehicle was tested on each fuel over three FTP and three UC tests. Measurements of regulated emissions were made for nitrogen oxides (NO_x), carbon monoxide (CO), total hydrocarbons (THC), nonmethane hydrocarbons (NMHC), methane (CH_4), and CO_2. Special attention was paid to gaseous air toxics, including carbonyl compounds (aldehydes and ketones) and volatile light aromatic hydrocarbons (benzene, toluene, ethylbenzene, and xylenes), and 1,3-butadiene. Particle emissions were characterized by defining PM mass, total particle number, particle size distributions, and black carbon concentrations. The results of this study will enhance our understanding of alcohol fuel formulations effects on gaseous and particle emissions from emerging vehicle technology.

(642a) Recovery of Phosphorus From Sewage Sludge and Sewage Sludge Ashes With the Thermo-Reductive Recophos Process

Schönberg, A., University of Leoben

Samiei, K., University of Leoben

Kern, H., University of Mining Leoben

Raupenstrauch, H., Montanuniversität Leoben

Originally presented on: 11/7/2013 8:30:00 - 8:55:00

The principle of the RecoPhos process is the fractioned extraction of phosphorus and heavy metals from sewage sludge and its ash at high temperatures under reducing conditions. The RecoPhos process transforms a problematic waste into several streams of secondary raw materials, leaving behind a minimum of solid waste. The aim of the planned RecoPhos project is to develop a completely new P-recycling process, which can avoid several of the known chemical-technical problems that are still limiting the possible success of phosphorus recycling.

<>

<>Introduction

There is no possibility to find a replacement for the resource phosphorus. It is an essential element for all life forms and its exploitable deposits are getting scarce quickly [1]. This and also the fact that phosphate mining and refining has severe bad effects on the environment, like enormous land use; and the production of vast amounts of dangerous wastes [2], will make it inevitable in the near future to recycle phosphorus from every suitable P-containing waste. Sewage sludge (coming from municipal wastewater treatment) is considered to be one of the most important alternative Phosphorus sources, as in this sludge all of the phosphorus discharged by humans is concentrated and can, if properly collected and treated, be re-used completely. Its inorganic dry matter (= ash content) has a phosphate content of 30 up to 70% of the content of phosphate rock produced from natural phosphate ores [3]. In the project, the process will be developed using modeling and lab scale experiments. Using these results, the basic design of a pilot scale plant will be carried out, a market study will be performed and the environmental impact of the process compared with the status quo will be evaluated in a Life Cycle Assessment (LCA) study [4].

<>The RecoPhos process

The core technique of the RecoPhos process, the so-called InduCarb reactor, works after the same chemical principle as the Woehler process [5], producing amongst others white phosphorus as a product.

$$Ca_3(PO_4)_2 + 3SiO_2 + 5C \rightarrow 3CaSiO_3 + 5CO + P_2$$

The above shown summary reaction runs in the inductive heated bed of carbon as illustrated schematically in figure 1.

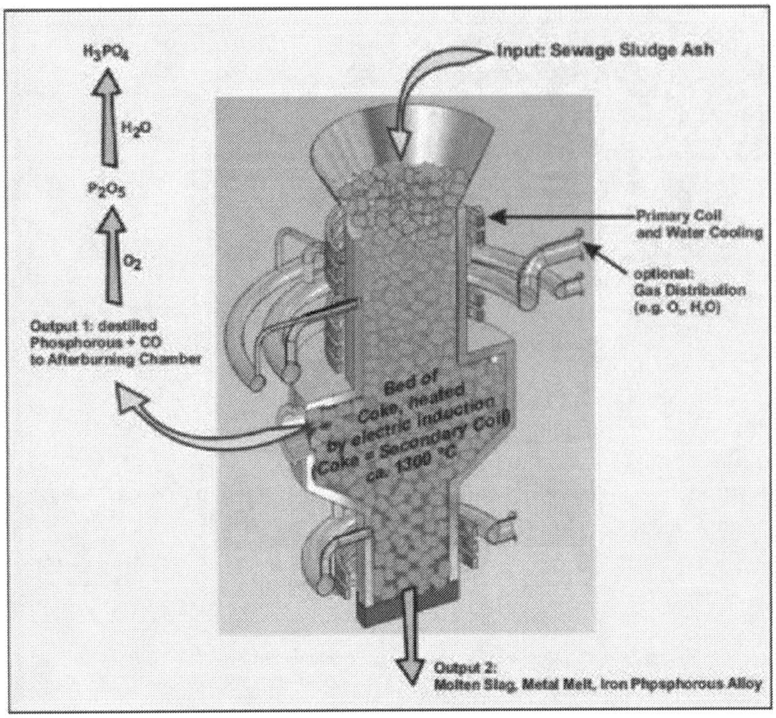

Figure SEQ "Abbildung" *Arabic 1: InduCarb reactor in the RecoPhos process

RecoPhos approaches the suitability of the following gas treatment aggregates very similar as used in waste and sludge incineration plants:

- Quench (temperature > 280 °C) for the removal of low boiling heavy metals and compounds as dust, leaving the phosphorus gaseous;
- Quench (temperature < 100°C) for the recovery of white phosphorus from the off-gas stream;
- Scrubber for the removal of acidic compounds and Hg;
- Dry and wet electrostatic precipitator for the removal of dusts and aerosols;
- Activated carbon adsorber and cloth filter or activated carbon cloth filter in appropriate positions;
- Afterburner, if the production of phosphoric acid is wished.

To be able to perform experiments, a fully operational continuous bench scale RecoPhos plant will be designed and constructed in the course of the project. This comprises the planning and basic engineering of the reactor as well as of periphery like frequency converters, off-gas cleaning and automation technology.

<>

<>Modeling and Simulation

The mathematical modeling and simulation consists of two parts: thermodynamic process simulation, and mathematical modeling of the inductive heated chemical reacting packed bed of carbon particles. The design will be constantly enhanced according to the experimental results. As a last step, using the most important results, the basic design of a pilot scale will be created.

The modeling of the InduCarb reactor also consists of two parts. At first the creation of a macroscopic model is necessary to get an overview of the occurrence in the reactor. Also it will be possible to get the dimensions for a reactor design. The aim of the model calculation is to find the necessary parameters for the dimensioning of the reactor. The correct choice of the reactor volume is decisive for the attainable conversion in the process. Furthermore statements relating to the design should be possible. The optimal height to diameter ratio of the reactor is an important statement here. Furthermore, modes of operation could be optimized in an existing reactor. Knowledge of the interactions between the processes in the reactor gains a deeper insight into the process. So you can make settings for targeted experiments.

The second part is a mesoscopic model that will be developed by MUL adapting on an existing transient, 2-dimensional heterogeneous model. The goal is to optimize the heat input, the product quality and the energy efficiency.

All over it is necessary to find out the key components and the reactions in the InduCarb reactor. Also a reducing of the surface of the carbon bed during operation has to be examined. Furthermore the input of additives should be checked (e.g. for minimizing the consumption of the reduction agent carbon or for an alternative energy supply to the reactor).

<>Summary

Recycling of Phosphorous from sewage sludge provides a sustainable solution for production of this finite resource essential for human nutrition. It will also be a solution to the environmental problems of Phosphate rich effluents from waste water treatment plants. To this end, the RecoPhos project employs the innovative InduCarb reactor design in which the ashes from sludge incineration plants will trickle down an inductively heated packed bed of Carbon. The reactions taking place inside the reactor for production of high quality Phosphorous are based on the same principle as the well-known Woehler process. The final design of the overall process will be based on the results of modeling and simulation in conjunction with experimental research at the pilot scale.

<>Acknowledgement

The RecoPhos research project has received funding from the European Union Seventh Framework Programme (FP7/2007-2013) under grant agreement no. 282856.

<>Literature

[1] Cordell, D.; Drangert, J.-O.; White, S. (2009). The story of phosphorus: Global food security and food for thought. Global Environmental Change 19: 292-305

[2] Sartorius C., von Horn J., Tettenborn F. (2011) Phosphorus recovery from Wastewater ? State-of-the-art and future potential, *International conference on Nutrient Recovery and Management 2011*, January 9-12, Miami, Florida, USA

[3] S. Donatello, D. Tong, C.R. Cheeseman. (2010). Production of technical grade phosphoric acid from incinerator sewage sludge ash (ISSA). Waste Management 30, 1634?1642

[4] RecoPhos Grand Agreement, Annex I ? Description of Work, 2011

[5] Corbridge, D.E.C. (1995). Phosphorus: An outline of its chemistry, biochemistry and uses. Elsevier. pp. 556.

(642b) Data Availability and Management for Process Sustainability Evaluation: Greenscope

Mukherjee, R., U.S. Environmental Protection Agency
Ruiz-Mercado, G. J., U.S. Environmental Protection Agency
Smith, R. L., U.S. Environmental Protection Agency
Gonzalez, M. A., U.S. Environmental Protection Agency
Originally presented on: 11/7/2013 8:55:00 - 9:20:00

Chemical products are obtained through processes that transform valuable resources into desired products, byproducts, and undesirable wastes and emissions. A particular product can be obtained from alternative processes involving varying amounts and types of resources, utilization and byproduct formation. Therefore, it is necessary to develop and identify the most sustainable process among many design alternatives.

The U.S. Environmental Protection Agency has created the GREENSCOPE tool (Ruiz-Mercado et al, 2013) (EXCEL® version) to guide decision-makers, designers, and stakeholders in developing more sustainable processes. GREENSCOPE calculates about 140 sustainability indicators and estimates chemical process sustainability in the areas of environment, efficiency, energy, and economics. This tool can be used for chemical process sustainability evaluation either in the early design or at the operational stage and aid in proper decision making to yield the most sustainable process possible.

GREENSCOPE process indicators are calculated using process data and chemical property data. The scope for successful implementation of chemical process sustainability relies on the availability and management of representative data from various sources. This includes chemical process data, classification lists, potency factors, and physicochemical, thermodynamic, and toxicological data. The complexity of data handling and management required for chemical process sustainability evaluation with GREENSCOPE has lead to the advent of a standalone software version. The standalone software tool is capable of handling challenges with data availability from various sources and heterogeneity of different chemical processes. This work describes development of the JAVA based GREENSCOPE tool for chemical process sustainability evaluation. This tool is designed to be integrated and available to other design tools such as simulator packages, optimization tools, sensitivity tools, etc.

In the GREENSCOPE JAVA version tool, a relational database for data management has been implemented to handle data from different sources. Both local and online databases are connected with GREENSCOPE. Currently, database creation, upload, modification, and access were performed using JAVA database connectivity functions. Different relational databases were created for proper handling of data using MySQL.

This presentation will describe the indicators used for sustainability evaluation of chemical processes. A detail description of different methods for obtaining required data and their handling methods for sustainability estimation with different indicators will be shown. It will also show how choices between different process design alternatives can be made using GREENSCOPE.

Ruiz-Mercado, Gerardo J., Michael A. Gonzalez, and Raymond L. Smith. “Sustainability Indicators for Chemical Processes: III. Biodiesel Case Study.” *Industrial and Engineering Chemistry Research*. **DOI:** 10.1021/ie302804x

(642c) Development of a Sustainable Phase Transfer Catalysed Continuous Process for a Pharmaceutical Intermediate

Teoh, S. K., Institute of Chemical & Engineering Sciences
Sa-ei, K., Institute of Chemical & Engineering Sciences
Noorulameen, M. S.,
Ng, Y. L., 1Institute of Chemical & Engineering Sciences
Sharratt, P. N.,
Wheelhouse, K.,
Rutter, A.,
Originally presented on: 11/7/2013 9:20:00 - 9:45:00

Three of the major environmental problems in the pharmaceutical industry are solvent-laden waste, catalyst-containing waste and waste produced from side or by reactions. Batch processing dominates pharmaceutical and fine chemical manufacturing. Compared to continuous processing, batch processing is inherently less mass and energy efficient. Continuous flow also provides the possibility of continuous solvent recovery even if a more dilute reaction be required to maintain mobility.

Phase Transfer Catalysis (PTC) is a long established procedure for the facilitation of reactions between two immiscible phases using quaternary ammonium salts. From the perspective of green chemistry, PTC provides environmental benefits in eliminating, reducing or replacement of environmentally incompatible dipolar aprotic solvent such as DMF, DMSO and THF. It could replace/eliminate stronger and/or highly toxic bases (NaH, organic base), increase conversion and/or selectivity, and hence reduce waste produced. We evaluate the synergetic environmental advantages of operating PTC in a continuous flow system.

The model chemical system used in our investigation is the heteroatom O-allylation of 3-phenyl-1-propanol with allyl bromide. A typical chemical system reported in the literature using 2-phenylethanol was conducted in plant scale batch wise at a reaction temperature of 65-75°C using NaOH micropearls, toluene (solvent) and tetrabutylammonium hydrogen sulfate (TBAH) reacted for at least 16 hours to give the product as a toluene concentrate (approximately 95 mol% product by GC). From our batch process development on 3-phenyl-1-propanol system, an effective amount of PTC accelerates the reaction by about 8 fold at room temperature. In the absence of PTC, the reaction temperature has to be elevated to achieve the required reaction rate and conversion. It was found that the model reaction could be operated at room temperature with a 0.028 mol eq. of Tetra-n-butylammonium bromide (TBAB). Using NaOH solution (50% w/w) instead of NaOH pellets, the reaction could perform as well (95% conversion in 1 hour with 97% product purity by GC) without the need of toluene and this also facilitates flow processing.

The model reaction has been demonstrated at scale both batch wise (≈ 2 liter) and continuously (≈ 3 liter/hour). In the continuous process, the reaction is operated counter-currently in a liquid-liquid extractor. The reaction is robust and takes about 100 mins to reach its steady state with a comparable reaction conversion to the batch process. The synergetic sustainability benefits of operating phase transfer catalyzed process in a continuous flow system are being compared to the traditional batch process from reaction to work-up. Sustainability metrics such as the efficiency of using chemical reagents/solvent/energy, the extent of aqueous/solids waste generation, E factor, equipment efficiency, and cost efficiency will be discussed.

(642d) Innovative Sustainable Materials & Processes

Enright, T., Xerox Research Center of Canada
Goredema, A.,
Originally presented on: 11/7/2013 9:45:00 - 10:10:00

Xerox Research Centre of Canada (XRCC) is the centre for global materials innovation in the Xerox Corporation. The mandate of XRCC is to develop all of the materials that are used in Xerox devices and to be on the forefront of materials that will enable new technologies. One of the primary goals of the centre is to deliver value to our customers through more sustainable and environmentally responsible materials and processes. This is being achieved through application of green chemistry and engineering from the conceptualization and design phase, through development and scale-up, and all the way through technology transfer to manufacturing. XRCC has commercialized a number of sustainable technologies, including functional materials that enable products to consume less energy, processes that minimize the overall amount of waste, and materials with higher amounts of bio-renewable content. A general overview of XRCC's sustainable product and process strategy will be presented, along with a highlight example that demonstrates how a multistep solvent-intensive process was redesigned to eliminate all waste solvents.

(642e) Dry Granulation of Blast Furnace Slag for Heat Recovery

Kofler, M., Montanuniversität Leoben

Doschek, K., Montanuniversität Leoben

Raupenstrauch, H., Montanuniversität Leoben

Originally presented on: 11/7/2013 10:10:00 - 10:35:00

Blast furnace slag is known as by-product from the production of hot metal at integrated steel and iron mills. Approximately 300 kg of this liquid slag with a temperature of about 1500 °C is tapped together with each ton hot metal. This means a worldwide annual production of about 400 million tons of blast furnace slag. Further processed, water granulated blast furnace slag can be produced, if the liquid slag is cooled down rapidly for solidification. State of the art is a closed system which uses lots of water for rapid cooling and solidification to produce a granulated blast furnace slag which has an amorphous structure. This product normally is dried and grinded for use as a binder in the cement production due to its latent hydraulic properties. The disadvantage of wet systems is on the one hand, the recooling of the circulating water and on the other hand the need for drying of the granulated blast furnace slag, due to the residual moisture up to 20 mass percent. Furthermore the sensible heat of about 1.5 GJ/t of slag is lost. [1]

Therefore a new lab rig has been installed by Siemens VAI in cooperation with Montanuniversität Leoben at the Chair of Thermal Processing Technology for research and development. In this project there are the co-partners voestalpine Stahl GmbH, ThyssenKrupp Steel Europe AG and FEhS-Institut für Baustoff-Forschung e.V. This project is funded by the German Federal Ministry of Economics and Technology (BMWi).

Figure 1: Lab rig at the University of Leoben, Chair of Thermal Processing Technology (Picture: Siemens VAI)

The lab rig is based on the ?Rotating Cup? principle. The liquid slag will be atomized by a cup and small particles spread away due to the occurring forces. First tests involving the ?Rotating Cup? principle were performed in Redcar at British Steel in the 90's. Further investigations took place at Vitkovice in the Czech Republic between 2002 and 2004. These tests were part of the ZeroWaste project of Siemens VAI. Both research activities did not focus on heat recovery due to the lack of political pressure at this time. It could be shown that dry granulation for producing glassy blast furnace slag is also feasible with the ?Rotating Cup?. [1]

Figure 2: Trials at Redcar (left) and Vitkovice (right) [1]

The new lab rig at the University of Leoben has been designed on basis of the results of the lab tests at FEhS-Institut and the modelling of Siemens VAI to find the best operating parameters for producing high quality granulated blast furnace slag and highest possible air outlet temperatures. Therefore about 300 kg of slag is molten in the ?Flash-Reactor? for each trial at the University's workshop. The liquid slag is tapped into special slag pots which are lined with refractory to prevent heat losses. Afterwards the slag pot is manipulated into a tilting unit, installed close to the granulator. The liquid slag feed into the granulator can be measured and controlled via load cells. The liquid slag will be delivered to the centre of the granulator passing a slag runner and a vertical refractory pipe. After atomizing the slag droplets cool down rapidly during their flight towards the granulator wall once spread away by the fast spinning cup. The surface of the particle has to solidify during this short flight time before falling into a developed fluidized bed made by granulated particles. [1]

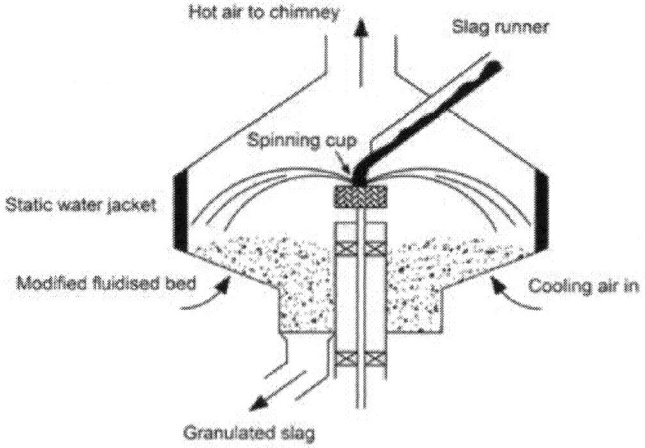

Figure 3: ?Rotating Cup? principle [1]

Furthermore the granulator wall is indirect cooled by a water jacket. Air will be introduced at the bottom of the device. A special air distribution insures a proper cooling of the particles, an agitated bed and no agglomeration. So a maximum heat exchange rate can be reached. Hot off gas can be added to raise the input air temperature for the granulation process. Hence a recirculation system can be simulated which should be used in industrial application. Flow straighteners at the top hinder any carryover of the particles, where the hot air leaves the granulator. [1]

Afterwards the air is cooled down by a spray tower before leaving the lab rig passing a droplet separator and the suction fan. The granulated slag particles can be discharged at the bottom of the granulator after each trial.

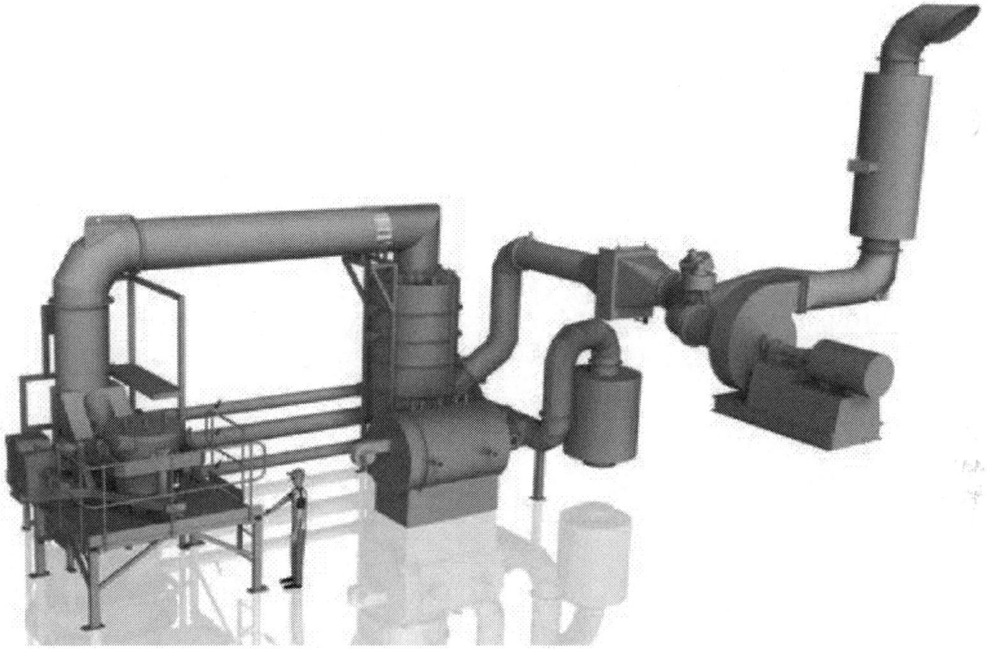

Figure 4: Isometric drawing of the lab rig

The whole lab rig is fully automated. More than 90 sensors are recording the process data like slag mass flow, different pressures, temperatures and other measurands. This helps analyzing each trial to find the best operating parameters for getting a proper product while having a maximum air outlet temperature. The main challenge considers ensuring rapid slag freezing to reach the transformation point of the slag very fast. Otherwise the product will not get fully glassy but partially crystalline which is unsuitable for further use in the cement industry. Cooling with air is much more difficult due to a substantially lower heat capacity compared with water used in conventional devices. Furthermore the air heats up while streaming through the agitated bed. So additionally the temperature difference worsens the setting process.

First campaigns showed promising results regarding glass content, grain size distribution and particle shape. Also a significant off gas temperature increase could be measured. With the obtained data, process optimization can be done and a scale up for further investigations can be realized in the future. Using dry slag granulation at integrated steel and iron mills will gain access to the sensible heat of the blast furnace slag. The recovered energy can be used for the production of steam or electricity or for other useful preheating processes. An energy potential for recovering 20 MW thermal energy or alternatively of about 6 MW electric power generation from a slag feed of one ton per minute can be calculated. Without the need of an upstream drying system for the blast furnace slag energy and as in consequence CO_2 can be saved. In summary dry granulation of blast furnace slag allows an environmental friendly and sustainable subsequent processing for one of the last big heat recovery potentials at a modern blast furnace. [1]

References

[1] McDonald, I., Long, E., Werner, A. & Most, D. (2010) Dry slag granulation - The environmental friendly way to making cement. In: IOM3 conference on Waste Recovery in Ironmaking and Steelmaking Processes. London: United Kingdom

(642f) Advances in Technology for Producing Alumina From Complex Resources in China

Zhang, Y.,

Originally presented on: 11/7/2013 10:35:00 - 11:00:00

The qualified bauxite ore in China to produce Alumina through Bayer process is limited. However, there are considerable complex resources containing significant amount of alumina in China, such as low grade bauxite, fly ash, red mud and nepheline, which can be used to produce alumina through an appropriate process. Compared with low grade bauxite and fly ash, red mud and nepheline, which contain 5-20% of sodium oxide and potassium oxide, could be used to economically extract alumina only through an alkaline sintering or leaching process, because the sodium oxide and the potassium oxide could be leached as valuable hydroxides. However, an acid leaching or sintering process is theoretically suitable to extract alumina from low grade bauxite and fly ash containing tiny sodium or potassium, meanwhile, the purification of the acid leachate is quite important for the development of new acid technology to produce qualified alumina. The improved technologies based on sintering process, alkaline and acid leaching processes to produce alumina from those complex resources in China, and especially, a Low Pressure Hydro-chemical Process (LPHP) leaching technology will be presented.

(664a) Evaluation of Spray Dried Oxygen Carriers Based On Manganese Ore and Ca(OH)2 for Chemical-Looping With Oxygen Uncoupling

Sundqvist, S., Chalmers University of Technology
Mattisson, T., Chalmers University of Technology
Leion, H., Chalmers University of Technology
Lyngfelt, A., Chalmers University of Technology
Originally presented on: 11/7/2013 12:30:00 - 12:55:00

Evaluation of spray dried oxygen carriers based on manganese ore and $Ca(OH)_2$ for chemical-looping with oxygen uncoupling

Sebastian Sundqvist[†,*], Tobias Mattisson[‡], Henrik Leion[†], Anders Lyngfelt[‡].

† Department of Chemical and Biological Engineering, Division of Environmental Inorganic

Chemistry, Chalmers University of Technology, S-412 96 Goteborg, Sweden

‡ Department of Energy and Environment, Division of Energy Technology, Chalmers University of

Technology, S-412 96 Goteborg, Sweden

*Corresponding author

The suitability of an oxygen carrier in chemical-looping combustion (CLC) or chemical-looping with oxygen uncoupling (CLOU) is determined by its lifetime and reactivity. Both are factors relates to the cost of the oxygen carrier in relation to the amount of carbon dioxide captured. There are likely losses of oxygen carrier from attrition in continuous operation and in the case with solid fuels there are additional losses expected together with the ash extraction. Hence there is a need to produce inexpensive oxygen carriers which requires the use of cheap raw materials, such as ores or industrial waste materials.

One type of oxygen carrier that has been shown to have reasonable reactivity is the perovskite structured calcium manganite, or $CaMnO_3$. The oxygen carrier has oxygen uncoupling properties, that is, it releases oxygen in the gas phase, the amount which is governed by the partial pressure of oxygen in the surroundings. In this project, the aim was to produce a $CaMnO_3$ particle from cheap raw materials such as manganese ore and commercial $Ca(OH)_2$ and test these with respect to parameters important for CLC and CLOU.

Several oxygen carriers of CaMnO3 have been produced utilizing different manganese ores as starting material. The manganese ores were mixed with commercial $Ca(OH)_2$ and each material was calcined at both 1200°C and 1275°C for 4h. Particles were then produced using spray-drying and sieved to obtain particles in the size range of 125-180 μm. The evaluation was conducted in a fluidized batch reactor, where the particles were alternatingly exposed to reducing and oxidizing conditions using methane CH_4 and syngas (50/50 H_2/CO) at 950°C. For the experiment 15 g of oxygen carrier was used with a flow of 450 ml/min of reducing gas. The oxygen carriers were also evaluated at different temperatures (850°C-1050°C) under inert conditions to measure the oxygen release (CLOU properties) and conversion of CH_4.

Most of the oxygen carriers have full conversion of syngas but the results for CH_4 conversion varies greatly between the different materials, with the average conversion ranging between 40-75% under the reducing period. However the release of oxygen during inert conditions is almost the same for all materials. The oxygen release also increases with temperature, and for some oxygen carriers the oxygen release rate doubles with a temperature increase of just 50°C.

(664b) Solid Fuel Conversion Via Chemical-Looping With Oxygen Uncoupling (CLOU) Using Co-Precipitation Derived Cu-Mn Oxygen Carriers

Hosseini, D., ETH Zurich
Imtiaz, Q., ETH Zurich
Kierzkowska, A., ETH Zurich
Muller, C. R., ETH Zurich
Originally presented on: 11/7/2013 12:55:00 - 13:20:00

Chemical-looping combustion (CLC) is an emerging CO_2 capture and storage (CCS) technology. In CLC oxygen derived from a solid oxygen carrier is used to completely combust a (hydro-) carbonaceous fuel to CO_2 and H_2O. Thus, in CLC CO_2 is inherently separated from the nitrogen in air. Conventional CLC for solid fuels requires a prior, relatively slow gasification step to produce a synthesis gas containing mostly CO and H_2. An alternative strategy is chemical-looping with oxygen uncoupling (CLOU) [1]. Here, oxygen is released *via* a decomposition reaction, with the combustion reaction of the solid fuel with molecular oxygen being very fast.

A suitable oxygen carrier for CLOU, typically a transition metal oxide, must possess a high oxygen storage capacity, thermal stability and fast decomposition kinetics at typical operating temperatures (800-1000 °C). In addition, resistance to attrition and agglomeration, cost and environmental impact have to be taken into account. With regards to equilibrium thermodynamics CuO and Mn_2O_3 are arguably the most attractive oxygen carriers for CLOU [1]. However, both copper and manganese have some drawbacks. For the CuO/Cu_2O system, agglomeration and a comparatively low oxygen partial pressure (at temperatures below 850 °C) are the main disadvantages, whereas Mn_2O_3/Mn_3O_4 has a low oxygen storage capacity and at temperatures above 900 °C Mn_3O_4 is thermodynamically stable and cannot be re-oxidized back to Mn_2O_3 using air.

In this study, mixed metal oxide oxygen carriers containing manganese and copper oxides were synthesized using a co-precipitation technique. The synthesized oxygen carriers were fully characterized and assessed with regards to solid fuel conversion under CLOU conditions. A comparison with the performance of the individual metal oxides revealed that the optimized Cu-Mn mixed oxides possessed an improved resistance to agglomeration, a higher oxygen carrying capacity, a high solid fuel conversion and an improved physical and chemical stability. These favorable characteristics make the newly developed Cu-Mn mixed oxides promising oxygen carriers for CLOU.

Reference

[1] Mattisson, T., Lyngfelt, A., Leion, H., Int. J. Greenhouse Gas Control 2009, 3 (1), 11-19

(664c) Using (MnxFe1-x)2SiO5 As Oxygen Carriers for Chemical-Looping With Oxygen Uncoupling (CLOU)

Mattisson, T., Chalmers University of Technology
Jing, D., Chalmers University of Technology
Van Noyen, J.,
Ryden, M., Chalmers University of Technology
Lyngfelt, A., Chalmers University of Technology
Originally presented on: 11/7/2013 13:20:00 - 13:45:00

Using $(Mn_xFe_{1-x})_2SiO_5$ as oxygen carriers for chemical-looping with oxygen uncoupling (CLOU)

Chemical-looping with oxygen uncoupling (CLOU) utilizes oxygen carrier materials which can release gas phase oxygen in the fuel reactor.Â This could be beneficial for the fuel oxidation process, especially when considering solid fuels, as the fuel will be converted through normal combustion, which is not the case with normal chemical-looping combustion.Â Copper oxide is the most widely investigated oxygen carrier for the CLOU technology, but recently a number of combined oxides have been proposed as feasible oxygen carriers.Â One especially interesting mixed oxide system is the Fe-Mn-O combination, which releases oxygen at relevant temperatures and partial pressures of oxygen.Â In earlier investigations by our research group at Chalmers, the behavior and functionality of this system has been shown to be highly dependent upon the molar ratio of Fe/Mn.Â However, the stability during operation has not been satisfactory, with considerable attrition found when used in continuous operation.Â Hence it is of interest to combine oxygen carriers of this system with active or inactive material in order to improve stability while keeping the reactivity intact.Â In this work, the effect of SiO_2 addition to the Fe-Mn-O system has been investigated. Â Three oxygen carriers based on the Fe-Mn-Si ternary were manufactured by spray-drying and evaluated with respect to parameters important for CLOU. Â The general composition of the material was $(Mn_xFe_{1-x})_2SiO_5$, with x=0.33, 0.5 and 0.67.Â After the spray-drying procedure the material was calcined at 1100 and 1200Â°C for 4h and sieved to a size fraction of 125-180 mm.Â The oxygen release properties and the reactivity with methane were investigated using a batch fluidized batch reactor made of quartz.Â The uncoupling properties were investigated in the temperature interval 850 ? 1050Â°C by exposing a sample of 15 g of particles to a flow of N_2, and then measuring the released oxygen as a function of time.Â For the investigations with methane as the reducing gas, the oxygen carrier particles were exposed to a CH_4 flow of 450 Nml/min in the fluidized bed for 20 s.Â The extent of reaction during the reduction was established by measurement of CO_2, CO and CH_4 in the outlet from the reactor. After reduction, the reduced particles were oxidized in 5% O_2.Â Â Â All of the materials showed oxygen release in the investigated temperature range, although there was not a huge variation with temperature. The $(Mn_{0.67}Fe_{0.33})_2SiO_5$ showed the highest release rate at 900Â°C, with 0.8% O_2 measured after 360 s reduction.Â The reactivity with methane increased with temperature for the $(Mn_{0.67}Fe_{0.33})_2SiO_5$ Â and $(Mn_{0.33}Fe_{0.67})_2SiO_5$ materials, and the average conversion approached 98% for the Mn-rich oxygen carrier.Â The oxygen carrier with a 1:1 molar ratio of Fe:Mn, i.e. $(Mn_{0.5}Fe_{0.5})_2SiO_5$, had remarkably high reactivity already at 850Â°C, with an average methane conversion of 90%, which decreased somewhat at 900 and 950Â°C.Â Â The investigated particles had a decent crushing strength, around 1 N, and little or no attrition was seen in the tests.Â From the current investigation it can be established that the addition of SiO_2 to the Fe:Mn system could be feasible, as the oxygen uncoupling properties seem to remain intact and the reactivity with fuel gas high at certain conditions.Â However, the mechanical stability of these type of materials need to be further investigated, to establish whether there is an advantage in comparison to pure Fe:Mn-oxides.

(664d) CO2 Utilization Via Chemical Looping Process

Galvita, V. V., Ghent University

Poelman, H., Ghent University

Marin, G. B., Ghent University

Originally presented on: 11/7/2013 13:45:00 - 14:10:00

Introduction

Chemical looping is one of several emerging technology options capable to reduce the carbon dioxide emissions and helping in a diverse range of applications for production of fuels, chemicals and electricity [1-3]. In a general chemical looping process, a metal oxide (MeO) is circulated between two reactors - the reducer and oxidizer. For a combustion process, MeO reacts in the reducer with the fuel to produce CO_2, H_2O and leaving metal (Me) or metal oxide in reduced state [1]. In the oxidizer it is re-oxidized to its initial state with air. Thus, fuel and air never mix and CO_2 is not diluted by nitrogen.

Other chemical looping schemes can produce hydrogen or synthesis gas. In chemical looping steam (CLSR) or dry reforming (CLDR) processes the fuel is again completely oxidized over the metal oxides (Eq. 1). The H_2O or CO_2 product is used as oxidant instead of air. The outlet from the looping reactor then consists of H_2 or CO.

A key issue in CLDR processes is the selection of an appropriate oxygen carrier material. The reactivity of the oxygen storage material, its cost and thermal stability are critical selection criteria. A broad range of metals was analyzed to determine the thermodynamic equilibrium limitations for redox cyclic processes [4-7]. While a number of metals/metal oxides (Mo, Cr, Zn, Co, Nb, Ce) give reasonable CO_2 reduction capacity, it is apparent that only iron yields high oxygen storage capacity from CO_2 (0.7 mol CO_2/ mol Fe) over a wide range of operating temperatures (600-1800°C).

Thus, the chemical looping conversion of CO_2 into CO or CO_2 and H_2O mixture to syngas over iron oxide as oxygen storage material has been one of the possible target technologies for CO_2 utilization. The ability of the oxygen storage material to maintain its high activity in repeated reduction/re-oxidation cycles is the most critical issue for the overall economic viability of the CO_2 splitting process.

The objective of the present study is to investigate the CO_2 re-oxidation efficiency of a series of CeO_2-Fe_2O_3 mixed oxides prepared by co-precipitation for the utilization of CO_2 in a chemical looping process. Experimental performance data are reported as well as structural characterization of the oxygen storage materials during reduction and CO_2 re-oxidation. In addition, the stability of performance was tested through prolonged use of the material.

Results

Mixed CeO_2-Fe_2O_3 materials with 0, 20, 50,70, 90 and 100 wt% CeO_2 were prepared for CO_2 transformation to CO by means of chemical looping. The samples were characterized with HRTEM, EDX, SAED, EEL and in situ XRD and tested in cycled reactions for activity and stability. The generation of CO from CO_2 through H_2-CO_2 redox cycles was investigated at 600°C to evaluate the effect of CeO_2 upon Fe_2O_3. Two different phases were observed in the CeO_2-Fe_2O_3 samples. For a CeO_2 content below 70wt%, a distinct Fe_2O_3 phase was always present. The second phase appeared in all mixed CeO_2-Fe_2O_3 samples and was identified as a nano-crystalline solid solution of iron in ceria with 20 nm crystallite size and an atomic ratio of Fe:Ce up to 1:5. In all mixed CeO_2-Fe_2O_3 samples, the solid solution crystallite size was considerably smaller compared to pure ceria, while a size decrease for Fe_2O_3 was noticed with increasing CeO_2 loading. The crystallographic structure of all CeO_2-Fe_2O_3 was studied under H_2 reduction and CO_2 oxidation, using time-resolved in situ X-ray diffraction. In a 90wt% CeO_2 sample, no iron related diffractions were present at the start, but the H_2-TPR treatment made Fe appear as phase segregating from ceria. For lower CeO_2 loadings, H_2-TPR reduced the separate Fe_2O_3 phase to Fe_3O_4, FeO and Fe. The solid solution phase was always partially reduced. In CO_2-TPO, re-oxidation of iron followed a one-step pathway to Fe_3O_4 around 500°C, while for the solid solution re-oxidation occurred at 450°C. The addition of CeO_2 to Fe_2O_3 had a beneficial effect upon the activity and stability of the material. Reduction-re-oxidation cycles could be repeated many times without significant loss of CO_2 conversion efficiency.

The activity and stability of the mixed CeO_2-Fe_2O_3 samples as compared to pure Fe_2O_3 and CeO_2 a test reaction with H_2 reduction and CO_2 re-oxidation was set up. The samples are reduced in 5%H_2 in Ar during 1 min. By contacting the reduced sample with CO_2, CO is produced. As an example, the observed space time yield of CO as a function of time for 20wt% CeO_2- Fe_2O_3 during the 2nd, 5th and 10th cycle is displayed in Fig. 1a.

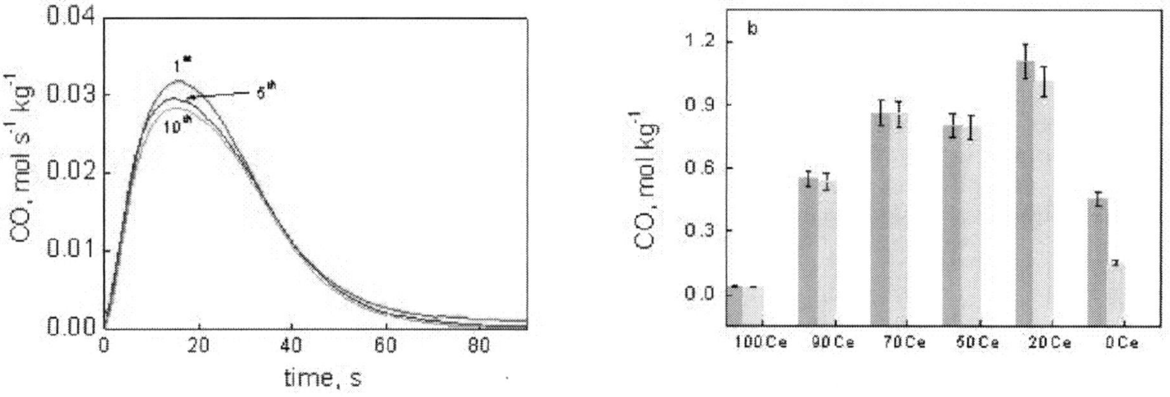

Figure 1. (a) Space time yield of CO during the re-oxidation phase with CO_2/He for 20wt% CeO_2-Fe_2O_3 after 1, 5 and 10 redox cycles; (b) CO produced after 1st (dark bar) and 10th cycle (light bar) over the xCe-Fe samples.

Fig. 1b shows the amount of produced CO after the 1st and 10th redox cycle for samples 0 to 100wt%CeO_2-Fe_2O_3. In both cycles, the product yield is higher for mixed materials than for pure ones. Especially the samples 20 to 70wt% CeO_2-Fe_2O_3 produce over 5 times more CO than Fe_2O_3, with the highest yield for 20 wt%CeO_2-Fe_2O_3. The amount of CO produced over CeO_2 was 10 times lower than for Fe_2O_3. Highest CO yield was obtained for 20wt% CeO_2-Fe_2O_3, while 70 and 50wt% CeO_2-Fe_2O_3 were the most stable combinations. The addition of ceria to Fe_2O_3 thus formed a solid solution phase, yielding improved activity. Moreover, the combination of both oxides efficiently suppressed the sintering effect, resulting in improved stability in chemical looping processes.

Acknowledgment

This work was supported by the 'Long Term Structural Methusalem Funding by the Flemish Government', the Fund for Scientific Research Flanders (FWO, project 3G004613), the Interuniversity Attraction Poles Programme, IAP7/5,- Belgian State – Belgian Science Policy.

References

1. J. Adanez, A. Abad, F. Garcia-Labiano, P. Gayan, and L.F. de Diego, Progress in Energy and Combustion Science 38 (2012) 215

2. L.-S. Fan, and F. Li, Ind. Eng. Chem. Res. 49 (2010) 10200.

3. C. Luo, Y. Zheng, N. Ding, Q. Wu, G. Bian, and C. Zheng, Ind. Eng. Chem. Res. 49 (2010) 11778.

4. M. Najera, R. Solunke, T. Gardner, and G. Veser, Chemical Engineering Research and Design 89 (2011) 1533.

5. A. Sim, N.W. Cant, and D.L. Trimm, Int. J. Hydrog. Energy 35 (2010) 8953-8961.

6. S. Abanades, Int. J. Hydrog. Energy 37 (2012) 8223.

7. J.R. Scheffe, M.D. Allendorf, E.N. Coker, B.W. Jacobs, A.H. McDaniel, and A.W. Weimer, Chemistry of Materials 23 (2011) 2030.

(664e) The Synthesis of Highly Efficient Ca-Based CO2 Sorbents Containing a Hierarchical Pore Structure

Broda, M., ETH Zurich

Müller, C. R., ETH Zurich

Originally presented on: 11/7/2013 14:10:00 - 14:35:00

In 2010 the global CO_2 emissions increased by more than 5 % compared to 2009, reaching 33 billion tonnes.[1] One possible, mid-term solution to reduce the release of CO_2 into the atmosphere is the implementation of CO_2 capture and storage (CCS) technologies.[2] However, the currently available CO_2 capture technology, i.e. amine scrubbing, is associated with a large penalty on plant efficiency. A promising, new CO_2 capture technology is the use of alkaline earth metal based CO_2 sorbents. Due to its (i) high theoretical CO_2 capture capacity of 0.78 g CO_2/ g sorbent, (ii) low cost and (iii) relatively vast abundance in naturally occurring minerals, e.g. limestone or dolomite, CaO is the most promising candidate in this class of materials. The selective capture and release of CO_2 is achieved by the following, theoretically reversible, reaction:

$$CaO(s) + CO_2(g) \ll CaCO_3\,(s),\ DH^0_{298K} = \pm\ 178\ kJ/\ mol \qquad (1)$$

However, CaO produced via the calcination of naturally occurring materials has a serious drawback, viz. the rapid decrease in the CO_2 capture capacity with repeated cycles of carbonation and calcination.[3] The carbonation reaction can be divided into two regimes.[4] It has been argued that the fast, kinetically controlled reaction stage, is associated with the filling of small pores. The importance of the available pore volume can be clearly seen by comparing the molar volume of the product, $CaCO_3$ ($M_V = 36.9\ cm^3/mol$), with that of CaO ($M_V = 16.7\ cm^3/mol$). Once the pores have been filled, the second, diffusion controlled reaction stage takes over, in which $CaCO_3$ is deposited on the partially carbonated CaO grains. Owing to the typically short contacting times in circulating fluidized beds, the reactor of choice for the commercial implementation of the calcium looping processes, only the CO_2 uptake achieved during the fast reaction stage will be of practical relevance. Since the Tammann temperature (temperature at which thermal sintering starts) of $CaCO_3$ (533 °C) is relatively low compared to the operating temperatures of 650-950 °C, the dramatic reduction in CO_2 uptake has been attributed to the reduction of surface area and the blockage of pores with diameter $d_{pore} <$ 100 nm. One potential strategy to increase the CO_2 capture stability of Ca-based sorbents is to stabilize CaO with high Tammann temperature matrix. However, work in this field has largely focused on the use of ?simple? synthesis techniques, which do not allow easily to adjust key structural properties, such as the pore volume. In this work we report the preparation of carbon gel templated, Ca-based CO_2 sorbents. Here, calcium nitrate and aluminium nitrate were used as the calcium and aluminium precursors, respectively, whereas resorcinol and formaldehyde were the precursors of the carbon gel. In addition, the diameter of the carbon sphere template was varied by changing the surfactant (CTAB) to resorcinol ratio. Upon removal of the carbon gel via calcination in air, hollow micro-spheres were obtained. The synthetic material was investigated using (i) scanning electron microscopy (SEM), (ii) surface area (BET) and pore volume (BJH) measurements, (iii) X-ray diffraction (XRD) measurements and (iv) mercury porosimetry measurements. The material synthesized possessed an excellent CO_2 capture capacity when compared to limestone or alternative CO_2 sorbents, such as hydrotalcites or metal organic frameworks. Furthermore, the current synthesis techniques allow the amount of Al_2O_3 required to stabilize the cyclic CO_2 capture capacity of the material to be reduced to only 5 wt.%.

References

<>1. Olivier, J. G. J.; Janssens-Maenhout, G.; Peters, J. A. H. W.; Wilson, J. *Long-term trend in global CO_2 emissions. 2011 report;* PBL Netherlands Environmental Assessment Agency: The Hague, **2011**

<>2. Herzog, H. What future for carbon capture and sequestration? *Environ. Sci. Technol.* **2001**, *35*, 148-153.

<>3. Grasa, G. S.; Abanades, J. C. CO_2 Capture Capacity of CaO in long series of carbonation/calcination cycles. *Ind. Eng. Chem. Res.* **2006**, *45*, 8846-8851.

<>4. Dennis, J. S.; Pacciani, R. The rate and extent of uptake of CO_2 by a synthetic, CaO-containing sorbent. *Chem. Eng. Sci.* **2009**, *64*, 2147-2157.

(664f) Li4SiO4-Based Novel Solid Fluidizable Sorbent for CO2 Capture:

Chowdhury, M. B. I., The University of Western Ontario
de Lasa, H., University of Western Ontario
Quddus, M. R., The University of Western Ontario

Originally presented on: 11/7/2013 14:35:00 - 15:00:00

This study investigates the cyclic use of several sorbents for CO_2 capture and comparative study of thermodynamic analysis to predict sorbent regeneration conditions and thermal levels where sorbent kinetics change from CO_2 absorption to CO_2 desorption. It is expected cycling processes with CO_2 capture and sorbent regeneration and alternative sorbents (e.g. lithium ortho-silicate based novel sorbent) could lead to significant energy savings with limited sorbent sintering. On this basis CO_2 absorption and desorption were quantified at various temperatures and CO_2 partial pressures. Temperature Programmed Carbonation (TPC) and Temperature Programmed Decarbonation (TPDC) were developed using temperature programmed equipment at CREC-UWO laboratories employing several sorbents subjected to a 5°C/min temperature ramp and with a gas stream containing a 10% CO_2 mole fraction. Calcium carbonate, lithium orthosilicate and a novel lithium orthosilicate modified sorbent are considered for these runs. TPC-TPDC runs confirm thermodynamic predictions for thermal inversion points. Furthermore, TPD-TPDC runs show that the novel lithium orthosilicate based sorbent provides a very stable and increased CO_2 sorption capacity over 10 absorption-regeneration cycles, while calcium carbonate displays a reduced CO_2 sorption capacity with cyclic operation. This fluidizable novel sorbent can significantly contribute towards CO_2 removal from flue gases emitted by power plants. The thermodynamic and TPC-TPDC analysis of the present study also provides a conceptual design from developing CO_2 capturing processes with alternative sorbents such as lithium ortho-silicate based sorbent.

(677a) Study Of Hydroden Production By Photofermentation From Different Organic ACID As Substrate

Oliveira, T. V., Federal University of Uberlandia
Oliveira, F. S., Federal University of Uberlandia
Sousa, T. M., Federal University of Uberlandia
Ferreira, J. S., Federal University of Uberlandia
Batista, F. R. X., Federal University of Uberlandia
Cardoso, V. L., Federal University of Uberlandia
Originally presented on: 11/7/2013 12:30:00 - 12:55:00

Hydrogen presents itself as an alternative fuel in the search for clean energy. One of the biological routes for hydrogen production is the photofermentation by purple non-sulfur photosynthetic bacteria. In general, these cells are able to synthesize hydrogen from organic acids. In this study, the potential of the H_2 production of ***Rhodopseudomonas palustris and Rhodobacter capsulatus*** strains was evaluated. The influence of organic acids (butyric, malic, lactic, propionic and acetic) as substrate was also studied at fixed ration of C/N of 35 mM : 1 mM, using sodium glutamate as nitrogen source. The results obtained from the fractional factorial design 2^{5-1}, based on variance analysis (ANOVA), showed that only the variables statistically significant at 90% confidence were malic acid and lactic acid for ***Rhodopseudomonas palustris*** and malic acid, lactic acid and acetic acid for ***Rhodobacter capsulatus***. The results showed that ***Rhodopseudomonas palustris*** presented higher productivity (26.01 μmol/L.h) when malic, lactic and butyric acids were combined as substrate resulting in a conversion of 0.1151 (mol of H_2/mol substrate). For ***Rhodobacter capsulatus***, the higher hydrogen production was verified when malic, lactic and propionic acids were combined (19.09 μmol/L.h) as substrate resulting in a conversion efficiency of 0.0068 (mol of H_2/mol substrate).

Acknowledgments: FAPEMIG, CNPq, CAPES and Vale S.A. by financial support.

(677b) Influence Of pH, Ammonium and Iron On The Hydrogen Production By Microbial Consortium

Romao, B. B., Federal University of Uberlandia
Santos, J. G., Federal University of Uberlandia
Oliveira, F. S., Federal University of Uberlandia
Ferreira, J. S., Federal University of Uberlandia
Batista, F. R. X., Federal University of Uberlandia
Cardoso, V. L., Federal University of Uberlandia
Originally presented on: 11/7/2013 12:55:00 - 13:20:00

Biological hydrogen production processes are known to be friendly and consume less energy compared to hydrogen production through the conventional route, such as thermochemical and electrochemical methods. The aim of this study was to investigate the influence of pH (4 to 7), ammonium sulfate (0.3 to 2 g/L) and iron sulfate (0.1 to 0.5 g/L) concentrations in the dark fermentation for hydrogen production by microbial consortium, using a composite design (2^3), on the substrate conversion (mol hydrogen per mol consumed substrate) and hydrogen productivity. Lactose (20 g/L), from milk whey permeate, was used as the carbon source. The fermentation medium was complemented with the following nutrients (g/L): 3 KH_2PO4, 7 K_2HPO_4, 1 $MgSO_4$, 3 yeast extract and 0.5 meat extract. The findings showed that the culture condition that resulted in higher conversion (6.73 mol H_2/ mol consumed substrate) was obtained to pH of 7, iron sulfate (0.5g/L) and ammonium sulfate (2g/L). In addition, a high level of the hydrogen productivity (73.7 mmolH_2/L.d) was obtained in the same condition.

Acknowledgments: FAPEMIG, CNPq, CAPES and Vale S.A. by financial support.

(677c) Monitoring of Gaseous Sulfur Species in a Modified Sulfur-Iodine Thermochemical Water Splitting Cycle in a Hydrophobic Ionic Liquid

Caple, K., Oregon State University
Originally presented on: 11/7/2013 13:20:00 - 13:45:00

Monitoring of Gaseous Sulfur Species in a Modified Sulfur-Iodine Thermochemical Water Splitting Cycle in a Hydrophobic Ionic Liquid

Kevin Caple and A.F.T. Yokochi

The development of a clean, simple, and renewable energy source is becoming one of the more prevalent topics amongst scientists, researchers, and politicians alike. Processes involving solar, wind, nuclear, chemical, and a smattering of other combinations of these have been studied and explored from numerous angles. Of the various processes studied, thermal based water splitting may be one of the most popularly studied, yet least commercially viable options available on the market today, due to drastic inefficiencies.

Some of these inefficiencies in thermal water splitting can be mitigated through what has been coined "thermochemical cycles" which are processes that involve several sequential and circular chemical reactions to take water and produce by hydrogen and oxygen while regenerating the initial reactants. Of these thermochemical cycles, one of the most commonly studied and most straightforward is the Sulfur-Iodine Cycle, which uses sulfur dioxide, iodine and a large excess of water to produce iodic and sulfuric acid. The two acids are then further processed into their initial reagents, as well as the desired hydrogen product. However, the S-I cycle has several drawbacks, including expensive catalysts for the acid treatments, material compatibility, and separation difficulty due to low percentage azeotropes.

In an attempt to moderate these drawbacks, we have developed a modification of the S-I cycle utilizing a previously undesired side reaction. To further alleviate any material compatibility issues, rather than run the reaction in water, an imidazolium based ionic liquid was chosen as a solvent for the following series of reactions to occur:

$$I_2 + SO_2 + 2H_2O \rightarrow 2HI + H_2SO_4 \text{ (Bunsen Reaction, Low Temp)}$$

$$8HI + H_2SO_4 \rightarrow 4I_2 + H_2S + 4H_2O \text{ (Side Reaction Producing H2S, Low Temp)}$$

$$H_2SO_4 \rightarrow SO_3 + H_2O \text{ (Sulfuric Acid Gasification, ~ 200°C)}$$

$$SO_3 \rightarrow SO_2 + .5O_2 \text{ (Reformation of Sulfur Dioxide, Low Temp)}$$

$$H_2S + 2H_2O \rightarrow SO_2 + 3H_2 \text{ (Steam Reformation, High Temp)}$$

This work focuses on the low temperature aspect of the cycle, specifically on the first two reactions involving the Bunsen Reaction that produces the two strong acids and the hydrogen sulfide production reaction, studying kinetic parameters, the extent of the cyclic reaction, and the amount of hydrogen sulfide produced.

(677d) The Water Splitting Kinetics of Two-Step Solar Thermochemical Process With CeO2

Arifin, D., University of Colorado Boulder
McDaniel, A. H., Sandia National Laboratories
Weimer, A. W., University of Colorado at Boulder
Originally presented on: 11/7/2013 13:45:00 - 14:10:00

Projected growth in global population and continued industrialization of developing countries will increase total world energy consumption by 50% in the next three decades. This increased demand for energy will be largely met by increased fossil fuel consumption, thereby increasing anthropogenic carbon in the atmosphere and further fuelling geopolitical conflicts over control of dwindling energy resources. Producing H_2 renewably by splitting H_2O in a solar-based thermochemical process is an attractive solution to both of these 21st century problems.

Due to the high thermal stability and fast kinetics, ceria is a favored material for the solar thermochemical process. In order to maximize the solar-to-hydrogen conversion efficiency, an understanding of the mechanistic pathway and a detailed engineering model of the water splitting process is necessary.

In this work, we report the results of a numerical analysis that separates experimental artifacts from material-specific H_2 production behavior yielding high quality fits to experimentally determined H_2 production rates over the entire reaction time domain. The results of this study show that the water splitting process is a surface mediated process analogous to a homogenous reaction, with low apparent activation energy.

(677e) Oxidation/Reduction Kinetics of Fe Particles in a ZiO2/ SiO2 Fluidized Bed

Greek, B., University of Florida
AuYeung, N., University of Florida
Imponenti, L., University of Florida
Klausner, J., University of Florida
Originally presented on: 11/7/2013 14:10:00 - 14:35:00

Reaction kinetics are crucial in the design of thermochemical reactors for hydrogen production. This paper investigates the reaction kinetics of various mass percentages of elemental iron (Fe) in a mixture of nonreactive particles, consisting of powdered zirconia (ZiO_2) and silica (SiO_2), when under a stream of H_2O. In addition, the oxidized bed was reduced under a stream of carbon monoxide (CO), and the reaction kinetics were characterized.. The overall cycling characteristics of an Fe-based system were investigated.. Temperatures well above the sintering temperature of the powdered iron particles alone were investigated due to the sintering inhibition provided by the nonreactive ceramic particles. During oxidation, a stream of H_2O was passed through the powder bed at temperatures ranging from 550-950°C. Reduction was performed using a stream of carbon monoxide (CO) at temperatures from 800-1000°C, with the minimum to avoid carbon depositing (due to the Boudouard reaction) in the powder bed. Products were measured via a mass flow meter and mass spectrometer during oxidation and reduction, respectively. Powder bed weight was used to close the reactant/product mass balance. Average hydrogen production rates over the first minute of reaction of 0.1-0.2 SLPM per gram of reactive material were achieved. This work characterizes the reaction kinetics of a readily available iron particle reactant in Fe-Fe_3O_4 redox reactions, using water as an oxidizing agent and CO as a reducing agent.

(677f) Hydrogen Fueling Stations: Optimization of Fill-Up Process

Olmos, F., UCLA
Manousiouthakis, V., University of California Los Angeles, Los Angeles
Originally presented on: 11/7/2013 14:35:00 - 15:00:00

One key challenge faced today by hydrogen fueling stations is to provide safe and efficient fill-ups for hydrogen fuel cell cars. Based on a self-consistent thermodynamic model of hydrogen, a dynamic model of the fill-up process, and a process simulator, strategies for the optimal control of this process can be devised. The simulator of the fill-up process is subject to optimization and control studies were global optimality is sought in order to identify the optimal fill-up strategy.

(704a) Sustainable Pathways for Recovering Materials and Energy From Waste

Sengupta, D., US Environmental Protection Agency
Smith, R. L., U.S. Environmental Protection Agency
Lee, C. C., U.S. Environmental Protection Agency
Takkellapati, S., U.S. Environmental Protection Agency
Originally presented on: 11/7/2013 12:30:00 - 12:49:00

Municipal Solid Waste (MSW), more commonly known as trash or garbage, consists of everyday items we use and then throw away, such as product packaging, grass clippings, furniture, clothing, bottles, food scraps, newspapers, appliances, paint, and batteries. This comes from our homes, schools, hospitals, and businesses. Waste generated by anthropogenic activity is a great source of carbon, and traditionally, it ends up in the landfill. In 2010, Americans generated about 250 million tons of trash and recycled and composted over 85 million tons of this material, equivalent to a 34.1 percent recycling rate. The purpose of this study is to evaluate various technologies to recover materials and energy from waste going to landfills. We explore two pathways, thermal and biological, to evaluate the recovery of materials from waste.

Thermal techniques of incineration of MSW reduce the volume of waste, and produce heat which can be used to produce steam and electricity. Gasification of MSW produces syngas, an important chemical intermediate, which can be converted into ethanol via catalytic methods and used as vehicle fuel. Biological techniques include separation of the MSW into an organic fraction and subsequent anaerobic digestion for obtaining biogas. Biogas can be cleaned and used for the production of electricity, or upgraded and used for natural gas applications. The organic fraction also resembles cellulosic feedstock, and can be fermented for the production of biofuels. Composting of the organic fraction is a useful way to obtain valuable materials for use in land applications.

Life cycle perspectives are used to evaluate each of these biological and thermal processes to calculate the cost, energy use, environmental releases, and societal issues for sustainability assessment of the process. The results of this research will help communities to decide on a sustainable technology for the treatment of their municipal solid waste. The research will also help them to develop strategies for waste collection, transport, materials and energy recovery, and green job creation in the community.

(704b) Simultaneous Reaction and Purification Process for Continuous Production of High Quality Biodiesel From Acid Oils

Shibasaki-Kitakawa, N., Tohoku University
Ihara, T., Tohoku University
Yonemoto, T., Tohoku University
Originally presented on: 11/7/2013 12:49:00 - 13:08:00

Introduction

Even today, the refined edible oils are mainly used as feedstocks for biodiesel, fatty aid ester, and this causes an increase in the production cost of biodiesel. The side-stream products, such as soapstocks, acid oils and deodorized distillates, obtained during edible oil refining have been suggested for use as alternative feedstocks for biodiesel. Especially, the crude oils of palm and rice bran have a high content of free fatty acid and, hence the amount of acid oils generated is large, about 10-15 wt% of refined oil output. There are some industrial uses for acid oils such as ink, but most acid oils are burned out as waste. The acid oils consist of about 95 wt% of free fatty acids (FFAs) and 5 wt% of acylglycerols (TAGs) and these components can be converted to fatty acid ester by esterification or transesterification using homogeneous acid catalyst. However, these reversible reactions do not go to completion, so that the separation and purification processes to remove the by-products and unreacted reactants are necessary to produce the high quality biodiesel in compliance with all international standards. These processes cause a further increase in the production cost of biodiesel and there is no commercially realizable process for biodiesel production from acid oils.

The purpose of this research is to construct a simple and economical production process of high quality biodiesel from waste acid oils using the ion-exchange resins as catalysts and adsorbents.

Materials and methods

The acid oil was donated by the rice bran oil company. Figure 1 shows a schematic diagram and photograph of our production process. Two columns were individually packed with the cation-exchange resin, Diaion PK208LH, and the anion-exchange resin, Diaion PA306S, respectively, and then connected in series. The temperature of each column was kept constant at 50 Â°C. The mixed solution at the molar ratio of methanol to the total fatty acid residue in the oil of 2:1 was supplied to the bottom of the first column at the constant flow rate. The effluent from the process was analyzed to check the quality of biodiesel.

Results and discussion

Table 1 shows the analytical result of standard tests for biodiesel quality of product without purification except for removing residual methanol by evaporation. The contents of product and residual reactants fully met the standard values. This indicated that free fatty acids and acylglyceroles were converted to biodiesel by esterification and transesterification with cation- and anion-exchange resin catalysts, respectively. The content of byproducts also met the standard values without downstream purification processes to remove them. This was because the by-products, water and glycerol were adsorbed on the anion-exchange resin and the effluent was free from them. Other properties for contaminants and fuel properties were found to satisfy with the standards.

Figure 2 shows the photographs of the feed oil and the effluents from the first and second columns at room temperature. The feed oil was liquefied by the esterificaion through the first column. The dark brown pigment was removed by the adsorption through the second column.

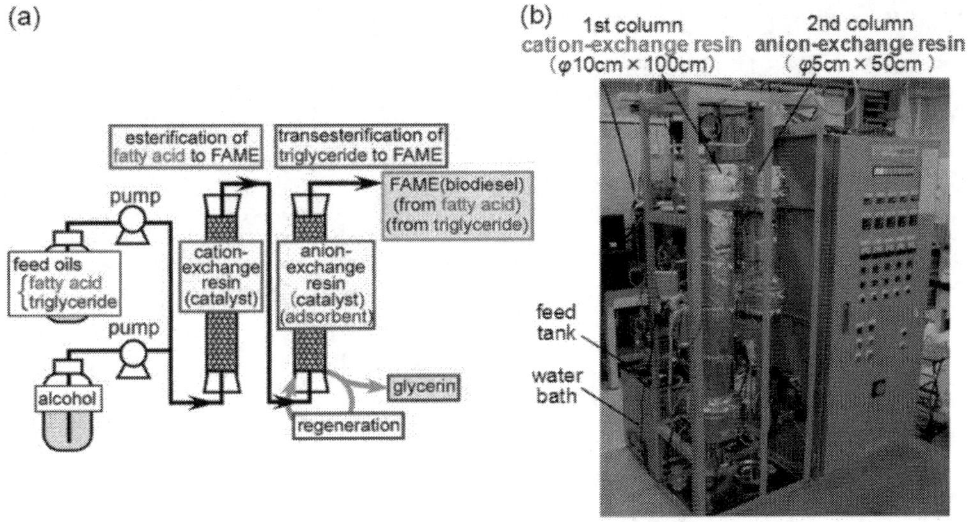

Fig.1 Schematic diagram (a) and photograph (b) of production process

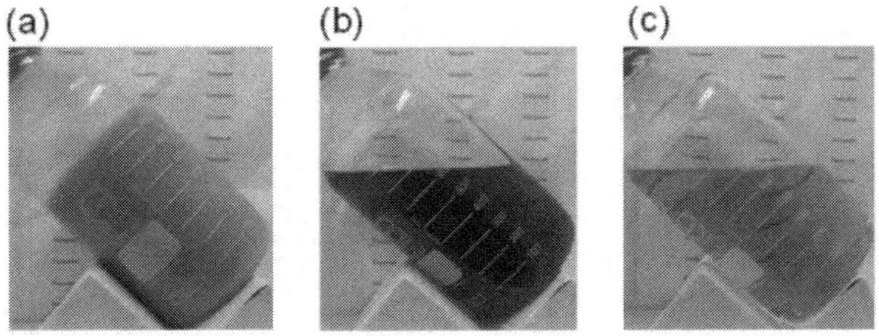

Fig.2 Photographs of feed oil (a) and effluents from the first (b) and second (b) columns at room temperature

Table 1 Analytical result of standard tests for biodiesel quality of product

Table 1 Analytical result of standard tests for biodiesel quality of product

properties		units	standards EN14214	product without purification
product	FAME	[wt%]	≥96.5	97.1
reactants	acid value (FFA)	[mg-KOH/g]	≤0.5	0.04
	monoglyceridel	[wt%]	≤0.8	0.48
	diglyceride	[wt%]	≤0.2	0.02
	triglyceride	[wt%]	≤0.2	0.03
byproducts	water	[mg/kg]	≤500	291
	free glycerol	[wt%]	≤0.02	0.00
	total glycerol	[wt%]	≤0.25	0.13
contaminants	sulfur	[mg/kg]	≤10	<5
	sulfated ash	[wt%]	≤0.02	<0.005
	phosphorous	[mg/kg]	≤4	<1
	carbon residue (at 10% distillation residue)	[wt%]	≤0.3	0.11
	total contamination	[mg/kg]	≤24	2
fuel properties	density at 15Â°C	[kg/m^3]	≥860, ≤900	880.4
	kinematic viscosity at 40Â°C	[mm^2/s]	≥3.5, ≤5.0	4.353
	flash point	[Â°C]	≥101	128.0
	cetane number	[-]	≥51	53.1
	iodin value	[-]	≤120	94.9
	methyl linolenate	[wt%]	≤12	1.2

(704c) Algal Biomass and Lipid Production From Wastewater Using a Phagotrophic Microalga

Li, C., The Univeristy of Akron

Originally presented on: 11/7/2013 13:08:00 - 13:27:00

The use of wastewater to culture microalgae can decrease the cost of microalgal-based biofuel production. Currently, studies have been focused on using photosynthetic microalgae with wastewater treatment. However, photosynthetic microalgae have low growth rates and their mass cultivation is limited to low cell concentrations in shallow ponds or complex narrow-diameter tubular photobioreactors due to the light penetration difficulty. It is also hard to maintain the selected high-lipid strains in open fields. In this study, we developed a new phagotrophic microalgae-based wastewater treatment process. By using the unique biological characteristic of the microalga with the engineering principles of continuous culture, a microbial population predominated by the microalga cells could be produced through the process. The lipid content of the produced culture was about 40% of its dry weight and could be potentially enhanced to about 80% with additional process manipulation. The cell concentration was not limited by light penetration. The microalga also showed strong ability in removing nutrients from wastewater. This microalgal lipid production process could potentially be incorporated as part of an overall wastewater treatment operation.

(704d) Novel Biomass-Supported Sorbents for Coal Combustion Desulfurization

Chu, S., University of Massachusetts-Amherst
Tompsett, G., University of Massachusetts
Yang, J., University of Massachusetts Amherst
Zhu, C., University of Massachusetts-Amherst
Mountziaris, T. J., University of Massachusetts
Dauenhauer, P., University of Massachusetts Amherst
Originally presented on: 11/7/2013 13:27:00 - 13:46:00

Coal is the largest source of energy for electricity generation in North America with over 200 years of supply available at the current consumption rate. At the same time, biomass is an attractive renewable fuel to supplement coal combustion. Although co-firing coal and biomass can reduce the total emissions per unit energy produced, more advanced technology will still be necessary to meet stricter emission regulations. A novel biomass-supported, calcium-based material named ReEngineered Feedstock™ (ReEF) has been developed by ReCommunity Inc. which combines waste paper and plastics from local sources integrating desulfurization sorbents into a fuel matrix. By directly injecting ReEF with coal into combustors, the biomass support burns to provide additional energy, thus releasing the calcium sorbent particles into the reactor to form solid, sulfated ash. The desulfurization performance of the ReEF containing 65 wt% biomass and 35 wt% high surface area calcium hydroxide are studied in this work in a quartz drop tube reactor. Characterization of ReEF including elemental analysis, SEM-EDS, XRD and TGA reveals the compositions of ReEF and the distribution of the calcium sorbent on the biomass support. The effects of reactor temperature, oxygen concentration and particle size on the desulfurization are studied to determine optimal ReEF design. The application of ReEF will impact communities across the country to achieve zero waste, thus improving air quality and helping local economies progress towards sustainable energy production.

(704e) Fuel Quality Roadmap for the Kingdom of Saudi Arabia Domestic Transportation Fuel

AlAlloush, S. S.,
Tukur, N. M., Saudi Aramco
Originally presented on: 11/7/2013 13:46:00 - 14:05:00

Abstract

Mobile sources have since been identified as the primary contributors of NOx, VOC, PM and CO emissions in KSA. Improvements in fuel quality for the existing vehicle fleet will provide immediate emission reductions of pollutants that are known to have detrimental effect on human health and the environment. Saudi Aramco took the proactive initiative to further improve air quality by developing a clean fuel quality roadmap for the Kingdom's domestic transportation fuel to achieve the following goals:

- improve the quality of the environment to provide clean air to protect human health for current and future generations;
- maintain a competitive position in global markets for continuing fuel export; and
- ensure fuel compatibility with future automotive engine technology.

The critical elements of the clean transportation fuels (CTF) program are a reduction in sulfur content in the gasoline and diesel products to 10 ppm, reduction in the gasoline benzene to 1 vol.% and a maximum gasoline aromatic content of 35 vol.%. As a consequence, CTF projects were initiated that will install new processing units at the company's domestic refineries to meet the new gasoline and diesel products requirements.

This presentation will focus on the primary specification changes for KSA fuels in the last 5 years and the estimated reduction in emissions and other benefits to be realized upon implementation of the roadmap.

(704f) Torrefied Biomass: Superior Fuel for Co-Firing in Pulverized Coal Fired Power Generation Facilities

Buchireddy, P., University of Louisiana at Lafayette
Guillory, J., University of Louisiana at Lafayette
Zappi, M., University of Louisiana at Lafayette
Originally presented on: 11/7/2013 14:05:00 - 14:24:00

Approximately 54% of the United States power supply is being generated by coal fired electric generators. Due to the abundance of coal reserves in the U. S., the use of coal will be predominant. However, the environmental concerns including greenhouse gas emissions have prompted several agencies to regulate the gaseous emissions. Among the techniques that were proposed to reduce these gaseous emissions and the costs associated with these gaseous emissions, biomass co-firing with coal has drawn huge attention from electric utility producers. Biomass co-firing is considered as one of the most practical strategic options that has a short to medium term implementation potential for renewable power production. It is one of the most attractive options that will help utility companies in complying with the greenhouse gas reductions and more importantly the renewable energy portfolio standards.

Biomass is considered carbon neutral, and has lower sulfur, nitrogen, ash, and heavy metals than coal. However, several disadvantages exist compared with coal, such as lower heating value, lower density and higher moisture content. These properties pose several challenges related to fuel preparation, handling, storage, transportation, milling, and feeding in coal fired power plants. Torrefaction is a thermal pretreatment technique and has a potential to address the above mentioned issues. This process involves heating biomass to temperatures of 230-300°C in the absence of an oxidizing agent and under atmospheric pressure. During this process, 70% of mass is retained in the solid product which contains 90% of the initial energy. The remaining 30% of biomass is converted to volatiles and gases that contain 10% of the biomass energy. Thus, energy densification can be achieved via torrefaction by a factor of 1.3 on mass basis. The product, torrefied biomass, has improved fuel properties in terms of increased mass energy density, possesses hydrophobic nature, and is more uniform, in comparison with unprocessed biomass. Also, the grindability characteristics of the torrefied biomass are greatly improved and are comparable with coal and petroleum coke. This allows for effective utilization of torrefied biomass in coal fired power plants using the existing burners with minor modifications.

Research is being performed on both laboratory scale and pilot scale, using various lignocellulosic materials including pine, willow, arundo, bamboo, cotton waste, and pecan shells at UL, Lafayette. The effect of various parameters including temperature, residence times, particle size, moisture content on torrefaction process is being evaluated, the results of which will be presented. Also, work is being performed on a pilot scale reactor that was fabricated by an industrial counterpart, which has a capacity of up to 30 lbs/hour. Testing is being carried out on the pilot scale reactor to improve the overall process efficiency. Mass and energy balances on the pilot scale reactor will be performed the results of which will be presented. Also, the effect of parameters tested on the energy density, bulk density, hydrophobicity, and grindability using modified hard grove grindability index (ASTM D409) will be presented and compared with representative samples of coal and petroleum coke.

(704g) The Osmotic Heat Engine: The Promise of High Power Density Closed Loop Pressure Retarded Osmosis

Anastasio, D., University of Connecticut
Arena, J. T., University of Connecticut
McCutcheon, J. R., University of Connecticut
Originally presented on: 11/7/2013 14:24:00 - 14:43:00

Pressure-retarded osmosis (PRO) is an engineered osmosis process that operates at pressures between forward and reverse osmosis. A highly concentrated draw solution is pressurized, and water from a clean feed source will permeate the membrane and increase the volume of the draw. As the volume changes at constant pressure, work is being done on the draw solution, and this energy can be harnessed by a hydroturbine. When operated in closed-loop, PRO is can function as an osmotic heat engine (OHE). If specially-designed osmotic agents are used, both working fluids can be recovered by using low-grade thermal energy, effectively converting this waste heat to useable electricity.

In this study, a cellulosic membrane was evaluated in closed-loop at a variety of temperatures, draw solution concentrations, and hydraulic pressures to determine how these parameters impact the flux of water and salt through the membrane. The selected conditions simulate OHE operation. Furthermore, the impact of these test conditions on membrane power density, and thus the maximum energy captured per unit of membrane area, was also examined. While the membranes were only tested up to 21 bar, it is clear that increasing salt concentration increases peak power density. This observation is appropriate with respect to theory. The experimental results also suggest that for constant draw concentration, increasing temperature or decreasing hydraulic pressure will increase water flux and power density. Under these same conditions, salt fluxes were reduced by up to a factor of three, suggesting that significant external concentration polarization is impacting the performance of the membrane. Maximum power density seen throughout the trial was 18 W/m^2, which was seen at 40 C with a 1.5 M sodium chloride draw solution.

(704h) A Participatory Pilot Study On Awareness of Biomass Pyrolysis: Feedback From Two Upper-Cumberland Counties

Murillo, J. D., Tennessee Technological University
Biernacki, J. J., Tennessee Technological University
Norris, L.,
Originally presented on: 11/7/2013 14:43:00 - 15:02:00

Government mandates and initiatives have prompted growth in all areas of the renewable energy sector in hopes of attaining national energy security, economic growth, and environmental sustainability. Of the various renewable energy resources, biomass-based energy is still the only meaningful route for producing liquid fuels, still in high-demand for transportation purposes. In addition to producing fuels, biomass-based processes are being honed to produce renewable chemicals for the lucrative and ever-expanding bioproducts industry, which is displacing many petroleum-based products. As technological advances in processing cellulosic biomass become a reality, a higher demand for various biomass feedstocks raises questions about supply logistics and sustainability. The new front-runners for biomass feedstock are waste products from agriculture, forestry, and municipalities; and energy crops, such as switchgrass, poplar, and camelina, which have to be grown, harvested and pre-processed. With all of the components that are factored into making biofuels from biomass, the most important and contentious questions are rarely raised. For instance, who will grow and provide these much needed feedstocks? And how will they be compensated? Somewhere along the logistical pathway to biofuels, the Nation's farmer, its rural community, and agriculture were left out of the biofuels equation. *To explore the current knowledge and awareness of biofuels, 28 farmers and 14 facilitators from local farming communities in Tennessee provided information through informal question and answer sessions, and in questionnaire format.*

A two-tiered participatory approach was taken to gather information from Scott and Putnam county farmers. The main goal of participatory research is to involve those who will be affected in all stages of the research; therefore, a focus group of 14 local conservationists was formed to uncover relevant questions and concerns that local growers may have with respect to changes brought about by a shift to a biofuels-centered agro-economy. The information gathered from the focus group was incorporated into a questionnaire administered at District Conservation meetings in Scott and Putnam counties following a brief presentation on biofuels given by the researcher. Many farmers were very aware of first generation biomass feedstock and fuels. However, the production of chemicals and non-ethanolic fuels were less familiar to the group. Many participants in both, the focus and community groups, were not familiar with pyrolysis as a biomass conversion process to fuel. The possibility of economic growth to their farming enterprise was the main driver for willingness to supply biomass residues and feedstock for biofuels production. Overall, this research showed that the farming community is open to the opportunity to participate if approached with transparency and equality.

(727a) Effect of Pore and Particle Size of Oxygen Carriers On Chemical-Looping Combustion and Reforming

Han, L., University of Connecticut

Zhou, Z., University of Connecticut

Bollas, G. M., University of Connecticut

Originally presented on: 11/7/2013 15:15:00 - 15:36:00

Effect of pore and particle size of oxygen carriers on chemical-looping combustion and reforming

Lu Han, Zhiquan Zhou, George M. Bollas

Abstract

Chemical-looping combustion (CLC) addresses the current CO_2 problem caused by fossil fuel utilization for power production. The main advantage of CLC over conventional power production technologies is that direct contact between air and fuel is circumvented, so that a pure stream of CO_2 is obtained without nitrogen dilution. An intermediate oxygen carrier (metal/metal oxide) is alternatively oxidized and reduced by the fuel and oxygen. Two successive reactions take place between two reactors shown in **Figure 1**. In this manner, CO_2 can be captured from the flue gases of the Fuel reactor without additional separation and high energy penalty. Chemical-looping reforming (CLR) is similar to CLC however a mixture of steam and fuel is sent to the Fuel Reactor to induce reforming reactions and produce syngas.

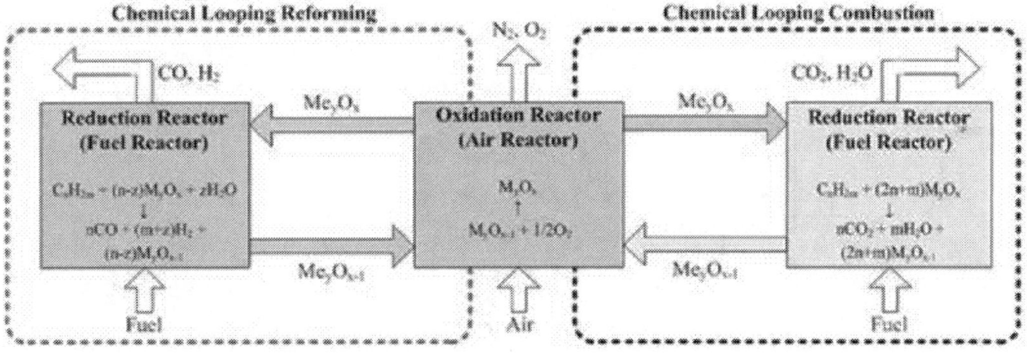

Figure 1: Basic concept of chemical-looping combustion and reforming processes

Research in the chemical-looping (CL) process has been focused on the development and testing of oxygen carrier particles. Favorable characteristics of high performance oxygen carriers include: high reactivity toward oxidation and reduction cycles, high melting temperature, low tendency of attrition and fragmentation, low toxicity, low cost and high combustion efficiency. Most reported work on CL has been accomplished using oxygen carrier particles where the active component (Cu, Fe, Ni, Mn) is formulated on an inert material [1]. This work focuses on reported studies utilizing Ni-based oxygen carriers for its high reactivity in the combustion of methane and catalytic properties.

A particle model is developed to study the effect of internal transport on the reaction rates. Intraparticle diffusion limitations need to be accounted for particles that are millimeters in size [2]. Due to the importance of Knudsen diffusion, internal transport resistance of different gases gives rise of concentration gradients inside the pores. Reaction rates are observed highest at the particle surface and decrease along the pore length. This knowledge can aid in oxygen carrier design to avoid over-estimation of the system performance when no mass transfer effects are considered. Incorporated into a reactor model, optimum particle dimensions are proposed to conduct CLR in a fixed-bed reactor that makes use of the intraparticle effects to maximize H_2 selectivity.

It is also worth to explore the change in particle morphology over many reduction and oxygen cycles in CLC and CLR. Investigators have reported increasing reactivity with usage associated with micro-cracking of the particle surface and migration of the Ni particles to the surface. With the particle model, the real changes to particle porosity, surface area, and pore size are accounted for, which influence the role of transport resistance on the performance. Observations made from experiments on the particle morphology are utilized to predict reactivity changes over cyclic tests.

Lastly the interaction between NiO and Al_2O_3 support is addressed because when present, the spinel can significantly lower the reactivity of the oxygen carrier. For instance, the reactivity of two supported NiO-oxygen carriers differing only by the calcination temperature of the support was measured in a thermogravimetric analyzer at 800°C in a H_2-environment (**Figure 2**). The Al_2O_3 support existed in the γ-phase in the low temperature pre-treatment and exhibited a phase transformation into α upon high temperature calcination. More instance of spinel formation occurred in the γ-Al_2O_3 support which resulted in comparably lower reactivity of the oxygen carrier. A kinetic scheme that combines the simultaneous reduction of NiO and $NiAl_2O_4$ is used in the particle model. Reduction data of Ni-based oxygen carriers with varying spinel phases are studied for the purpose of assessing the impact of spinel on kinetic rates. Kinetic parameters derived from the particle model are presented and model-experiment deviations are discussed.

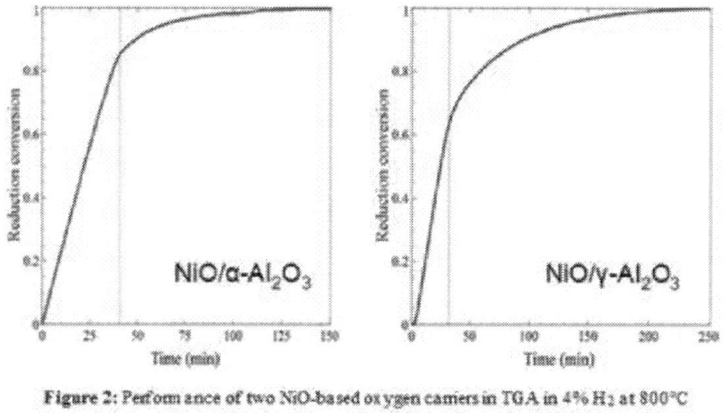

Figure 2: Performance of two NiO-based oxygen carriers in TGA in 4% H_2 at 800°C

Acknowledgement: This material is based upon work supported by the National Science Foundation under Grant No. 1054718.

References:

[1] M. Johansson, T. Mattisson, M. Rydén, and A. Lyngfelt, ?Carbon Capture via Chemical-Looping Combustion and Reforming Chemical-Looping Combustion,? in *International Seminar on Carbon Sequestration and Climate change*, 2006, no. October, pp. 24?27.

[2] S. Noorman, F. Gallucci, M. van Sint Annaland, and J. a. M. Kuipers, ?A theoretical investigation of CLC in packed beds. Part 1: Particle model,? *Chemical Engineering Journal*, vol. 167, no. 1, pp. 297?307, Feb. 2011.

(727b) Thermal Evaluation of Chemical Looping Process for Hydrogen Production Using Bivalent Metal Added Iron Oxide As An Oxygen Carrier

Cho, W., Korea Institute of Energy Research
Kang, K., Korea Institute of Energy Research
Park, C., Korea Institute of Energy Research
Jeong, S. U., Korea Institute of Energy Research
Lee, D., KAIST
Seo, M. W., Korea Institute of Energy Research
Kim, C. H., KIER
Bae, K. K., Korea Institute of Energy Research
Kim, S. D., KAIST
Originally presented on: 11/7/2013 15:36:00 - 15:58:00

The chemical looping process for hydrogen production co-generates hydrogen and electricity with simultaneous capture of carbon dioxide. This process is composed of interconnected three reactors: a fuel reactor where an oxygen carrier is reduced by fuel, a steam reactor where hydrogen is produced by water splitting, and air reactor where air is combusted. The iron oxide is used as a suitable oxygen carrier from thermodynamic and chemical equilibrium point of view. The many studies were conducted to improve the reactivity of the iron oxide such as preparation methods, sintering temperatures, supporters, etc. In addition, the mixing the bivalent metal (M = Cu, Ni, Mo) to the iron oxide has been studied to enhance the reactivity of the redox reaction. In this study, the bivalent-added iron oxides showed the kinetic improvement, higher steam conversion to hydrogen, and the lower carbon deposition. The chemical looping process was simulated and the thermal evaluation was conducted based on the experimental data, varying the oxygen carriers. The several cases were compared varying the temperatures and the extent of solid conversion in the fuel and steam reactors.

(727c) Modeling of Cu Oxidation in Air Reactors of Chemical Looping Combustion Systems

Goldstein, E. A., Stanford University
Mitchell, R. E., Stanford University
Gur, T. M., Stanford University
Originally presented on: 11/7/2013 15:58:00 - 16:19:00

An oxidation model is developed that describes *Cu* (copper) oxidation kinetics to *CuO* (cupric oxide) for applications in the air reactor of chemical looping combustion (*CLC*) systems. The model, which is based on the Wagner theory of oxidation, is developed by considering the fluxes of interstitial and vacancy defects in the cation sublattice of Cu_2O (cuprous oxide) and *CuO*. Defects in the anion sublattice are a priori included in the model, however are subsequently found to be too small in concentration and diffuse too slowly to contribute significantly to the oxidation of Cu. The oxidation model presented here is capable of predicting Cu oxidation rates and relative thicknesses of the resulting Cu_2O and *CuO* oxide layers as a function of the partial pressure of oxygen in the ambient and temperature.

Experiments were performed in a thermogravimetric analyzer (*TGA*) at conditions relevant to air reactors of *CLC* systems to obtain data that was used to characterize parameters in the model. Experiments were performed on copper spheres having diameters in the range 15 to 25 μm, at temperatures between 500 and 900 °C and at 1 atm. The gaseous environments contained 0.3%, 6% and 21% O_2, by volume, with the balance N_2. Based on this model, the temperature and oxygen partial pressure dependencies of the oxidation rate as well as overall rate constants were determined and compared to published results for Cu oxidation. A sensitivity analysis was performed in order to assess the influence of defect diffusivities and formation energies as well as particle diameter and surface area on oxidation rates. Based on our understanding of defects and their role in Cu oxidation, insights regarding the design of copper particles for use in CLC have been made. Assuming that oxidation is the rate-limiting process in *CLC*, the recirculation rate and copper loadings have also been estimated when CuO is used to oxidize a typical coal.

(727d) An Overview of Chemical-Looping Reduction in Fixed-Bed and Fluidized Bed Reactors Focused On Oxygen Carrier Utilization and Reactor Efficiency

Zhou, Z., University of Connecticut
Han, L., University of Connecticut
Du, S., University of Connecticut
Fischer, A., University of Connecticut
Bollas, G. M., University of Connecticut
Originally presented on: 11/7/2013 16:19:00 - 16:41:00

Chemical-looping combustion is a method for the oxidation of hydrocarbons with in-situ CO_2 capture. The basic concept of the process involves two interconnected reactors, a Reducer and an Oxidizer, with an oxygen carrier (OC) Â¨C a metal/metal oxide circulating between the two. The main principle in chemical-looping is the oxidation of hydrocarbons with a metal oxide in a N_2-free environment. In this study, a transient non-ideal reactor model was developed for two types of Reducers: fixed and fluidized bed [1,2]. The models capture the most important dynamic and spatial phenomena in CLC reduction. The mass balance of 9 species and the energy balance are simulated in both models by the same kinetics mechanisms and reaction constants.

In this work, Ni-based CLC tests in fixed-bed and spouted-bed CLC reactors operated in the University of Connecticut are discussed and the simulation of different fixed and fluidized bed unit data reported in the literature and data of this study are presented. Comparison between experimental and predicted results reveals that the process models and kinetics network proposed are successful in predicting chemical-looping combustion in fixed- and fluidized bed reactors. Through model-assisted comparison between the two types of reactors, some substantial issues in the CLC system performance are discussed. For instance, the comparison of carbon formation (Figure 1) between the two types of reactors (reactor configurations of Iliuta et al. [3]) shows that carbon deposition is not significant in the fluidized bed reactor; whereas, it is of concern at higher conversions in a fixed-bed reactor. Based on the developed reactor models, fixed-bed scale-up calculations are carried out. Spatial and transient pressure and temperature variations are shown and analyzed. Other issues, like challenges in operating CLC at minimum oxygen carrier loading (therefore, smaller Reducer size), OC loading required for high CH_4 conversion and OC conversion, time of starting of significant carbon formation at a given OC loading, and CO_2 selectivity of fluidized-bed and scale-up fixed-bed units will be discussed.

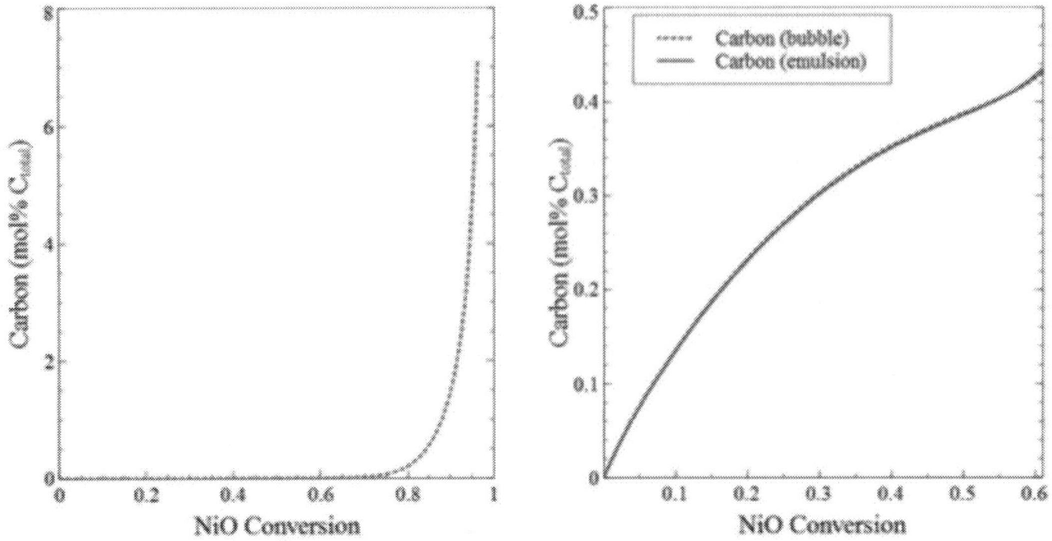

Figure 1: *Carbon formation of NiO based OC chemical-looping reduction process in (a) fixed-bed and (b) fluidized bed reactors[3].*

Acknowledgement: This material is based upon work supported by the National Science Foundation under Grant No. 1054718.

[1] D. Kunii, O. Levenspiel, Bubbling bed model: model for the flow of gas through a fluidized bed, I & EC Fundamentals. 7 (1968) 446Â¨C452.

[2] D. Kunii, O. Levenspiel, Bubbling bed model for kinetic processes in fluidized beds: Gas-Solid Mass and Heat Transfer and Catalytic Reactions, I & EC Process Design and Development. 7 (1968) 481Â¨C492.

[3] I. Iliuta, R. Tahoces, G.S. Patience, S. Rifflart, F. Luck, Chemical-looping combustion process□: kinetics and mathematical modeling, AIChE J. 56 (2010) 1063Â¨C1079.

(727e) Modeling A Scaled 300 kWth Circulating Fluidized Bed Reactor With Barracuda

Hamilton, M. A., University of Utah
Lighty, J. S., University of Utah
Originally presented on: 11/7/2013 16:41:00 - 17:02:00

MODELING A SCALED 300 kW_{th} CIRCULATING FLUIDIZED BED REACTOR WITH BARRACUDA

CHEMICAL LOOPING COMBUSTION WITH OXYGEN UNCOUPLING

ABSTRACT

As global warming is continuing to gain recognition as a problem, new technologies for capturing greenhouse gasses such as CO_2 are being introduced.Â The aim of these techniques is to capture CO_2 before it enters the atmosphere thus ameliorating its deleterious effects on global temperature and climate.Â Chemical looping combustion with oxygen uncoupling (CLOU), a subset of chemical looping combustion (CLC), allows for isolation of carbon dioxide (CO_2) as an almost pure gaseous stream, suitable for compression and storage.Â This is done by using two interconnected fluidized beds (CFB) with a metal oxygen carry capable of transferring oxygen from an air reactor to a fuel reactor.Â The flue gas of the fuel reactor contains a highly concentrated stream of CO_2 and H_2O.Â After condensing H_2O the CO_2 can be easily captured.Â This eliminates any direct efficiency loss from the combustion process.Â The goal of this study was to study fluid dynamics of a 300 kW_{th} CFB using computation particle fluid dynamic software Barracuda VR, then scaling this reactor to a cold flow model and demonstrating with Barracuda that similar fluid dynamics could be obtained.

The scaling was performed similar to Chalmers University of Technologies which was adapted from Levenspiel's text Fluidization Engineering.Â This method uses dimensionless π-groups to determine the similarities of the beds.Â Then from these an overall scaling value is determined and used to adjust the geometric dimensions of the CFB.Â The time scale dimension is determined from raising the overall scaling value to the Â½ power.Â The particle density and fluid density ratio must be maintained for the cold and hot set up.Â The hot system has a gas density of 0.236 kg/m^3 and a particle density of 2140 kg/m^3.Â From these scaling parameters it was determined that helium (0.167 kg/m^3) was the required fluidization gas and that a particle with density of 1450 kg/m^3 would be required.Â The particle of the hot system will fall within the Geldart A group and the particle in the cold system will fall in the Geldart A group.Â Due to the price and difficulty of obtaining helium it was also determined that the scaled reactor would be run using air at 20 °C as the fluidizing gas.Â The models to be meshed in Barracuda were created using Solidworks 2012.

Using Barracuda the following transient data was obtained and analyzed: oxygen carrier volume fraction (at points in the air reactor, at points in the air reactor riser, at point in the fuel reactor, and loop seals); the gas pressure at different points of the reactors.Â The oxygen carrier mass was monitored at the outlet of the air reactor riser and the cyclone outlet.Â The average residence time of the oxygen carrier was determined in the air reactor and fuel reactor.Â The reactors were visual inspected using movies created in Barracuda.Â These were compared for the hot reactor, cold reactor with helium as the fluidizing gas, and the cold reactor with air as the fluidizing gas.

From the Barracuda simulations it was determined that the hot and both cold reactor systems function with similar fluidization patterns. Â It is recommended that the cold flow model should be constructed to study the effects of static electricity and other unknown inputs.Â Then kinetics can be added to Barracuda to fully study the hot system.

(727f) The Ohio State University Chemical Looping Processes: 25 Kwth Sub-Pilot Performance Studies for Natural Gas Conversion and 250 Kwth Pilot Unit Developments

Kathe, M., The Ohio State University
Tong, A., The Ohio State University
Zeng, L., The Ohio State University
Wang, D., The Ohio State University
Bayham, S.,)William G. Lowrie Department of Chemical and Biomolecular Engineering, The Ohio State University, Columbus, OH
McGiveron, O., Ohio State University
Fan, L. S., The Ohio State University
Originally presented on: 11/7/2013 17:02:00 - 17:24:00

The increasing focus on developing alternative fossil fuel conversion technologies is fueled by growing concerns on controlling carbon emissions from traditional technologies in a cost-effective manner. Amongst the newer technologies, chemical looping has been established as a promising candidate exemplified by larger scale demonstrations world-wide. Chemical looping processes utilize metal based oxygen carriers to inherently produce a concentrated CO_2 stream coupled with efficient heat-integration which serves to reduce costs and justify potential commercial scale application. The Ohio State University (OSU) has developed multiple product flexible iron-based chemical looping processes for utilizing fossil fuels like coal and natural gas in a unique moving bed reactor design. OSU's chemical looping technologies are driven by the thermodynamic motivation of using higher oxygen carrying capacity of iron-oxide. This enables the process to enlist the flexibility of converting a variety of fuels like coal, natural gas and biomass to generate electricity, hydrogen, syngas or a combination of products with zero or negative CO_2 emissions. The iron-oxide based oxygen carriers have been developed to sustain hundreds of redox cycles without loss in reactivity and structural integrity. Amongst the OSU processes, the Syngas Chemical Looping (SCL) process has been developed for converting gaseous fuels to high purity CO_2 and demonstrated for more than 350 hours on a sub-pilot (25 kW_{th}) scale. With support of the Advanced Research Projects Agency – Energy (ARPA-E) of the United States Department of Energy (USDOE) a 250 kW_{th} high-pressure SCL pilot unit is being constructed at the National Carbon Capture Center. The pilot unit is designed to handle a 250 kW_{th} syngas produced from a commercial Kellogg, Brown & Root transport gasifier to generate high purity hydrogen coupled with a concentrated CO_2stream. This will validate the chemistry and operational capability on a commercially applicable scale, thereby completing an important step in the pathway to commercialization.

This presentation will summarize multiple parametric studies which elucidate the interplay of factors like fuel residence time, oxygen-carrier to fuel ratio and temperature for optimizing the natural gas conversion and thereby the process efficiency in the SCL sub-pilot system. The presentation will also include theoretical reactor modeling studies coupled with an overview and comprehensive commercial techno-economic analyses of other OSU chemical looping technologies. Finally, the design development and the progress of the pilot plant construction will be presented.

(727g) Combustion Of Polyurethane By Chemical Looping Process In A Moving Bed Reactor Using Iron-Based Oxygen Carrier

Ku, Y., National Taiwan University of Science and Technology
Chiu, P. C., National Taiwan University of Science and Technology
Wu, H. C., National Taiwan University of Science and Technology
Tseng, Y. H., National Taiwan University of Science and Technology
Kuo, Y. L., National Taiwan University of Science and Technology
Originally presented on: 11/7/2013 17:24:00 - 17:45:00

Polyurethane (PU) as a thermal setting plastic with high recovery cost was employed as fuel for chemical looping combustion. The heat value of PU was estimated to be 0.28 kJ/g based on the ultimate analysis of PU by an elemental analyzer. Alumina supported iron-based composite oxygen carriers were prepared for combustion with PU by a moving bed reactor. PU was gasified and reformed by gasification reactants, CO_2/O_2, to be CO and H_2 as fuels prior to combustion with iron-based composite oxygen carriers at 900°C. By moving bed reactor operation, CO_2 yield and PU conversion were both in the moving bed reactor. Long-term operation for PU combustion was lasted for 24 hours, and 100% of CO_2 yield was stably maintained. PU combustion with iron-based oxygen carrier in moving bed operation was technically feasible; therefore, chemical looping combustion using solid waste as fuel can be considered as an alternative technology for solid waste treatment.

(735a) Detailed Kinetic Modeling of Selective Catalytic Reduction of NO With C3H6 On Cu-Chabazite Monolithic Catalyst

Raj, R., University of Houston

Harold, M. P., University of Houston

Balakotaiah, V., University of Houston

Originally presented on: 11/7/2013 15:15:00 - 15:30:00

Detailed Kinetic Modeling of Selective Catalytic Reduction of NO with C_3H_6 on Cu-chabazite Monolithic Catalyst

Richa Raj, Michael P. Harold and Vemuri Balakotaiah

Department of Chemical & Biomolecular Engineering, University of Houston, Houston, TX, United States.

Abstract

NO_x storage and reduction (NSR) and selective catalytic reduction (SCR) are the two commercialized aftertreatment technologies for the reduction of NO_x emissions from lean?burn engines The combined LNT-SCR systems are emerging as promising technology for NO_x reduction as NO_x can be stored during lean phase and NH_3 and hydrocarbons (C_3H_6, C_2H_4 etc.) released during rich phase can be stored on SCR catalyst and these stored reductants can reduce the NO_x coming out from LNT during lean phase. The storage of hydrocarbons and their role as co-reductants of NO_x lead to an intricate system. Hence it is important to study the HC-SCR chemistry. A systematic steady state and transient experimental studies were performed to elucidate the mechanistic pathways involved in NO_x reduction with C_3H_6 [1].

The steady state experiments were performed for different feed compositions for $CO+O_2$, $C_3H_6+O_2$ and $C_3H_6+NO+O_2$ reaction systems. Fig.1a shows that C_3H_6 light-off curves for C_3H_6 oxidation for varying O_2 concentration (0.5-5%) in the feed. Similarly, fig.1b shows the C_3H_6 conversion with temperature for different C_3H_6 concentration (200-500 ppm). Fig.2a and 2b show the C_3H_6 and NO conversion in temperature range of 200-550^oC for feed with 500 ppm C_3H_6, 500 ppm NO and 1% and 5% O_2 respectively. A steady state detailed kinetic model for NO reduction with C_3H_6 on Cu-chabazite (Cu-SSZ13) monolith catalyst is developed based on bench scale flow reactor experiments The following mechanistic sequence is proposed based on experimental studies:

R1: $C_3H_6 + S \leftrightarrow C_3H_6\text{-}S$

R2: $O_2 + S \leftrightarrow O_2\text{-}S$

R3: $O_2\text{-}S + S \leftrightarrow 2O\text{-}S$

R4: $C_3H_6\text{-}S + O\text{-}S \rightarrow C_3H_4\text{-}S + H_2O$

R5: $C_3H_4\text{-}S + O\text{-}S \rightarrow C_3H_4O\text{-}S$

R6: $C_3H_4O\text{-}S + 4O\text{-}S \rightarrow 3CO\text{-}S + 2H_2O$

R7: CO ?S$\leftrightarrow$ CO + S

R8: $CO\text{-}S + O\text{-}S \leftrightarrow CO_2 + 2S$

R9: $C_3H_4O\text{-}S + NO \rightarrow C_2H_2\text{-}NCO\text{-}S + H_2O$

R10: $C_2H_2\text{-}NCO\text{-}S + 3O\text{-}S \rightarrow 3CO\text{-}S + N\text{-}S + H_2O$

R11: $N\text{-}S + N\text{-}S \leftrightarrow N_2 + 2S$

. Kinetic Parameters estimated capture the experimental trends and meet the thermodynamic constraints. The spatially 2D (1D + 1D) mathematical model of the catalytic monolith reactor coupled with the detailed kinetic model is used to study the complex reaction system. The kinetic model developed capture the main reaction pathways and predicted conversion (CO, C_3H_6 and NO) and product distribution (CO, CO_2 and N_2) are in good qualitative and quantitative agreement with the experimental data. The model predicts for the light off curves for three reaction systems $CO+O_2$, $C_3H_6+O_2$ and $C_3H_6+NO+O_2$ for different feed compositions. The kinetic model is used to determine the trends in species surface coverages to explain the experimentally observed trends in selectivities.

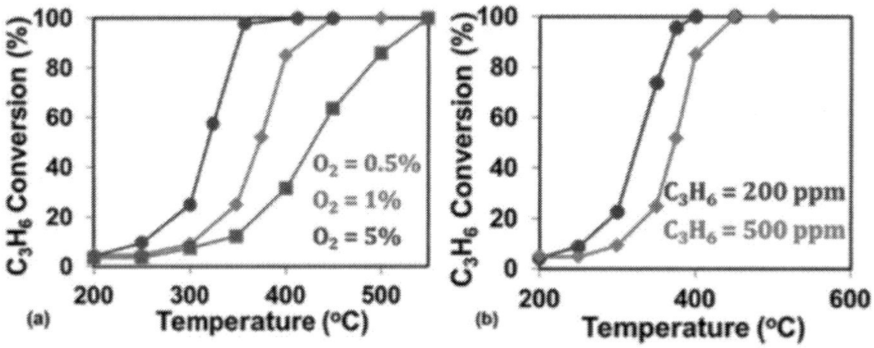

Figure 1 (a) Light- off curves of C_3H_6 for different oxygen concentrations in the feed (C_3H_6 = 500 ppm, O_2 = 0.5 or 1 or 5%, Balance gas: Ar), (b) Light- off curves of C_3H_6 for different C_3H_6 concentrations in the feed (C_3H_6 = 200 or 500 ppm, O_2 = 1 %, Balance gas: Ar),

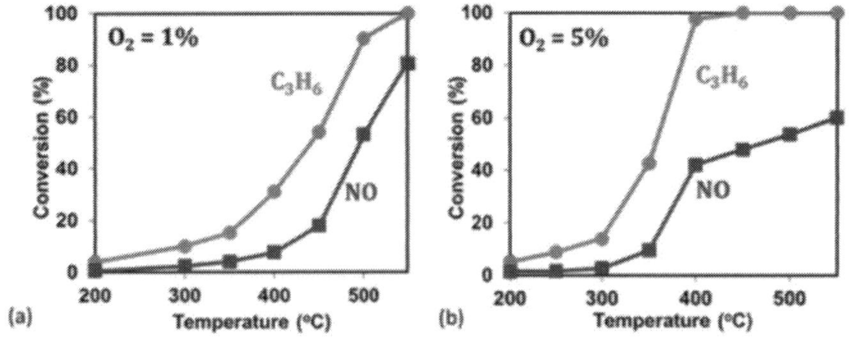

Figure 2(a) C_3H_6 and NO conversion with temperature (C_3H_6 = 500 ppm, NO = 500 ppm, O_2 = 1%, Balance gas: Ar) (b) C_3H_6 and NO conversion with temperature (C_3H_6 = 500 ppm, NO = 500 ppm, O_2 = 5%, Balance gas: Ar)

Keywords: SCR, Kinetic Modeling, Cu-chabazite, Propylene

References:

[1] R. Raj, M.P. Harold, V. Balakotaiah, Industrial & Engineering Chemistry Research, revisions pending (2013)

(735b) A Detailed Microkinetic Model for SO2 Oxidation From Diesel Engine Exhaust

Sharma, H., University of Connecticut
Mhadeshwar, A. B., University of Connecticut
Ramprasad, R., University of Connecticut
Originally presented on: 11/7/2013 15:30:00 - 15:45:00

A detailed microkinetic model for SO_2 oxidation from diesel engine exhaust

Hom N. Sharma[1], Ashish B. Mhadeshwar[1,x] and Rampi Ramprasad[1, 2,*]

[1]Department of Chemical and Biomolecular Engineering, University of Connecticut, Storrs, CT

[2]Material Science and Engineering, University of Connecticut, Storrs, CT

*rampi@ims.uconn.edu

[x] Currently at ExxonMobil Research & Engineering, Annandale, NJ

Currently used ultra-low sulfur diesel (USLD) contains up to 15 ppm sulfur, consistent with the United States Environmental Protection Agency (US-EPA) regulations [1]. In typical diesel engine exhaust, this results into ~1 ppm of sulfur in the oxide form [2], which can deactivate the diesel oxidation catalyst (DOC) after long term exposure as well as result in increased particulate matter (PM) emissions [3-4]. The noble metals (Pt/Pd) based DOC is one of the most expensive components of the emissions aftertreatment system. Despite the extensive research conducted on DOC performance, the interaction of DOC with sulfur oxides in the exhaust is not clearly understood. In this work, first we will discuss the development of a detailed 24 elementary-step microkinetic model for Pt-DOC that accounts for SO_2 oxidation. The detailed mechanism development is carried out using several parameter estimation techniques, such as semi-empirical Unity Bond Index-Quadratic Exponential Potential (UBI-QEP), Transition State Theory (TST), quantum mechanical Density Functional Theory (DFT), and temperature programmed experiments in literature. Model predictions for catalytic oxidation of SO_2 in fixed bed and monolith reactors under practically relevant operating conditions will be presented (see Figure). The microkinetic model is validated against experimental data and reaction pathways are analyzed. Next, we explore the SO_x ($x = 0 - 4$) interactions and reaction pathways of SO_2 oxidation on Pt and Pd surfaces using first principles DFT. Our results show good agreement with the experimental observations. The elementary-step surface reaction mechanism and DOC-SO_x interactions could be utilized for improved design and control of the aftertreatment system as well as for appropriate DOC sizing.

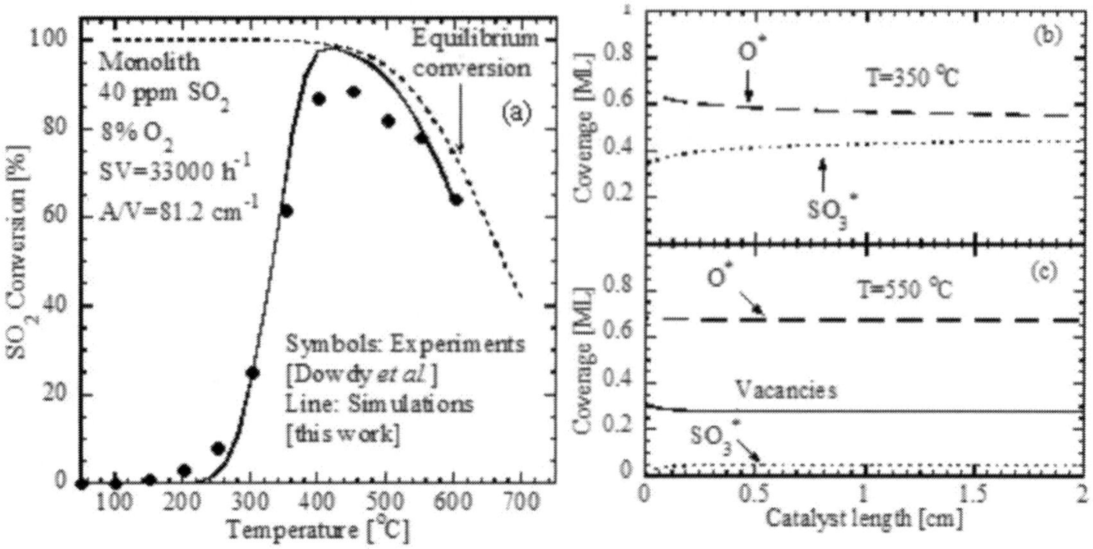

Figure: *Panel a: performance of the microkinetic model [5] for* SO_2 *oxidation on Pt. Symbols represent experimental data; solid lines represent our simulations; and dashed lines represent the calculated equilibrium conversion. Panel b: simulated axial coverage profiles at 350 Â°C.Â Panel c:Â simulated axial coverage profiles at 550 Â°C.*

References

[1] Â Environmental Protection Agency, Federal Register. Vol. 71, No. 83 (2006).

[2] Â O. KrÃ¶cher, M. Widmer, M. Elsener, D. Rothe, Ind. Eng. Chem. Res. 48 (2009) 9847-9857.

[3] Â M. Lemmetty, T. RÃ¶nkkÃ¶, A. Virtanen, J. Keskinen, L. Pirjola, Aerosol Sci. Technol. 42 (2008) 916-929.

[4] Â T. Kolli, M. Huuhtanen, A. Hallikainen, K. Kallinen, R. Keiski, Catal. Lett. 127 (2009) 49-54.

[5] Â H.N. Sharma, S.L. Suib, A.B. Mhadeshwar, in: Novel Materials for Catalysis and Fuels Processing, ACS, 2013, In press.

(735c) Reaction Characteristics of Sncr De-NOx in Moderate Temperature Range Based Urea Spiked Hydrazine Reductant

Chen, H., Thermal & Environmental Engineering Institute, Tongji Unversity
Hong, L., Thermal & Environmental Engineering Institute, Tongji University
Chen, D. Z., Thermal & Environmental Engineering Institute, Tongji University
Originally presented on: 11/7/2013 15:45:00 - 16:00:00

Large amounts of NOx are discharged from waste incinerator and sintering furnace in iron and steel industry. Some suitable reagent such as $N_2H_4.H_2O$ can reduce NOx efficiently through selective non-catalytic reduction (SNCR) process within temperature range of 500~600Â°C, however adding urea into hydrazine hydrate would reduce the cost of the reagents.

Thermal decomposition of urea and hydrazine hydrate in N_2 and flue gas, the SNCR de-NOx characteristics of the urea/hydrazine hydrate mixture, the influence of factors such as flue gas temperature, oxygen concentration, normalized stoichiometric ratio (NSR) and proportion of urea in hydrazine hydrate were investigated.

It has been found that urea starts decomposing around 200Â°C and completes the decomposition around 500Â°C with NH_3 emission as an index. Under N_2 atmosphere, hydrazine hydrate decomposes around 200Â°C and NH_3 emission increases with the temperature rising. While in flue gas, very little NH_3 emission can be detected when hydrazine hydrate decomposes. The appearance of NOx from the mixture oxidation increases with the rising of temperature.

The effective temperature window of urea / hydrazine hydrate mixture is 450~550Â°C, much lower than that of urea and even lower than that of $N_2H_4.H_2O$. By adding 16.7% of urea into hydrazine hydrate the effective temperature window slightly moved to lower temperature side.

Through the sensitive analysis it was found that by adding urea into hydrazine hydrate, these following reactions can promote De-NOx process:

NH_2CONH_2=>NH_3+HNCO

H+O_2+M<=>HO_2+M,

H_2O_2+M<=>2OH+M,

NH_2+NO<=>N_2+H_2O

$2N_2H_3$<=>N_2H_4+N_2H_2

N_2H4(+M)<=>2NH_2(+M).

(735d) Kinetic Model Study in a Photo-CREC Reactor for Photocatalytic Hydrogen Production

de Lasa, H., University of Western Ontario
Escobedo, S. Jr., Western University
Serrano, B., Universidad Autonoma de Zacatecas.
Originally presented on: 11/7/2013 16:00:00 - 16:15:00

Kinetic Model Study in a Photo-CREC Reactor for Photocatalytic Hydrogen Production

Salvador Escobedo[a], Benito Serrano[b], Hugo de Lasa[a,*]

[a] *Western University, London, N6A 3K7, Canada*

[b] *Universidad Autonoma de Zacatecas, Zacatecas, 98000, Mexico*

**Corresponding author: hdelasa@eng.uwo.ca*

Introduction

A new type of photocatalysts have been studied in different photocell configurations and showed one of the best alternatives to generate environmental-friendly hydrogen. The goal of this study is to demonstrate the feasibility of photocatalytic water splitting for hydrogen production through a Photo-CREC reactor, using a sacrificial agent, acid pH and an optimum modified platinum loaded semiconductor of DP25 (TiO_2) [1]. It is also the objective of this study to analyze the byproducts and the kinetic modeling of this photochemical reaction process [2].

Experimental

A modified platinum Degussa P25 (TiO_2) was impregnated by incipient wetness impregnation method. This step was followed by calcinations and reduction. A 2.73 eV reduced band gap was observed for the modified semiconductor [1] (see Figure 1). The optimum pH and platinum impregnated photocatalyst were studied in a Photo-CREC-Water II (see Figure 2). A kinetic modeling study is represented by the photocatalytic formation of hydrogen, addressing the reaction network of ethanol diluted in water (see Figure 3).

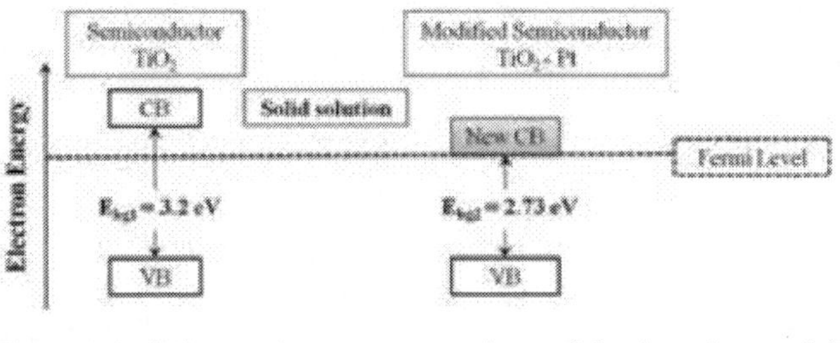

Figure 1. Schematic representation of the band gap shifting by the use of an optimum modified photocatalyst of DP25-1 wt% Pt [1]

Results and Discussion

Impregnation of Pt on TiO_2 leads to enhanced particle size distribution with reduction of particle agglomeration, with better distribution of charges on the semiconductor [1]. On the other hand, it is shown that Pt did not affect the anatase and rutile phases as well as the specific surface area of the DP25. Figure 2 reports hydrogen production rates as a function of contact time under free oxygen conditions and pH equals to 4. Runs were developed with excess of Ethanol as $OH^?$ scavenger reagent. Hydrogen free of oxygen was produced in all cases with 1wt% Pt on Degussa P25 yielding the best hydrogen production rates (see Table 1).Moreover, it is proven that Ethanol helps averting h^+ and e^- recombination, with the highest 7.86% quantum yields [3] obtained between 1-6 hours of irradiation [1].

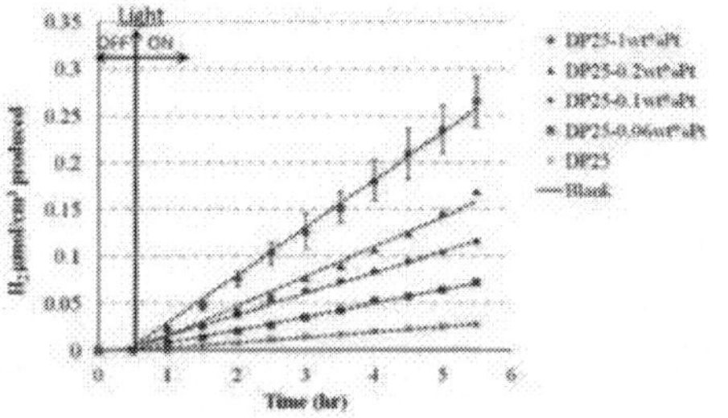

Figure 2. Hydrogen Production under pH=4, Ethanol 2%v/v and Atmosphere of Argon.

Table 1 Reaction Rate at different Loading of Pt on TiO_2 and pH=4

Catalyst	Reaction rate [a] (mol h^{-1} g_{cat}^{-1})	Quantum Yield (% φ)
		H_2 Production
DP25	36	0.7
DP25 0.06wt% Pt	89	1.8
DP25 0.1wt% Pt	142	2.9
DP25 0.2wt% Pt	200	4.1
DP25 1wt% Pt	383	7.9

[a.] Reaction conditions: 298 K, 1 atm.

The results of photocatalytic hydrogen production came along with useful hydrocarbons and CO_2. The study of those hydrocarbons and CO_2 as byproducts is essential on this photochemical reaction system to understand the reaction network suggested by the authors (see Figure 3),

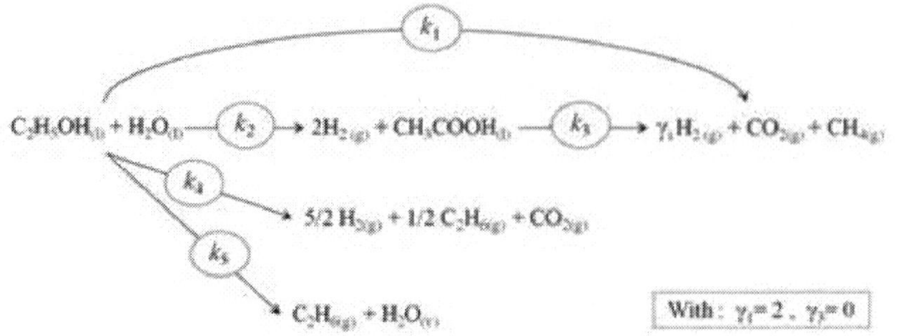

Figure 3. Reaction network for kinetic modelling study

The main assumptions of the proposed heterogeneous kinetic model are stated as a system of differential equations. This valuable approach is used to establish the kinetic model of the reaction network presented above.

The kinetic model adopted in this study includes the Langmuir-Hinshelwood approach. The results of the kinetic model include the kinetic constants and the adsorption constants involved in the generation of hydrogen, the by-products and the intermediate species.

References

1. S. Escobedo, B. Serrano and H. de Lasa, *Appl. Catal. B: Environ.* accepted April (2013)

2. H. de Lasa, B. Serrano and M. Salaices, ?Photocatalytic Reactor Engineering?, *Springer Science* (2005), pp 187; ISBN 0-387-23450-0

3. B. Serrano, A. Ortiz, J. Moreira and H. de Lasa, *Ind. Eng. Chem. Res.*48 (2009) 9864

(735e) Abatement of Diesel Soot Under Various Catalyst/Soot Contact Conditions

Su, C., University of Notre Dame

Originally presented on: 11/7/2013 16:15:00 - 16:30:00

When investigating metal oxides in graphite oxidation, McKee for the first time, pointed out that the activity of an oxide depends more on the localized interaction between catalyst and carbon than on the thermodynamic properties of oxides. The former was latterly referred as "contact condition" in diesel soot oxidation, and it is widely accepted that the reported performances in literature for soot oxidation catalysts are strongly affected by "contact condition", exhaust composition and the nature of the soot used in the experiments. It is thus often difficult to compare the performances of these catalysts without assessing various factors affecting the catalytic performance. For this reason, during the evaluation of a novel K-containing glass soot oxidation catalyst [1], present study investigated in an attempt to understand the effects of various contact conditions on soot oxidation.

Konstandapoulus et al. developed a two-layer soot oxidation model to describe the effect of catalyst coating of porous filter taking explicitly into account the soot-catalyst contact. Catalytic zone was defined as a region over which a spatial "field of catalyst activity" exists, and soot particles oxidize within the "field of activity" are considered in contact with the catalyst. Non-catalytic zone is formed by soot particles which form a "queue" on top of the filled-up catalyst affected layer. In present study, the content of SOF/VOF in engine soot is not negligible, and soot oxidation TGA curves indicated that catalyst also plays a role on SOF/VOF oxidation by decreasing SOF/VOF ignition temperature by 250 oC. An additional type of soot oxidation presenting SOF/VOF in soot particle was defined. Consequently, catalytic soot oxidation on coated wire mesh can be subdivided to three zones, namely, zone I for SOF/VOF oxidation, zone II for catalytic soot oxidation (by O2), and zone III for thermal soot oxidation.

For implementation of continuous DPF regeneration, this present model can applied by a balance temperature analysis while soot accumulation and soot oxidation reaching equilibrium. Due to high soot capacity with tight contact, Zone II of catalytic soot oxidation on wire mesh was employed and its comparison with non-catalytic soot oxidation was analyzed.

(735f) Gklj;lklk

Chanda Nagarajan, X.,

Originally presented on: 11/7/2013 16:30:00 - 16:45:00

Limit of Liability/Disclaimer of Warranty: While the publisher and author have used their best efforts in preparing this book, they make no representations or warranties with respect to the accuracy or completenessofthe
contentsofthis bookandspecificallydisclaimanyimpliedwarranties ofmerchantability
or fitness for a particular purpose. No warranty may be created or extended by sales representatives or writtensalesmaterials.Theadviceand strategies containedhereinmaynotbesuitableforyour situation.
Youshouldconsultwithaprofessionalwhereappropriate.Neitherthepublishernorauthorshallbeliable
for any loss of profit or any other commercial damages, including but not limited to special, incidental, consequential, or other damages.
For general information on our other products and services or for technical support, please contact our Customer Care Department within the United States at (800) 762-2974, outside the United States at (317) 572-3993 or fax (317) 572-4002.
Wiley also publishes its books in a variety of electronic formats. Some content that appears in print may not be available in electronic formats. For more information about Wiley products, visit our web site at www.wiley.com.

Limit of Liability/Disclaimer of Warranty: While the publisher and author have used their best efforts in preparing this book, they make no representations or warranties with respect to the accuracy or completenessofthe
contentsofthis bookandspecificallydisclaimanyimpliedwarranties ofmerchantability
or fitness for a particular purpose. No warranty may be created or extended by sales representatives or writtensalesmaterials.Theadviceand strategies containedhereinmaynotbesuitableforyour situation.
Youshouldconsultwithaprofessionalwhereappropriate.Neitherthepublishernorauthorshallbeliable
for any loss of profit or any other commercial damages, including but not limited to special, incidental, consequential, or other damages.
For general information on our other products and services or for technical support, please contact our Customer Care Department within the United States at (800) 762-2974, outside the United States at (317) 572-3993 or fax (317) 572-4002.
Wiley also publishes its books in a variety of electronic formats. Some content that appears in print may not be available in electronic formats. For more information about Wiley products, visit our web site at www.wiley.com.

Limit of Liability/Disclaimer of Warranty: While the publisher and author have used their best efforts in preparing this book, they make no representations or warranties with respect to the accuracy or completenessofthe

(735g) System-Wide Modeling of HCCI Engine With After-Treatment

Chanda Nagarajan, P., IIT Madras

Aghalayam, P., Indian Institute of Technology Madras

Originally presented on: 11/7/2013 16:45:00 - 17:00:00

System-Wide Modeling of HCCI Engine with After-Treatment

C N Pratheeba[a] and Preeti Aghalayam[b]

Department of Chemical Engineering,

Indian Institute of Technology Madras, Chennai - 600036, India

Automotive engines are synonymous with power and emissions. The stringent norms imposed by legislations drive the design of IC engines. Typical engine-out emissions contain unburned fuel, CO, CO_2, particulate matter and NOx. Many after treatment methods help in curtailing emissions and adhering to emission norms. Engine design also has a major role in reducing emissions. With more stringent emission norms, optimal operation can only be envisaged when the engine and the after-treatment device are considered in conjunction. This paper focuses on system-wide transient models for HCCI engines with the after-treatment options being Exhaust Gas Recirculation (EGR) and Selective Catalytic Reduction (SCR) unit.

EGR refers a part of the exhaust stream is recycled to the engine along with air-fuel intake[1]. EGR helps in NOx reduction by reducing the peak in-cylinder temperature and air required for combustion. The engine-out NOx is then selectively reduced using NH_3 in a catalytic converter to yield N_2, this is known as Selective Catalytic Reduction (SCR). This operates at lower temperatures than its SCNR (Selective Non Catalytic Reduction) counterpart, and thus presents many advantages.

In this work, the Engine + Aftertreatment Device units are to be simulated using the commercial reaction analysis software called DARS. The engine model chosen to represent HCCI is a Stochastic Reactor Model in DARS.

A zero dimensional stochastic model is used for modelling the HCCI engine. The Stochastic model is conceived as a probability density function (PDF) consisting of a discrete number of particles. The number of particles is a measure of the accuracy and computational load.

The mass density function (MDF) is the discretized form of PDF, contains the user-defined number of particles. The number of particles is a measure of accuracy and computational load. The transient MDF is assumed to be a partially stirred tank reactor with the following equation containing the flow, source and mixing terms[2]:

$$\frac{\partial}{\partial t}F_\phi(\psi,t)=\frac{\partial}{\partial\psi_{S-1}}\left\{V\frac{1}{Cp}\frac{dp}{dt_{air}}F_\phi(\psi,t)\right\}+\frac{C_\phi\beta}{\tau}\left[\int_{\Delta\phi}F_c(\psi-\Delta\psi,t)F_c(\psi+\Delta\psi,t)-F_\phi(\psi,t)\right]+$$

$$\frac{\partial}{\partial\psi_{S+1}}\left\{V\frac{1}{Cp}\frac{dp}{dt_{mix}}F_c(\psi,t)\right\}-\left(\frac{\partial}{\partial\psi_{S+1}}\left(\frac{1}{Cp}\sum_{i=1}^{S}h_i\frac{M_i}{\rho}\omega(\phi)F_c(\psi,t)\right)-\sum_{i=1}^{S}\frac{\partial}{\partial\psi_i}\left(\frac{M_i}{\rho}\omega(\phi)F_c(\psi,t)\right)\right\}+$$

$$\frac{\partial}{\partial\psi_{S+1}}\left\{V\frac{1}{Cp}\frac{dp}{dt_{chemistry}}F_\phi(\psi,t)\right\}-\frac{\partial}{\partial\psi_{S+1}}\left(\frac{h_gA}{Cp}(\psi_{S+1}-T_w)F_\phi(\psi,t)\right\}+\frac{\partial}{\partial\psi_{S+1}}\left(V\frac{1}{Cp}\frac{dp}{dt_{heattransfer}}F_c(\psi,t)\right)$$

The terms in the above equation represent change due to piston movement, mixing, chemical reaction and heat transfer, respectively.

The conservation equations for the SCR[3] (including intrinsic catalytic reaction and external mass transfer resistance) are as follows:

$$\frac{\partial}{\partial t}(C_i) + \frac{\partial}{\partial z}(\upsilon C_i) = -k_c a\left[C_i - C_i^*\right]$$

$$k_c a\left[C_i - C_i^*\right] + \sum_{j=1}^{nrxn} \nu_{ij} r_j = 0$$

$$\frac{\partial T_g}{\partial t} + \upsilon \frac{\partial T_g}{\partial z} = \frac{h_s a}{\rho_g C_{pg}}\left[T_s - T_g\right] + \frac{a}{\rho_g C_{pg}} \sum_{j=1}^{nrxn} (-\Delta H_r) r_j$$

$$\frac{\partial T_s}{\partial t} = \frac{\lambda_s}{\rho_s C_{ps}} \frac{\partial^2 T_s}{\partial z^2} - \frac{h_s a}{\rho_r C_{pr}}\left[T_s - T_g\right] + \frac{a}{\rho_g C_{pg}}\left\{\alpha \sum_{j=1}^{nrxn} (-\Delta H_r) r_j\right\}$$

The above equations are Partial Differential Equations (PDEs). They are solved in DARS using the operator splitting technique. The initial conditions that are to be specified for the engine module are the fuel and the oxidant concentration. The engine physical parameters and its rpm, Woschni's heat transfer coefficients, the number of particles for MDF and percentage of EGR are specified. The kinetics used here is the microkinetic combustion model for n-heptane developed by Zeuch et al [4] ? consisting of 121 species and 381 reversible reactions.

The SCR model requires the engine-out exhaust and ammonia dosage to be specified as inlet condtions. The reduced kinetics for the intrinsic reaction, developed by Tsinoglou and Koltsakis[5] will be used here.

The solutions of the above HCCI equations yield the concentration profiles, temperature and pressure profiles as a function of Crank Angle Displacement (CAD), at engine-out. The profiles are validated with experimental data with similar engine parameters. The exhaust profiles with and without EGR are compared, first.

Presented in Figure (1) are the NOx profiles from the engine out emissions, obtained in our preliminary simulations. The mole fractions of NOx in the engine-out exhaust are plotted for different EGR namely 0%, 5% and 10%. The case of No-EGR has significant NOx in the exhaust stream. Even the 5% EGR reduces NOx considerably. This is due to the reduction in In-cylinder peak temperature as shown in Table (1). In-cylinder Peak Pressure variation is insignificant, though a boost compressor can be added for making up the lost pressure head, if required.[6]

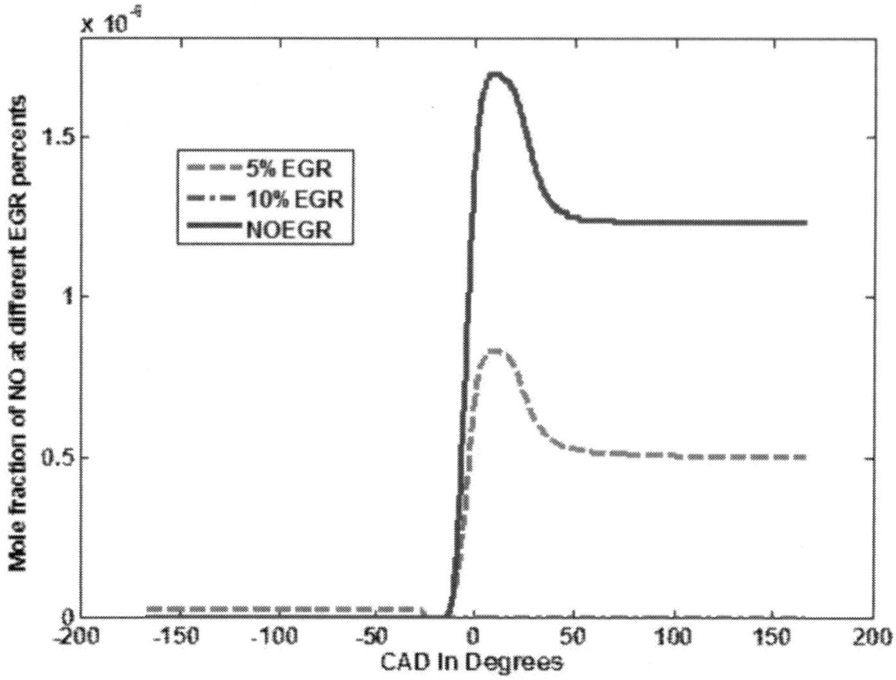

Fig 1. Comparison of NOx Profiles Vs CAD at various EGR fractions

Table 1: In-cylinder Peak Temperature and Pressure Vs EGR

	Temperature in kelvin	**Pressure in Mpa**
0 % EGR (No EGR)	1833	10.44
5 % EGR	1795	10.27
10% EGR	1609	10.14

In the next step, the transient engine-out profiles will be used as inlet profiles for the SCR device, and by solving the SCR equations, profiles of various species of relevance in the tail-pipe will be evaluated. Independent validations of the Engine & SCR units will be undertaken as well.

The integrated approach proposed above has the potential to provide clues for better engine operation and optimal control of tail-pipe emissions. Since the HCCI engine can be a possible candidate for achieving very low concentrations of NOx when coupled with after treatment options, such modeling efforts will have good benefits for the future.

Keywords:

Emissions, Exhaust Gas Recirculation, Selective Catalytic reduction.

References:

1. Heywood JB. Internal Combustion Engine Fundamentals, Mc Graw Hill; 1988.

2. DARS Manual Book 3 Engine In-cylinder Models. 2012

3. Kaisare NS, Lee JH, Fedorov AG. Hydrogen generation in a reverse flow microreactor: 1. Model formulation and Scaling. *AIChEJ.* 2005; 51, 8: 2254-2264.

4. Zeuch T, Moreac G, Ahmed SS, Mauss FA. Comprehensive skeletal mechanism for the oxidation of n-heptane generated by chemistry-guided reduction. *Combust Flame* 2008; 1555: 651-674.

5. Tsinoglou D, Koltsakis G. Modelling of the selective catalytic NOx reduction in diesel exhaust including ammonia storage. *P I Mech Eng D-J Aut. 2007;* 221: 117 - 133.

6. http://www.dieselnet.com/tech/engine_egr_control.php

---

[a] Presenting Author, cnpratheeba@gmail.com

[b] Corresponding Author, preeti@iitm.ac.in

(735h) A New Approach for NOx Control Based On the Chelate Absorption-Biofilm Electrode Integrated System

Xia, Y.,
Xu, B., Insititute of Industrial Ecology and Environment
Zhou, Y., Taihu Basin Authority
Li, W., Insititute of Industrial Ecology and Environment
Originally presented on: 11/7/2013 17:00:00 - 17:15:00

Nitric oxides (NO_x) are hazardous air pollutants that may cause acid rain, acid fog, photochemical smog, health risks of flora and fauna, etc. More and more attention has been paid to the control of NO_x. To date, only Selective Catalytic Reduction (SCR) and Selective Non-Catalytic Reduction (SNCR) obtained industrial applications widely. However, researchers never stopped in studying efficient, economic and green approach for NO_x control. Chemical absorption-biological reduction integrated system is one of the promising approaches which well integrates biological treatment with physicochemical methods. In order to break through the limitation of biological reduction rate, electrochemical action was introduced into this system in recent years. Biofilm electrode method combines biofilm process with electrochemical process, and enhances the reduction of Fe(II)EDTA-NO and Fe(III)EDTA by the cathodic hydrogen production.

Based on the biofilm electrode method, a chelate absorption–biofilm electrode integrated approach was innovatively utilized for the removal of nitrogen monoxide (NO) from the simulated flue gas. The feasibility of this system and parameters that may affect its performance were investigated. It was hoped that this study could provide theoretical foundation and basic data for the industrial application of this new approach.

A reliable chelate absorption–biofilm electrode integrated system was built up. The simulated flue gas with an inlet gas flow rate of 1 L/min contained 500 ppm of NO, 1%~10% of O_2, 15% of CO_2 and N_2. An absorption tower was used to absorb the simulated flue gas and a three-dimensional biofilm electrode reactor was used to regenerate the absorption liquid. Denitrifying bacteria (*Pseudomonas sp.* DN-2) and iron-reducing bacteria (*Escherichia coli* FR-2) were mixed cultivated to form biofilm. After inoculation process, the integrated system could maintain steady removal efficiency (90%) during long-term continuous operation. The electric current has an obvious impact on the NO removal efficiency during startup phase, and a relative small effect during steady state.

Effects of several key parameters such as gas residence time, inlet concentration of NO and O_2, and concentration of Fe(II)EDTA in the absorption liquid on the NO removal efficiency were also investigated in the integrated system. The inlet concentration of NO and O_2 had little effects on the NO removal, but a relatively obvious effects on the concentration of Fe(II)EDTA. The fall in the concentration of Fe(II)EDTA showed the greatest when concentration of inlet O_2 was increased. The decrease of gas residence time had the greatest impact on the NO removal efficiency.

Compared to the chemical absorption-biological reduction integrated system, this new approach has several evident advantages. The reduction rates of Fe(II)EDTA-NO and Fe(III)EDTA are enhanced, tolerance of O_2 in the flue gas is improved, processing load of NO is increased. Therefore, chelate absorption–biofilm electrode integrated system could provide better performance in the NO removal and larger processing load than the biological packed tower in the almost same or even harder conditions.

(735i) Preparation and Catalytic Performance of Au/3DOM Co3O4 for the Oxidative Removal of CO and BTX

Liu, Y.,
Deng, J., Beijing University of Technology
Dai, H., Beijing University of Technology
Originally presented on: 11/7/2013 17:15:00 - 17:30:00

Preparation and catalytic performance of Au/3DOM Co_3O_4 for the oxidative removal of CO and BTX

Yuxi Liu, Jiguang Deng, Hongxing Dai*

Department of Chemistry and Chemical Engineering, College of Environmental and Energy Engineering, Beijing University of Technology, Beijing 100124, PR China. E-mail: hxdai@bjut.edu.cn.

Co_3O_4 is catalytically active for the oxidation of CO and volatile organic compounds (VOCs) [1,2]. The Co_3O_4 material with a mesoporous structure possesses high surface area and developed porous structure, hence favoring the improvement in catalytic performance. In recent years, supported gold catalysts attract much attention. The factors, such as Au particle size and support nature, influence the catalytic activities of supported Au catalysts. Here, we report the preparation and catalytic applications of three-dimensionally ordered mesoporous (3DOM) Co_3O_4-supported gold catalysts for the oxidation of CO and BTX (benzene, toluene, and xylene).

3DOM Co_3O_4 and xAu/3DOM Co_3O_4 (x = 3.7-9.0 wt%) catalysts were prepared using the nanocasting strategy with KIT-6 as hard template and bubble-assisted polyvinyl alcohol-protected reduction method[3] with $NaBH_4$ as reducing agent, respectively. Techniques, such as XRD, BET, TEM, XPS, and H_2-TPR, were used to characterize the physicochemical properties of the materials, and their catalytic activities were evaluated for the oxidation of CO and BTX. The feedstock was 1% CO + 99% air (space velocity (SV) = 60 000 mL/(g h)) or 1000 ppm BTX + O_2 + N_2 (balance) (SV = 20 000 mL/(g h)).

XRD results indicate that the 3DOM Co_3O_4 and xAu/3DOM Co_3O_4 samples exhibited a cubic crystal structure. The BET surface areas of 3DOM Co_3O_4 and xAu/3DOM Co_3O_4 were 94 and 91-93 m^2/g, respectively. TEM results indicate that the 3DOM Co_3O_4 and 6.5Au/3DOM Co_3O_4 samples displayed a good quality of 3DOM architecture, a pore size of ca. 4 nm, and an Au particle size of 2-5 nmÂ°≤XPS and H_2-TPR results reveal that the surface Co^{2+}/Co^{3+} and O_{ads}/O_{latt} molar ratios and low-temperature reducibility of 6.5Au/3DOM Co_3O_4 were higher or better than those of 3DOM Co_3O_4. Catalytic activity evaluation results show that the 6.5Au/3DOM Co_3O_4 sample performed the best, with the $T_{90\%}$ values being -45, 189, 138, and 162 °C, lower than those of 3DOM Co_3O_4 by 115, 50, 67, and 39 °C, respectively. Therefore, it is concluded that the excellent catalytic performance of 6.5Au/3DOM Co_3O_4 was associated with its larger surface area, developed pore structure, higher oxygen adspecies concentration, and better low-temperature reducibility as well as strong Au-3DOM Co_3O_4 interaction.

The work was financially supported by the NSF of China (Nos. 20973017 and 21077007).

References

[1] X. Xie, Yong Li, Z.Q. Liu, M. Haruta, Wenjie Shen. *Nature*, **2009**, *458*, 746-749.

[2] L.H. Hu, Q. Peng, Y.D. Li. *J. Am. Chem. Soc.*, 2008, *130*, 16136Â¨C16137.

[3] M. Comotti, W.C. Li, B. Spliethoff, F. SchÂ¨Â¹th. *J. Am. Chem. Soc.*, **2006**, *128*, 917-924.

(735j) Kinetic Mechanism Investigation On NOx Removal With Hydrazine Hydrate At Moderate to High Temperatures

Hong, L., Thermal & Environmental Engineering Institute, Tongji University
Chen, D. Z., Thermal & Environmental Engineering Institute, Tongji University
Wang, D.,
Originally presented on: 11/7/2013 17:30:00 - 17:45:00

The moderate temperature selective non-catalytic reduction (SNCR) process is needed in many complicate practices. In this research kinetic mechanism of NO_x reduction using hydrazine hydrate in SNCR process was studied and verified by the experimental results, the dominant radicals and reactions were confirmed and key parameters such as residence time and normalized stoichiometric ratio (NSR) in the SNCR process have been obtained through sensitivity analysis. The results of kinetic calculations based on the proposed mechanism showed that the effective temperatures for SNCR De-NOx using hydrazine hydrate were bimodal distributed with the optimum temperatures being around 893K and 1248K respectively, and the lower temperature window fell in the range of 850-1000 K; while the experimental results proved the bimodal distribution of effective temperatures with the peak values being around 896 K and 1241 K respectively and the lower temperature window was within the range of 837-957 K for the hydrazine-based SNCR process. Chemkin/Fluent coupling simulation was adopted to eliminate the discrepancy between the results of experiments and perfectly stirred reactor (PSR) model, and the calculation results from Chemkin/Fluent coupling simulations and the experimental data matched well with respect to reaction behaviors versus temperature. Through sensitivity analysis it had been found that the most effective reaction contributing to NO reduction was the decomposition reaction of N_2H_4 to NH_2 and the most helpful radical was NH_2. At temperature below 900 K, increasing of residence time and NSR could raise the efficiency of NO reduction, and the increased residence time would expand the temperature window to lower side significantly; however, if the reaction temperature is higher than 900 K, increase of residence time or NSR has almost none effect on NO reduction. All of these results showed that hydrazine hydrate corresponds to a much lower temperature window than the traditional reagents and it will be widely applied to those SNCR processes for NO_x reduction where a high temperature reactor is not available.

(764a) Sulfolenes As Sustainable Substitutes for Dimethyl Sulfoxide

Urena-Benavides, E. E., Georgia Institute of Technology
Boigny, A. J., Georgia Institute of Technology
Randall, Z. J., Georgia Institute of Technology
Pollet, P., Georgia Institute of Technology
Liotta, C. L., Georgia Institute of Technology
Eckert, C. A., Georgia Institute of Technology
Fisk, J. S., The Dow Chemical Company
Holden, B. S., The Dow Chemical Company
Originally presented on: 11/7/2013 15:15:00 - 15:35:00

Dimethyl sulfoxide (DMSO) is a widely used solvent for chemical manufacturing processes; however it often involves energy intensive product purification methods. Recovery of the solvent is cost prohibitive, resulting in large amounts of DMSO containing waste streams that often times have to be incinerated. Sulfolenes have similar solvent properties as DMSO, but in contrast, can be easily removed and recycled. In fact, upon mild heating (110 to 135 °C), sulfolenes undergo a retro-cheletropic reaction yielding sulfur dioxide and the corresponding diene; these two volatile products are easily recombined to reform and recycle the solvent.

Some examples of chemical processes effectively carried in piperylene sulfone and/or butadiene sulfone include nucleophilic substitution reactions, copper catalyzed oxidation of alcohols and hydrolysis of β-pinene. For all cases, the "built in" switchability of the solvent was exploited to enhance the production scheme involving efficient reaction and facile separation. Process improvements include in-situ acid catalysis, facile product purification, efficient catalyst and solvent recovery schemes, etc.

In this presentation we will review the advances done in developing sustainable processes using switchable sulfolene solvents as substitutes for DMSO. A scope of their applications will be discussed in comparison with traditional dipolar, aprotic solvents such as DMSO. Moreover, details will be provided regarding solvent recovery methods and recyclability.

(764b) Production of Ethylene and Methanol From Dry Natural Gas and Crude Oil: A Deep Dive Sustainability Assessment

Upadhyayula, V. K. K., Oak Ridge Institute of Science and Education (ORISE)
Masanet, E., Northwestern University
Warren, J.,
Sujit, D.,
Graziano, D. J., Argonne National Laboratory
Riddle, M.,
Conzelmann, G.,
Originally presented on: 11/7/2013 15:35:00 - 15:55:00

Chemical feedstocks (mainly Ethylene, Methanol and NH_3) are conventionally produced from crude oil but deriving the same from natural gas is an option under active perusal by chemical and petrochemical industries today. Since natural gas is an abundantly available and a much cleaner source than crude oil, chemical feedstock production routes using natural gas are expected to be more sustainable than crude oil process path ways. Two major production routes are available for converting dry natural gas to chemical feedstocks. The first one is the indirect conversion route where syngas produced as an intermediate product gets converted to chemicals. The second one is direct conversion route where natural gas gets directly transformed into chemicals. Indirect conversion of natural gas to chemicals can be achieved in two ways: (a) optimization of syngas production through process intensification where advanced chemical engineering principles are applied to design a process pathway that is not only compact and safe but also reduces the energy and environmental footprint of syngas synthesis which is subsequently used to produce chemicals; (b) improving the efficiency of natural gas to syngas conversion process by employing advanced catalytic materials. On the other hand in direct conversion of chemicals from natural gas (no syngas intermediate step) is attained either by employing membrane technology or catalytic reactor systems. While natural gas to chemicals routes are hypothesized to be more sustainable than crude to chemicals pathways, these claims can be validated when life cycle (cradle to gate) impacts of producing chemicals from natural gas and crude can be compared. To accomplish this objective, a deep dive analysis is conducted and sustainability potential of selected Ethylene and Methanol production pathways of natural gas and crude oil are evaluated. The cradle to gate energy and environmental burden associated with respective production routes, process economics and material substitution effects are studied and the results will be discussed in the presentation.

(764c) Process Ecoefficiency of Methanol and Dmc Production From CO2: A Comparative Analysis

Alves, R. M. B., Braskem S.A.
Ribeiro, C., UFRJ - Universidade Federal do Rio de Janeiro
Medeiros, J. L., UFRJ - Universidade Federal do Rio de Janeiro
Coutinho, P. L. A., Braskem S.A.
Araujo, O., UFRJ - Universidade Federal do Rio de Janeiro
Originally presented on: 11/7/2013 15:55:00 - 16:15:00

Environmental performance of products and processes has gained importance as the industry is trying to minimize its environmental impacts. Of special concern to society, are climate effects associated to the rise of CO_2 atmospheric concentration. To mitigate emissions effects, processes are searched that use CO_2 as raw material. It is worth noting that an emission mitigating process should necessarily be a commodity as large production scale are required. With this orientation, methanol and DMC are promising alternative products: the first is a very attractive additive to diesel, and it blends with gasoline to yield a high octane fuel with low Reid vapor pressure. Methanol, on the other hand, is considered as a promising energy alternative besides displaying a wide range of industrial applications. However, in order to claim that a given technology is clean, its ecoefficiency should be addressed with quantitative sustainability indicators. In this sense, the present work selects a set of environmental metrics along with safety and economical (CAPEX and OPEX) indicators. A comparative analysis is presented for two potentially green processes: (i) methanol synthesis; and (ii) DMC synthesis. Furthermore, the later synthesis is approached in two alternatives, a direct and an indirect reaction path. In the methodology used, process simulation and optimization are employed to compare process performances at maximum yields.

(764d) An Innovative Process for the Production of Polymer Grade Propylene From Propane

Julka, V., GRT Inc.
Gadewar, S. B., GRT, Inc.

Originally presented on: 11/7/2013 16:15:00 - 16:35:00

The availability of natural gas from shale basins has had a profound impact on the U.S. Petrochemical Industry by decreasingthe costs of both raw materials and energy. The price of US natural gas has declined from $12.50/MBTU in 2008 to approximately $3.00/MBTU in 2012, with prices expected to decline even further.[1]Â Shale gas has also resulted in the availability of tremendous quantities of low cost natural gas liquids (ethane, propane and butane) recovered as byproducts of natural gas processing which have the potential to transform the chemical industry.Â It has already made a significant impact on the ethylene industry, as a number of producers have switched the feedstock for their steam crackers from naphtha to ethane for higher yield and lower production costs.

Propylene is one of the largest commodity chemicals in the world with an annual worldwide demand of ~ 70 million metric tons, and projected growth at over 6% per year.Â Propylene is primarily produced (~94%) as a co-product of naphtha cracking (ethylene is the primary product) and FCC units in refineries (gasoline is the primary product).Â As ethylene manufactures switch feedstock from naphtha to ethane, combined with a reduced gasoline demand, co-production will not be able to satisfy the growing demand for propylene.Â It is estimated that by 2020, ~ 20% of propylene demand will be satisfied by on-purpose technologies.Â This shortfall in supply, coupled with the availability of inexpensive propane from shale gas, has resulted in significant interest in the manufacture of on-purpose propylene from propane.Â The only commercially available technology for achieving this is propane dehydrogenation (PDH) from UOP (Oleflex) and Lummus Technology (CATOFIN).

As a result of a multi-year research and development effort, GRT Inc., has developed a proprietary technology for the sustainable production of high purity (polymer grade) propylene using propane as the feedstock.Â In the GRT process, propane is first reacted with bromine under relatively mild conditions to produce propyl bromide and HBr.Â Propyl bromide is subsequently dehydrobrominated in the presence of a catalyst to produce propylene and HBr.Â All of the HBr produced in the process is recovered and reacted with oxygen from air over a catalyst to form bromine and water.Â The water produced is discharged from the process, while the bromine is recycled to the bromination step.

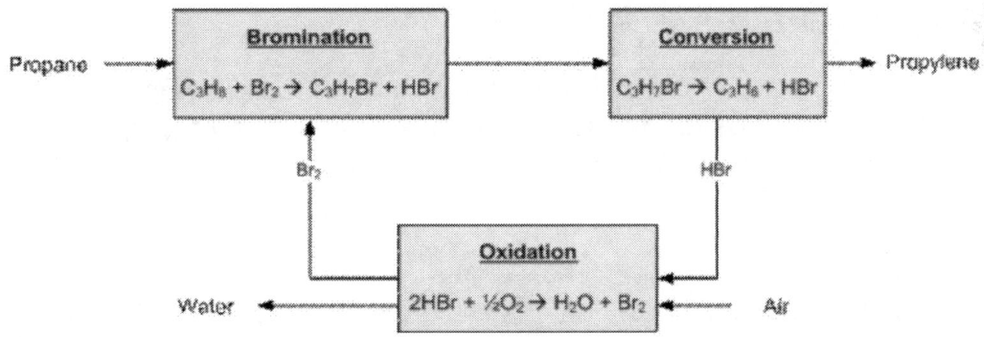

Based on capital and operating cost estimation performed by a top-tier Engineering Company, compared to current state of the art propane dehydrogenation (PDH) processes, the GRT process has a number of important advantages:

? Highly selective production of propylene; no significant byproducts

? Simple energy efficient separations; no propane/propylene splitter or cold box

? Minimal coke formation ? infrequent de-coking and simple reactor design

? No refrigeration

? ~ 9% improvement in feedstock conversion efficiency (98% vs. 90%)

(764e) Gas to Olefins: Sustainable Development of Shale Gas?

Huh, H.,
Davis, B. J., The Cooper Union
Originally presented on: 11/7/2013 16:35:00 - 16:55:00

The development of shale gas plays has received much recent attention due to their huge potential contribution to the United States economy. The economic output of shale gas is remarkable because a cheap and abundant supply of natural gas benefits not only direct industries that produce natural gas but also indirect industries that use natural gas to manufacture other products. The induced economic impact is also significant, as increased earnings correspond to increased expenditures, stimulating monetary circulation. However, the environmental issues associated with the shale gas development have raised extensive public concern. While it is true that extraction activities have increased the amount of greenhouse gases emitted to the atmosphere and that hydraulic fracturing requires extensive use and disposal of water, unconventional natural gas is nonetheless a much cleaner fuel as compared to coal and somewhat cleaner than petroleum-derived gasoline. Regulations at the federal and the state levels, in addition to proper management practices, should enable safe and reliable shale gas development.

A downstream opportunity of shale gas development that aims to meet growing worldwide propylene demand and supply shortage is Gas-to-Olefins (GTO) technology. This process consists of four stages: Syngas production, dimethyl ether (DME) production, hydrocarbon production from DME, and product fractionation.

In this presentation, we will outline our projection for the economic value of a plant which converts 100 million standard cubic feet of methane to polymer grade propylene (43.4% yield) and gasoline (3.5% yield). The annual production of propylene would be 440 thousand metric tons and the sales of the gasoline by-product would also generate revenue. A preliminary economic analysis for the proposed plant design showed that the total capital investment would be $462 million and the investment would accumulate $1.6 billion by the end of the 15 year plant life. The capital investment could be paid back in slightly over three years after startup and the internal rate of return was projected to be 22%. Reducing operating costs through more thorough plant optimization would make this project an even more promising investment opportunity.

(764f) Polyol-Ester Synthesis Via Transesterification of Jatropha-Based Methyl Ester With Trimethylolpropane

Yusup, S., Universiti Teknologi Petronas
Nik Mohamad Kamil, R.,
Mohamed, M., Universiti Teknologi PETRONAS
Originally presented on: 11/7/2013 16:55:00 - 17:15:00

Environmental awareness triggers the increase in demands for synthetic esters derived from vegetable oils and polyhydric alcohols as a replacement for petroleum oils. Biodegradable polyol esters are known for its excellent features in many applications such as automotive, aviation, refrigeration, compressor, offshore drilling, etc. [1,2]. Polyol esters are products of transesterification of fatty acids (FAs) or FA esters with polyhydric alcohol such as trimethylolpropane (TMP), neopentylglycol (NPG) and pentaerythritol (PE). Jatropha seeds oil can be processed to produce high quality end product, with properties like: low acidity, good oxidation stability as compared to soybean oil, low viscosity as compared to castor oil and better cold properties as compared to palm oil [3]. Jatropha oil methyl ester (JOME) was chosen as starting material for the synthesis of polyol ester. Polyol esters synthesis can be carried out either via chemical or enzymatic reactions using various types of catalysts. In this study, the experimental design for polyol ester production by transesterification of JOME and TMP was performed using Taguchi method, employing the L9 orthogonal array and S/N concept for higher the better characteristics for the conversion obtained. Four major controllable factors of the polyol esters synthesis, namely, temperature, JOME:TMP molar ratio, pressure and catalyst amount were investigated at three different levels. The temperature levels were 423 K, 453 K and 483 K. The level of molar ratio of JOME:TMP were 2.5:1, 4:1 and 7.5:1. The pressure was studied at 0.5, 25 and 50 kPa and the catalyst amounts were 0.5, 1.0 and 1.5 wt%. The advantage of using Taguchi design is that the method allows several effects of factors to be simultaneously determined effectively and efficiently.

Transesterification of Jatropha oil was carried out using the two-step acid/alkali-catalyzed transesterification process due to the presence of high free fatty acid (FFA) in the oil. The synthesis of Jatropha oil trimethylolpropane (TMP) ester was performed in a 250 ml three-neck flask equipped with a thermometer, a sampling port, a reflux condenser and a second condenser at lower temperature (283 K) located after the first condenser. The condenser was connected to a vacuum line equipped with relief valve, accumulator and a vacuum trap. The flask was filled with 100 g of JOME and a known amount of TMP. The solution was stirred constantly.The weight of TMP was determined based on the required molar ratio and the calculated mean molecular weight of JOME. The mixture was then heated to the reaction temperature. Then, sulphuric acid catalyst was added. The vacuum was gradually applied to the system until it reached the desired pressure and maintained at the same pressure, up to 5 h. Nine experiments composed in the L9 Taguchi method were replicated and performed randomly. TheJOME conversion was reported as total conversion of JOME in % to trimethylolpropane esters (mono-, di-, and tri-esters of trimethylolpropane) [4,5]. The polyol esters analysis was performed by gas chromatography using method described by Yunus et al. [4] on Shimadzu Gas Chromatography Model GC-2010. From the JOME conversion results, signal to noise ratio (S/N ratio) was calculated using larger-the-better quality characteristicfor each of the nine experiments. S/N ratio is the ratio of the mean (signal) to the standard deviation (noise) or representing the magnitude of the mean of a process compared to its variation. It provides a measure of the impact of noise factors on performance. The average performance of temperature, pressure, molar ratioof JOME:TMP and catalyst amount at different levels with respectto signal to noise ratio will be presented. The polyol ester conversion increases with increase in temperature, with a rapid conversion from 423 K to 453 K followed by slow conversion from 453 K to 483 K. These observations were due to endothermic nature of the reaction which increases the polyol ester concentration along with increase in temperature. Increase in pressure from 0.5 kPa to 25 kPa results in rapid increase of polyol ester conversion. A slight decrease in polyol ester conversion was observed when the pressure was further increased to 50 kPa. Additional experiment was conducted to further assess the effect of pressure on polyol ester conversion. A conversion of greater than 80% can be achieved when the pressure is between 20 and 50 kPa. The polyol ester conversion increases with increase in molar ratioof JOME:TMP from 2.5:1 to 4:1 but almost constant when it was further increased to 7.5:1. There is slight decreased in polyol ester conversion at JOME:TMP molar ratio of 7.5:1. Increasing the molar ratio up to 7.5:1 does not significantly improve the polyol ester conversion. Furthermore, the large excess of unconsumed JOME would require more energy to remove them. Yunus et al. [4] observed that polyol ester conversion increased from 83% to 86% as the ratio of POME:TMP was increased from 3.5:1 to 3.7:1 and at3.9:1 ratio of POME:TMP, further improvement on polyol ester conversion was observed. A 98% w/w polyol ester was successfully synthesized at POME:TMP molar ratio of 3.9:1, 403 K, pressure2 kPa and sodium methoxide catalyst 0.8%.The effect of catalyst loading on polyol ester conversion is shown in Figure 1. The polyol ester conversion increased with increase in catalyst amount up to 1 wt%, but decrease when catalyst amount was further increased to 1.5 wt%. A similar finding was reported by Uosukainen et al. (1998) where the optimum catalyst weight percentage is in the range of 0.1–2.0% [5]. The order of importance of the variables was found to be pressure, temperature, JOME:TMP molar ratio and catalyst. The optimum reaction conditionswere identified to be 483 K, 25 kPa; 1.0% w/w catalyst and JOME:TMP molar ratio of 4:1. At this optimum condition the JOMEconversion obtained was 91.5%.The value of average polyol ester produced from the experiment was then compared with the estimated value for confirmation runs[6]. It was found that the estimated optimum value and experimental value were at 94 wt% and 92 wt% respectively with 2.3% deviation. In polyol ester synthesis using TMP and fatty acid methyl ester (FAME), various experimental conditions have been applied (or reported) by different authors [4–6]. It was noticed thatoperation at reduced pressure which ranged from 25 kPa up to2 kPa, were applied during the reaction for continuous removalof methanol. This indicated the importance of reduced pressure for the transesterification reaction of TMP and FAME. This study discussed the application of Taguchi method to establish the optimal set of conditions for the synthesis of Jatropha oil-based polyol ester. Statistical results showed that the order of importance of the variables was found to be pressure, temperature, JOME:TMP molar ratio and catalyst. The analysis of the confirmation experiment for polyol esters production has shown that Taguchi parameter design can successfully verify the optimum process parameters for polyol ester synthesis. The optimum reaction conditions were identified to be 483 K, 25 kPa; 1.0% w/w catalyst and JOME:TMP molar ratio of 4:1. At this optimum condition the JOME conversion obtained was 91.5%. It was observed that the difference between the value of the predicted polyol ester production and the actual polyol ester production from the confirmation experiments is 2.3%.

(764g) Studies On the Stability Behavior of a Depolymerization Reaction Based On Modified Denbigh Scheme for the Reclamation of Waste Tires

Dutta, A., Groep T - Leuven Engineering College (Associatie KU Leuven)
Constales, D., Ghent University
Beny, G., Groep T - Leuven Engineering College (Associatie KU Leuven)
Sharma, K. R., Prairie View A & M University
Originally presented on: 11/7/2013 17:15:00 - 17:35:00

It is estimated that 70% of the annual production of polybutadiene (PBd) goes into tire manufacture. As the value of the monomer in the waste tires is lot higher than the value attained from using waste tires as fuel, reclamation is considered to be more valuable than incineration. It is believed that a systematic study of the thermodynamics, involving both reaction equilibria and phase behavior, can lead to a better understanding of PBd. Once optimal pressure, temperature and catalyst conditions of the reactor are obtained at the depolymerization step, it will add value by resource recovery.

In this study, a state space model representing a depolymerization reaction is used to understand the stability behavior of a depolymerization reaction based on a modified Denbigh scheme of reactions. The Denbigh reaction scheme has two reactions in series and two reactions in parallel, involving five distinct species. The different instabilities that may arise from the system are studied. This helps in the final design and optimization of yield of the valuable material in the tires for reclamation, thereby leading to a green sustainable economy.

(774a) A Novel Fluidized Bed Hydrator for Steam Reactivation of Calcium-Based CO2 Sorbent

Wang, A.,
Wang, D., The Ohio State University
Deshpande, N., The Ohio State University
Phalak, N., The Ohio State University
Wang, W., The Ohio State University
Fan, L. S., The Ohio State University
Originally presented on: 11/8/2013 8:30:00 - 8:55:00

Calcium-based solid sorbents has been widely applied for the removal of acid gases, such as sulfur dioxide (SO_2), sulfur trioxide (SO_3), and hydrochloric acid (HCl), from coal-fired flue gas. Amid possible carbon dioxide (CO_2) regulations, scientists have explored ways to extend this technology for CO_2 removal especially from sizable point sources at low cost. Here at the Ohio State University (OSU), researchers have developed a 3-step CaO-based carbonation-calcination reaction (CCR) process that operates in the range of 450 – 950 °C for the simultaneous removal of both CO_2 and SO_2 from the post-combustion flue gas. The decay in sorbent reactivity for CO_2 capture over multiple reaction cycles poses a significant challenge to large-scale cyclic carbonation-calcination process. Thermal sintering at elevated temperatures and the subsequent loss in pore volume both contribute to decreasing CO_2 capture capacity. Intermediate hydration has been suggested to be one of the most promising approaches to overcome this limitation by regenerating the particle pore volume and surface area. However, most existing hydration studies have been performed at laboratory scale using thermogravimetric techniques, often investigating conditions not amenable to the 3-step CCR process. In addition, a reactor design specifically suited to high temperature steam hydration is warranted. Experimental data from such a reactor can offer insights to the mass transfer limitations associated with the gas-solid reaction of the extremely cohesive calcium hydroxide ($Ca(OH)_2$) particles under fluidized bed conditions. For the first time, this study reports the design of a bench-scale high-temperature steam hydrator for reactivation of CaO sorbent. The hydrator, consisting of a fluidized-bed reactor with additional internals such as an agitator and a cooling coil, was evaluated using cold-flow tests. Based on the results obtained, a high temperature reactor unit was constructed. The high-temperature steam hydration tests were performed at isothermal conditions in the range of 300-500 °C and 0.6-1.0 atm steam partial pressure. In particular, at an average temperature of 473 °C and P_{H2O} = 1 atm, the 30 min hydration conversion exceeded 70%. Sorbent reactivity toward CO_2 was substantially recovered by steam hydration and the correlation follows a linear trend. Lastly, the cooling coil design demonstrated a proof-of-concept for extracting the heat from this highly exothermic high temperature (>450 °C) reaction. This further expands the application of such reactor design beyond CO_2 capture, for other applications such as chemical heat pumps.

(774b) Dynamic Modeling of a Solid-Sorbent CO2 Capture System

Modekurti, S., West Virginia University
Bhattacharyya, D., West Virginia University
Zitney, S. E., National Energy Technology Laboratory
Miller, D. C., U.S. Department Of Energy, National Energy Technology Laboratory
Originally presented on: 11/8/2013 8:55:00 - 9:20:00

A one-dimensional (1D), two-phase, non-isothermal, dynamic model of a solid sorbent-based CO_2 capture system has been developed in Aspen Custom Modeler (ACM). The system consists of a two-stage bubbling fluidized bed (BFB) adsorber-reactor for CO_2 capture, a single stage moving bed regenerator for the sorbent regeneration, a multi-stage integral-gear centrifugal compressor system for CO_2compression, a pneumatic transport system for solids transport between the adsorber and regenerator, and the balance of the plant. In addition, a heat recovery system has been modeled for the regenerator which includes pre- and post-heat exchangers using steam/boiler feed water as the heating/cooling medium. The solid sorbent is NETL 32D developed at National Energy Technology Laboratory (NETL). The system of partial differential algebraic equations (PDAEs) has been generated using conservation equations, hydrodynamics, and reaction kinetics along with the appropriate boundary and initial conditions.

Flue gas from a supercritical pulverized coal (SCPC) plant enters at the bottom of the lower stage adsorber and regenerated sorbent enters the upper stage adsorber. Embedded heat exchangers have been considered for cooling of the bed using water as the cooling medium. The CO_2-rich sorbent is pneumatically transported to the regenerator through a pre-heater. An embedded heat exchanger in the regenerator maintains the unit at near-isothermal conditions. The regenerated sorbent passes through a low pressure steam generator that helps to recover heat for improving the overall efficiency. The medium selected for transporting the solid sorbent between the adsorber and regenerator is the one that minimizes the auxiliary power requirement without violating the overall target for CO_2 capture. The CO_2 rich gas stream from the regenerator is compressed to pipeline pressure by a multi-stage integral-gear compressor. A glycol absorption system is used to satisfy the pipeline specification for water in the CO_2stream. For off-design operation of the compressors, surge detection and control algorithms are developed.

The PDAE system with the appropriate boundary conditions is solved in ACM using the well-known method of lines which involves discretization of the spatial domain and dynamic integration of the time domain. The transient responses of various process variables such as CO_2 capture, flue gas and sorbent temperatures, and loading of the sorbent are studied by simulating typical disturbances such as change in the flue gas temperature, flowrate, and composition. A lower level control system is designed to maintain the overall CO_2 capture at the desired value in the face of such disturbances.

(774c) Comparative Environmental Life Cycle Assessment of CO2 Capture Systems Based On Ionic Liquid and Monoethanolamine Solvents

Farahipour, R.,
Karunanithi, A. T., University of Colorado Denver
Originally presented on: 11/8/2013 9:20:00 - 9:45:00

Carbon capture, storage and utilization (CCUS) is increasingly being viewed as a global climate mitigation option. Central to a viable CCUS system is the development of cost-effective methods for removal of CO_2 from flue-gas streams of power plants and other stationary CO_2 sources. In addition to economic viability it is important to consider the life cycle impacts of CCUS methods in order to evaluate the net benefits. Moreover, in addition to climate mitigation we would need to consider other issues such as increased life cycle impacts on human and/or ecosystem health. Amin-based (e.g., monoethanolamine) absorption systems for CO_2 capture is a fairly mature technology. However, they can be very energy intensive and less attractive from an environmental stand point due to emissions of the used volatile solvent components. In recent years, because of their low volatility, ionic liquids show significant promise as a CO_2 absorption solvent for post-combustion CO_2 capture. In this work, we present process simulation results of post-combustion CO_2 capture with monoethanolamine (MEA) and ionic liquid bmim [PF6]. The heat and material balances derived from these simulations are utilized to construct the life cycle inventory for the capture process. In addition data related to other upstream processes such as solvent (MEA and ionic liquid) production and downstream processes such as CO_2 compression, transportation, storage and/or utilization is used to perform a life cycle assessment with an aim to compare the environmental impacts of absorption systems based on these two different solvents.

(774d) Whole-Chain CCS System Modelling: Enabling Technology to Help Accelerate Commercialisation and Manage Technology Risk

Matzopoulos, M., Process Systems Enterprise Ltd.
Ramos, A., Process Systems Enterprise, Inc.
Originally presented on: 11/8/2013 9:45:00 - 10:10:00

The commercial implementation of CCS still faces significant challenges. Many of these arise from the fact that the whole chain – and, eventually, whole CO_2 transportation network – needs to be considered as a single system in order to make design and operation decisions that satisfactorily address the commercial imperatives and risk requirements of the various stakeholders along the chain. Even a quick analysis shows that design and operating decisions at the power plant can have a significant effect on storage providers at the other end of the chain, and vice versa. A systems modelling approach is essential, but there are currently no tools that can satisfactorily provide this capability over the whole CCS chain.

In order to address this, a £3m project was commissioned and co-funded by the ETI and project participants, who comprise E.ON, EDF, Rolls-Royce, Petrofac (via its subsidiary CO2DeepStore), Process Systems Enterprise (PSE) and E4tech. The project is aimed at delivering a robust, fully integrated tool-kit that can be used by CCS stakeholders across the whole CCS chain that will be released as a commercially-supported software product at the end of the project.

Because of the requirement for many stakeholders to be able to model areas of the chain beyond their own specific processes (for example, a key requirement for power generators is the ability to investigate the effects on their operation of amine and compression systems attached to their plants), the tool-kit will provide process models for the whole chain. This will also allow it to be used by groups who require a whole-chain view, such as engineering companies or government departments who need to quantify policy decisions.

The tool-kit includes models for conventional generation (pulverised coal and combined cycle gas turbine), new generation (gasification and oxyfuel), solvent-based carbon capture, compression, transmission and injection. The individual process models are mostly 'medium-to-high fidelity': multicomponent streams; equilibrium methods for vapour-liquid separation; comprehensive compressor models allowing multi-stage, multi-section compressors with manufacturers' curves; and distributed pipeline models for construction of pipeline networks that can take elevation into account. The capture models use rate-based techniques for accurate quantification of chemical and physical capture in order to quantify energy penalties accurately and allow meaningful analysis of transient operations.

Using the system modelling tool-kit it will be possible to look at single areas such as amine plants in detail; investigate partial-chain operations – for example, power generation, capture and compression; or analyse interactions across the whole chain and eventually the whole CO2 transmission network with multiple sources and multiple sinks.

The tool-kit also caters for interfacing of external software packages such as E.ON's PROATES, in order to simplify interoperability analysis while allowing companies to preserve existing workflows where necessary. In addition an advanced custom modelling capability means that it is easy to add models of new processes and combine them with existing flowsheeting components.

This presentation describes the status of developments to date in whole chain CCS system analysis and plans to roll the tool-kit software out to the wider 'CCS community' following the current testing phase.

(774e) Optimization of CO2 Capture, Utilization and Sequestration (CCUS) Supply Chain Networks

Hasan, M. M. F., Princeton University
Boukouvala, F., Princeton University
Floudas, C. A., Princeton University
Originally presented on: 11/8/2013 10:10:00 - 10:35:00

CO_2 capture, utilization and storage (CCUS) is an enabling technology toward reducing CO_2 emissions from stationary sources which include power plants, cement production plants, iron and steel plants, refineries, petrochemicals and gas processing plants. More than 60% of the total CO_2 emissions in the United States are attributed to stationary sources [1]. CCUS connects CO_2 sources and utilization/storage sites through three sequential operations: capture, compression, and transportation. CO_2 is first captured and compressed at the source plant, and then transported via pipeline to a site where it is utilized or injected for sequestration for geological storage.

It is critical to integrate capture, compression, transportation, utilization and sequestration in an optimal fashion. However, standalone development of CCUS components have made the design of large-scale CCUS networks a complex and challenging task. In this work, we present a mixed-integer optimization (MILP) model for the synthesis of CCUS supply chain networks for nationwide and regional CO_2 management. We design the CCUS supply chain network by considering simultaneous selection of source plants, capture technologies, capture materials, and locations of utilization and sequestration sites for the first time. We consider real geographic locations and reliable storage estimates from the available database. Furthermore, our cost model includes the costs of capture, compression, transportation and injection for sequestration, and also revenues from CO_2 utilization. The capture and compression costs are derived using advanced modeling, simulation and optimization techniques for a wide range of flue gas compositions and flow rates [2-3], which are crucial to incorporate stationary sources of different types in a large CCUS network. We apply our model to determine the CCUS supply chain network with minimum net expenditure for a given CO_2 reduction level.

Our model selects the optimal nationwide CCUS supply chain network from 3317 stationary sources, 491 saline formations, 1837 unmineable coal areas, and 161 utilization locations (oil and gas reservoirs). We also allow four leading capture technologies (absorption, membrane, pressure swing adsorption (PSA) and vacuum swing adsorption (VSA)), and 9 top materials (2 solvents, 4 zeolites [4-5], and 3 membranes) to choose from for CO_2 capture. Results suggest that it is possible to capture 50% of the current CO_2 emissions from the stationary sources in the United States at a cost less than \$30 for each ton of CO_2. Results for the CCUS in South Central United States, and Texas are also favorable. We demonstrate that it is crucial to consider both material and technology selections as integral parts of a large-scale CCUS network design.

References:

[1]. DOE/NETL. Carbon Sequestration Atlas of the United States and Canada, 2012.
[2]. Hasan, M. M. F.; Baliban, R. C.; Elia, J. A.; Floudas, C. A. modeling, simulation, and optimization of postcombustion CO_2 capture for variable feed concentration and flow rate. 1. Chemical absorption and membrane processes. Ind. Eng. Chem. Res. 2012a, 51, 15642-15664.
[3]. Hasan, M. M. F.; Baliban, R. C.; Elia, J. A.; Floudas, C. A. modeling, simulation, and optimization of postcombustion CO_2 capture for variable feed concentration and flow rate. 2. Pressure swing adsorption and vacuum swing adsorption processes. Ind. Eng. Chem. Res. 2012b, 51,15665-15682.
[4]. Hasan, M. M. F.; First, E. L.; Floudas, C. A. Cost-effective CO_2 capture based on in silico screening of zeolites and process optimization. Submitted for publication, 2013.
[5]. U.S. Provisional Patent Application #61/761,436 and #61/765,284.

(808a) CO2 Capture By Gas Hydrate Formation: Combined Effect of Thermodynamic and Kinetic Promoters

Daraboina, N., Denmark Technical University
von Solms, N., Technical University of Denmark (DTU)
Originally presented on: 11/8/2013 12:30:00 - 12:55:00

Carbon capture by gas hydrate crystallization is a promising technology currently being considered in the battle to mitigate climate change. However, the requirement of both high pressure and slow formation rates are serious limitations which need to be resolved prior to implementing this technology at demonstration scale. Various additives (e.g.: THF, TBAB, cyclopentane and others) are known to promote hydrate formation thermodynamically by reducing the hydrate formation pressure. Furthermore there are also kinetic promoters such as surfactants which can increase the hydrate crystallization rate based on the concentration. The appropriate combination of thermodynamic and kinetic promoters has the potential to move the technology towards practical application.. In this work, the hydrate-based process for carbon dioxide capture was tested in a rocking cell (RC-5; Systemtechnik, Germany) apparatus. Various combinations of surfactant (kinetic promoters: SDS and CTAB) and THF (thermodynamic promoter) in liquid water were assessed in order to find the optimum operating conditions.

(808b) CO2/Brine/Rock Interactions Under CO2 Sequestration Conditions

Soong, Y., National Energy Technology Laboratory (NETL), Office of Research and Development, Department of Energy

Howard, B. H., National Energy Technology Laboratory

Hedges, S. W., U.S. Department of Energy

McIntyre, D., National Energy Technology Laboratory

Warzinski, R. P., U.S. Department of Energy, National Energy Technology Laboratory

Irdi, G.,

Haljasmaa, I.,

Crandall, D., National Energy Technology Laboratory (NETL), Office of Research and Development, Department of Energy

Originally presented on: 11/8/2013 12:55:00 - 13:20:00

CO_2/brine/rock interactions under CO_2 sequestration conditions

Y. Soong*, B. H. Howard, S. W. Hedges, D. McIntyre, R. Warzinski, G. Irdi, I. Haljasmaa[1], and D. Crandall[1]

U.S. DOE, National Energy Technology Laboratory,

P.O. Box 10940, Pittsburgh, PA 15236

[1] URS Co.

It has indicated by Intergovernmental Panel on Climate Change Study that injection of CO_2 into confined geological formations, given their potentially massive carbon storage capacity and widespread geographic distribution, represents one of the most promising options for mitigating anthropogenically-induced global climate change. The Mount Simon sandstone in the Midwest region of the U.S is recognized as a promising candidate host reservoir for carbon sequestration. The Mount Simon is a deep saline aquifer and is a primary reservoir target for large scale carbon dioxide injection tests due to its proximity to CO_2 sources, favorable depth, thickness, permeability, porosity, and the presence of the overlying Eau Claire Formation as a seal. The Midwest Geological Sequestration Consortium (MGSC) has selected the Mount Simon as the reservoir at its major demonstration project. While the Mount Simon Sandstone has ideal reservoir characteristics in some areas, there are significant variations in porosity, permeability, and mineralogy.

Numerous studies have been conducted to investigate changes in host rock properties when exposed to CO_2. Luquot and Gouze conducted CO_2-enriched fluid flow through a carbonate core sample under in situ sequestration conditions (T = 100 °C and P = 12 MPa). Under these conditions, a decrease in permeability and porosity is seen that is linked to precipitation of Mg-rich calcite. Gouze and Luquot also conducted an X-ray microtomography characterization study of porosity, permeability and reactive surface changes during the dissolution of pure calcite in a brine-CO_2 mixture. It showed that the increase in permeability is due to the decrease of the tortuosity for homogeneous dissolution, whereas it is due to the combination of tortuosity decrease and hydraulic radius increase for heterogeneous dissolution. Jove Colon et al., studied the permeability, porosity and reactive surface area evolution during dissolution of nonfractured, clay-free Fontainebleau sandstone cores using a flow-through percolation reactor. The experiments were performed at 80°C with 0.1 M NaOH solution. The permeability was found to vary depending on where the dissolution occurred: in the pore void or the pore throat. Liu et al., conducted a simulation study of CO_2 sequestration in the Mount Simon sandstone via TOUGHREACT. They indicated that a strongly acidified zone (pH 3-5) forms in the areas affected by the injected CO_2 and consequently causes extensive secondary mineral precipitation (calcite, magnesite, ankerite, alunite and anhydrite) and dissolution of feldspars. Thus, the injection of CO_2 into a deep saline aquifer affects porosity since it can cause both mineral dissolution and precipitation in the formation. To predict if there will be any changes to the reservoir during sequestration, it is critical to understand the behavior of the authigenic minerals that line pore spaces when they are in contact with the CO_2 saturated brine.

An experimental study was conducted to assess the impacts of injected CO_2. Core samples from the Mount Simon formation were placed in CO_2 saturated brines at sequestration temperature of 85 °C and pressure of 23.8 MPa in the presence of synthetic Illinois Basin brine for a period of six months in a set of 1.3 liter high-pressure vessels (17-4PH-1150 stainless steel, 10.2 cm I.D. x 16.5 cm depth, manufactured by Thar Technologies, Inc.). One sandstone sample with a porosity of 7.9% was obtained at depth of 1,769.7 m from a well located in Vermillion County, IN. The other sample with a porosity of 1.4% was collected at depth of 2634.2 m from a well located in Knox County, IN. The major mineral composition for the first sample is 78% quartz, 15% Feldspar and with less than 2% microcline and trace minerals. The primary mineral composition of the second sample obtained from Knox County is 70% quartz, 22% Feldspar, and around 4% illite/muscovite. Micro CT, CT, XRD, SEM, petrography, porosity, and permeability analyses were conducted prior to the six month exposure. Upon completion of the experiment, the analyses were repeated.

The permeability of the Vermillion sandstone after the six month exposure to CO_2 and brine was 0.85 mD compared to 1.6 mD for the fresh sample. On the other hand, the permeability of Knox sandstone was 69 nD compared to 50 nD for the fresh sample. The results suggest that the exposure altered the pore characteristics and thus the permeability of the sample. Micro CT analysis as well as SEM backscattered images were used to probe the Vermillion and Knox Co. sandstone before and after the six month exposure to the CO_2/brine environment. It can be seen that mineral dissolution and mineral precipitation occurred after the samples were exposed to CO_2/brine for six months. Mineral precipitation occurred in the Vermillion sample pores and cracks blocking flow and resulting within the observed 50% decrease in permeability. For the case of Knox sandstone, the mineral dissolution occurred in pore though the precipitation was observed in the exterior of the sample. This might result in increasing in permeability. In addition, from the analyses of the reacted brines within the reactor, it indicated the possibly of Feldspar dissolution.

When CO_2 is injected into a saline formation, acidic conditions form in the brine in the nearby portion of the reservoir thus reducing the pH of the brine in the vicinity of the injection system. Minor minerals in the sandstone, such as K-feldspar, calcite, kaolinite, microcline etc., that are stable in the native environment but not stable at lower pH can dissolve. Eventually, the increase of cation concentration produced by mineral dissolution can supersaturate the fluids with respect to other minerals. If this occurs, precipitation of new minerals can result in a change of porosity and permeability which could affect both CO_2 storage capacity and fluid movement. The issue is further complicated by the evolution of pore geometry due to these geochemical reactions that may affect the overall hydrologic properties of the host rock and therefore the performance of the unit as an effective reservoir.

The decreasing permeability found in Vermillion sandstone was probably due to Feldspar dissolution, migration and precipitation altering the sandstone pore/crack structure. For the case of sandstone from Knox county, the permeability increased by 38%. It might be due to the Feldspar dissolution in the pore and mineral precipitation at exterior surface of the sandstone. Investigations are continuing to more clearly elucidate the mechanisms and geochemistry of this system, especially at the pore structure scale.

(808c) Novel CO2 Capture and Carbonation Process With Waste Concrete

Shuto, D.,
Yamasaki, A., Seikei University
Hongo, T., Seikei University
Iizuka, A., Tohoku University

Originally presented on: 11/8/2013 13:20:00 - 13:45:00

A novel CO_2 capture and carbonation process is developed. In this process, wastes containing calcium, such as waste cement particles or steel slug, are used as a calcium source for carbonation reaction with CO_2 emitted from point sources. The process is composed of several steps. CO_2 in the flue gas is captured through absorption process with alkaline solution such as sodium hydroxide to form sodium carbonate and sodium bicarbonate. The calcium content in the waste cement particles is extracted with acid such as nitric acid or acetic acid to form solution containing calcium. Mixing of these two solutions leads to the precipitation of calcium carbonate: CO_2 fixation via carbonation reaction. Electrodialysis is applied for the regeneration of acid and alkaline from the remained solution after filtration of the precipitant. Thus, the net input of the process is CO_2 in the flue gas and waste cement particles, and the output is calcium carbonate. The process feasibility was examined through laboratory -scale experimental studies under various operation conditions. Based on the experimental results, the process was optimized and evaluated in terms of the power consumption and cost for the CO_2 recovery, and compared with conventional processes. Net CO_2 sequestration can be achieved by the process under certain conditions.

(808d) Direct Capture and Conversion of CO2 From Flue Gas Via Ex-Situ Mineral Carbonation Using Chemical and Biological Catalysts

Swanson, E. J., Columbia University
Brady, P. V., Sandia National Laboratories
Park, A. H. A., Columbia University
Originally presented on: 11/8/2013 13:45:00 - 14:10:00

Converting carbon dioxide from flue gas directly into a solid magnesium carbonate presents a compelling option for the mitigation of power plant emissions. The carbonate product is environmentally benign and thermodynamically stable, thus creating a safe and reliable alternative to geological storage. However, given the low partial pressure of carbon dioxide in flue gas and the low reactivity of magnesium bearing minerals, achieving the kinetics required to make such a process feasible is challenging. In addition, magnesium carbonates formed at low temperature tend to incorporate significant waters of hydration, which increases transportation requirements and limits utilization of by-products. Each of these issues was addressed in the design of a multistep process for the conversion of magnesium silicates to carbonates as a carbon capture and storage option. The mineral dissolution reaction is accelerated using a mixture of organic chelating agents to produce a concentrated magnesium solution. This acceleration appears to occur through a stabilization effect on the pores formed as the silicate dissolves. The concentrated magnesium solution produced during the dissolution of the silicate is used to drive the precipitation of carbonates directly from flue gas. Through strict control over the chemistry of a simultaneous absorption and precipitation process, the phase of magnesium carbonate formed can be controlled. In addition, the mass transfer rate of carbon dioxide from flue gas into the precipitating liquor is accelerated using biological catalysis, which is shown to provide significant savings in process equipment size and operating parameters. Finally, the process as a whole is demonstrated to have a negative overall carbon balance through life cycle assessment.

(808e) A Green Approach to Mitigate CO2 Using Mixed Microbial Culture

Raghuvanshi, S., Birla Institute of Technology & Science, Pilani
Gupta, S., Birla Institute of Technology & Science, Pilani
Mishra, S.,
Originally presented on: 11/8/2013 14:10:00 - 14:35:00

Carbon dioxide is one of the GHG's (Green House Gases) & responsible for Global warming.. As of July 2011, CO_2 in the earth's atmosphere is at a concentration of 392 ppm by volume which is 35% higher as compared to the mid-1800s. Carbon dioxide, which is estimated to contribute 50-60% of the Global warming in the last century, is majorly released during the combustion of fossil fuel. The CO_2 concentration in earth atmosphere can be reduced by biological, chemical and technological techniques available through either mitigating or sequestering emissions. In the recent years, biological route has been identified as one of the most viable options available for the mitigation of CO_2. There are six carbon fixing pathway's. Out of which, the Calvin cycle (Calvin-Benson-Bassham) is used by photoautotroph's to fix CO_2. There are microbes which are ubiquitous in nature and thrive in extreme environment classified as chemoautotrops. They also classified as Extremophiles. The requirement of industries is to develop a economic, efficient and effective system which not only fix CO_2 but also capable to withstand in the severe industrial environment. Mixed culture is proved to be a superior candidate, as the consortium of different engineered species develop a symbiotic relationship which helps each other not only to survive in harsh conditions but also curb down CO_2. The another possibilities are simultaneous waste water treatment, production of Methane, and bio-hydrogen production. The present work deals in identification of microbes present in sludge of waste water treatment plant having CO_2 fixation capabilities. It also gives the insight of the relation which exists between natural selected mixed culture and to understand the capabilities of CO_2 fixation. The batch study has been conducted to measure the capabilities of mixed culture to mitigate CO_2 for five different V/V % concentration of CO_2 4, 12, 16, 22 and 25. It was observed that CO_2 is completed consumed by microorganisms in 4 days. Mitigation rate for different concentration of CO_2, bio-mass growth rate and resulting product has been analyzed on per hour basis and per day basis. Mixed culture is found to be superior candidate than photoautotrophic organisms for CO_2 mitigation. They show a promising future to develop a low cost, compact size closed system for *in-situ* CO_2 mitigation.

AUTHOR INDEX

Abbasian, Javad ... 6
Abdel-Wahab, Ahmed ... 134
Adewuyi, Yusuf G (Debo) ... 167, 174
Adhikari, Shreya ... 308
Adhikari, Sushil ... 359
Agarwal, Sushant ... 249
Agblevor, Foster ... 394
Aghalayam, Preeti ... 488
Agnihotri, Dileep ... 284
Agrawal, Rakesh ... 399
Ahmad, Iftekhar ... 343
Ahunbay, M. Goktug ... 417
Akinmolayan, Folashade ... 44
Akseli, Ahu Aydogan ... 16
Al Shoaibi, Ahmed ... 246
Alalloush, Saeed S. ... 463
Alanazi, Yousef ... 199
Alayedi, Ibrahim ... 229
Alibabaei, Leila ... 289
Allen, David T. ... 23, 82, 110, 187
Allen, Mark G. ... 100
Allen, Ray W. K. ... 351
Allen, Sue Ann Bidstrup ... 100
Almquist, Catherine B. ... 366
Alnouri, Sabla ... 43
Alves, Rita M. B. ... 498
Amos, Delaina A. ... 218
Amrose, Susan ... 213
Amundsen, Ted ... 361
Amundsen, Ted J. ... 81
Anastasio, Daniel ... 465
Andrade-Espinosa, Guillermo ... 371
Anid, Nada Marie ... 28
Appuhamilage, Nilusha Sudasinghe ... 419
Ara, Mahbuba ... 39, 105
Arastoopour, Hamid ... 6
Araújo, Ofélia ... 498
Arce, Pedro E. ... 191, 319
Arena, Jason T. ... 465
Arie, Arenst Andreas ... 107
Arifin, Darwin ... 454
Arlt, Wolfgang ... 66
Armutlulu, Andac ... 100
Asa-Awuku, Akua ... 86, 427
Ascue, Katia ... 5
Asghar, Hafiz M. A. ... 77
Aston, Victoria J. ... 408
Austin, Sharon ... 23, 187
Auyeung, Nicholas ... 455
Ayyakkannu, Manivannan ... 104
Azargohar, Ramin ... 242
Bae, Ki-Kwang ... 469
Bai, Fengwu ... 273
Bai, Qing ... 374
Bajpai, Rakesh ... 364
Bakshi, Bhavik R. ... 17, 307, 387
Balakotaiah, Vemuri ... 206, 231, 477
Balasubramanian, Rajasekhar ... 126, 370
Bandyopadhyaya, Rajdip ... 184
Banholzer, William ... 30
Bao, Zhenan ... 330
Barbour, Natalia ... 186
Barford, John ... 368
Barghi, Seyedhamed ... 352
Barnes, Jack ... 214
Barroso, Bruno L. ... 194
Ba-Shammakh, Mohammed S. ... 94
Batchelor, Bill ... 134
Batista, Fabiana R. X. ... 451, 452
Bayham, Samuel ... 475
Beck, Sara ... 15
Belharouak, Ilias ... 58
Bellona, Christopher ... 192
Benavides, Pahola T. ... 379
Beny, Guido ... 503
Berberich, Jason A. ... 366
Berber-Mendoza, Maria Selene ... 371
Bernal-Jácome, Luis Armando ... 371
Bertazzoli, Rodnei ... 194
Bessette, Roxanne ... 176
Bhandari, Vinay M. ... 140, 149
Bharathiraja, Balasubramanian ... 116
Bhattacharyya, Debangsu ... 505
Bhavsar, Saurabh ... 407
Biddinger, Elizabeth J. ... 5
Biernacki, Joseph J. ... 466
Bingham, Jon-Paul ... 178
Bird, Jeffrey ... 16
Bisht, Varsha ... 271
Biswas, Pritam ... 184
Blaney, Lee ... 177
Blinde, Paul ... 228
Blowers, Paul ... 229, 382
Boateng, Akwasi A. ... 401, 402
Bobba, Christopher ... 342
Bohn, Christopher ... 344
Boigny, Afrah J. ... 496
Bolaños-Reynoso, Eusebio ... 424
Bolintineanu, Dan S. ... 217
Bollas, George M. ... 467, 471
Boningari, Thirupathi ... 196, 208
Booshehri, Amin Yoosefi ... 117
Bothun, Geoffrey ... 342
Boukouvala, Fani ... 508
Boyd, Christopher M. ... 90
Bradley, Kenneth ... 252
Brady, Patrick V. ... 515
Brendel, Brittany ... 421
Brennecke, Joan F. ... 413
Bricka, R. Mark ... 364

AUTHOR INDEX

Broda, Marcin 448
Brown, Gordon E. 243
Brown, Nigel W. 1, 77
Brown, Tobin 256
Bryant, Steven 311
Buchanan, James H. 241
Buchireddy, Prashanth 364, 464
Burns, Mark 186
Burton, Edward 374
Cabezas, Heriberto 379, 384
Cai, Tianxing 230, 318
Cairncross, Richard 343
Camacho, Lucy Mar 50
Camargo, Gabriel 423
Campen, Andrew K. 1
Cano, Natalia Andrea 391
Cao, Jinlei 133
Capecelatro, Jesse S. 356
Caple, Kevin 453
Cappelle, Malynda A. 50
Cappillino, Patrick 67
Cardoso, Vicelma L 451, 452
Cargnello, Matteo 350
Carlier, Dany 56
Carpenter, John R. 400
Carter, William 85
Case, Paige 358
Castranova, Vince 378
Castro, Jose J. 55
Cavanagh, Eduardo J. 386
Cerda, Katherine V. 191
Chan, Sui Yung 68
Chang, Wounjim 158
Chang, Yuan 225, 388
Chapman, Chris 159
Chatterton, Bill 49
Chawla, Ramesh 323
Cheali, Peam 305
Chen, De-Zhen 481, 495
Chen, Gen 36
Chen, Hui 481
Chen, J. Paul 180, 181, 202
Chen, Liang-Yih 296
Chen, Lijie 273
Chen, Liping 260
Chen, Liwen 9, 227
Chen, Si Cong 72
Chen, Wanghua 260
Chen, Yanguang 403
Chen, Yuan 97
Chen, Yuanxin 162
Cheng, Peng 240
Cheng, Yi 118
Chiu, Ping-Chin 476
Cho, Daehwan 38
Cho, Seonghyun 27
Cho, Wonchul 469
Choudhary, Surbhi 346, 348
Chowdhury, Muhammad B. I. 450
Chozhavendhan, Sivasankaran 116
Christofides, Panagiotis D. 212
Chu, Sheng 462
Chun, Jaehun 3
Chung, Tai Shung 68, 70
Chung, Tai-Shung 72, 166
Chyou, Yau Pin 247
Clark, Blythe 67
Clark, Douglas S. 190
Cleveland, Vincent 178
Cocker, David 85
Coggon, Matthew 88
Cohen, Joel 221
Cohen, Yoram 8, 80, 212, 252, 286, 287, 336
Constales, Denis 503
Conzelmann, Guenter 497
Cooper, David G. 256
Copeland, John R. 332
Cortés, Daniel 25, 383
Coutinho, Paulo L. A. 498
Crandall, Dustin 511
Croguennec, Laurence 56
Crossley, Steven 395
Cruz, Yair 55
Cruze, Nathan 387
Cui, Wenquan 239
Cui, Yi 420
Culp, Jeffrey 328
Cumming, Denis J. 351
Currier, Neal 238
Dahlin, Robert S. 49
Dai, Fei 192
Dai, Hongxing 493
Dai, Sheng 51
Dalai, Ajay K. 242
Dan, Seungkyu 27
Dang, Thu Thi 369
Daous, Muhammad A. 131
Daraboina, Nagu 510
Dass, Sahab 346, 348
Datta, Moni Kanchan 349
Dauenhauer, Paul 462
Dauenhauer, Paul J. 355
Davis, Benjamin John 500
Davis, Tom 50
Dayton, David C. 400
De Angelis, Maria Grazia 414
De Lasa, Hugo 404, 450, 482
De Lasa, Hugo I. 411
De Wit, Jos S. 70
De, Mahuya 242

AUTHOR INDEX

Deb, Aniruddha 101
Debelak, Kenneth A. 188
Degenstein, John 399
Delaire, Caroline 213
Delgass, W. Nicholas 399
Demokritou, Philip 221
Deng, Baolin 283
Deng, Da 41, 105, 198
Deng, Jiguang 493
Deng, Shuguang 419
Depaoli, David W. 51
Derk, Raymond 378
Deschamps, Michaël 56
Deshpande, Niranjani 504
Desisto, William J. 358
Desjardins, Olivier 356
Ding, Chang 278
Dinu, Cerasela Zoica 249, 250, 339, 373, 377
Diwekar, Urmila D. 379
Djire, Abdoulaye 101
Dong, Chenbo 339, 373
Dong, Xiuqin 124
Dorado, Christina 401, 402
Doschek, Klaus 435
Dougless, Austin 229, 382
Dowding, Matthew 421
Drioli, Enrico 166
Druffel, Thad 218
Du, Shoucheng 471
Dungen, Barry 419
D'Unger, Dan 251
Dunn, Russell F. 188
Durbin, Thomas 86, 427
Duri, Bushra 112
Dutta, Abhishek 503
Dutta, Prabir K. 162
Dyson, Anna 16
Dziubla, Thomas D. 200
Dzombak, David A. 6
Eason, Tarsha 384
Eckert, Charles A. 496
Edgar, Thomas F. 31
Ehrman, Sheryl H. 57
Eidawud, Reem 339
Ekerdt, John G. 288
Eldawud, Reem 249, 373
Elder, Rachael H. 351
El-Farra, Nael H. 19
El-Halwagi, Mahmoud 43
Ellmaier, Lisa 362
El-Shazly, Ahmad 131
Enick, Robert M. 333
Enright, Tom 434
Epling, William 210, 234
Epstein, Scott A. 87
Erythropel, Hanno C. 256
Escobedo, Salvador 482
Esposito, Daniel V. 291
Ettireddy, Padmanabha Reddy 208
Evangelisti, Sara 159
Evanko, Brian W. 408
Fagnant, Daniel 413
Fan, L. S. 475, 504
Fan, Mingxia 274
Fanarraga, Mónica L. 372
Fane, Anthony G. 121
Fang, Jun 283
Fann, Steven 52
Farahipour, Reza 506
Farfan, Gabriela 243
Faria, Jose 8
Farmer, Justin 400
Farnoud, Amir M. 376
Farrell, Anthony 323
Faxon, Cameron 82
Fei, Ling 34
Feiner, Roland 362
Feng, Hao 248
Feng, Xuelan 118
Ferreira, Juliana S. 451, 452
Fiegel, Jennifer 376
Fischer, Ari 471
Fisher, Richard P. 15
Fisk, Jason S. 496
Floudas, Christodoulos A. 508
Fogwell, Thomas W. 4
Forero, Carlos A. 423
Foust, Thomas D. 356
France, Bryan 252
Franclemont, Joshua 192
Freeman, Benny D. 284
Frenkel, Val 287
Fröhlich, Eleonore 253
Fu, Feng-Jiang 72
Fujiki, Junpei 334
Fujiwara, Kakeru 337
Gadewar, Sagar B. 499
Gadgil, Ashok 213
Gaglione, Anthony 264
Gallego, Dario 391
Galleher, Connor 190
Galvita, Vladimir V. 445
Ganesh, Panchapakesan 37
Gangadhar, Gayathri 71, 156
Gangadharan, Preeti 227
Gao, Larry 212
Garcia, Bruno S. 123
Garrido, Rene 353
Gary, Brannon 330
Gass, Samuel 221

AUTHOR INDEX

Gaston, Sharyn....186
Gatica, Jorge E....308
George, Saji....252
Gernaey, Krist V....305
Gerwe, Brian....229, 382
Ghassemi, Abbas....12
Ghosh, U. K....151
Ghoshal, Aloke Kumar....279
Giacomelli, Jason J....214
Giarola, Sara....301
Gibbemeyer, Erin L....17
Gjoka, Eriugen....5
Glass, Jeffrey T....289
Glover, T. Grant....241
Goel, Prem K....387
Goff, George S....413
Goh, Shuwen....121
Goldstein, Eli A....470
Gomes, Mariany L. M....194
González, Jesus....372
Gonzalez, Michael A....380, 432
Gonzalez, Yuri....423
Gonzalez-Garcia, Pedro....424
Gonzalez-Mejia, Alejandra....384
Gopalaswami, Nirupama....261
Goredema, Adela....434
Graves, David B....190
Graziano, Diane J....497
Greek, Benjamin....455
Greenlee, Lauren F....285
Grem, Izabel C....123
Grenville, Richard K....214
Grieco, Scott....317, 365
Grossmann, Ignacio E....303
Gu, Han....212
Gu, Hongbo....98
Gu, Qiang....236
Gu, Sai....397
Guerrero-Gallego, Juan....324
Guillén, Gonzalo....383
Guillén-Gosálbez, Gonzalo....25, 390
Guillory, John....364, 464
Guimerá, Roger....25
Guiver, Michael D....414
Gullapalli, Sravani....211
Gumidyala, Abhishek....395
Gunugunuri, Krishna Reddy....331
Guo, Jiang....98
Guo, Zhanhu....98, 106, 201
Gupta, Rakesh K....249
Gupta, Suresh....71, 156, 516
Gür, Turgut M....470
Guri, Sonia....158
Gutierrez, Angela Maria....200
Hagino, Hiroyuki....85
Hale, William....421
Halitoglu-Velioglu, Sadiye....417
Haljasmaa, Igor....511
Hall, Maurice....45
Hamilton, Matthew A....473
Hammer, Nicole L....353
Hammond, Paula T....40
Hamzehlouyan, Tayebeh....234
Han, Kai....62
Han, Lu....467, 471
Hanes, Rebecca....387
Hansen, Nathaniel S....38
Hanson, Kenneth....289
Hanumantha, Prashanth Jampani....104
Hara, Nobuo....416
Hara, Shigeki....416
Haraya, Kenji....416
Harold, Michael P....206, 232, 477
Hart, Ashley E....251
Hasan, M. M. Faruque....508
Hattar, Khalid....67
Hayes, Thomas....284
Hayner, Cary M....61, 62
He, Cheng-Chun....155
He, Guanghu....351
He, Jiajun....330
He, Jianzhong....278
He, Ke....177
Hedges, S. W....511
Hekmaty, Michelle....67
Held, Kerstin....266
Helmer, Bradley J....70
Henrichsen, Matthew....237
Heo, Gookyoung....85
Her, Sungyun....258
Heras, Nuria De Las....77
Hernandez, Juan....324
Hernandez-Rivera, Samuel....262
Hesse, Marc....311
Higbee, Robert W....214
Hill, Justin J....96
Hilt, J. Zach....200
Hinks, Mallory L....87
Hiremath, R. S....140
Hlebak, Josh....400
Ho, Tsung-Yeh....296
Ho, W. S. Winston....162
Hoertz, Paul....289
Holden, Bruce S....496
Holsen, Thomas....192
Hong, Dae Ho....104
Hong, Liu....481, 495
Hongo, Teruhisa....367, 514
Honma, Chiemi....161
Hossain, Mohammad M....411

AUTHOR INDEX

Hossain, Mohammad Mozahar 404
Hosseini, Davood 442
Hou, Ariel 264
Hough, Blake 355
Howard, Bret H. 511
Hrma, Pavel 3
Hsieh, Ming-Kai 6
Hsuan, Grace 343
Huang, Ching Ying 247
Huang, Xiaofei 284
Huang, Yinlun 197, 219
Huang, Yuxiong 119
Huelsman, Chad M. 263
Huh, Heejae 500
Hussain, Syed N. 77
Hussein, Abdulmenan 298
Hutchins, Margot 228
Hutchinson, Todd M. 214
Hyder, Md. Nasim 40
Ighere, Jude 323
Ihara, Toru 458
Ilzuka, Atsushi 367, 514
Imponenti, Luca 455
Imtiaz, Qasim 409, 442
Intoci, Nicholas 421
Inturi, Siva Nagi Reddy 196
Irdi, Gino 511
Ishizuka, Masanori 69
Jacobs, Benjamin 67
Jain, Akshay 126, 370
Jamieson, Heather 281
Jarvis, Mark 356
Jensen, Cory D. 189
Jentoft, Rolf 395
Jeong, Seong Uk 469
Jeong, Soyeon 209
Jew, Adam D. 243
Jia, Charles Q. 274
Jia, Shao-Yi 182
Jiang, Lan Ying 74
Jiang, Wanlin 203
Jiang, Wenjuan 341
Jimenez, Laureano 383
Jiménez-Esteller, Laureano 390
Jing, Dazheng 443
Jones, Christopher 67
Jones, Christopher W. 332
Jones, Mark E. 30
Jonsson, Haflidi 88
Joo, Yong L. 38
Joo, Yong Lak 64
Jovanovic, Goran N. 76
Judkoff, Ron 13
Julka, Vivek 499
Jung, Bahngmi 134
Jung, Ji-Eun 243
Kadakia, Karan 104, 349
Kajale, Atul 48
Kalanyan, Berc 289
Kalra, Vibha 95
Kamani, Haarika 341
Kamasamudram, Krishna 210, 238
Kamegawa, Takashi 332
Kamil, Ruzaimah Nik Mohamad 501
Kan, Eunsung 173, 178
Kang, Byung Joon 27
Kang, Kyoungsoo 469
Kang, M. G. 344
Kanjilal, Baishali 421
Kansha, Yasuki 69
Kapoor, Kriti 31
Karakitsios, Spyros 91
Karavalakis, Georgios 86, 427
Karimi, Shahrad 327
Karunanithi, Arunprakash T. 223, 224, 385, 506
Kasemset, Sirirat 284
Kashon, Michael L. 339
Kathe, Mandar 475
Kavian, Reza 40
Kearns, Josh 15
Keller, Arturo A. 119
Kelley, Steve 359
Kent, Paul R. C. 37
Kermenidou, Marianthi 91
Kern, Hannes 429
Khan, Md Arif 174
Kierzkowska, Agnieszka 442
Kiggans, Jim 51
Kim, Chang Hee 469
Kim, Do Han 289
Kim, Do Heui 209
Kim, Hyosang 173
Kim, Jae Jeong 335
Kim, Jangwoo 64
Kim, Jihyun R. 173
Kim, Jong 31
Kim, Kyung Woo 38
Kim, Sang Done 469
Kim, Woonhoe 53
Kim, Yo-Han 335
Kim, Yong Seok 38
Kirschling, Teresa L. 285
Kitchens, Christopher L. 251, 342
Klausner, James 455
Knecht, Christopher J. 214
Knutsson, Pavleta 406
Kobayashi, Daisuke 161
Kocherlakota, Lakshmi Suhasini 412
Kofler, Markus 435
Koong, Len Foong 368

AUTHOR INDEX

Koronaios, Peter....333
Kota, Arun S....231
Kotrba, Adam....208
Kreider, Peter....199
Ku, Young....476
Kumar, Ashok....234, 238
Kumar, Rahul....320
Kumar, Ramanujam Praveen....116
Kumta, Prashant N....104, 349
Kung, Harold H....61, 62
Kung, Mayfair C....61, 62
Kuo, Yu-Lin....476
Kuroda, Chiaki....161
Kuwahara, Yasutaka....332
Kwan, Joseph Zheng Xin....121
Lackey, Dana....313
Lal, Banwari....271, 277
Lam, Khee Poh....14
Lam, Koon Fung....368
Lammers, Peter....419
Langrock, Alex....57
Lapkin, Alexei....299
Leask, Richard L....256
Leazer, John....21
Lechman, Jeremy B....217
Lee, Albert....284
Lee, Chang Ha....335
Lee, Chun C....457
Lee, Doh C....290
Lee, Doyeon....469
Lee, Hyoungill....248
Lee, Jae W....65
Lee, Jinwoo....33
Lee, Kwanghee....258
Lee, Youngmin....291
Lehr, Claus-Michael....253
Lei, Lecheng....135, 146
Leion, Henrik....406, 440
Leitinger, Gerd....253
Lettieri, Dave....228
Lettieri, Paola....159
Lewandowski, Allan....15
Li, Christopher....343
Li, Cong....461
Li, Fanxing....403
Li, Guiming....124
Li, Guoting....180, 181, 202
Li, Haitao....222
Li, Hao....135
Li, Hua....284
Li, Junhui....234
Li, Wei....492
Li, Wenqiang....114
Li, Xiaosen....309
Li, Yan....297
Li, Ying....63
Li, Yonghui....124
Li, Zhongjian....146
Li, Zhuo....18
Li'An, Hou....183
Lighty, Joann S....473
Lim, Nonnie....317
Lim, Young-Il....369
Lin, Ronghong....120
Lind, Mary Laura....281
Linden, Karl G....15
Lindner, Jeffrey S....2
Lindsay, Abby....24
Linke, Patrick....43
Liotta, Charles L....496
Liu, H. Haven....336
Liu, Jie....289
Liu, Lianjun....63
Liu, Rong....252
Liu, Xiang....114
Liu, Ying....340
Liu, Yun-Quan....357
Liu, Yuxi....493
Liu, Zhengang....126
Livingston, Andrew G....163
Llop, Maria....25
Lo, S. L....155
Lobban, Lance....395
Loeb, Christopher....76
Lohcharoenkal, Warangkana....377
Lou, Helen H....9, 227
Lowry, David....339
Lowry, Gregory V....338
Lu, Diannan....309
Lu, Wei....118
Luebke, David....333
Luo, Hongmei....34, 36
Luss, Dan....231
Lv, Dong-Can....357
Lv, Jiayu....260
Lyngfelt, Anders....440, 443
Ma, Jing....124
Ma, Yue....180
Macedonio, Francesca....166
Machala, Zdenko....190
Mahanta, Narahari....154, 202
Maheshwari, Utkarsh....276
Mahle, John J....241
Mahoney, Ryan B....15
Mahoney, Sharlee....374
Majumder, Subhajit....71, 156
Mallinson, R. G....395
Mandal, Ajoy K....271, 277
Manousiouthakis, Vasilios....456
Mansur, Claudia E....123

AUTHOR INDEX

Mao, Lei 274
Mao, Yuanbing 99
Maric, Milan 256
Marin, Guy B. 445
Martín, Mariano 303
Masanet, Eric 225, 228, 388, 497
Masquelier, Christian 56
Massiot, Dominique 56
Mathews, Alexander P. 52, 169
Matranga, Christopher 328
Matsumoto, Hideyuki 161
Mattisson, Tobias 440, 443
Matzopoulos, Mark 507
Maurmann, Leila 52
Mayes, Richard T. 51
Mba, Jean-Marcel Ateba 56
Mbachu, Chiedozie 75
Mbah, Jonathan C. 262
McCool, Brian C. 8
McCutcheon, Jeffrey R. 465
McDaniel, Anthony H. 454
McDaniel, Martin D. 288
McEvoy, Joanne Gamage 239
McGinn, Paul 237
McGiveron, Omar 475
McIntyre, Dustin 511
McKay, Gordon 368
McNeill, V. Faye 89
Mededovic, Selma 192
Medeiros, José L. 498
Medellín-Castillo, Nahúm Andrés 371
Meeks, Noah D. 245
Mefford, O. Thompson 251
Mehrkesh, Amirhossein 223
Meighan, Terence 378
Mejic, Dragan 15
Ménétrier, Michel 56
Meng, Fanxu 220
Meng, Jiajia 359
Meng, Tiejun 105
Meng, Xinghua 198
Merlano, Aura 324
Messinger, Robert J. 56
Mesta, Adam 229, 382
Meyer, Thomas 289
Mhadeshwar, Ashish B. 204, 479
Mi, Baoxia 72
Miller, Andrew 289
Miller, Daniel J. 284
Miller, David C. 6, 29, 505
Miller, Dennis J. 398
Minmin, Liu 183
Mishra, Anurag 378
Mishra, Somesh 516
Mitchell, Reginald E. 470
Mizuno, Hiroyuki 69
Mlusu, Jacob 323
Modekurti, Srinivasarao 505
Moffat, Thomas P. 291
Mohamed, Mustakimah 501
Mohamed, Shamseldin 411
Molina, Alejandro 324
Moon, Il 258
Moorer, Kiara 262
Moraes, Maria De Lourdes L. 194
More, Amey 407
Moscow, Prukinthraj 116
Mostafa, Mohamed 7
Mountziaris, T. J. 462
Mousavi, Bahador 224, 385
Muhich, Christopher L. 408
Mukherjee, Rajib 432
Mullen, Charles 401
Mullen, Charles A. 402
Müller, Christoph R. 409, 442, 448
Müller, Karsten 66
Muppaneni, Tapaswy 419
Murillo, Jessica D. 466
Nagarajan, Pratheeba Chanda 488
Nagarajan, X. Chanda 486
Nair, Dr. Chandra 381
Najera, Michelle 374
Nakao, Shunsuke 85
Nakhla, George 168
Nangia, Shikha 341
Negishi, Hideyuki 416
Nel, Andre E. 252
Newsome, Brad J. 200
Ng, K. Y. Simon 41, 198
Ng, Ng Lee 90
Ng, Simon 105
Ng, Simon K. Y. 39
Ng, Yi Ling 433
Nguyen, Nhan 23, 187
Nguyen, Tu 218
Nicol, Katherine T. 96
Nikolaki, Spyridoula 91
Nirmalakhandan, Nagamany 419
Nizkorodov, Sergey A. 87
Noorulameen, Mohammad Salih 433
Norris, Lachelle 466
Noshadi, Iman 421
Nulwala, Hunaid 333
Nyman, Marianne 16
Oberg, Aaron A. 398
Oliva, Diego Gabriel 390
Oliveira, Felipe Segismundo 451, 452
Oliveira, Marizilda E. 194
Oliveira, Thamayne V. 451
Olmos, Fernando 456

AUTHOR INDEX

Ong, Rui Chin 70
Ono, Toshisato....190
Orlov, Alexander....297
Osorio, Andres 324
Otake, Katsuto 161
Overney, René M. 412
Oversby, Cori 15
Paasikallio, Ville....394
Pacsi, Adam 10
Pakzadeh, Behrang 49
Palazoglu, Ahmet 19
Panda, Kishora K....169
Pande, Priyanka 101
Papadikis, Konstantinos....397
Park, Ah-Hyung Alissa....515
Park, Chusik....469
Park, Jaedeuk....258
Park, Sunkyu....359
Parnas, Richard 421
Parsons, Gregory N....289
Pascual, Janire 383
Paul, Chen J. 154
Paul, Donald R. 284
Paulsen, Alex D. 355
Pavlovich, Matthew J....190
Peláez, Daniel 324
Pellegrino, John....165
Peng, Qing 289
Penner-Hahn, James....101
Pepiot, Perrine....356
Perez, Gonzalo Vilela....406
Perng, Charles....232
Pesado, Lidia Cervantes-....424
Peters, Dr. Robert W. 18, 48, 167, 176, 186
Peters, Robert W....7, 313
Peterson, Eric....255
Peterson, Gregory W. 241
Petty, Brian....46
Pfaendtner, Jim....355
Phalak, Nihar 504
Pham, Tiep Hoang 412
Phillips, Valerie....2
Pierce, David 3
Pirbazari, Massoud....53, 215
Pliego-Bravo, Y. S. 424
Poelman, Hilde....445
Pokorny, Richard....3
Pollet, Pamela....496
Pommerenck, Jordan....199
Pommerenck, Justin 199
Poston, James....104
Powell, Brian A. 251
Powell, Joseph B....22
Powell, Kody....31
Pratsinis, Sotiris E....221, 337
Pretzer, Lori A....211
Prigiobbe, Valentina....311
Privitera, Marc 42
Pucher, Hannes....362
Pucher, Peter 362
Pyrgiotakis, Georgios....221
Qadir, Riffat Maqbool 171, 172
Qi, Li....85
Qiu, Guanglei....78
Quddus, Mohammad R....404, 411, 450
Queirós, Yure G. 123
Qullin, Steven....331
Raghuvanshi, Smita 71, 156, 516
Rahardianto, Andi....287
Rahardianto, Anditya 8, 80, 212, 286
Raj, Richa 206, 477
Raju, Muralikrishna 37
Rallo, Robert....252
Ramakrishnan, Girish 297
Ramarao, Bandaru V....365
Ramasubramanian, Kartik....162
Ramírez, Juan....324
Ramos, Alfredo....507
Ramprasad, Ramamurthy....204, 479
Ranade, Vivek V....140, 149
Randall, Zachary J....496
Rasmussen, Paul G....101
Rastegary, Jalal 12
Raupenstrauch, Harald....266, 429, 435
Ravindran, Varadarajan 53, 215
Ray, Amiya Kumar 151
Ray, Madhumita B. 79, 168
Razzak, Shaikh....411
Redding, Adam....317
Reddy, Harvind Kumar....419
Reddy, K Suresh Kumar....246
Reed, Morgan 255
Reeves, Stephen A. 308
Regupathi, Iyyasamy....116
Resasco, Daniel E....395
Ribeiro, Camila 498
Ribeiro, Fabio H. 399
Rico, Enrique....55
Riddle, Mathew 497
Riedthaler, Melissa J....308
Rincon, Sahra 423
Rindfusz, Kurt....264
Roberts, Edward P. L. 1, 77
Robinson, David B....67
Robinson, Kenza....111
Roblegg, Eva....253
Rodrigues, Christiane A....194
Rodriguez-Fernández, Lidia....372
Rojanasakul, Yon 373, 377, 378
Romanin, Vince 228

AUTHOR INDEX

Romão, Betania B. 452
Roszczenko, Janusz....214
Rugel, Caleb 229, 382
Ruiz-Hernández, Antonio....25
Ruiz-Mercado, Gerardo J. 380, 432
Runde, Wolfgang 413
Rupp, Erik C. 243
Rutter, Andrew 433
Ryden, Magnus....443
Sa-Ei, Kanicha....433
Safari, Iman....6
Saha, Rajib K 79
Sahimi, Muhammad 352
Sales-Pardo, Marta 25
Salley, Dr. Steven 39
Samee, Masoud....215
Samiei, Kasra 429
Sampara, Chaitanya 234
Sanaei, Shabnam....389
Santiano, Branden 173
Santos, Julia G....452
Saraeian, Alireza....12
Sargent, Linda M. 339, 373
Sarigiannis, Dimosthenis 91
Sarkar, Shubhajit....168
Sarma, Priyangshu M....271, 277
Sarti, Giulio C. 414
Sasaki, Takeshi....367
Satrio, Justinus A. 353
Satsangi, Vibha Rani 346, 348
Savage, Phillip E. 263
Savelski, Mariano J....386
Schartz, Daniel....355
Schaub, Tanner....419
Scheffe, Jonathan 408
Schindler, Bryan J. 241
Schneider, Marc....253
Schönberg, Andreas....429
Schunk, P. Randall....217
Schwaiger, Nikolaus....362
Schwartz, Nicholas....361
Schwebel, Georg....406
Schweitzer, Dirk....402
Selvaratnam, Thinesh....419
Sengupta, Debalina 457
Seo, Myung Won....469
Serrano, Benito....482
Severance, Michael A....162
Shah, Nilay....301
Shao, Ting....118
Shaofeng, Li....113
Shao-Horn, Yang 40
Sharifi, F....344
Sharma, Dipika 346
Sharma, Hom 204, 479
Sharma, Kal Renganathan 503
Sharma, Ketki....51
Sharma, Pooja 151
Sharma, Vinit....204
Sharratt, Paul N. 433
Shen, Chunmiao 127
Shen, Jingmei....62
Shen, Peichuan....297
Shende, Anuradha....193
Shende, Rajesh....298
Shende, Rajesh V. 193
Shewchuk, Stanley R. 242
Shi, Fan 328
Shibasaki-Kitakawa, Naomi....458
Shimada, Yuichiro....161
Shipley, Sarah....256
Shirazi, Saeid Aghahossein....12
Shonnard, David R. 23, 32, 187
Shono, Atsushi 161
Short, Daniel....86, 427
Shrivastav, Rohit....346, 348
Shuto, Daiki 514
Siebenhofer, Matthäus....170, 362
Sierra, Fabio 423
Sievers, Carsten....332
Sin, Gürkan....305
Singh, Bina....271, 277
Singh, Nirupama 348
Slater, C. Stewart 386
Sleightholme, Alice E. S. 101
Smirniotis, Panagiotis 196, 208, 331
Smith, Dave....242
Smith, Laura T. 2
Smith, Raymond L. 380, 432, 457
Snell, Kristi 402
Snowberger, Sebastian....177
Song, Hao 219
Soong, Yee....511
Sorensen, Eva....44
Sorokhaibam, Laxmi Gayatri....149
Sorooshian, Armin 88
Sotiriou, Georgios A. 221, 337
Sousa, Thiago M. 451
Sparks, Kevin....402
Spinks, Thomas 208
Srinivasakannan, C 115, 246
Srinivasan, M. P. 126, 370
Srivastava, Ranjan 421
Stack, Daniel....330
Stafford, Randy 238
Steiner, Lisa 362
Stemmler, Elizabeth....358
Stenstrom, Michael K....80
Stevenson, Craig....333
Stillwell, Ashlynn S....47

AUTHOR INDEX

Stratton, Gunnar.....192
Stuart, Paul R.....389
Stueckle, Todd.....377, 378
Su, Changsheng.....237, 485
Su, Dong.....297
Su, Jincai.....72
Sui, Dandan.....203
Suidan, Makram.....196, 384
Sujit, Das.....497
Summers, R. Scott.....15
Sun, Bingbing.....375
Sun, Dan.....397
Sun, Jingwen.....272
Sun, Shi-Peng.....68
Sun, Xing.....99
Sun, Yong-Li.....182
Sundqvist, Sebastian.....440
Sunkara, Mahendra.....218
Sunol, Aydin K.....222
Sunol, Sermin G.....222
Surawanvijit, Sirikarn.....287
Swanson, Edward J.....515
Szalai, Veronika.....344
Tagliaferri, Carla.....159
Takahashi, Tomoki.....161
Takkellapati, Sudhakar.....457
Talin, A. Alec.....291
Tan, Timothy T. Y.....143
Tan, Xiaoyao.....127
Tantekin-Ersolmaz, S. Birgül.....417
Tao, Ling.....422
Tatarchuk, Bruce.....236, 240
Tavlarides, Lawrence L.....120
Teichmann, Daniel.....66
Teng, Hsisheng.....292
Teoh, Soo Khean.....433
Teubl, Birgit.....253
Theregowda, Ranjani B.....6
Thirumalai, Pranav.....248
Thompson, John.....287
Thompson, John F.....286
Thompson, Levi T.....101
Thornhill, Nina F.....44
Thwaites, Fred.....388
Tijaro-Rojas, Rocio.....319
Tillie, Charles F.....308
Tilotta, David.....359
Tilton, Robert D.....338
Ting, Yen-Peng.....78
To, John.....330
Toghiani, Rebecca K.....2
Tompsett, Geoffrey.....462
Tong, Andrew.....475
Tong, Shitang.....274
Tong, Zi.....162
Tran, Chau.....95
Tseng, Yao-Hsuan.....476
Tsotsis, Theodore T.....352
Tsouris, Costas.....51
Tsutsumi, Atsushi.....69
Tuan, Hsing-Yu.....35
Tuan, Vu Anh.....335
Tukur, Nasiru M.....463
Ugaz, Victor M.....220
Upadhyay, Sumant.....346
Upadhyayula, Venkata K. K.....497
Urban, Robert A.....307
Ureña-Benavides, Esteban E.....496
Uttarwar, Rohan.....197, 219
Vahdat, Nader.....262
Vaillancourt, Austin.....49
Vakharia, Varun.....162
Valiente, Rafael.....372
Van Duin, Adri C. T.....37
Van Genuchten, Case.....213
Van Noyen, Jasper.....443
Varadarajan, Anandavalli.....360
Vega, Lourdes F.....158
Velásquez, Hector Ivan.....391
Velazquez, Juan C.....211
Velikokhatnyi, Oleg.....349
Vendra, Venkat Kalyan.....218
Venkatakrishnan, Vinod Kumar.....399
Veser, Götz.....374, 407
Villegas, Juan C.....372
Vollen, Jason.....16
Von Solms, Nicolas.....510
Voorhies, Wayne.....419
Vopicka, Ondrej.....414
Vu, Diep.....86, 427
Vutukuri, Joesph.....364
Wagner, Andrew L.....81
Walker, Michael E.....6, 228
Waller, Aaron.....281
Walters, Keisha B.....360
Wan, Shaolong.....395
Wang, Alan.....504
Wang, Chao.....345
Wang, Chunsheng.....57
Wang, Dawei.....475, 504
Wang, Di.....210
Wang, Du.....495
Wang, Duo.....357
Wang, Jun.....127, 182
Wang, Lixin.....105
Wang, Liying.....377, 378
Wang, Qianli.....146
Wang, Shanquan.....278
Wang, Wei-Cheng.....422
Wang, Wentan.....118

AUTHOR INDEX

Wang, William 504
Wang, Xiaonan 19
Wang, Yanbin 133
Wang, Yi Yang 229
Wang, Zhen 88
Wang, Ziqiang 114
Ward, Barbara J. 15
Warman, Samantha 333
Warren, Bethany 85
Warren, Joshua 497
Warzinski, Robert P. 511
Watanabe, Masayuki 367
Waters, Christopher 395
Webber, Michael E. 47
Wei, Huige 98
Wei, Li 97
Wei, Suying 98, 106, 201
Wei, Yun 125
Weimer, Alan W. 15, 408, 454
Wheeler, M. Clayton 358
Wheelhouse, Katherine 433
Wickenden, David A. 1
Wilcox, Jennifer 243, 330
Winslett, Matthew 48
Wirth, Stacy M. 338
Woisetschläger, Dieter 170
Wong, Michael S. 211
Woo, Sunghyun 258
Worden, Robert M. 340
Wu, Hsuan-Chih 476
Wu, Qiyuan 297
Wu, Song-Hai 182
Wyczalkowski, Wojciech 214
Xi, Aihong 342
Xia, Tian 252, 375
Xia, Yinfeng 492
Xiangru, Fan 199
Xie, Liping 127
Xin, Feng 203
Xiong, Ran 65
Xu, Bailong 492
Xu, Lu 90
Xu, Qiang 230, 318
Xu, Rong 117
Xu, Yuming 327
Yacob, Tesfayohanes W. 15
Yamaguchi, Takeo 416
Yamasaki, Akihiro 367, 514
Yamashita, Hiromi 332
Yan, Binhang 118
Yang, Bin 135
Yang, Jae-Kyu 119
Yang, Jin 462
Yao, Yuan 225, 388
Yaseneva, Polina 299
Ydstie, B. Erik 14
Yeh, Te-Fu 292
Yelvington, Paul E. 361
Yetayew, Alemayehu 50
Yezerets, Aleksey 238
Yiacoumi, Sotira 51
Yin, Huidan 281
Yin, Jun 64
Yin, Xiaohong 203
Yogo, Katsunori 334
Yokochi, Alexandre F. T. 199
Yonemoto, Toshikuni 458
Yoon, En Sup 27
Yoon, Hyungjoon 258
Yoon, Yi 258
Yoshimune, Miki 416
Youn, Seunghee 209
Yu, Dingshan 97
Yu, Miao 145
Yu, Yang 180, 181, 202
Yu, Yue Yang 61, 62
Yu, Zita L. T. 80
Yuan, Fang-Wei 35
Yuan, Liang 357
Yuan, Ying 237
Yuan, Zhenhong 108
Yusup, Suzana 501
Yüzbasi, Nur Sena 409
Zan, Ting-Ting 182
Zappi, Mark 464
Zappi, Mark E. 364
Zastrow, Dustin 96, 361
Zastrow, Dustin J. 81
Zeng, Liang 475
Zhang, Chao 79, 192
Zhang, Haiyuan 252
Zhang, Jin Song 121
Zhang, Junshe 65
Zhang, Li 210
Zhang, Ruiling 310
Zhang, Xiaofeng 58
Zhang, Y. 179
Zhang, Yifei 439
Zhang, Yuzhong 127
Zhang, Zisheng 239
Zhao, Cunyu 63
Zhao, Guohua 133
Zhao, Hongying 133
Zhao, Huilei 63
Zhao, Jie 14
Zhao, Lin 162
Zhao, Shen 297
Zhao, Xin 61
Zhao, Yanan 162
Zheng, Bangxing 124

AUTHOR INDEX

Zheng, Liu **309**
Zhou, Xingfu **125**
Zhou, Ya **492**
Zhou, Zhiquan **467, 471**
Zhu, Cheng **462**
Zhu, Jia Ming **74**
Zhu, Jiahua **106, 201**
Zhu, Jian **41**
Zimmer, Andreas **253**
Zitney, Stephen E. **505**
Ziv, Guy **307**
Zmiewski, Alexander **353**
Zokoe, James **237**
Zou, Chenxi **127**
Zydney, Andrew L. **164**